Undergraduate Texts in Mathematics

Editors
S. Axler
F.W. Gehring
K.A. Ribet

Springer
New York
Berlin
Heidelberg
Barcelona
Budapest
Hong Kong
London
Milan
Paris
Santa Clara
Singapore
Tokyo

Undergraduate Texts in Mathematics

Anglin: Mathematics: A Concise History and Philosophy.
Readings in Mathematics.

Anglin/Lambek: The Heritage of Thales.
Readings in Mathematics.

Apostol: Introduction to Analytic Number Theory. Second edition.

Armstrong: Basic Topology.

Armstrong: Groups and Symmetry.

Axler: Linear Algebra Done Right.

Beardon: Limits: A New Approach to Real Analysis.

Bak/Newman: Complex Analysis. Second edition.

Banchoff/Wermer: Linear Algebra Through Geometry. Second edition.

Berberian: A First Course in Real Analysis.

Brémaud: An Introduction to Probabilistic Modeling.

Bressoud: Factorization and Primality Testing.

Bressoud: Second Year Calculus.
Readings in Mathematics.

Brickman: Mathematical Introduction to Linear Programming and Game Theory.

Browder: Mathematical Analysis: An Introduction.

Buskes/van Rooij: Topological Spaces: From Distance to Neighborhood.

Cederberg: A Course in Modern Geometries.

Childs: A Concrete Introduction to Higher Algebra. Second edition.

Chung: Elementary Probability Theory with Stochastic Processes. Third edition.

Cox/Little/O'Shea: Ideals, Varieties, and Algorithms. Second edition.

Croom: Basic Concepts of Algebraic Topology.

Curtis: Linear Algebra: An Introductory Approach. Fourth edition.

Devlin: The Joy of Sets: Fundamentals of Contemporary Set Theory. Second edition.

Dixmier: General Topology.

Driver: Why Math?

Ebbinghaus/Flum/Thomas: Mathematical Logic. Second edition.

Edgar: Measure, Topology, and Fractal Geometry.

Elaydi: Introduction to Difference Equations.

Exner: An Accompaniment to Higher Mathematics.

Fine/Rosenberger: The Fundamental Theory of Algebra.

Fischer: Intermediate Real Analysis.

Flanigan/Kazdan: Calculus Two: Linear and Nonlinear Functions. Second edition.

Fleming: Functions of Several Variables. Second edition.

Foulds: Combinatorial Optimization for Undergraduates.

Foulds: Optimization Techniques: An Introduction.

Franklin: Methods of Mathematical Economics.

Gordon: Discrete Probability.

Hairer/Wanner: Analysis by Its History.
Readings in Mathematics.

Halmos: Finite-Dimensional Vector Spaces. Second edition.

Halmos: Naive Set Theory.

Hämmerlin/Hoffmann: Numerical Mathematics.
Readings in Mathematics.

Hijab: Introduction to Calculus and Classical Analysis.

Hilton/Holton/Pedersen: Mathematical Reflections: In a Room with Many Mirrors.

Iooss/Joseph: Elementary Stability and Bifurcation Theory. Second edition.

Isaac: The Pleasures of Probability.
Readings in Mathematics.

(continued after index)

George E. Martin

The Foundations of Geometry and the Non-Euclidean Plane

 Springer

George E. Martin
Department of Mathematics and Statistics
State University of New York at Albany
1400 Washington Avenue
Albany, New York 12222
U.S.A.

Mathematics Subject Classification (1991): 51-01, 51-03.

This book was originally published by Intext Educational Publishers.

Library of Congress Cataloging in Publication Data

Martin, George Edward, 1932–
    The foundations of geometry and the non-Euclidean
plane.

    (Undergraduate texts in mathematics)
    Reprint. Originally published: New York : Intext
Educational Publishers, 1975.
    Includes index.
    1. Geometry—Foundations. 2. Geometry, Non-Euclidean.
I. Title. II. Series. III. Series: Intext series in
mathematics.
QA681.M34   1982   516′.1   82-728

Printed in the United States of America
9 8 7 6 5 4 (Corrected fourth printing, 1998)

ISBN 0-387-90694-0 Springer-Verlag New York Heidelberg Berlin
ISBN 3-540-90694-0 Springer-Verlag Berlin Heidelberg New York SPIN 10650750

To Margaret

Contents

Preface

This book is a text for junior, senior, or first-year graduate courses traditionally titled Foundations of Geometry and/or Non-Euclidean Geometry. The first 29 chapters are for a semester or year course on the foundations of geometry. The remaining chapters may then be used for either a regular course or independent study courses. Another possibility, which is also especially suited for in-service teachers of high school geometry, is to survey the the fundamentals of absolute geometry (Chapters 1–20) very quickly and begin earnest study with the theory of parallels and isometries (Chapters 21–30). The text is self-contained, except that the elementary calculus is assumed for some parts of the material on advanced hyperbolic geometry (Chapters 31–34). There are over 650 exercises, 30 of which are 10-part true-or-false questions.

A rigorous ruler-and-protractor axiomatic development of the Euclidean and hyperbolic planes, including the classification of the isometries of these planes, is balanced by the discussion about this development. Models, such as Taxicab Geometry, are used extensively to illustrate theory. Historical aspects and alternatives to the selected axioms are prominent. The classical axiom systems of Euclid and Hilbert are discussed, as are axiom systems for three- and four-dimensional absolute geometry and Pieri's system based on rigid motions.

The text is divided into three parts. The *Introduction* (Chapters 1–4) is to be read as quickly as possible and then used for reference if necessary. The formal axiomatic development begins in Chapter 6 of Part One, *Absolute Geometry* (Chapters 5–25). Chapter 5 contains a list of 15 models that are used throughout Part

One in discussing the relative consistency and independence of the axioms used in building our system. Isometries are introduced as soon as they are useful. In fact, the existence of the reflections is shown to be equivalent to the familiar SAS axiom. Chapter 25 shows that our five axioms for absolute geometry together with one of the equivalents of Euclid's Parallel Postulate (Theorem 23.7 gives 26 such equivalents) form a categorical system. Section 25.1 contains a detailed survey of the contents of Part Two, *Non-Euclidean Geometry* (Chapters 26–34). Although Part Two concentrates on hyperbolic geometry, many of the results have direct application to Euclidean geometry as well.

The classification of the isometries of the hyperbolic plane and, as a corollary, the classification of the isometries of the Euclidean plane appear in Chapter 29 of Part Two. In order to be sure of covering this important material in a one-semester or a two-quarter course it is suggested that Chapter 20 be finished halfway through the course. Chapters 10, 11, 15, and even 25 might be assigned as outside reading, postponed, or omitted. On the other hand, Chapter 30 should be included in such a course if time allows. (For a semester course meeting three times a week, the author uses the following schedule where exam days and reading days are omitted: 1–3, 4, 5, 6, 7, 8, 9, 9, 12, 13, 14, 16, 16, 17, 18, 19, 19, 20, 21, 21, 22, 22, 23, 23, 23, 24, 24, 26, 26, 27, 27, 28, 28, 28, 29, 29, 29.)

Special acknowledgment is heartily granted to my colleague Hugh Gordon, who made many very helpful suggestions when he was teaching from the preliminary version of this book. I am grateful to Mary Blanchard, who typed the manuscript. Finally, I wish to express appreciation to the Cambridge University Press for permission to quote the statements of the definitions, axioms, and theorems of *Book* 1 from its definitive publication on Euclid: *The Thirteen Books of Euclid's Elements* by T. L. Heath.

Foreword to the Student

"Thales, well known for his control of oil through a monopoly on the olive presses, today announced the invention of a means for obtaining knowledge. He calls the process *deduction.*" So began the front page story of the *Miletus Times* dated July 3, 576 B.C. An accompanying article reported the reactions of Oracle Joe to the invention. The utterances of Oracle Joe were deemed mysterious, as usual, and were quoted verbatim as follows: "Lines. O.J. sees parallel lines. Some seem more parallel than others in the hyperbolic plane. That's Non-Euclidean geometry. Just last week O.J. predicted that in a couple hundred years in a city near Egypt a guy named Euclid would make a big deal about parallel lines in a book that will endure as long as the stories of Homer. Euclid will use deduction. This deduction thing will hurt the oracle business, but the advice of oracles will be sought even into the Age of Aquarius. O.J. now sees tables, chairs, and beer mugs. Yes, it will be well over two thousand years and in worlds yet to be discovered before the implications and limitations of deduction begin to be fully realized. Non-Euclidean geometry will play an important role in all this. O.J. is never wrong—and is now open on Saturnday." With that we end the fantasy in this book but not, perhaps, the fantastic. (We shall see rectangles relegated to the domain of unicorns and pentagons with five right angles.)

There are many ways to distinguish between Euclidean and non-Euclidean geometry. The business about parallel lines is only one of the interrelated aspects whose totality is called *the theory of parallels.* To understand the theory of parallels we must begin our geometry almost from scratch. Thus we shall avoid the various traps that have ensnared mathematicians of the greatest genius. Also, the dynamics of building an axiom system very similar to but, in the end, vastly different from Euclid's are as exciting as any mystery novel. The story behind non-Euclidean geometry is one of the fascinating chapters in man's search for knowledge. In this text you will learn something of this story as well as the mathematical theory itself. For an appreciation of either, some understanding of the other is required. For those of you who may become teachers and feel non-Euclidean geometry is irrelevant, we quote the geometer Felix Klein: "After all, it is in order for the teacher to know a little more than the average pupil."

The following method is suggested for a quick, rough self-evaluation of your mastery of a particular chapter. After you have studied a chapter, answer each part of the True-or-False exercise in turn without allowing yourself to look ahead or to change an answer. Then score yourself, using the *Hints and Answers* section in the back of the book. If you missed a question because you forgot a definition from the theory, the *Index* will help you find the definition.

The author hopes that you enjoy your study of the theory of parallels.

INTRODUCTION

The Introduction contains the prerequisites to our study of the foundations of geometry. In order to begin Part One, it is sufficient that the following questions be understood and answered: What is an equivalence relation on a set? What is a one-to-one mapping from one set onto another? What does it mean to say that an axiom system is consistent, independent, or categorical? The Introduction answers these specific questions and contains enough additional material so that almost every reader will encounter something new. It is recommended that these first four chapters be read as quickly as possible and then used for reference later if necessary.

CHAPTER 1

Equivalence Relations

1.1 LOGIC

We agree that a statement is either *true* or *false* (Law of the Excluded Middle) but not both (Law of Noncontradiction). Our use of "not," "and," "or," "if . . . then . . . ," and "iff" in relation to arbitrary statements p and q is explained by the *truth tables* in Table 1.1, where "T" stands for true and "F" for false. In mathematics "or" is always used in the inclusive sense. The conditional $p \Rightarrow q$ may be read in any one of the following equivalent ways:

1 If p then q.
2 q if p.
3 p only if q.
4 q or not p.
5 p is a sufficient condition for q.
6 q is a necessary condition for p.

The sentence "p implies q" means that the conditional "if p then q" is true. To say "(if p then q) and (if q then p)," we merely say "p if and only if q" and write "p iff q" or "$p \Longleftrightarrow q$."

Related to the conditional "if p then q" are its *converse* "if q then p" and its *contrapositive* "if not q then not p." It should be easy to think of a conditional which is true but whose converse is false. On the other hand, a conditional is true if and only if its contrapositive is

TABLE 1.1

p	q	not p	p or q	p and q	if p then q	p iff q
T	T	F	T	T	T	T
T	F	F	T	F	F	F
F	T	T	T	F	T	F
F	F	T	F	F	T	T

true. One way of convincing yourself of this is to observe that the following are all equivalent: (1) If not q, then not p. (2) (Not p) or not (not q). (3) (Not p) or q. (4) q or not p. (5) If p, then q. Another way is to check the truth table in Table 1.2, where the numbers at the bottom indicate the order in which the columns were entered in constructing the table.

You intuitively know the meaning of the two quantifiers that are used in basic logic. One is the *existential quantifier,* which may be denoted by any one of the following: there exists, there exist, there is, there are, for some. The other is the *universal quantifier,* which may be denoted by any one of the following: for any, for all, each, every. Actually, the universal quantifier may be logically defined in terms of the existential quantifier and negation. For example, if p denotes some proposition about the integers, then "for all integers, p" means the same thing as "there does not exist an integer such that not p." One thing to look out for is that the little words $a, an,$ and *the* are often hidden quantifiers in English. For example, "The diameters of a circle intersect at a point" contains three quantifiers and means that *for any* circle *there exists* a point such that *each* diameter of that circle passes through that point.

Consider the statement "If N is a positive integer, then $N^2 - 79N + 1601$ is a prime." To prove this statement it would not be sufficient to show that $N^2 - 79N + 1601$ is a prime for several values of N. Even to show that you get a prime for the first seventy-nine positive integers is not a proof of the statement. Actually, the statement is false as $N^2 - 79N + 1601 = 41^2$ when $N = 80$. Note that one case where the statement is false proves that the statement is false! In other words, it only takes one *counterexample* to disprove a statement.

TABLE 1.2

p	q	$(p \Rightarrow q)$	iff	$(\text{not } q)$	$\Rightarrow$	$(\text{not } p))$
T	T	T	T	F	T	F
T	F	F	T	T	F	F
F	T	T	T	F	T	T
F	F	T	T	T	T	T
1	2	3	7	4	6	5

1.2 SETS

Most of us have heard that a *set* is a collection of elements. "$x \in A$" means that x is an element of set A; "$x \notin A$" means that x is not an element of set A. The statement that set A is a *subset* of set B is written "$A \subset B$" and means $x \in A$ only if $x \in B$. The set of all positive integers is a subset of the set of all integers. Some sets can be exhibited explicitly. For example, the set of odd digits is $\{1, 3, 5, 7, 9\}$. Often it is impractical or impossible to list the elements of a set. If **R** is the set of all real numbers, we may denote the set of all positive reals by "$\{x | x \in \mathbf{R}, x > 0\}$" and read "the set of all elements x such that x is a real number and x is greater than zero."

Let A and B be sets. The *union, intersection, difference,* and *Cartesian product* of A and B are defined, respectively:

$$A \cup B = \{x | x \in A \quad \text{or} \quad x \in B\},$$
$$A \cap B = \{x | x \in A \quad \text{and} \quad x \in B\},$$
$$A \setminus B = \{x | x \in A \quad \text{but} \quad x \notin B\},$$
$$A \times B = \{(x, y) | x \in A, \quad y \in B\}.$$

Since "but" means "and" in mathematical logic, we see that $A \setminus B$ is the set of all elements of A that are not also elements of B. Note that $A \times B$ is just the set of all ordered pairs such that the first element is in A and the second element is in B.

If A and B are sets with no element in common, then A and B are *disjoint*. In this case we write "$A \cap B = \varnothing$." So $\varnothing$ is the set which contains no elements and is called the *empty set* or *null set*. The empty set is a subset of every set. Two sets *intersect* if they are not disjoint.

If L and R are sets, then $L = R$ iff $L \subset R$ and $R \subset L$. One may exercise his ability to use "and" and "or" by proving the following distributive laws, where A, B, C are sets:

$$(A \cup B) \cap C = (A \cap C) \cup (B \cap C),$$
$$(A \cap B) \cup C = (A \cup C) \cap (B \cup C).$$

We may wish to speak of a set of sets. In this case the elements of the set are subsets of some other set. For example, $\{\{1, 2, 3\}, \{3, 4, 5, 6\}\}$ is a set with exactly the two elements $\{1, 2, 3\}$ and $\{3, 4, 5, 6\}$. Note that for general element S, we have $S \neq \{S\}$. In particular, $\varnothing \neq \{\varnothing\}$

since $\{\varnothing\}$ has one element. Although

$$\{\{(x, y) | x \in \mathbf{R}, \quad y \in \mathbf{R}, \quad ax + by + c = 0\} | a, b, c \in \mathbf{R}, \quad a^2 + b^2 \neq 0\}$$

is a rather formidable looking set, it is really something familiar. First of all, note that it is a set of sets. "$a^2 + b^2 \neq 0$" is a short way of saying real numbers a and b are not both zero. Thus, element $\{(x, y) | x \in \mathbf{R}, y \in \mathbf{R}, ax + by + c = 0\}$ is the set of all ordered pairs (x, y) of real numbers that satisfy the nondegenerate real linear equation $ax + by + c = 0$. Geometrically, an element is the set of all points on some line in the Cartesian plane. Thus, thinking of a line as a set of points, our formidable looking set is the set of all lines in the Cartesian plane.

1.3 RELATIONS

If D and C are sets and $G \subset D \times C$, then the ordered triple (D, C, G) is a *relation* between D and C. The letters stand for *domain, codomain,* and *graph.* If $D = C = S$, then we say "relation on S" rather than "relation between S and S." For an example of a relation, if D is the set of points of a plane, C is the set of lines in the plane, and G is the set of all ordered pairs (P, l) such that point P is on line l, then (D, C, G) is the relation called *incidence* between the points and lines of the plane. For another example, containment is a relation on 2^S, where 2^S is the set of all subsets of set S. Here $D = C = 2^S$ and $(A, B) \in G$ iff $A \subset B$ for subsets A and B of S.

Given set S, we shall define a very important type of relation on S. Relation (D, C, G) such that $D = C = S$ is an *equivalence relation* on S if for all elements a, b, c in S: (a) $(a, a) \in G$, (b) $(a, b) \in G \Rightarrow (b, a) \in G$, (c) $(a, b), (b, c) \in G \Rightarrow (a, c) \in G$. Perhaps this will look more familiar if we let $\sim = (D, C, G)$ and write "$a \sim b$" and say "a wiggle b" iff $(a, b) \in G$. Then, $\sim$ is an equivalence relation on S iff for all $a, b, c \in S$ the following axioms are satisfied:

R: (Reflexive Law) $a \sim a$,

S: (Symmetric Law) $a \sim b \Rightarrow b \sim a$,

T: (Transitive Law) $a \sim b, b \sim c \Rightarrow a \sim c$.

For a simple example of an equivalence relation on S, let $S = \{1, 2, 3\}$ and $G = \{(1, 1), (2, 2), (3, 3)\}$. Since $a \sim b$ iff $a = b$ in this example, the axioms are easily checked. The example shows that the set of the three axioms for an equivalence relation is *consistent,* that is, no contradiction can be derived from this set of axioms.

When $\sim$ is an equivalence relation on a set, "$a \sim b$" is generally read "a is equivalent to b." However, in specific cases a more specialized phrase may be used, such as "is parallel to," "is congruent to," or "is similar to." We shall give several examples of equivalence relations.

Example 1 Equality is the most familiar equivalence relation. Let S be an arbitrary nonempty set and $G = \{(a, a) | a \in S\}$. For the relation of equality, an element is equivalent only to itself.

Example 2 Let S be an arbitrary nonempty set and $G = S \times S$. This is the other extreme from equality. In this equivalence relation any element is equivalent to every element. These first two examples are said to be the trivial equivalence relations on a set S.

Example 3 Let $\mathbf{z}$ be the set of integers. Define an equivalence relation on $\mathbf{z}$ by $a \sim b$ if $a - b$ is even. So $(a, b) \in G$ iff a and b are either both even or both odd.

Example 4 Let $\mathbf{z}$ be the set of integers. Define an equivalence relation on $\mathbf{z}$ by $a \sim b$ iff $a = b = 0$ or $ab > 0$.

Example 5 Parallelness is an equivalence relation on the set of lines in the Euclidean plane, i.e., $a \sim b$ iff $a \parallel b$ when a and b are lines. (A line is *parallel* to itself and any other line which it does not intersect.)

Example 6 Congruence is an equivalence relation on the set of triangles in the Euclidean plane.

Example 7 Similarity is an equivalence relation on the set of triangles in the Euclidean plane. (Recall that two triangles are similar if they have corresponding angles congruent.)

We have already noted that the set of axioms for an equivalence relation is consistent. Let's show that the set of axioms is also *independent,* that is, no one of the three axioms is a consequence of the other two. We can do this by constructing three relations on a set S where a given axiom does not hold but the remaining two axioms do hold. Let $S = \{1, 2, 3\}$ and

$G_1 = \{(1, 1), (2, 2), (1, 2), (2, 1)\}$,

$G_2 = \{(1, 1), (1, 2), (1, 3), (2, 2), (2, 3), (3, 3)\}$,

$G_3 = (S \times S) \setminus \{(1, 2), (2, 1)\}$.

Although the relation on S defined by G_1 is symmetric and transitive, the relation is not reflexive as $(3, 3) \notin G_1$. Although the relation on S defined by G_2 is reflexive and transitive, the relation is not symmetric as $(1, 2) \in G_2$ but $(2, 1) \notin G_2$. Although the relation on S defined by G_3 is reflexive and symmetric, the relation is not transitive as $(1, 3), (3, 2) \in G_3$ but $(1, 2) \notin G_3$. These three relations show that the set of axioms for an equivalence relation is independent.

When proving a given relation is an equivalence relation, it is sufficient to prove that the relation is reflexive, symmetric, and satisfies the rule: if a, b, c are distinct, $a \sim b$, and $b \sim c$, then $a \sim c$. To see that the transitive law always holds under these assumptions, we have only to prove that the transitive law holds when the three elements are not distinct. But this is a trivial observation. (In case $a = b$: $a \sim a, a \sim c$ implies $a \sim c$ trivially. In case $a = c$: $a \sim b$ and $b \sim a$ implies $a \sim a$ by the reflexive law. In case $b = c$: $a \sim c, c \sim c$ implies $a \sim c$ trivially.)

Suppose $\sim$ is an equivalence relation on nonempty set S. For each element a in S we define the *equivalence class of a* to be $[a]$ where $[a] = \{x | x \in S, x \sim a\}$. Obviously, an equivalence class is a subset of S. Since $a \sim a$, we have $a \in [a]$ and $[a] \neq \varnothing$. So every element in S is in some equivalence class. We want to show that no element of S is in two distinct equivalence classes, that is, two distinct equivalence classes are disjoint. We shall prove $[a] \cap [b] \neq \varnothing$ implies $[a] = [b]$. By hypothesis we let $c \in [a] \cap [b]$. Then $c \in [a]$ and $c \in [b]$. So $c \sim a$ and $c \sim b$ by definition of $[a]$ and of $[b]$. But then $a \sim c$ and $c \sim b$ by the symmetric law. Thus, by the transitive law $a \sim b$ and, by the symmetric law, $b \sim a$. Now we are ready to prove $[a] = [b]$. We shall first show $[a] \subset [b]$. Suppose $x \in [a]$, then $x \sim a$ by definition of $[a]$. But, since $x \sim a$ and $a \sim b$, we have $x \sim b$ by the transitive law and $x \in [b]$ by definition of $[b]$. Hence, $[a] \subset [b]$. Similarly, if $y \in [b]$, then $y \sim b$. But $y \sim b$ and $b \sim a$ implies $y \sim a$ or $y \in [a]$. Hence, $[b] \subset [a]$. Thus $[a] = [b]$, as desired. Altogether we have shown: *The set of equivalence classes of an equivalence relation on a nonempty set S is a partition of the set S into disjoint nonempty subsets. Every element of S is in exactly one equivalence class.*

Letting P_i be the set of equivalence classes under the equivalence relation given in Example i above, you should obtain the following results. P_1 is the set of all one element subsets of S; $P_1 = \{\{a\} | a \in S\}$. P_2 contains exactly one element S itself; $P_2 = \{S\}$. P_3 has two elements: the set of even integers and the set of odd integers. P_4 has three elements: $\{0\}$, the set of all positive integers, and the set of all negative integers. The elements of P_5 are called *parallel pencils*. So a parallel pencil consists of all the lines parallel to a given line. In the Euclidean plane, there are of course an infinite number of parallel pencils, one corresponding to every line through some fixed point.

1.4 EXERCISES†

● **1.1** Construct counterexamples to show that the following are not valid arguments: (a) If q is true and p implies q, then p is true. (b) If p is false and p implies q, then q is false.

● **1.2** Assume the following three statements are all true and prove the converse of each one where a, b, c, d are real numbers: (i) If $c = d$, then $a = b$. (ii) If $c > d$, then $a > b$. (iii) If $c < d$, then $a < b$.

1.3 Define two nontrivial equivalence relations on $\{1, 2, 3, 4\}$.

● **1.4** If a and b are integers such that 5 divides $a - b$, then number theorists say that a is *congruent* to b modulo 5. Show that congruency modulo 5 is an equivalence relation on the set of integers and describe the equivalence classes.

● **1.5** True or False?

(a) "not $(p$ or $q)$" means "(not p) or q."

(b) "not $(p$ or $q)$" means "(not p) and (not q)."

(c) "not $(p$ and $q)$" means "(not p) and (not q)."

(d) $A \cup B = B \cup A$.

(e) $A \cap B = B \cap A$.

(f) $A \setminus B = B \setminus A$.

(g) $A \times B = B \times A$.

(h) $A \cap B = A$ iff $A \cup B = B$.

(i) Containment is an equivalence relation on the set of all subsets of a set.

(j) In the Euclidean plane two parallel pencils are disjoint.

● **1.6** Describe $\{(x, y) \mid x, y \in \mathbf{R}, \ 0x + 0y + 0 = 0\}$ and $\{(x, y) \mid x, y \in \mathbf{R}, \ 0x + 0y + 2 = 0\}$ as sets of points in the Cartesian plane.

1.7 Show "p and q" is equivalent to "not $((\text{not } p)$ or $(\text{not } q))$."

1.8 Show "not (if p then q)" is equivalent to "p and not q."

---

†The bullet ● before an exercise indicates that there is some reference to that exercise in the *Hints and Answers* section. The starred exercises throughout the book range from those that might be difficult for some students to those that will be very difficult for any student or any instructor.

1.9 If n is a nonnegative integer and set S has exactly n elements, then how many elements does 2^S have?

1.10 Show that every partition of a nonempty set into disjoint nonempty subsets determines an equivalence relation on the set.

*1.11 How many equivalence relations are there on a set of n elements?

*1.12 Consider the statement "All Cretans are liars," made by the Cretan philosopher Epimenides in the sixth century B.C.

GRAFFITI

A pride of lions. A school of fish. A knot of toads. A gaggle of geese.
A labor of moles. A gam of whales. A leap of leopards.
An exaltation of larks.

This statement is false.

The Greek alphabet

Letters		Names	Letters		Names	Letters		Names
A	α	alpha	I	ι	iota	P	ρ	rho
B	β	beta	K	κ	kappa	Σ	σ ς	sigma
Γ	γ	gamma	Λ	λ	lambda	T	τ	tau
Δ	δ	delta	M	μ	mu	Y	υ	upsilon
E	ϵ	epsilon	N	ν	nu	Φ	ϕ	phi
Z	ζ	zeta	Ξ	ξ	xi	X	χ	chi
H	η	eta	O	o	omicron	Ψ	ψ	psi
Θ	θ	theta	Π	π	pi	Ω	ω	omega

CHAPTER 2

Mappings

2.1 ONE-TO-ONE AND ONTO

Recall that a *rational number* is a real number of the form a/b where a and b are integers with $b \neq 0$. A *complex number* is of the form $x + yi$ where x and y are real numbers and $i^2 = -1$. (More on complex numbers in Section 3.1.) Many mathematicians use the following symbols, given with their meanings:

∀	for any, for every, for all
∃	there exists, there exist
∋	such that
!	unique
z	the integers
Q	the rationals
R	the reals
C	the complex numbers

Let p be some statement about the elements of sets A and B. Note that "$\forall x \in A \exists y \in B \ni p$" and "$\exists y \in B \ni \forall x \in A, p$" mean different things. (For example, let p be "$y = x + 1$" and $A = B = \mathbf{z}$.) The negation of "$\forall x \in A, p$" is "$\exists x \in A \ni$ not p." It follows that the negation of "$\exists x \in A \ni p$" is "$\forall x \in A$, not p." Hence the negation of "$\forall x \in A \exists y \in B \ni f(x) = y$" is "$\exists x \in A \ni \forall y \in B, f(x) \neq y$."

If D and C are sets, then f is a *function* or *mapping from D into*

C if for every element x in D there is a unique element $f(x)$ in C. If f is a mapping from D into C and for each element y in C there is an element x in D such that $f(x) = y$, then f is said to be *onto* or *surjective*. If f is a mapping and $f(x) = f(y)$ implies $x = y$, then f is said to be *one-to-one* or *injective*.

Perhaps you have noticed that a mapping is another type of relation: if D and C are sets and $G \subset D \times C$, then the relation (D, C, G) is called a *mapping from D into C* if every element of the domain D occurs exactly once as the first element of an ordered pair in the graph G. Before looking at the next sentence below, try your hand at using the symbols introduced above to express the condition that relation (D, C, G) be a mapping. You should convince yourself that each of the following expresses the condition:

(a) $\forall x \in D \, \exists! \, y \in C \ni (x, y) \in G$.

(b) $D = \{x | \exists! \, y \in C \ni (x, y) \in G\}$.

For f as a mapping from D into C write "$f : D \to C$" and read "f maps D into C." If (a, b) is in the graph of mapping f, we do not write "afb" as we did for an equivalence relation but rather "$f : a \mapsto b$" or "$f(a) = b$" and say that f maps a to b, that b is the image of a under f, that the value of f at a is b, or that f of a is b. Note the difference between f and $f(a)$: f is a mapping while $f(a)$ is an element in the codomain of f. You are probably most familiar with mappings from **R** into **R** where the function is defined by a formula in a variable element of the domain, e.g., $f(x) = x^2$ or $f(x) = \cos x$.

The square function from **R** into **R** demonstrates that an element of the codomain may be the image of more than one element of the domain, as $f(x) = x^2$ implies $f(2) = 4 = f(-2)$. We have several equivalent ways of saying that this sort of thing does not happen for function f with graph G:

1 f is one-to-one.
2 f is injective.
3 f is an injection.
4 $f(x) = f(y)$ implies $x = y$.
5 Distinct elements have distinct images.
6 $(x, z), (y, z) \in G \Rightarrow x = y$.

Associated with mapping $f : D \to C$ is its *range* R where $R = \{y | \exists x \in D \ni f(x) = y\} = \{f(x) | x \in D\}$. The range of a mapping is a subset of the codomain of the mapping. The square function from **R** into **R** demonstrates that every element of the codomain need not be an element of the range, as there is no real number x such that $f(x) =$

$x^2 = -1$. We have several equivalent ways of saying that this sort of thing does not happen for function f with range R where $f = (D, C, G)$:

1 f is onto.
2 f is surjective.
3 f is a surjection.
4 $R = C$.
5 Every element in C is the image of some element in D.
6 $C = \{y|\; \exists x \in D \ni (x, y) \in G\}$.

Clearly there is an over-abundance of language used to describe mappings. However, you should be familiar with all of it. Some people prefer the function:one-to-one:onto language while others prefer the mapping:injection:surjection language. We shall use both interchangeably.

Let's practice using our symbols and review the definitions. If D and C are sets and $G \subset D \times C$, then (D, C, G) is a *relation*. If $\forall x \in D \exists! y \in C \ni (x, y) \in G$, then the relation is a *function* with range R where $R = \{y|\; \exists x \in D \ni (x, y) \in G\}$. Further, if $\forall y \in R \exists! x \in D \ni (x, y) \in G$, then the function is an *injection*; if $\forall y \in C \exists x \in D \ni (x, y) \in G$, then the function is a *surjection*.

When the domain and codomain of a function f are both **R**, we have a nice way of picturing a function using the Cartesian plane. We let the x-axis represent the domain and the y-axis represent the co-domain. The graph consists of all points $(x, f(x))$. The range consists of all numbers $f(x)$. See Figure 2.1, illustrating $f: \mathbf{R} \to \mathbf{R}$, $f: x \mapsto e^x$. Also, in the case $f: \mathbf{R} \to \mathbf{R}$, we can give a geometric interpretation of

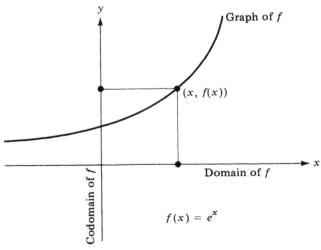

FIGURE 2.1

an injection and of a surjection. Function f is one-to-one if every line parallel to the x-axis intersects the graph at most once; function f is onto if every line parallel to the x-axis intersects the graph at least once.

Examples of formulas defining functions $f:\mathbf{R}\to\mathbf{R}$ where f is neither an injection nor a surjection: $f(x)=x^2$, $f(x)=2$, $f(x)=\cosh x = \frac{1}{2}(e^x+e^{-x})$.

Examples of formulas defining functions $f:\mathbf{R}\to\mathbf{R}$ where f is an injection but not a surjection: $f(x)=e^x$, $f(x)=\arctan x$, $f(x)=\tanh x = (e^x-e^{-x})/(e^x+e^{-x})$.

Examples of formulas defining functions $f:\mathbf{R}\to\mathbf{R}$ where f is a surjection but not an injection: $f(x)=x(x-1)(x+1)$, $f(x)=x^3/(1+x^2)$, $f(x)=x\sin x$.

Examples of formulas defining functions $f:\mathbf{R}\to\mathbf{R}$ where f is both an injection and a surjection: $f(x)=2x+3$, $f(x)=x^3$, $f(x)=\sinh x = \frac{1}{2}(e^x-e^{-x})$.

The functions sinh, cosh, and tanh defined in the examples above are called *hyperbolic trig functions*. If $x=\cosh\theta$ and $y=\sinh\theta$ for any real number θ, then $x^2-y^2=1$. Further properties of the hyperbolic trig functions can be found in any calculus text or Section 31.2.

Let $f:\mathbf{Z}\to\mathbf{R}$, $f:x\mapsto x+3$ and $g:\mathbf{Z}\to\mathbf{Z}$, $g:x\mapsto x+3$. Although f

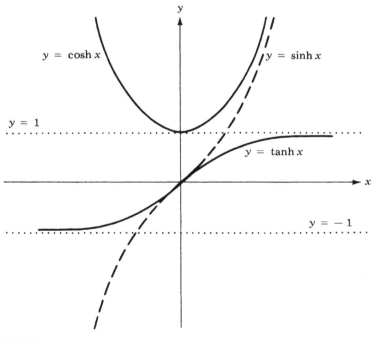

FIGURE 2.2

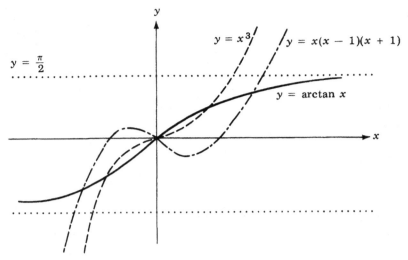

FIGURE 2.3

and g have the same domain **z** and the same graph G where $G = \{(x, x+3)|x \in \textbf{z}\}$, we note that $f \neq g$ as f and g have different codomains. Observe that g is a surjection while f is not a surjection.

A mapping that is both an injection and a surjection is called a *bijection*. We shall use this word often! A bijection is a mapping that is both one-to-one and onto. If S and T are sets, saying that there is a *one-to-one correspondence* between S and T means that there exists a bijection from S onto T. For example, to see that there is a one-to-one correspondence between all the integers and the even integers, consider the mapping f from **z** into the even integers defined by $f(x) = 2x$. For another example, the existence of the mapping g from the set of positive integers into the set of all integers defined by $g(x) = \frac{1}{2}x$ if x is even and $g(x) = -\frac{1}{2}(x-1)$ if x is odd proves that there is a one-to-one correspondence between the set of positive integers and the set of all integers. Only for infinite sets is it possible that there exists a one-to-one correspondence between a set and a proper subset of the set.

Occasionally one wants to consider what a mapping does to some particular subset of its domain. If $f: D \to C$ and $A \subset D$, then the *restriction of f to A* is $g: A \to C$ defined by $g(x) = f(x)$ for x in A. Loosely speaking, g just copies f for a smaller domain. Of course g and f may have different properties since they are different mappings when $A \neq D$.

Associated with a mapping f from set D into set C is a certain mapping f_{*} which maps subsets of D to subsets of C, namely, $f_{*}: 2^{D} \to 2^{C}$ where $f_{*}(T) = \{f(t)|t \in T\}$ if $T \subset D$. In particular, $f_{*}(D)$ is the range of f and $f_{*}(\varnothing) = \varnothing$. Although f and f_{*} are clearly different functions, we shall follow the customary abuse of language and write "f" in place of "$f_{..}$." This convention is possible because there is no likelihood of confusion. To illustrate where this convention is used in geometry, suppose f is a mapping from the points into the points of the Euclidean plane and T is some set of points such as a line or a triangle. It is probably clear what "$f(T)$" should mean; $f(T)$ is the set of all points $f(t)$ such that t is a point of T. For example, a translation f of the Euclidean plane is really a mapping from the points onto the points. Associated with f is $f_{..}$ which maps, say, line l to line m. Our convention allows us to say that f maps l to m; we write "$f(l) = m$" rather than the more formal "$f_{..}(l) = m$."

2.2 COMPOSITION OF MAPPINGS

Given mappings $f:D \to C$ and $g:B \to A$ such that the range of f is a subset of the domain of g, we can define mapping $gf:D \to A$ by $gf:x \mapsto g(f(x))$. So $gf(x) = g(f(x))$. The reason for the requirement $f(D) \subset B$ is clear, as otherwise $g(f(x))$ is not defined. This mapping gf is called the *product* or *composition of f followed by g*. See Figure 2.4. For example, suppose $f:\mathbf{Z} \to \mathbf{Z}$ is defined by $f(x) = x^2$; $g:\mathbf{R} \to \mathbf{R}$ is defined by $g(x) = \sin x$; and $k:\mathbf{R} \to \mathbf{R}$ is defined by $k(x) = x^2$. Then, gf is the mapping $gf:\mathbf{Z} \to \mathbf{R}$ defined by $gf(x) = \sin x^2$, but fg is not even defined (e.g., $fg(1) = f(\sin 1)$ is not defined since $\sin 1$ is not an integer). The order of f and g in "gf" is important! In our example gk and kg are mappings from $\mathbf{R}$ into $\mathbf{R}$ where $gk(x) = \sin x^2 = \sin (x^2)$ and $kg(x) = \sin^2 x = (\sin x)^2$. Since $\sin x^2$ and $\sin^2 x$ are not equal for every real number x, we have $gk \neq kg$.

It is a simple exercise to prove once and for all that *composition of mappings is associative*, i.e., mappings $h(gf)$ and $(hg)f$ are equal when they are defined. Both mappings have the domain of f and the codomain of h. So, to show that the mappings are equal, we must show

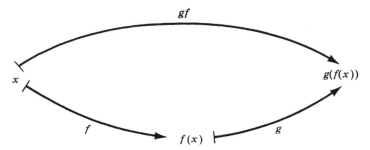

FIGURE 2.4

that the two mappings also have the same graph. That is, we must show $[h(gf)](x) = [(hg)f](x)$ for each x in the domain of f. Indeed, for each x in the domain of f, we have

$$[h(gf)](x) = h(gf(x)) = h(g(f(x)))$$
$$= hg(f(x)) = [(hg)f](x).$$

We have the desired result. (The first equality follows from the definition of the composition of gf followed by h. The second equality follows from the definition of the composition of f followed by g. The third equality follows from the definition of the composition of g followed by h. The last equality follows from the definition of the composition of f followed by hg.)

If f and g are both injections and gf is defined, then gf is also an injection. To show this we must prove $gf(x) = gf(y)$ only if $x = y$. But, $gf(x) = gf(y)$ implies $g(f(x)) = g(f(y))$ by definition of gf. Then, $f(x) = f(y)$ since g is an injection. Finally, since f is an injection, $x = y$ and we are done.

A bijection from a set S onto itself is often called a *permutation* on S. The least exciting permutation on S is $\iota : S \rightarrow S, \iota : x \mapsto x$, called the *identity mapping on* S. (Note that "ι" is not really "i" with a dot missing but is the Greek letter iota.) It is quickly observed that ι is the unique permutation on S such that $f\iota = f = \iota f$ for every permutation f on S.

Is the product of two permutations on set S always a permutation on S? Yes! Since permutations on S are injections, we have already seen that their product is an injection. To show the product of two permutations f and g on S is also a surjection, we must show that for every y in S there is an x in S such that $gf(x) = y$. Since g is onto, if $y \in S$ there exists t in S such that $g(t) = y$; since f is onto there exists x in S such that $f(x) = t$. Hence, $gf(x) = g(f(x)) = g(t) = y$. We have shown that the product of two permutations on set S is a permutation on S.

Suppose f is a relation and $f = (S, S, G)$. An element of G is an ordered pair (x, y) with x and y in S. Now f is a mapping iff every element of S appears exactly once as a first element of an ordered pair in G. Mapping f is a permutation on S iff every element of S appears exactly once as a second element of an ordered pair in G (onto iff at least once, and one-to-one iff at most once). Hence, if f is a permutation on S and $f = (S, S, G)$, then f' is a permutation on S where $f' = (S, S, G')$, $G' \subset S \times S$, and $(y, x) \in G'$ iff $(x, y) \in G$. Then, $ff'(y) = f(f'(y)) = f(x) = y = \iota(y)$ and $f'f(x) = f'(f(x)) = f'(y) = x = \iota(x)$. We see that if f is a permutation on S, then f' is the unique permutation on S such that $ff' = \iota = f'f$ where ι is the identity mapping on S.

For easy reference, we list the results that we have obtained about permutations on a set S:

1 The product of two permutations on S is a permutation on S, and products of permutations on S are associative.

2 There is a unique permutation ι on S such that $f\iota = f = \iota f$ for each permutation f on S.

3 If f is a permutation on S, then there exists a unique permutation f' on S such that $ff' = \iota = f'f$.

2.3 EXERCISES

● **2.1** Which of the following functions are injections and which are surjections?

$f_1 : \mathbf{R} \to \mathbf{R}, \qquad f_1 : x \mapsto 2x;$

$f_2 : \mathbf{Z} \to \mathbf{Z}, \qquad f_2 : x \mapsto 2x;$

$f_3 : \mathbf{Z} \to \mathbf{R}, \qquad f_3 : x \mapsto 2x;$

$f_4 : \mathbf{R} \to \mathbf{R}, \qquad f_4 : x \mapsto x^3;$

$f_5 : \mathbf{Z} \to \mathbf{Z}, \qquad f_5 : x \mapsto x^3;$

$f_6 : \mathbf{Z} \to \mathbf{R}, \qquad f_6 : x \mapsto x^3.$

● **2.2** Let f, g, h be mappings from $\mathbf{R}$ into $\mathbf{R}$ where $f(x) = x - 1, g(x) = 2x$, and $h(x) = x^3$. Find $fg(x)$, $gf(x)$, $fgh(x)$, $hgf(x)$, and $gfh(x)$.

● **2.3** Fill in the missing words: If $f : D \to C$, then f is a mapping _____ D _____ C. That $f(a) = b$ may be expressed in several ways: (1) the value _____ f _____ a is b, (2) f maps a _____ b, (3) b is the image of a _____ f. If $f(D) = C$, then f is _____. If $D = C = f(D)$ and f is _____, then f is a permutation _____ D.

● **2.4** Find a counterexample to: If $\forall x \in B \, \exists y \in A \ni (x, y) \in D$ where $D \subset A \times B$, then (B, A, D) is a mapping.

2.5 Give an example where function g is a restriction of function f, g is an injection, and f is not an injection.

● **2.6** True or False?
Assume only that f is a mapping from D into C.

(a) $\forall\, x \in D\, \exists!\, y \in C \ni f(x) = y.$

(b) $\forall\, x \in D\, \exists\, y \in C \ni f(x) = y.$

(c) $\forall\, y \in C\, \exists\, x \in D \ni f(x) = y.$

(d) $\forall\, y \in C\, \exists!\, x \in D \ni f(x) = y.$

(e) $\forall\, y \in D\, \exists\, x \in C \ni f(x) = y.$

(f) $f\!:\!a \mapsto b$ iff $f(a) = b.$

(g) $g\!:\!A \to B$ iff $g(A) = B.$

(h) "One-to-one" is short for "one-to-one correspondence."

(i) A permutation is a bijection.

(j) A bijection is a permutation.

● **2.7** Let set S have n elements, where n is a positive integer. How many mappings are there from S into S? How many of these are injections? How many are surjections? How many are bijections?

2.8 Let S be a finite set and $f\!:\!S \to S$. Show that f is an injection iff f is a surjection.

● **2.9** Let $f\!:\!D \to C$ have graph G. Find a counterexample to: $(x, y) \in G$ implies $(y, x) \in G$ only if f is a permutation on D.

● **2.10** Let $f\!:\!D \to D$ have graph G. Prove: $(x, y) \in G$ implies $(y, x) \in G$ only if f is a permutation on D. Give an example where f is not the identity on D.

2.11 Give examples of three distinct functions having the same graph.

● **2.12** Define three functions from **Z** into **Z** which are onto but not one-to-one.

2.13 State the negative of the following where $\sim$ is any relation between sets A and B: (a) $\forall\, x \in A\, \exists!\, y \in B \ni x \sim y.$ (b) $\exists!\, x \in A \ni x \sim y \,\forall\, y \in B.$ (No fair just adding "not" in front.)

*2.14 When is $(D, C, \varnothing)$ a mapping?

*2.15 Show that $f\!:\!\mathbf{R} \to \mathbf{R}$ where $f(x) = x^3 - 2x$ is a surjection but not an injection, while $g\!:\!\mathbf{Z} \to \mathbf{Z}$ where $g(x) = x^3 - 2x$ is an injection but not a surjection.

GRAFFITI

I have no fault to find with those who teach geometry. That science is the only one which has not produced sects; it is founded on analysis and on synthesis and on the calculus; it does not occupy itself with probable truth; moreover it has the same method in every country.

Frederick the Great

In my opinion a mathematician, in so far as he is a mathematician, need not preoccupy himself with philosophy—an opinion, moreover, which has been expressed by many philosophers.

Lebesgue

Motto of the Pythagoreans:

Number rules the universe.

This skipping is another important point. It should be done whenever a proof seems too hard or whenever a theorem or a whole paragraph does not appeal to the reader. In most cases he will be able to go on and later on he may return to the parts which he skipped.

Artin

Statement	Negation
not p	p
p and q	[not p] or [not q]
p or q	[not p] and [not q]
q or not p	p and not q
if p, then q	p and not q
if not q, then not p	p and not q
$\forall x, P(x)$	$\exists x \ni$ not $P(x)$
$\exists x \ni P(x)$	$\forall x,$ not $P(x)$
$\forall x,$ if $P(x)$, then $Q(x)$	$\exists x \ni P(x)$ and not $Q(x)$
$\exists x \ni$ if $P(x)$, then $Q(x)$	$\forall x \ni P(x),$ [not $Q(x)$]

CHAPTER 3

The Real Numbers

3.1 BINARY OPERATIONS

Let A, B, C, D, and S be sets. If the relation (D, C, G) is a mapping and $D = A \times B$, then the relation is a *binary operation from A and B into C*. We shall have use here only for the special case where $A = B = C$. Thus, for our purposes, a *binary operation on set S* is simply a mapping from $S \times S$ into S. If binary operation $*$ maps (a, b) to c, then we write "$a * b = c$."

Example 1 Perhaps the most familiar binary operation is addition on **z**, the integers. For every ordered pair (a, b) of integers there is a unique integer $a + b$.

Example 2 Another very familiar binary operation is multiplication on **R**, the reals. For every ordered pair (a, b) of real numbers there is a unique real number ab. As "+" always denotes ordinary addition on subsets of **R**, "·" always denotes ordinary multiplication on subsets of **R**. However, the symbol denoting multiplication is often suppressed, so that instead of "$a \cdot b$" we write "ab."

Binary operation $*$ on set S is *associative* if $*$ satisfies the associative law and is *commutative* if $*$ satisfies the commutative law:

Associative Law: $\qquad a * (b * c) = (a * b) * c \qquad$ for all a, b, c in S.

Commutative Law: $\qquad a * b = b * a \qquad$ for all a, b in S.

Addition on the integers and multiplication on the reals are both associative and commutative binary operations.

Example 3 Subtraction on **z** is a binary operation that is neither associative nor commutative. If a and b are integers, then $a-b$ is a unique integer. One counterexample is sufficient to demonstrate that subtraction is not associative: $1-(2-3)=2$, $(1-2)-3=-4$, but $2 \neq -4$. One counterexample demonstrates that subtraction is not commutative: $1-2=-1$, $2-1=1$, but $-1 \neq 1$.

Example 4 For real numbers a and b, let $a * b = a^2 + b^2$. Then $*$ is an example of a binary operation on **R** that is commutative but not associative.

Example 5 Define binary operation $*$ on **R** by $a * b = |a|b$. Then $*$ is a binary operation that is associative but not commutative. (Exercise 6.2.) Recall that $|a| = a$ if $a \geqq 0$ but $|a| = -a$ if $a < 0$. So $|a|$, called the *absolute value* of a; is always nonnegative.

Example 6 Another example of an associative binary operation that is not commutative is composition of permutations on a set A where A has at least three elements. Notice that the binary operation of composition is not on the set A itself but rather on the set of permutations on A.

A definition is an agreement to substitute a simple term or symbol for more complex terms or symbols. This is precisely how we are going to treat the word "group." To say that ordered triple (S, $*$, e) is a *group* means:

1 $*$ is an associative binary operation on set S.
2 e is the unique element of S such that $a * e = a = e * a$ for all a in S.
3 If a is in S, then there exists unique a' in S such that $a * a' = e = a' * a$.

We call $*$ the *multiplication* of the group. The element e is called the *identity* of the group and a' is called the *inverse* of a.

Note that group multiplication need not be commutative. If the multiplication does satisfy the commutative law, then the group is said to be a *commutative group* (or an *Abelian group*). In passing we might also note that the English language is not commutative. For example, (3) above states "$\forall a \in S \; \exists! a' \in S \ni a * a' = e = a' * a$" and *not* "$\exists! a' \in S \ni a * a' = e = a' * a \; \forall a \in S$."

Example 7 The most familiar example of a group is $(\mathbf{Z}, +, 0)$. Here the group multiplication is ordinary addition on the integers, 0 is the additive identity, and the additive inverse of a is $-a$. Check axioms 1, 2, and 3 above with $S = \mathbf{Z}$, $* = +$, $e = 0$, and $a' = -a$. Likewise, $(\mathbf{Q}, +, 0)$ and $(\mathbf{R}, +, 0)$ are groups where in each case the inverse under ordinary addition of a is $-a$.

$\mathbf{Z}^*$, $\mathbf{Q}^*$, and $\mathbf{R}^*$ are the nonzero elements of $\mathbf{Z}$, $\mathbf{Q}$, and $\mathbf{R}$, respectively. (The star used as a superscript should not be confused with a star used to denote a binary operation.) $\mathbf{Z}^+$, $\mathbf{Q}^+$, and $\mathbf{R}^+$ are the positive elements of $\mathbf{Z}$, $\mathbf{Q}$, and $\mathbf{R}$, respectively.

Example 8 $(\mathbf{Z}, \cdot, 1)$ is not a group as 2 does not have a multiplicative inverse in $\mathbf{Z}$ since $\frac{1}{2}$ is not an integer. $(\mathbf{R}, \cdot, 1)$ is not a group since 0 does not have a multiplicative inverse in $\mathbf{R}$. $(\mathbf{Z}^+, -, 0)$ is not a group as subtraction is not even a binary operation on $\mathbf{Z}^+$.

Example 9 It should be easy to check that each of $(\mathbf{Q}^+, \cdot, 1)$, $(\mathbf{Q}^*, \cdot, 1)$, $(\mathbf{R}^+, \cdot, 1)$, and $(\mathbf{R}^*, \cdot, 1)$ is a group. In each case the inverse of a is $1/a$.

Example 10 The last thing we did in Section 2.2 was to show that $(P, \circ, \iota)$ is a group where P is the set of all permutations on a nonempty set S, $\circ$ is composition of mappings, and ι is the identity mapping on S.

Let's prove one theorem about groups. The *Left Cancellation Law* states that if $(S, *, e)$ is a group and $a * x = a * y$ for a, x, y in S, then $x = y$. To prove this statement, we first note that since a is in S there exists a' in S such that $a' * a = e$. Then $a' * (a * x) = a' * (a * y)$. Using the associative law, we obtain $(a' * a) * x = (a' * a) * y$. Since $a' * a = e$, we now have $e * x = e * y$. Because e is the identity, we are left with $x = y$, as desired. The *Right Cancellation Law* states that if $(S, *, e)$ is a group and $x * a = y * a$ for a, x, y in S, then $x = y$. We leave the proof of this fact to Exercise 3.3.

We might observe that in just a few lines we have proved the left cancellation law for *all* groups. There are more groups than anybody knows about. Yet for each group the left cancellation law holds. Although our little result is not earthshaking, it does demonstrate the power of modern abstract mathematics.

There are volumes and volumes written about groups. However, our use of this group theory is limited to the definition and the cancellation laws. The word "group" now has a technical meaning and should no longer be used as a general collective noun. A mathematician might possibly refer to a gaggle of geese as a group of geese, since geese are not *usually* considered mathematical objects, but a mathematician would never be caught referring to the collection of odd integers as the group of odd integers.

A *rational* number is any number of the form a/b where a and b are integers but $b \neq 0$. The words "ratio" and "reason" come from the same stem. **Q** is the set of all rational numbers. When we add or multiply rational numbers together we always obtain a rational number. We know $(\mathbf{Q}, +, 0)$ and $(\mathbf{Q}^*, \cdot, 1)$ are groups. The binary operations of addition and multiplication on **Q** are related to each other through the *left distributive law* and the *right distributive law:*

L: $a \cdot (b+c) = (a \cdot b) + (a \cdot c)$ for all a, b, c in **Q**.

R: $(a+b) \cdot c = (a \cdot c) + (b \cdot c)$ for all a, b, c in **Q**.

The number system $(\mathbf{Q}, +, \cdot, 0, 1)$ is called the *field of rationals.*

The very early Greeks thought that all numbers had to be rational numbers. The whole of religion and philosophy of the early Pythagorean school was based on this supposed fact. It came as quite a shock to find that the diagonal of a square with sides of length 1 could not be expressed as a quotient of integers. In other words, there do not exist integers a and b such that $\sqrt{2} = a/b$. To prove this fact, one begins by assuming $\sqrt{2} = a/b$ where a and b are integers and a/b has already been reduced to its lowest terms. Then a and b are not both even. After squaring the equation, one goes on to deduce that a and b are both even. The contradiction proves that the original assumption must be false. The details of this historically famous proof are left for Exercise 3.1.

Considering the set of all real numbers and the usual operations, we know that $(\mathbf{R}, +, 0)$ and $(\mathbf{R}^*, \cdot, 1)$ are commutative groups. The binary operations of addition and multiplication on **R** are related to each other through the distributive laws. The number system $(\mathbf{R}, +, \cdot, 0, 1)$ is called the *field of reals.*

Generalizing the idea of the field of rationals and the field of reals, we say $(S, +, \cdot, 0, 1)$ is a *field* iff

1   $(S, +, 0)$ is a commutative group.
2   $(S^*, \cdot, 1)$ is a commutative group.
3   $a \cdot (b+c) = (a \cdot b) + (a \cdot c)$ for all a, b, c in S.

The group in **1** is called the *additive group* of the field, and the identity of this group is called the *zero*. The group in **2** is called the *multipli-*

cative group of the field where $S^{*} = S \setminus \{0\}$, and the identity of this group is called the *unity*. Statement **3** is called the *left distributive law*. That $(b+c) \cdot a = (b \cdot a) + (c \cdot a)$ for all a, b, c in S, follows immediately from the left distributive law and the commutativity of multiplication.

Obviously, the field of rationals $(\mathbf{Q}, +, \cdot, 0, 1)$ and the field of reals $(\mathbf{R}, +, \cdot, 0, 1)$ are fields. Two other fields, given in Examples 11 and 12 below, enter into later discussions.

Example 11 *The Field of Complex Numbers* A *complex number* is a number of the form $x + yi$ where x and y are real numbers and $x_1 + y_1 i = x_2 + y_2 i$ iff $x_1 = x_2$ and $y_1 = y_2$. $\mathbf{C}$ is the set of all complex numbers. Addition and multiplication are defined on $\mathbf{C}$ as follows:

$$(x_1 + y_1 i) + (x_2 + y_2 i) = (x_1 + x_2) + (y_1 + y_2) i,$$
$$(x_1 + y_1 i)(x_2 + y_2 i) \quad = (x_1 x_2 - y_1 y_2) + (x_1 y_2 + x_2 y_1) i.$$

$(\mathbf{C}, +, 0)$ is a commutative group where $0 = 0 + 0i$ and the additive inverse of $x + yi$ is $(-x) + (-y) i$. $(\mathbf{C}^{*}, \cdot, 1)$ is a commutative group where $\mathbf{C}^{*} = \mathbf{C} \setminus \{0\}$, $1 = 1 + 0i$, and the multiplicative inverse of $x + yi$ is $(x/(x^2 + y^2)) + (-y/(x^2 + y^2)) i$. The distributive laws hold. If you are not already familiar with the complex numbers, it is sufficient for our purposes to know that they exist and that $(\mathbf{C}, +, \cdot, 0, 1)$ is a field. As the field of reals contains the field of rationals, so the field of complex numbers contains the field of reals where real number r is identified with the complex number $r + 0i$.

Let $(S, +, \cdot, 0, 1)$ and $(S', +', \cdot', 0', 1')$ be fields F and F', respectively. If $f: S \to S'$ is a bijection from S onto S' such that f preserves addition, meaning $f(a + b) = f(a) +' f(b)$ for all a and b in S, and such that f preserves multiplication, meaning $f(a \cdot b) = f(a) \cdot' f(b)$ for all a and b in S, then f is called an *isomorphism* from F onto F'. It follows necessarily that $f(0) = 0'$ and $f(1) = 1'$. As a bijection, f determines a one-to-one correspondence between the elements of the fields. The existence of an isomorphism from F onto F' means that F and F' are abstractly the same. When F and F' are actually the same, the isomorphism is called an *automorphism*. The field of complex numbers provides an example of a nonidentity automorphism. Define $f: \mathbf{C} \to \mathbf{C}$ by $f(x + yi) = x + (-y) i$. The mapping f is called the *conjugate map*, and $f(z)$ is usually denoted by $\bar{z}$ for z in $\mathbf{C}$. That the conjugate map is a bijection on $\mathbf{C}$ is quickly checked. That the conjugate map is an automorphism on $\mathbf{C}$ then follows from the easily proved identities $\overline{z_1 + z_2} = \bar{z}_1 + \bar{z}_2$ and $\overline{z_1 z_2} = \bar{z}_1 \bar{z}_2$ for all z_1 and z_2 in $\mathbf{C}$.

The square root of $z\bar{z}$ is called the *modulus* or *absolute value* of complex number z and is denoted by $|z|$. So $|x + yi| = (x^2 + y^2)^{1/2}$ when x and y are real.

The equation $z^2 = -1$ has no solution in **R** but has a solution i in **C** where $i = 0 + 1i$. Complex numbers are usually introduced in high school algebra so that all quadratic equations with real coefficients have solutions. A complex number $x + yi$ with $y \ne 0$ is often called an *imaginary number*. As the very words "rational" and "irrational" indicate an earlier view of numbers, so the words "real" and "imaginary" indicate how numbers were considered in the last century. These words have a technical meaning today, independent of the insight they provide into the history of mathematics. As we would not question the rationality of a person just because that person used an irrational number such as $\sqrt{2}$, we should be aware that $1 + 2i$ is no more real or imaginary, in the everyday use of these words, than is -3. Since negative numbers are no longer called *fictitious numbers* as they once were, since negative numbers are introduced in grade school, and since imaginary numbers are not introduced until high school if at all, today's college student is usually surprised to learn that negative numbers and imaginary numbers were widely accepted at about the same time.

The requirements for a field demand that a field must contain at least two elements, namely, the zero and the unity. For our fourth example of a field, we see that there is a field with only these two elements.

Example 12 *The Field of Two Elements* The entire addition and multiplication tables for a field with exactly two elements is given in Table 3.1. For this unity 1 and this addition $+$, we have the somewhat peculiar fact that $1 + 1 = 0$. So $-1 = 1$ in this little field. The multiplicative group contains only the one element 1. Although the tables in Table 3.1 can be logically deduced from the requirements for a field and the assumption that $S = \{0, 1\}$, there is no need for us to do so. It takes just a minute to check that the three requirements for a field are satisfied. The sky will not fall if we consider this field only as an amusing toy.

TABLE 3.1

$+$	0	1
0	0	1
1	1	0

$\cdot$	0	1
0	0	0
1	0	1

3.2 PROPERTIES OF THE REALS

For our purposes, we may consider the *real numbers* to be the (positive, negative, or zero) infinite decimals. Since a rational number is a quotient of two integers, it follows from the algorithm for long division and the formula for the sum of an infinite geometric series that, of the real numbers, it is exactly the rationals that have a repeating infinite decimal. (Rationals of the form $10^n a$, where a and n are integers with $a \neq 0$, have two infinite decimal representations, one terminating in repeating 9 and one terminating in repeating 0. So, if each real number is to have exactly one infinite decimal representation, we discard all the infinite decimals that terminate in repeating 9.) An *irrational* is a real number that does not have a repeating infinite decimal representation. Every real number is either rational or irrational.

It is almost certain that Pythagoras, who was born about 572 B.C., was not aware that $\sqrt{2}$ is irrational. One legend attributes the discovery of the irrationality of $\sqrt{2}$ to Hippasus about 470 B.C.; another legend tells that Hippasus was drowned by his fellow Pythagoreans for disclosing this secret outside the brotherhood. In any case, the scandal within logic caused by the incommensurables (irrationals) which jeopardized the theory of proportion was resolved by Eudoxus about 370 B.C. Eudoxus' work is preserved in Book V of Euclid's *Elements*. The Pythagorean idea that all (real) numbers eventually depend on the integers for their definition was vindicated by the work of Richard Dedekind in 1872. Dedekind (1831–1916), following in the footsteps of Eudoxus, was among those who first gave a rigorous definition of the real numbers. A thorough understanding of the real numbers is only a hundred years old!

Dedekind defined an *infinite set* to be any set such that there is a one-to-one correspondence between the set and some proper subset of the set. Another way of saying this is that set S is an infinite set iff there is a mapping $f : S \rightarrow S$ which is one-to-one but not onto. The set $\mathbf{Z}^+$ of positive integers is an infinite set since $f : \mathbf{Z}^+ \rightarrow \mathbf{Z}^+, f : n \mapsto n^2$ is such a mapping. The existence of this one-to-one correspondence between the set of positive integers and its proper subset consisting of the squares was actually observed by Galileo (1564–1642). However, the possibility of making some important use of Galileo's observation was not realized for two hundred and fifty years.

If there is a one-to-one correspondence between sets A and B, then A and B are said to have the *same cardinality*. So two sets have the same cardinality iff there is a bijection from one onto the other. By Galileo's observation mentioned above, the set of all positive integers and the set of all the squares of the positive integers have the same cardinality. A set S and $\mathbf{Z}^+$ have the same cardinality iff there is an infinite sequence of terms from S such that each element of S occurs exactly once as a term of the infinite sequence. (An infinite sequence is, after all, only a mapping whose domain is $\mathbf{Z}^+$.) That $\mathbf{Z}$ and $\mathbf{Z}^+$ have the same cardinality is proved by reference to the infinite sequence 0, 1, -1, 2, -2, 3, -3,. . . . That is, $f:\mathbf{Z}^+ \to \mathbf{Z}$ defined by $f(2n) = n$ and $f(2n+1) = -n$ is a bijection.

TABLE 3.2

$1/_1$;

$1/_2$, $2/_1$;

$1/_3$, $2/_2$, $3/_1$;

$1/_4$, $2/_3$, $3/_2$, $4/_1$;

$1/_5$, $2/_4$, $3/_3$, $4/_2$, $5/_1$;

$1/_6$, $2/_5$, $3/_4$, $4/_3$, $5/_2$, $6/_1$;

. . . .

To show $\mathbf{Z}^+$ and $\mathbf{Q}^+$ have the same cardinality, first think of the infinite array suggested by Table 3.2, where in the nth row are listed all the fractions p/q with p and q positive integers such that $p+q = n+1$. Since every positive rational number has a unique representation p/q in reduced form and appears in some row of the array, an infinite sequence of positive rationals where each occurs exactly once can be constructed by taking the rows of the array in turn but omitting those fractions that are not reduced. The infinite sequence is

$1/_1$, $1/_2$, $2/_1$, $1/_3$, $3/_1$, $1/_4$, $2/_3$, $3/_2$, $4/_1$, $1/_5$, $5/_1$, $1/_6$, $2/_5$, $3/_4$, $4/_3$, $5/_2$, $6/_1$, $1/_7$, $3/_5$, $\cdots$

Defining $g(m)$ to be the mth term of this infinite sequence gives a bijection g from $\mathbf{Z}^+$ onto $\mathbf{Q}^+$. (Giving a formula for g is not easy and is left for students of the theory of numbers; we are quite happy to know that g exists.) Sandwiching in zero and the negative rationals, we obtain an infinite sequence of all the rationals where each rational occurs exactly once. More formally, $h:\mathbf{Z}^+ \to \mathbf{Q}$ where $h(1) = 0, h(2m) = g(m)$, and $h(2m+1) = -g(m)$ is a bijection from $\mathbf{Z}^+$ onto $\mathbf{Q}$. Hence $\mathbf{Z}^+$ and $\mathbf{Q}$ have the same cardinality.

Dedekind's friend, the great Georg Cantor (1845–1918), studied infinite sets and developed transfinite arithmetic. This was the beginning of what is now called *set theory*. We cannot go into the astounding results of this work here. However, stemming from Cantor's work there eventually arose contradictions in mathematics. It was again a scandalous matter for logic, this time leading to the establishment of the several modern mathematical schools of thought. Although the resulting problems have not been totally resolved to this day, the effect of the scandal was to leave mathematics greatly enriched. Mathematics is truly a phoenix. Cantor may be compared with his contemporary Sigmund Freud (1856–1939); for although much of their early groundwork has been discarded, each of these giants opened radically new worlds for others to explore.

One of Cantor's results is that the set of positive integers and the set of real numbers do not have the same cardinality. It was shocking to the Pythagoreans to learn that $\sqrt{2}$ is not rational; it was almost as shocking to mathematicians at the end of the last century to learn that not all infinite sets have the same cardinality. Suppose there were a bijection g from $\mathbf{Z}^+$ onto $\mathbf{R}$, then $f : \mathbf{Z}^+ \to I$ with $f(n) = \frac{1}{2}(1 + \tanh g(n))$ would be a bijection from $\mathbf{Z}^+$ onto I, where I is the set of real numbers between 0 and 1. We shall show that $\mathbf{Z}^+$ and $\mathbf{R}$ do not have the same cardinality by showing that there does not exist any mapping from $\mathbf{Z}^+$ *onto* I. Assume the contrary, that f is some mapping from $\mathbf{Z}^+$ onto I. We shall now obtain a contradiction. Let $f(n)$ have digit d_m^n in its mth place as a nonterminating infinite decimal. See Table 3.3. Let d be the infinite decimal $0.d_1 d_2 d_3 d_4 \ldots$ where $d_n = 2$ if $d_n^n \neq 2$ and

TABLE 3.3

$f(1) = 0.d_1^1 d_2^1 d_3^1 d_4^1 \cdots$

$f(2) = 0.d_1^2 d_2^2 d_3^2 d_4^2 \cdots$

$f(3) = 0.d_1^3 d_2^3 d_3^3 d_4^3 \cdots$

$\cdots$

$f(n) = 0.d_1^n d_2^n d_3^n d_4^n \cdots d_n^n \cdots$

$\cdots$

$d_n = 3$ if $d_n^n = 2$. Since d and $f(n)$ differ in their nth places, $f(n) \neq d$ for every positive integer n. Then, since d is in I, it follows that f is not onto. This proof, due to Cantor, is one of the most famous proofs in mathematics.

We know $\mathbf{Z}^+$ and $\mathbf{Q}$ have the same cardinality. Assuming $\mathbf{Q}$ and $\mathbf{R}$ had the same cardinality, it would follow that $\mathbf{Z}^+$ and $\mathbf{R}$ have the same cardinality, which contradicts Cantor's theorem. Hence, there does not exist a one-to-one correspondence between $\mathbf{Q}$ and $\mathbf{R}$.

The properties of order for the field of real numbers are considered next. In general, a field is *ordered* if there exists a subset P of elements satisfying the following three properties.

O1 a and b in P implies $a + b$ in P.
O2 a and b in P implies ab in P.
O3   For each element a in the field, exactly one of the following holds: $a = 0$, $a \in P$, $-a \in P$.

Since $-1 = 1 \neq 0$ in the field of two elements, contrary to (O3), it follows that the field of two elements cannot be an ordered field. Taking P to be the set of positive real numbers, it is easily seen that the field of real numbers is an ordered field. For this reason, given any field with a subset P satisfying the three requirements, the elements of P are said to be *positive*. You should be able to guess what $|x|$ means where x is an element of any ordered field; $|x| = x$ if x is positive or zero, and $|x| = -x$ if $-x$ is positive.

For an ordered field with given set P of positive elements, relation $>$ is defined on the elements of the field by $a > b$ iff $a - b$ is positive. In particular, $x > 0$ iff x is positive. Also, relation $<$ is defined by $b < a$ iff $a > b$, where "$<$" is read *less than* and "$>$" is read *greater than*. Ten properties of the relation $>$ follow (Exercise 3.4):

    1 For elements a and b, exactly one of the following holds: $a > b$, $a = b$, or $b > a$.

    2 $a > b$ and $b > c$ implies $a > c$.

    3 $a > 0$ and $b > 0$ implies $ab > 0$.

    4 $a > b$ implies $a + c > b + c$ for every element c.

    5 $a > 0$ iff $-a < 0$; $a < 0$ iff $-a > 0$.

    6 $a > b$ and $c > d$ implies $a + c > b + d$.

    7 $a > 0$ and $b > c$ implies $ab > ac$.

    8 $a < 0$ and $b > c$ implies $ab < ac$.

    9 $a \neq 0$ implies $a^2 > 0$.

  10 $|t| < a$ iff $-a < t < a$.

For the field of complex numbers, we have $+1 = 1^2$ and $-1 = i^2$. So (9) contradicts (O3) for complex numbers, as $+1$ and -1 can't both be positive. Hence the field of complex numbers is not an ordered field. This explains why it is senseless to ask which of $2 + 3i$ and $3 + 2i$ is greater than the other.

An ordered field may or may not have the following property: If $B > 0$ and $t > 0$, then there is a positive integer n such that $nt > B$. (For any field, nt means the sum $t + t + \cdots + t$ with n terms.) This property is called *Archimedes' axiom* and is named after Archimedes (287–212 B.C.). The axiom was probably known to Eudoxus. Anyway, before Archimedes, Euclid had expressly stated the axiom in considering the ratio of two magnitudes. The import of the axiom is that no matter how big B is and no matter how tiny t is there is an integer n such that nt is greater than B. This is a simple idea but very subtle. An ordered field that does not satisfy Archimedes' axiom is said to be *non-Archimedian*. Although admittedly fascinating, these fields with *infinitely small* and *infinitely large* elements are not essential to our work.

The field of real numbers is an Archimedean ordered field. Thinking of B and t as positive infinite decimals, there is an integer n such that $nt > B$ where n is an integral power of 10. In this case n has the effect of moving the decimal point in t far enough to the right to obtain a real number greater than B. Similar arguments show that for any real number a there exist integers n and m such that $n < a < m$. Also the set of rationals is *dense* in the set of reals, meaning that between any two real numbers there is a rational number. This last property together with Cantor's theorem that the reals and the rationals do not have the same cardinality may point out the necessity of having more than an intuitive definition of the real numbers.

A field is *Pythagorean* if $1 + a^2$ is a square for every element a; an ordered field is *Euclidean* if every positive element is a square. Unlike the field of rationals, the field of reals is Pythagorean and Euclidean as well as Archimedean. However, there is one property that distinguishes the reals from all other ordered fields. This is the least upper bound property that you may or may not remember from calculus.

Let F be the set of elements from an ordered field, and let S be a nonempty subset of F. If there is an element b in F such that $x \leqq b$ for all x in S, then b is an *upper bound* of S. Further, if b is less than any other upper bound of S, then b is called the *least upper bound* of S or the *supremum* of S and we write $b = \text{lub } S$. An ordered field is *complete* if every nonempty set of elements having an upper bound has a least upper bound. There is also the corresponding idea that if $c \leqq x$ for every x in nonempty subset T of F, then c is a *lower bound* of T. Further, if c is greater than any other lower bound of T, then c is called the *greatest lower bound* of T or the *infimum* of T and we write $c = \text{glb } T$. If T is a nonempty set of elements from a complete ordered field and T has a lower bound, then $\text{glb } T = -\text{lub } R$ where $R = \{-x | x \in T\}$. Of course, the greatest lower bound must be less than or equal to the least upper bound when they both exist. Considering the set of all rationals whose square is less than 2, we see that the ordered field of rationals is not complete.

Let's show that a complete ordered field is necessarily Archimedean. Assume, to the contrary, that some complete ordered field has positive elements t and B such that $nt \leqq B$ for every integer n. Then B is an upper bound of the set S of all elements nt with n an integer. Since the field is assumed to be complete, we may let $b = \text{lub } S$. Then $(m + 1)t \leqq b$ for every integer m. So $mt \leqq b - t$ for every integer m. Hence $b - t$ is an upper bound of S. Thus $b \leqq b - t$ and $t > 0$, a contradiction. Therefore, our assumption was incorrect, and every complete ordered field is Archimedean.

Every real number is the least upper bound of a set of rationals.

For if c is a real number, let S be the set of all rationals r such that $r < c$. Then c is certainly an upper bound of S. Also, since between any two reals there is a rational, it follows that c is the least upper bound of S. This is the whole idea behind Dedekind's definition of a real number. A *Dedekind cut* is a nonempty proper subset C of the rationals such that (1) $x \in \mathbf{Q}$, $c \in C$, $x < c$ implies $x \in C$ and (2) $x \in C$ implies there exists $y \in C$ such that $x < y$. Starting with the rationals and defining the real numbers to be the Dedekind cuts, one can go on to define $+$, $\cdot$, and $<$ to obtain a complete ordered field. This is not easy.

We might also mention Cantor's definition of a real number. A *Cauchy sequence* of rationals is a sequence $\{a_n\}$ of rational numbers such that for every positive rational e there is an integer N such that $|a_m - a_n| < e$ whenever n and m are both greater than N. Cauchy sequences $\{a_n\}$ and $\{b_n\}$ of rationals are said to be equivalent if for every positive rational e there is an integer N such that $|a_n - b_n| < e$ whenever $n > N$. Starting with the rationals and defining the real numbers to be the equivalence classes of Cauchy sequences of rationals, one can go on to define $+$, $\cdot$, and $<$ to obtain a complete ordered field. This is not easy.

It is not terribly difficult to argue that the ordered field of real numbers is complete, if you consider the real numbers to be defined as infinite decimals. What is somewhat difficult to show is that the infinite decimals form a field in the first place. Whether you start with infinite decimals, Dedekind cuts, or equivalence classes of Cauchy sequences of rationals, a rigorous development of the real numbers is not trivial. One knows that these approaches give the same abstract result since it can be shown that any two complete ordered fields are isomorphic. So, up to isomorphism, there is one complete ordered field, the reals.

3.3 EXERCISES

3.1 Show that $\sqrt{2}$ is irrational.

3.2 Let $*$ and $\#$ be the binary relations defined on the set of real numbers by $a * b = a^2 + b^2$ and $a \# b = |a| b$. Show that $*$ is commutative but not associative, while $\#$ is associative but not commutative.

3.3 Prove the right cancellation law for groups.

● **3.4** Prove the ten properties listed in the text for ordered fields.

● **3.5** True or False?

(a) If a, x, y are in a field and $ax = ay$, then $x = y$.

(b)   A group may have exactly one element.

(c) $|ab| = |a| \cdot |b|$ for all real numbers a and b.

(d)   If ρ, σ, τ are elements of group $(S, \cdot, \iota)$ such that $\rho\sigma = \tau$ and $\rho^2 = \iota$, then $\sigma = \rho\tau$.

(e)   If S is a subset of T and S is an infinite set, then T is an infinite set.

(f)   If two sets have the same cardinality, then the sets are infinite sets.

(g)   If two sets are infinite sets, then the two sets have the same cardinality.

(h) $0 < x < y < 1 < z$ implies $y^2 < x^2 < 1 < z^2$ for real x, y, z.

(i)   Both the ordered field of rationals and the ordered field of reals are Archimedean.

(j)   Both the ordered field of rationals and the ordered field of reals satisfy the least upper bound property.

● **3.6**   The set of rooms of a rather large motel has the same cardinality as $\mathbf{Z}^+$. One night all the rooms were full when one more customer pulled up to the manager's office. Without turning anyone out or making people double up, the manager rearranged the guests to accommodate the newcomer. How?

3.7   For the real numbers, there is only one possible set P that satisfies the three requirements for a field to be ordered.

3.8   Show that if $>$ is any relation on the elements of a field and if $>$ satisfies the first four of the ten properties listed in the text, then the field has a set P satisfying O1, O2, and O3.

3.9   Show $(\mathbf{Z}, \ast, 0)$ and $(\mathbf{Q}^\ast, \#, 1)$ are non-Abelian groups where $m \ast n = m + (-1)^m n$, $x \# y = xy$ if $x > 0$, and $x \# y = x/y$ if $x < 0$.

● **3.10**   Is the English language associative?

3.11   Show that all the real numbers of the form $a + b\sqrt{2}$ where a and b are rational form a field under the usual operations of addition and multiplication.

3.12   Prove that between every two real numbers there is a rational number and that between every two rational numbers there is an irrational number.

● **3.13**   Let $((a, b), c)$ be in the graph of some relation. Why do we write "$(a, b) \ast c$" if $\ast$ is an equivalence relation, "$\ast (a, b) = c$" if $\ast$ is a mapping, but "$a \ast b = c$" if $\ast$ is a binary operation?

3.14 Show that the only automorphism of the field of the reals is the identity mapping. Explain why a mathematician might say "a group of autos over a field" not thinking about a collection of cars parked in some country lot.

3.15 Prove $|a + b| \leq |a| + |b|$ for reals a and b.

3.16 Prove an ordered field is Archimedean iff $x > 0$ implies there is an integer n such that $nx > 1$; prove an ordered field is Archimedean iff $a \geq 0$, $b > 0$, and $na \leq b$ for every integer n implies $a = 0$.

3.17 For a study of the process of evolution in mathematics with particular attention to the concept of number, see *Evolution of Mathematical Concepts: an Elementary Study* by R. L. Wilder (Wiley, 1968). A thorough study of the foundations of the real numbers may be found in the second edition of Wilder's *Introduction to the Foundations of Mathematics* (Wiley, 1965).

*3.18 Show that **R** and **C** have the same cardinality.

*3.19 Prove any two complete ordered fields are isomorphic.

*3.20 Show that if two infinite decimals are equal, then they are equal to $10^n a$ for some integers a and n.

*3.21 Do there exist ordered fields which are Pythagorean but not Euclidean?

*3.22 Find a field besides the reals which is both Euclidean and Pythagorean.

GRAFFITI

z: *The German word for* integer *is* Zahl.

No one shall expel us from the paradise which Cantor has created for us.

Hilbert

He is unworthy of the name of man who is ignorant of the fact that the diagonal of a square is incommensurable with its side.

Plato

CHAPTER 4

Axiom Systems

4.1 AXIOM SYSTEMS

An *axiom system* or *postulate system* consists of some undefined terms and a list of statements, called *axioms* or *postulates,* concerning the undefined terms. One obtains a *mathematical theory* by proving new statements, called *theorems,* using only the axioms and previous theorems. Definitions are made in the process in order to be more concise. Aesthetically it may be preferable to give the list of axioms all at once. This may be impractical, however, as some of the axioms often depend on definitions and theorems resulting from earlier axioms. Usually one does not construct an axiom system from scratch. It is common to assume at least a language, a logic, and some set theory.

In order to point out a language convention used in this text, consider the following four sentences:

1   P and Q are points.
2   P and Q are two points.
3   P and Q are two distinct points.
4   P and Q are distinct points.

The meanings of (1), (3), and (4) should be clear. Statements (3) and (4) say the same thing, assuming one can count to 2. Unlike (3) and (4), statement (1) allows for the two possibilities that either P and Q are distinct or else $P = Q$. Now, does (2) mean the same thing as (1) or as (3)? Unfortunately, different mathematicians will give different answers. What is worse, and totally inexcusable, is to use (2) for both (1) and (3). Without further ado, we declare that (2) and (3) mean the same thing. Statement (3) will be used in place of (2) only for emphasis.

"There are three letters in the English alphabet" is a true statement. If you want "three" to mean "exactly three" rather than "at least three," you must say so. We have already mentioned that "or" is always used in the inclusive sense in mathematics. Another modern mathematical convention is the use of "equals" only in the sense of "is exactly the same thing as." The old fashion use of "equal" for "equivalent (in some sense)" should be avoided. When we write "$a = b$" we mean that "a" and "b" are names for the same object.

The logic and set theory that we shall assume as prerequisites are given in Chapter 1.

Some concepts that are applicable in general to an axiom system are given next. We have already encountered some of these. They are listed below for easy reference but are best learned from seeing them used in context.

An axiom system is *consistent* if there is no statement such that both the statement itself and its negation are theorems of the axiom system. One of the ways of showing that an axiom system is consistent is to assign meanings to the undefined terms of the axiom system in such a way that the axioms then become true statements. This may not be easy as true statements are hard to come by in this world. If the undefined terms of a given axiom system are assigned meanings from a second axiom system (e.g., Euclidean geometry or the real number system) such that the axioms of the first axiom system are theorems of the second axiom system, then the result is a *model* of the first axiom system. In this case we say that the first axiom system is *relatively consistent* with the second, as any inconsistency in the first axiom system would be reflected as an inconsistency in the second axiom system. Often relative consistency is all we can hope for, as Gödel has shown that there is no internal proof of consistency for a system that involves infinite sets. See Exercise 4.12.

In an axiom system, an axiom is *independent* if it is not a theorem following from the other axioms. Whereas consistency or relative consistency is an absolute requirement for any worthwhile axiom system, independence is not. For obvious pedagogical reasons, a simple looking theorem that has a long and difficult proof is often taken as an axiom in an elementary text.

Models of an axiom system are *isomorphic* if there is a one-to-one correspondence between their elements which preserves all relations. That is to say the models are abstractly the same, only the notation is different. If every two models of an axiom system are isomorphic, then the axiom system is *categorical*. It must not be assumed that categoricalness is always desirable. Indeed, there is great economy in proving theorems in a noncategorical axiom system because the theorems are then true statements for every model of the axiom system. As an example, once you have shown that the three axioms of a group are true statements for a set with a binary operation, then you immediately know literally thousands and thousands of true statements since all the theorems of group theory hold without further proof.

4.2 INCIDENCE PLANES

If $(\mathscr{P}, \mathscr{L}, \mathscr{F})$ is a relation such that $\mathscr{P}$ and $\mathscr{L}$ are disjoint, then the relation is an *incidence plane*. If this doesn't look like a geometric axiom system, let's start again. We take "point" and "line" as undefined terms. We have four axioms. Axiom A: The class of all points is a set $\mathscr{P}$; Axiom B: The class of all lines is a set $\mathscr{L}$; Axiom C: $\mathscr{P} \cap \mathscr{L} = \varnothing$; Axiom D: $\mathscr{F} \subset \mathscr{P} \times \mathscr{L}$. Axiom C requires that a point and a line be different. For specific types of incidence planes $(\mathscr{P}, \mathscr{L}, \mathscr{F})$, further requirements are made on the graph $\mathscr{F}$. Incidence planes have their own notation to express the fact that an ordered pair is in the graph. Since we are doing geometry, it ought to sound like geometry! (Is that backwards?) Thus, the following are equivalent for an incidence plane $(\mathscr{P}, \mathscr{L}, \mathscr{F})$:

1 $(P, l) \in \mathscr{F}$.
2 (P, l) is a flag.
3 Point P and line l are incident.
4 Point P is on line l.
5 Line l is on point P.
6 Line l passes through point P.
7 Line l is through point P.

So, if $(\mathscr{P}, \mathscr{L}, \mathscr{F})$ is an incidence plane, then $\mathscr{P}$ is the set of points, $\mathscr{L}$ is the set of lines, and $\mathscr{F}$ defines incidence between points and lines. We shall frequently use "off" for "not on." Further, if l and m are lines such that there is no point incident with both lines or if $l = m$, then we say that l is *parallel* to m or $l \parallel m$. Obviously, $l \parallel m$ implies $m \parallel l$.

For illustrative purposes only, consider $(\mathscr{P}, \mathscr{L} \ \mathscr{F})$ where $\mathscr{P} = \{A, B, C, D\}$, $\mathscr{L} = \{k, l, m, n\}$, and $\mathscr{F} = \{(B, k), (C, l), (D, l)\}$. See Figure 4.1. This incidence plane has the peculiarity of having a point which is not on any line and lines which pass through no point. Also, all the lines are parallel. Passing from the ridiculous to the sublime,

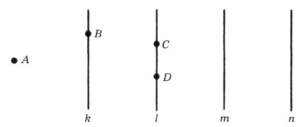

FIGURE 4.1

we leave this example to consider an example from analytic geometry.

If $\mathscr{L} = \{\{(x, y)|x, y \in \mathbf{R},\ ax + by + c = 0\}|a, b, c \in \mathbf{R},\ a^2 + b^2 \neq 0\}$, $\mathscr{P} = \{(x, y)|x, y \in \mathbf{R}\}$, and $\mathscr{F}$ is defined by set inclusion, meaning $((x, y),\ l) \in \mathscr{F}$ for $l \in \mathscr{L}$ iff $(x, y) \in l$, then $(\mathscr{P}, \mathscr{L}, \mathscr{F})$ is the familiar incidence plane called the *Real Cartesian Incidence Plane*. In this description, line $\{(x, y)|x, y \in \mathbf{R},\ ax + by + c = 0\}$ is a set of points satisfying a nondegenerate real linear equation in x and y. The plane is named after René Descartes (1596–1650). The Real Cartesian Incidence Plane is an example of the first of three types of incidence planes that we shall consider.

Axiom System I   An *affine plane* is an incidence plane such that

> AXIOM 1   If P and Q are two points, then there exists a unique line through P and Q.

> AXIOM 2   If P is any point off line l, then there exists a unique line through P that is parallel to l.

> AXIOM 3   There exist four points such that no three are on any line.

Considering only incidence and with the usual interpretation of "point" and "line," the Euclidean plane is an affine plane. Thus the axiom system for affine planes is relatively consistent with Euclidean plane geometry. Any inconsistency that could be deduced from the axiom system for affine planes would give an inconsistency in the Euclidean plane. For this simple looking axiom system we can actually prove consistency by giving a finite model $(\mathscr{P}, \mathscr{L}, \mathscr{F})$, where $\mathscr{P} = \{A, B, C, D\}$, $\mathscr{L} = \{\{A, B\}, \{A, C\}, \{A, D\}, \{B, C\}, \{B, D\}, \{C, D\}\}$, and $\mathscr{F}$ is determined by set inclusion, i.e., P on l iff P in l. See Figure 4.2. In this geometry there are exactly four points and exactly six lines! Is the line $\{A, C\}$ perpendicular to the line $\{B, D\}$? This is a trick question. The word "perpendicular" is a technical word that has not been defined. At this point the question makes as much sense as to

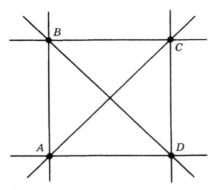

FIGURE 4.2

ask whether the point A is blue. Actually, since lines $\{A, C\}$ and $\{B, D\}$ have no point in common, the two lines are parallel by definition. If you thought the two lines were perpendicular, it is probably because you were misled by Figure 4.2. Beware: Figures help, but they may mislead! Since you may quickly check that this four point geometry is an affine plane, no inconsistency can be deduced from the axioms for an affine plane.

Geometers have their own word, *collineation*, for an isomorphism between incidence planes. A collineation from incidence plane $(\mathcal{P}_1, \mathcal{L}_1, \mathcal{F}_1)$ onto incidence plane $(\mathcal{P}_2, \mathcal{L}_2, \mathcal{F}_2)$ consists of a bijection $f: \mathcal{P}_1 \to \mathcal{P}_2$ and a bijection $g: \mathcal{L}_1 \to \mathcal{L}_2$ such that $(P, l) \in \mathcal{F}_1$ iff $(f(P), g(l)) \in \mathcal{F}_2$. Since there is obviously no one-to-one correspondence between all the points of the Euclidean plane considered as an affine plane and the points of the affine plane with just four points, we see that not all models of an affine plane are isomorphic. Therefore, the axiom system for affine planes is not categorical.

If one line in an affine plane has exactly n points then so does every line and the total number of points in n^2. Determining the possible values for n has been an open problem for many years.

Each of the axioms for an affine plane is independent of the other two. To show that Axiom 1 is independent, we need an incidence plane $(\mathcal{P}, \mathcal{L}, \mathcal{F})$ such that Axiom 2 and Axiom 3 hold but Axiom 1 fails. For such a model take $\mathcal{P}$ to be the set of points in Cartesian three-space, $\mathcal{L}$ to be all planes perpendicular to an axis, and $\mathcal{F}$ given by the usual incidence of Cartesian three-space. Once you get over any prejudice you might have that a plane in one geometry cannot be a line in some other geometry, it is trivial to check that this model has the desired properties.

Skipping Axiom 2 for the moment, Axiom 3 is seen to be independent by considering $(\mathcal{P}, \mathcal{L}, \mathcal{F})$ where $\mathcal{P}$ is an arbitrary set, $\mathcal{L} =$

$\{\mathscr{P}\}$, and $\mathscr{F} = \mathscr{P} \times \mathscr{L}$. Since there is exactly one line (which passes through every point), Axiom 1 must hold but Axiom 3 necessarily fails. The purpose of Axiom 3 is to omit trivial incidence planes. Note that Axiom 2 says *if* point P is off line l then something happens. Since there are no points off the only line in this geometry, Axiom 2 is never denied. We say that Axiom 2 holds *vacuously*.

We now turn to the important Axiom 2: If point P is off line l, then there exists a unique line through P that is parallel to l. You have no doubt seen it before. We shall have *much* more to say about this postulate later. For the moment, we want to show that the axiom is independent in our axiom system for affine planes. We need an incidence plane where there are at least four points of which no three are on one line and where every two points are on a unique line but such that Axiom 2 fails. The negation of Axiom 2 merely requires the existence of some particular point P_0 off some particular line l_0 such that there is not a unique line passing through P_0 and parallel to l_0. So there must be either no line through P_0 that is parallel to l_0 or there must be at least two lines through P_0 that are parallel to l_0. Let $\mathscr{P}_1 = \{A, B, C, D, E, F, G\}$, $\mathscr{L}_1 = \{\{A, B, F\}, \{A, C, E\}, \{A, D, G\}, \{B, C, D\}, \{B, E, G\}, \{C, F, G\}, \{D, E, F\}\}$, and $\mathscr{F}_1$ determined by point P in $\mathscr{P}_1$ is on line l in $\mathscr{L}_1$ iff P is in l. See Figure 4.3. It is quickly checked that Axiom 1 and Axiom 3 hold in incidence plane $(\mathscr{P}_1, \mathscr{L}_1, \mathscr{F}_1)$. In this seven point and seven line geometry, Axiom 2 fails because there are no parallel lines. Every two lines intersect in a unique point! We have now shown that the three axioms for affine planes are independent. We also have a model of our second type of incidence plane, defined next.

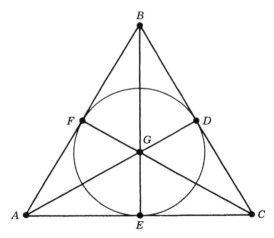

FIGURE 4.3

Axiom System 2   A *projective plane* is an incidence plane such that

AXIOM 1   If P and Q are two points, then there exists a unique line through P and Q.

AXIOM 2′   If l and m are two lines, then there exists a unique point on l and m.

AXIOM 3   There exist four points such that no three are on any line.

The existence of the finite projective plane $(\mathcal{P}_1, \mathcal{L}_1, \mathcal{F}_1)$ given above demonstrates that the axiom system for projective planes is consistent. There should be no question that Axiom 2′ is independent. To show that this axiom system is not categorical, we give a model $(\mathcal{P}_2, \mathcal{L}_2, \mathcal{F}_2)$ of a projective plane which is not isomorphic to a finite projective plane. Let O be any fixed point in Euclidean three-space. The elements of $\mathcal{P}_2$ are the Euclidean lines through O. The elements of $\mathcal{L}_2$ are the Euclidean planes through O. For $P \in \mathcal{P}_2$ and $l \in \mathcal{L}_2$, define $(P, l) \in \mathcal{F}_2$ iff in Euclidean three-space P is in l. Once you have suppressed any prejudice of what a point and a line should be, it is easily seen that $(\mathcal{P}_2, \mathcal{L}_2, \mathcal{F}_2)$ is a projective plane. For, if P and Q are two points, then P and Q lie on a unique line, since in Euclidean three-space two lines through O determine a unique plane through O. Also, if l and m are two lines, then l and m pass through a unique point, since in Euclidean three-space two planes through O determine a unique line through O. Any difficulty you might have in comprehending this model is psychological (it's dumb to say a line is a point!) or semantical ("line" is used with two meanings, as elements of $\mathcal{P}_2$ and as elements of $\mathcal{L}_2$). A common way around this is to use the adjectives "old" and "new." Then a new point is an old line, and two new points lie on a unique new line since two old lines through O determine a unique old plane through O. Any geometry isomorphic to $(\mathcal{P}_2, \mathcal{L}_2, \mathcal{F}_2)$ is the *real projective plane*.

The real projective plane $(\mathcal{P}_2, \mathcal{L}_2, \mathcal{F}_2)$ contains a copy of the incidence structure of the Euclidean plane. Consider the geometry determined by throwing away some fixed new line l and all the new points that were on l. See Figure 4.4. Admittedly, the resulting geometry does not look like a Euclidean plane at first glance. Let E be any Euclidean plane parallel to l and off O. The subgeometry is isomorphic to E. There is an obvious one-to-one correspondence between the remaining new points and all the Euclidean points of E and a one-to-one correspondence between the remaining new lines

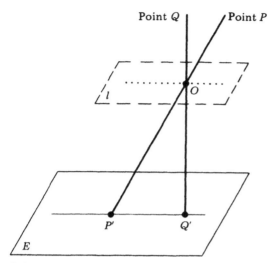

Point Q   Point P

FIGURE 4.4

and all the Euclidean lines in E. These correspondences are determined by set intersections as in Figure 4.4.

It is a fact that every affine plane can be extended to some projective plane. Let any affine plane $(\mathscr{P}, \mathscr{L}, \mathscr{F})$ be given. We shall construct a projective plane $(\mathscr{P}_3, \mathscr{L}_3, \mathscr{F}_3)$ that contains $(\mathscr{P}, \mathscr{L}, \mathscr{F})$ as a subgeometry. For each $n \in \mathscr{L}$ define $P_n = \{l|l \in \mathscr{L}, \ l \| n\}$, and let $l_x = \{P_n | n \in \mathscr{L}\}$. Since parallelism is an equivalence relation on the set of lines in an affine plane (Exercise 4.1), $P_n = P_m$ iff $n \| m$. So l_x is the set of all parallel pencils P_n of $(\mathscr{P}, \mathscr{L}, \mathscr{F})$, where P_n consists of all the old lines parallel to n. Let

$$\mathscr{P}_3 = \mathscr{P} \cup l_x, \qquad \mathscr{L}_3 = \mathscr{L} \cup \{l_x\},$$

and

$$\mathscr{F}_3 = \mathscr{F} \cup \{(P_n, n)|n \in \mathscr{L}\} \cup \{(P_n, l_x)|n \in \mathscr{L}\}.$$

Thus, all the old points are new points, and all the old lines are new lines. The set of new points consists of all the old points and all the old parallel pencils. To the set of old lines we have added only one new line l_x. For each old line n, we have added one new point P_n on n, and all the new points that are not old points have been put on the one new line l_x. To understand this model requires intellectual powers stronger than any old prejudices about what a point and a line are; there is no reason that a set of parallel lines in one geometry can't be a point in some other geometry. The verification of the fact that $(\mathscr{P}_3, \mathscr{L}_3, \mathscr{F}_3)$ is actually a projective plane is left as Exercise 4.3. If $(\mathscr{P}, \mathscr{L}, \mathscr{F})$ is a Euclidean plane, then $(\mathscr{P}_3, \mathscr{L}_3, \mathscr{F}_3)$ is the real projective plane, i.e., isomorphic to $(\mathscr{P}_2, \mathscr{L}_2, \mathscr{F}_2)$ above (Exercise 4.8).

Axiom 2 for an affine plane requires that there be exactly one line that is parallel to line l and passes through point P when P is off l. Axiom 2′ for a projective plane requires that there be exactly zero lines that are parallel to line l and pass through point P when P is off l. Axiom 2″ below requires that there be two lines that are parallel to line l and pass through point P when P is off l. Axioms 2, 2′, and 2″ are called *parallel postulates*.

Axiom System 3   A *hyperbolic plane* is an incidence plane such that:

AXIOM 1   If P and Q are two points, then there exists a unique line through P and Q.

AXIOM 2″   If P is any point off line l, then there exist two lines through P that are parallel to l.

AXIOM 3′   There exist four points such that no three are on any line; every line has a point on it.

If a person were marooned for many many years on the proverbial uninhabited desert island, then it is conceivable that he might possibly consider verifying that $(\mathscr{P}, \mathscr{L}, \mathscr{F})$ is a hyperbolic plane when $\mathscr{P}, \mathscr{L}$, and $\mathscr{F}$ are defined as follows. The ten digits are the points: $\mathscr{P} = \{0, 1, 2, 3, 4, 5, 6, 7, 8, 9\}$. The set $\mathscr{L}$ of lines consists of the twenty-five numbers 10, 15, 16, 20, 23, 24, 36, 39, 45, 47, 59, 67, 78, 80, 89, 128, 137, 149, 257, 269, 340, 358, 468, 560, and 790. $\mathscr{F}$ is defined by saying that point P is on line l iff P occurs as a digit of l. This is an example of a finite hyperbolic plane. In general, an incidence plane $(\mathscr{P}, \mathscr{L}, \mathscr{F})$ is said to be *finite* if both $\mathscr{P}$ and $\mathscr{L}$ have a finite number of elements.

If $\mathscr{P} = \{(x, y) \mid x, y \in \mathbf{R}, x > 0, y > 0\}$, $\mathscr{L} = \{\{(x, y) \mid (x, y) \in \mathscr{P}, ax + by + c = 0\} \mid a, b, c \in \mathbf{R}, a^2 + b^2 \neq 0\} \setminus \{\varnothing\}$, and $\mathscr{F}$ is defined by set inclusion, then $(\mathscr{P}, \mathscr{L}, \mathscr{F})$ is the subgeometry of the Real Cartesian Incidence Plane obtained by restricting ourselves to the first quadrant. We shall call this the *Quadrant Incidence Plane* or Q_1 (see Figure 4.5a). Replacing $\mathscr{P}$ by $\{(x, y) \mid x, y \in \mathbf{R}, y > 0\}$ we have Q_2, the *Halfplane Incidence Plane* (see Figure 4.5b); replacing $\mathscr{P}$ by $\{(x, y) \mid x, y \in \mathbf{R}, x > 0$ or $y > 0\}$ we have Q_3, the *Missing-Quadrant Incidence Plane* (see Figure 4.5c).

You should quickly convince yourself that the Quadrant Inci-

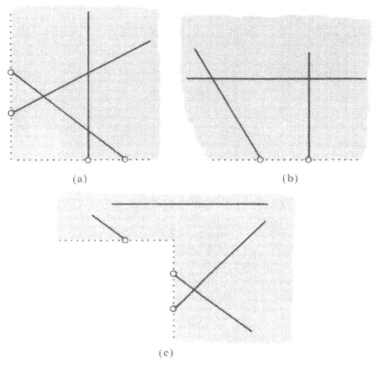

(a)

(b)

(c)

FIGURE 4.5

dence Plane Q_1 is a hyperbolic plane. See Figure 4.6. Using the points on the axes of the Real Cartesian Incidence Plane, even though these are not points in Q_1 itself, it is seen that if P is a point in Q_1, l is a line in Q_1, but (P, l) is not a flag in Q_1, then there are actually an infinite number of lines through P that are parallel to l. Since Axiom 2″ requires only that there be two such lines, the axiom is certainly satisfied.

The Halfplane Incidence Plane Q_2 is not a hyperbolic plane because in Q_2 there exists point P off line l such that there is a unique line through P that is parallel to l. See Figure 4.7. However, Q_2 is not an affine plane either since there exists point P off line l such that there are two lines through P that are parallel to l. In fact, if in Q_2 point P is off line l, then there is either exactly one or else an infinite number of lines passing through P that are parallel to l.

The Missing-Quadrant Incidence Plane Q_3 is neither a hyperbolic plane nor an affine plane. See Figure 4.8. Of course, neither Q_2 nor Q_3 is a projective plane as distinct parallel lines exist in each. To see that Q_3 is not isomorphic to Q_2, we need to find some incidence

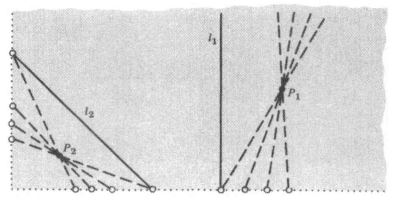

FIGURE 4.6

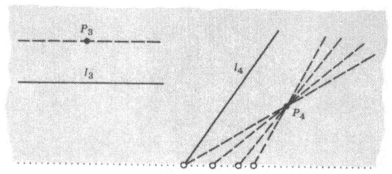

FIGURE 4.7

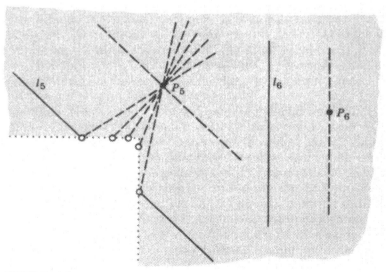

FIGURE 4.8

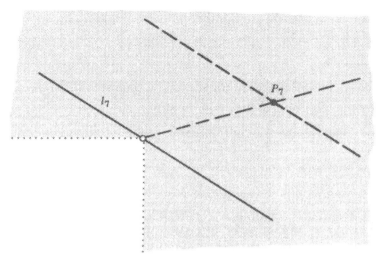

FIGURE 4.9

property of one plane that does not hold for the other. In Q_3 there exists a point P off a line l such that there are exactly two lines through P that are parallel to l. See Figure 4.9. If there were a collineation from Q_3 onto Q_2, then P and l would have to be mapped, respectively, to point P' in Q_2 and line l' in Q_2 such that there would be exactly two lines in Q_2 passing through P' that were parallel to l'. Since we have already noted that in Q_2 there is either exactly one or an infinite number of lines that are parallel to a given line and pass through a point off the given line, it follows that there can be no collineation from Q_3 onto Q_2. None of the incidence planes Q_1, Q_2, or Q_3 is isomorphic to the other.

4.3 EXERCISES

4.1 Show that parallelism is an equivalence relation on the set of lines of an affine plane but parallelism is not an equivalence relation on the set of lines of a hyperbolic plane.

● **4.2** Show that Axioms 1, 2′, and 3 are independent in the axiom system for projective planes.

● **4.3** Show that the incidence plane $(\mathscr{P}_3, \mathscr{L}_3, \mathscr{F}_3)$ constructed in the text from any affine plane $(\mathscr{P}, \mathscr{L}, \mathscr{F})$ is actually a projective plane.

● **4.4** True or False?

(a) The set $\{1, 2, 3, 4\}$ has three elements.

(b) If P and Q are two points, then possibly $P = Q$.

(c) $8/2 = 4$.

(d) "$8/2$" = "4."

(e) Today, "postulate" and "axiom" mean the same thing.

(f) A point may be a star, a rock, a flower, or a bird.

(g) Any two models of a consistent axiom system are isomorphic.

(h) Any worthwhile axiom system must be consistent.

(i) Any worthwhile axiom system must be categorical.

(j) If a statement is true for one model of an affine plane, then the statement is a theorem for any affine plane.

4.5 Read "Modern Axiomatic Methods and the Foundations of Mathematics" by Jean Deudonné (pages 251–266) in *Great Currents of Mathematical Thought* Vol. II, Edited by F. LeLionnais (Dover, 1971).

● **4.6** Give an example of a categorical axiom system.

● **4.7** Give a model of an incidence plane where three points determine a line.

● **4.8** Show that $(\mathscr{P}_3, \mathscr{L}_3, \mathscr{F}_3)$ is isomorphic to $(\mathscr{P}_2, \mathscr{L}_2, \mathscr{F}_2)$ when $(\mathscr{P}_3, \mathscr{L}_3, \mathscr{F}_3)$ is derived, as in the text, from a Euclidean plane.

4.9 Let l be a fixed line in any projective plane $(\mathscr{P}, \mathscr{L}, \mathscr{F})$. Show that $(\mathscr{P}_4, \mathscr{L}_4, \mathscr{F}_4)$ is an affine plane where $\mathscr{P}_4 = \mathscr{P} \setminus \{P | P \text{ on } l\}$, $\mathscr{L}_4 = \mathscr{L} \setminus \{l\}$, and $\mathscr{F}_4 = \mathscr{F} \setminus \{(P, l) | P \text{ on } l\}$.

4.10 Show that if $(\mathscr{P}, \mathscr{L}, \mathscr{F})$ is a projective plane, then $(\mathscr{L}, \mathscr{P}, \mathscr{F}')$ is a projective plane where $(l, P) \in \mathscr{F}'$ iff $(P, l) \in \mathscr{F}$.

4.11 Discussion questions: What is a point? What is a line?

*__4.12__ Read *Gödel's Proof* by Ernest Nagel and James R. Newman (New York University Press, 1958), or read "Goedel's Proof" in *The World of Mathematics* by James R. Newman (Simon and Schuster, 1956).

*__4.13__ Given any incidence plane $(\mathscr{P}, \mathscr{L}, \mathscr{F})$, show that $(\mathscr{P}, \mathscr{L}', \mathscr{F}')$ is isomorphic to $(\mathscr{P}, \mathscr{L}, \mathscr{F})$ iff for no two lines in $(\mathscr{P}, \mathscr{L}, \mathscr{F})$ is the set of points on one line equal to the set of points on the other where

$$\mathscr{L}' = \{\{P | (P, l) \in \mathscr{F}\} | l \in \mathscr{L}\}$$

and

$$\mathscr{F}' = \{(P, l') | P \in l', l' \in \mathscr{L}'\}.$$

So, if different lines of an incidence plane have different sets of points on them, we may assume $\mathscr{L} \subset 2^{\mathscr{P}}$ without loss of generality.

*4.14 Show that if an affine plane has a finite number of points, then there exists an integer n such that the number of points is n^2, the number of lines is $n(n+1)$, there are exactly n points on every line, and there are exactly $n+1$ lines through every point.

*4.15 Read "The Role of the Axiomatic Method" by R. L. Wilder in *The American Mathematical Monthly* Vol. 74 (1967), pp. 115–127.

GRAFFITI

A mathematical point is the most indivisible and unique thing which art can present.

John Donne

A line is not made up of points.

Aristotle

Why are you so sure parallel lines exist?

Believe nothing, merely because you have been told it, or because it is traditional, or because you have imagined it.

Gutama Buddha

Part One

ABSOLUTE GEOMETRY

Our study of the foundations of geometry begins with an examination of the common ground between non-Euclidean geometry and Euclidean geometry. This common ground is called absolute geometry and is independent of any assumption about parallel lines. In constructing this part of non-Euclidean geometry, we necessarily learn about the structure of Euclidean geometry as well. Throughout Part One we are most concerned with the actual development of an axiom system for the absolute plane. In building our structure, we are as much interested in the absence of certain propositions in the theory as the presence of others. We are never in the position of pretending we do not know something! Many models, including the Cartesian plane, are used to illustrate the growth of our axiom system. After selecting our five axioms for the absolute plane, we are forced to consider the theory of parallels.

CHAPTER 5

Models

5.1 MODELS OF THE EUCLIDEAN PLANE

The words "point" and "line" are usually undefined when studying the Euclidean plane in high school. Later every point is named in the usual fashion by a unique ordered pair of real numbers, called *coordinates*, and every ordered pair of real numbers is the name of some point. See Figure 5.1. The lines are then shown to be exactly the sets of all points with coordinates (x, y) that satisfy an equation $ax + by + c = 0$ for real numbers a, b, c with not both a and b zero. This introduction of coordinates enables one to use algebraic methods to solve geometric problems.

Now, taking a different approach, we construct a geometry by *defining* a *point* to *be* an ordered pair of real numbers and every ordered pair of real numbers to *be* a point. Before, $(2, 3)$ was the *name* of a point; now, $(2, 3)$ *is* a point. Further, *lines* are *defined* to be exactly the sets of all points (x, y) such that x and y satisfy an equation $ax + by + c = 0$ for real numbers a, b, c with not both a and b zero. At this point we have the *Real Cartesian Incidence Plane*. Then, *distance* from (x_1, y_1) to (x_2, y_2) is *defined* to be the real number $[(x_2 - x_1)^2 + (y_2 - y_1)^2]^{1/2}$. We'll forego actually going on to define *angle* and *angle measure*. The result of all this is the geometry called the *Cartesian plane*. There are no geometric axioms here; one can immediately start proving theorems based on the axioms and theorems of the real numbers. Saying there is no difference between the high school Euclidean plane and the Cartesian plane is almost correct. Indeed, the whole

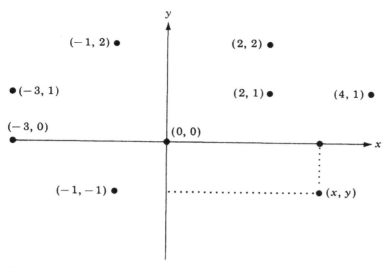

FIGURE 5.1

idea is that the Cartesian plane is a model of the Euclidean plane! It is no exaggeration to state that the Cartesian plane is the most useful model ever devised by man.

The Cartesian plane is named after René Descartes (1596–1650), the founder of modern philosophy. Descartes' *La Géométrie* appeared in 1637 as the third appendix of his *Discours (A Discourse on the Methods of Correct Reasoning and Seeking Truth in the Sciences.)* Algebra and geometry were directly combined for the first time in a published work. Actually Pierre de Fermat (1608–1665) had accomplished the same thing a couple of years earlier, but Fermat did not publish the work. As Fermat and Descartes independently began the development of analytic geometry, so Newton and Leibniz independently began the development of the calculus. Gottfried Wilhelm Leibniz (1646–1716) first published work on the calculus, but Isaac Newton (1647–1727) did his work earlier. An intelligible account of Newton's methods of calculus finally appeared in 1704 as an appendix to his *Opticks.* The same book contained a second appendix on enumerating curves of third degree. It is in this second appendix that the use of negative numbers, as well as positive numbers, for coordinates first appears in any systematic way. Negative numbers have been called *absurd numbers, false numbers,* and *fictitious numbers* at various times. An explicit presentation of the material usually found in the first dozen pages of any modern book on analytic geometry finally appeared in 1797 in the text *Traité de calcul* by Sylvestre Francois Lacroix (1765–1843). Certainly the fundamental assumption that associates the geometry of Euclid and the algebra of the real numbers is the one-to-one correspondence between the points on a Euclidean line and the set of real numbers. The real numbers were not placed on a logical foundation until 1872, two hundred and thirty-five years after Descartes' initial work. The Cartesian plane, as we know it, did not appear overnight, as do mushrooms.

In the following chapters *we are not going to pretend ignorance of the Cartesian plane!* Nor are we going to be so ignorant as to pretend knowledge about things we do not know. If we were pressed to give some definition of the Euclidean plane now, we could say the Euclidean plane is anything that is isomorphic to the Cartesian plane.

Another model of the Euclidean plane is the *Gauss plane*. Here the set of points is the set of all complex numbers. For example, $1 - i$, 2, i, and $2 + 3i$ *are* points. For the lines we take the sets of all points Z that satisfy an equation $BZ + \overline{BZ} + C = 0$ where B and C are complex numbers with $B \neq 0$ and C real. If $z_1 = x_1 + y_1 i$ and $z_2 = x_2 + y_2 i$ with x_1, x_2, y_1, y_2 real, then the distance from z_1 to z_2 is defined to be $|z_2 - z_1|$. The Gauss plane is a model of the Euclidean plane because the Gauss plane is isomorphic to the Cartesian plane. The mapping which takes (x, y) to $x + yi$ for all real x, y is a bijection from the set of points of the Cartesian plane onto the set of points of the Gauss plane. It can be checked that this mapping induces a collineation, taking the line in the Euclidean plane with equation $ax + by + c = 0$ to the line in the Gauss plane with equation $BZ + \overline{BZ} + C = 0$ where $B = a - bi$ and $C = 2c$. Since this mapping also preserves distance, it follows that the Cartesian plane and the *Gauss plane* are isomorphic.

The Gauss plane, which is obviously named after Carl Friedrich Gauss (1777–1855), is sometimes called the *Cauchy plane* after Augustin Louis Cauchy (1787–1857), who popularized complex numbers. The plane is also known as the *Argand diagram* as Jean Robert Argand (1768–1823) had previously noted in 1806 that the complex number $x + yi$ can be represented by the point (x, y). This supposedly concrete representation of a complex number was very influential in the acceptance of the so-called imaginary numbers. By historical accident, this plane of complex numbers is *not* called the *Wessel plane,* although Caspar Wessel (1745–1815) had published the correspondence between complex numbers and points of the Euclidean plane in 1798.

The next model of the Euclidean plane is described quite informally. This model, as well as all the remaining models in this section, is given for the sole purpose of stretching your imagination. Once looked at, these models may be safely forgotten. We start with a rectangular sheet of paper. Let's agree that the paper approximates a piece of the Euclidean plane. (That should be amusing, considering that the Euclidean plane was devised to be a system which described

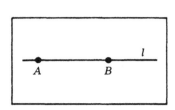

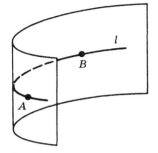

FIGURE 5.2   **FIGURE 5.3**

reality.) On the paper we draw a line l through two points A and B which are six units apart (see Figure 5.2). We then stand the paper on a desk so that a longer side touches the desk in an arc of a parabola. The surface in space represented by our sheet of paper is a Euclidean plane *provided* we interpret "point," "line," "distance," and "angle measure" exactly as they were before we bent the paper. For example the distance from A to B in Figure 5.3 is still defined to be 6.

For another model of the Euclidean plane, we start by observing that $f(x) = e^x$ defines a bijection from the set of all reals onto the set of positive reals. Using this fact, we can map all the points of the Cartesian plane in a one-to-one fashion onto the points of the first quadrant of the Cartesian plane by the mapping α which sends (x, y) to (e^x, e^y). The points of this model are defined to be the ordered pairs of positive real numbers. The lines of this model are defined to be exactly those sets of points that are the images of the lines of the Cartesian plane under the mapping α.

So lines in our model have equations $x = e^a$ or $y = x^m e^b$ coming from the lines of the Cartesian plane with equations $x = a$ or $y = mx + b$, respectively. For example, the set of all points (x, y) in the model

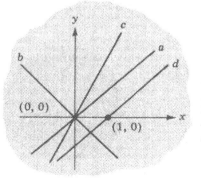

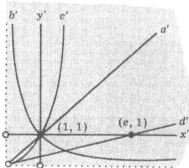

FIGURE 5.4

such that $xy=1$ is a line in this model. See Figure 5.4. Of course, for the model to be a model of the Euclidean plane we *copy* the distance and angle measure as well. So the distance from (x_1, y_1) to (x_2, y_2) in our model is the real number $[\ln^2(x_2/x_1) + \ln^2(y_2/y_1)]^{1/2}$ as can be checked by observing that α sends $(\ln x, \ln y)$ to (x, y). Without being told about the map α, it might take some time to recognize that this model is indeed isomorphic to the Cartesian plane.

Finally, we indicate four more models of the Euclidean plane that might appeal to those who really like to get their hands on things. We shall need to know that $g(x) = \tanh x$ and $h(x) = (2/\pi)\arctan x$ each define bijections from the set of all reals onto the set of reals between -1 and $+1$. We let $\mathscr{P}$ be the set of points in the Cartesian plane that are in the interior of the square with equations $|x+y| + |x-y| = 2$; the square has vertices $(1, 1)$, $(-1, 1)$, $(-1, -1)$, and $(1, -1)$. (See Figure 5.5.) The mapping β which sends (x, y) to $(\tanh x, \tanh y)$ is a bijection from the points of the Cartesian plane onto $\mathscr{P}$. We define $\mathscr{L}$, the set of lines of our model, to be such that β determines a collineation. In other words, we *copy* lines as the images of the lines in the Cartesian plane under the mapping β. If we also copy distance and angle measure from the Cartesian plane, then the result is a model of the Euclidean plane. (The equations for lines and the formulas for distance and angle measure are horrid.) Another model of the Euclidean plane having $\mathscr{P}$ as its set of points can be obtained by defining lines, distance, and angle measure such that β' is an isomorphism where β' sends (x, y) to $((2/\pi)\arctan x, (2/\pi)\arctan y)$.

Is the Euclidean plane rectangular? The preceding string of words with a question mark at the end is not a question; it doesn't make any sense. However, for those who like "round" models and know about polar coordinates, let $\mathscr{P}$ now be the points in the interior of the unit circle in the Cartesian plane. The circle has equation $x^2 + y^2 = 1$. (See Figure 5.6.) The mapping γ sending the point in the Cartesian plane with polar coordinates (r, θ) to the point with polar coordinates $(\tanh r, \theta)$ is a bijection from the points of the Cartesian plane onto $\mathscr{P}$.

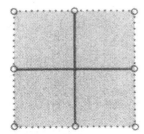

FIGURE 5.5

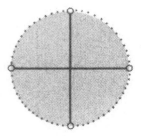

FIGURE 5.6

The same can be said for the mapping γ' sending the point with polar coordinates (r, θ) to the point with polar coordinates $((2/\pi)$ arc-tan $r, \theta)$. Then as for the previous two models, one can define lines, distance, and angle measure to obtain a model of the Euclidean plane such that γ is an isomorphism from the Cartesian plane onto the model. Likewise, still another model is obtained by making the definitions such that γ' is an isomorphism.

5.2 MODELS OF INCIDENCE PLANES

There are several incidence planes $(\mathscr{P}, \mathscr{L}, \mathscr{F})$ that are referred to in later chapters. These are listed together for easy reference. In each case the elements of the set $\mathscr{L}$ of lines are subsets of the set $\mathscr{P}$ of points. The graph $\mathscr{F}$ is always assumed to be determined by set inclusion. So a line is a set of points, and point P is incident with line l iff point P is an element of line l. The list begins with our old friend the Cartesian plane, but here we are content to restrict ourselves to incidence. Later we shall add distances *(plural!)* to this incidence plane.

Model 1 *The Real Cartesian Incidence Plane* Points are defined to be the ordered pairs of real numbers; $\mathscr{P} = \{(x, y) \mid x, y \in \mathbf{R}\}$. A line is the set of all points (x, y) that satisfy some equation $ax + by + c = 0$ where $a, b, c \in \mathbf{R}$ and not both a and b are zero. Conversely, every such set is a line. This model is certainly an affine plane. If $x_1 \neq x_2$, then the line thru (x_1, y_1) and (x_2, y_2) is said to have *slope* $(y_2 - y_1)/(x_2 - x_1)$. A line with equation $y = mx + b$ has slope m.

Model 2 *The Rational Cartesian Incidence Plane* Points are defined to be the ordered pairs of rational numbers; $\mathscr{P} = \{(x, y) \mid x, y \in \mathbf{Q}\}$. A line is the set of all points (x, y) that satisfy some equation $ax + by + c = 0$ where $a, b, c \in \mathbf{Q}$ and not both a and b are zero. Conversely, every such set is a line. If (x_1, y_1) and (x_2, y_2) are two points, then the points determine the unique line having equation $(y_2 - y_1)x + (x_1 - x_2)y + (x_2 y_1 - x_1 y_2) = 0$. Also, as in Model 1, lines with equations $a_1 x + b_1 y + c_1 = 0$ and $a_2 x + b_2 y + c_2 = 0$ are parallel iff $a_1 b_2 = a_2 b_1$ and otherwise intersect in point

$$\left(\frac{b_1 c_2 - b_2 c_1}{a_1 b_2 - a_2 b_1}, \frac{a_2 c_1 - a_1 c_2}{a_1 b_2 - a_2 b_1} \right).$$

It is not improper to think of Model 2 as Model 1 with a lot of holes poked in it.

Model 3 *The Complex Cartesian Incidence Plane* Points are defined to be the ordered pairs of complex numbers; $\mathcal{P} = \{(x, y) | x, y \in \mathbf{C}\}$. A line is the set of all points (x, y) that satisfy some equation $ax + by + c = 0$ where $a, b, c \in \mathbf{C}$ and not both a and b are zero. Conversely, every such set is a line. Model 3 should not be confused with the Gauss plane, which is isomorphic to Model 1. The same formulas that were given for Model 2 also apply here. One might even think of Model 1 as Model 3 with a lot of holes poked in it.

The first three models are all affine planes. Each of these planes is determined by some field. Actually every field determines an affine plane, formed by replacing the real numbers in Model 1 by elements from that field. The formulas given in Model 2 still apply for an arbitrary field.

Model 4 *The Space Incidence Plane* The points and lines are those of ordinary Euclidean three-space, where the lines are thought of as sets of points. If you like, think of the three-dimensional Cartesian coordinate system with the usual three axes. If you object by saying this is not a plane, then you are probably prejudiced by previous experience with the word *plane*. It can be shown that Model 4 is isomorphic to a subgeometry of Model 3. Model 4 is really a very nice example of an incidence plane. In fact, according to our definitions, the Space Incidence Plane is a hyperbolic plane.

Model 5 *The Quadrant Incidence Plane* Points are the ordered pairs of positive real numbers; $\mathcal{P} = \{(x, y) | x, y \in \mathbf{R}^+\}$. A line is the nonempty set of all points (x, y) that satisfy some equation $ax + by + c = 0$ where $a, b, c \in \mathbf{R}$ and not both a and b are zero. Conversely, every such set is a line. This model is compared with Model 6 and Model 7 at the end of Section 4.2.

Model 6 *The Halfplane Incidence Plane* Points are the elements of $\mathcal{P}$ where $\mathcal{P} = \{(x, y) | x \in \mathbf{R} \text{ and } y \in \mathbf{R}^+\}$. The lines are defined as in Model 5. This model is compared with Model 5 and Model 7 at the end of Section 4.2

Model 7 *The Missing-Quadrant Incidence Plane* Points are the elements of $\mathcal{P}$ where $\mathcal{P} = \{(x, y) | x, y \in \mathbf{R}, \text{ and } x \in \mathbf{R}^+ \text{ or } y \in \mathbf{R}^+\}$. The lines are defined as in Model 5. This model is compared with Model 5 and Model 6 at the end of Section 4.2

Model 8 *The Missing-Strip Incidence Plane* Points are the elements of $\mathcal{P}$ where $\mathcal{P} = \{(x, y) | x, y \in \mathbf{R}, \text{ and } x \leq 1 \text{ or } x > 2\}$. The lines are defined as in Model 5. This model is like the previous three models in that a set of points is removed from the Real Cartesian Incidence Plane. Model 8 contains all points (x, y) of Model 1 except those for which $1 < x \leqq 2$.

Model 9 *The Cubic Incidence Plane*   Points are the same as for Model 1; $\mathscr{P} = \{(x, y) | x, y \in \mathbf{R}\}$. A line is either the set of all points (x, y) that satisfy some equation $y = (ax + b)^3$ with $a, b \in \mathbf{R}$ or else the set of all points (x, y) that satisfy some equation $x = c$ with $c \in \mathbf{R}$. Conversely, every such set is a line. Some of the lines of this geometry are cubic curves in Model 1. Nevertheless, they are lines here. Some lines are indicated in Figure 5.7. That two points determine a unique line is left for Exercise 5.7.

Model 10 *The Moulton Incidence Plane*   Points are the same as for Model 1; $\mathscr{P} = \{(x, y) | x, y \in \mathbf{R}\}$. A line is the set of all points (x, y) that satisfy one of the following three types of equations where $a, b, m \in \mathbf{R}$:

$x = a,$

$y = mx + b \qquad \text{with} \quad m \leqq 0,$

$y = \begin{cases} mx + b & \text{if} \quad x \leqq 0 \\ \frac{1}{2}mx + b & \text{if} \quad x > 0 \end{cases} \quad \text{with} \quad m > 0.$

Conversely, every such set is a line. So those lines of Model 1 that have

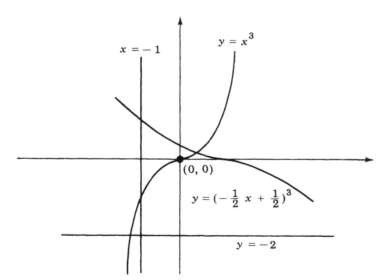

$x = -1$

$y = x^3$

$(0, 0)$

$y = \left(-\frac{1}{2} x + \frac{1}{2}\right)^3$

$y = -2$

FIGURE 5.7

either no defined slope, negative slope, or zero slope are lines in this model. The points of such a line satisfy some equation of the first two types. The remaining lines of Model 10, given by some equation of the third type, might be called *bent lines* in Model 1. However, these are lines for this model. (See Figure 5.8.) Note that the set of all points (x, y) satisfying equation $y = 3x + 4$ is not a line! This model, which is always encountered in the study of projective planes, was given in 1902 by the American mathematician Forest R. Moulton.

Do distinct points (x_1, y_1) and (x_2, y_2) lie on a unique line in the Moulton Incidence Plane? We may suppose $x_1 \leqq x_2$. If $x_1 = x_2$ or $y_2 \leqq y_1$, then the unique line through the two points has the same equation as in Model 1. If x_1 and x_2 are either both positive or both negative, then it should not be difficult to find the equation of the unique line through the two points. Suppose now that $x_1 < 0 < x_2$ and $y_1 < y_2$. Then a line through the two points must have an equation of the third type and pass through $(0, b)$ for some b. Borrowing the idea of slope from Model 1, we see that it is necessary and sufficient to have $m = (b - y_1)/(0 - x_1)$ and $\frac{1}{2}m = (y_2 - b)/(x_2 - 0)$. From these equations it follows that m and b are uniquely determined. Thus, when $x_1 < 0 < x_2$ and $y_1 < y_2$, the unique line through (x_1, y_1) and (x_2, y_2) has equation of the third type where $m = 2(y_2 - y_1)/(x_2 - 2x_1)$ and $b = (x_2 y_1 - 2x_1 y_2)/(x_2 - 2x_1)$. Hence two points always determine a unique line. A moment's reflection will show that, given point P and line l, there is a unique line parallel to l that passes through P. Therefore, the Moulton Incidence Plane is an affine plane.

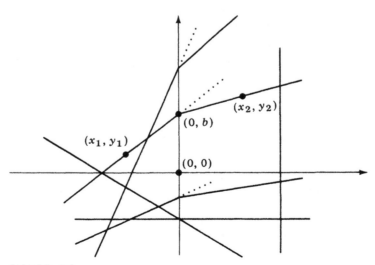

FIGURE 5.8

Model 11   *The Poincaré Incidence Plane*   Points are the elements of $\mathscr{P}$ where $\mathscr{P} = \{(x, y) \mid x, y \in \mathbf{R}$ and $x^2 + y^2 < 1\}$. A line is either the set of all points (x, y) that satisfy some equation $(x - a)^2 + (y - b)^2 = a^2 + b^2 - 1$ with $a, b \in \mathbf{R}$ such that $a^2 + b^2 > 1$ or else the set of all points (x, y) that satisfy some equation $ax + by = 0$ with $a, b \in \mathbf{R}$ such that $a^2 + b^2 \neq 0$. Conversely, every such set is a line. So the points are exactly those points we think of as being in the interior of the unit circle in the Cartesian plane. In the Cartesian plane, equation $(x - a)^2 + (y - b)^2 = a^2 + b^2 - 1$ describes the circle with center (a, b) and radius r where $1^2 + r^2 = a^2 + b^2$. Recalling that two circles are orthogonal in the Cartesian plane iff their tangents are perpendicular at a point of intersection, it follows (see Figure 5.9) that the circle described by the equation is orthogonal to the unit circle. In the Cartesian plane, equation $ax + by = 0$ describes a line through $(0, 0)$. Therefore, a line in the Poincaré Incidence Plane is either the set of all points in the Cartesian plane that lie in the interior of the unit circle and on a circle orthogonal to the unit circle or else the set of all points in the Cartesian plane that lie in the interior of the unit circle and on a Cartesian line through $(0, 0)$. See Figure 5.10. The Poincaré Incidence Plane is a very important example of a hyperbolic plane and is named after the great mathematician Henri Poincaré (1854–1912).

Model 12   *The Poincaré Halfplane Incidence Plane*   Points are the same as for Model 6; $\mathscr{P} = \{(x, y) \mid x \in \mathbf{R}, y \in \mathbf{R}^+\}$. However, here a line is either the set of all points (x, y) that satisfy some equation $(x - a)^2 + y^2 = r^2$ with $a \in \mathbf{R}$ and $r \in \mathbf{R}^+$ or else the set of all points (x, y) that satisfy some equation $x = a$ with $a \in \mathbf{R}$. Conversely, every such set is a line. See Figure 5.11. Model 12 is a hyperbolic plane.

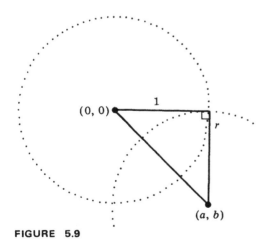

FIGURE 5.9

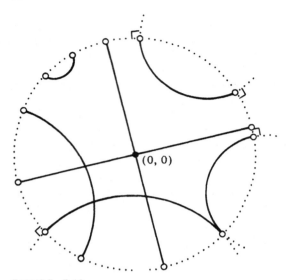

FIGURE 5.10

Model 13 *The Cayley–Klein Incidence Plane* Points are the same as for Model 11; $\mathscr{P} = \{(x, y) \,|\, x, y \in \mathbf{R}, x^2 + y^2 < 1\}$. However, here a line is the nonempty set of all points (x, y) that satisfy some equation $ax + by + c = 0$ with a, b, $c \in \mathbf{R}$ but not both a and b zero. Conversely, every such set is a line. See Figure 5.12. It is very easy to see that this model is a hyperbolic plane. The model is named after both Arthur Cayley (1821–1895) and Felix Klein (1849–1929). We shall see a lot more of this particular model.

Model 14 *The Sphere Incidence Plane* The points are the Euclidean

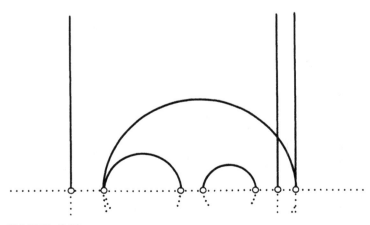

FIGURE 5.11

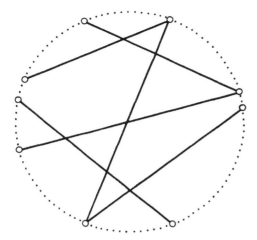

FIGURE 5.12

points on a Euclidean sphere, and the lines are the great circles thought of as sets of points. (A great circle is a circle on the sphere whose center is the center of the sphere.) This model is different from all others that have been considered in that here it may take *three* points to determine a line. There are an infinite number of great circles passing through both the north pole and the south pole. Such opposite points on a sphere are called *antipodal points.*

Model 15 *The Riemann Incidence Plane* The set of points is the set of all pairs of antipodal points of the Euclidean unit sphere. Taking the sphere to be the unit sphere in Cartesian three-space, a point is then a pair $\{(x, y\, z), (-x, -y, -z)\}$ where $x^2 + y^2 + z^2 = 1$. A line is the set of all pairs of antipodal points which lie on a fixed great circle, and for each great circle such a set is a line. Although this model is related to Model 14, here two points do determine a unique line. This model, named after Bernard Riemann (1826–1866), is a real projective plane (Exercise 5.10).

Henceforth M1, M2, . . . , and M15 will stand for Model 1, Model 2, . . . , and Model 15, respectively.

5.3 EXERCISES

● **5.1** For the incidence plane determined by the field of two elements, the points are defined to be the ordered pairs of elements of the field. A line is the set of all points (x, y) that satisfy some equation $ax + by + c = 0$ where a, b, c are elements of the field but not both a and b are 0. Conversely, every such set is a line. Now, find all the points and all the lines. Have you seen this geometry before?

● **5.2** Is the following incidence plane isomorphic to a familiar incidence plane? The set of points is exactly the same as the set of points of M8, the Missing-Strip Incidence Plane. A line is the set of all points (x, y) that satisfy any one of the following two types of equations where $a, b, m \in$ **R**:

$$x = a \quad \text{with} \quad a \leq 1 \text{ or } a > 2;$$

$$y = \begin{cases} mx + b & \text{if } x \leq 1 \\ mx + b - m & \text{if } x > 2. \end{cases}$$

Conversely, every such set is a line.

5.3 Show that M13, the Cayley–Klein Incidence Plane, is a hyperbolic plane.

● **5.4** In M10, the Moulton Incidence Plane, find the equations of the six lines determined by the four points $(-1, -1)$, $(1, -1)$, $(2, 1)$, and $(1, 3)$.

● **5.5** True or False?

(a) In the Gauss plane the distance from $2 + 3i$ to $7 - 9i$ is 13.

(b) In M1, the Real Cartesian Incidence Plane, the line with equation $y = x$ intersects the set of points (x, y) such that $x^2 + y^2 = 1$.

(c) In M2, the Rational Cartesian Incidence Plane, the line with equation $y = x$ intersects the set of points (x, y) such that $x^2 + y^2 = 1$.

(d) In M1, the Real Cartesian Incidence Plane, the line with equation $y = 5$ intersects the set of points (x, y) such that $x^2 + y^2 = 1$.

(e) In M3, the Complex Cartesian Incidence Plane, the line with equation $y = 5$ intersects the set of points (x, y) such that $x^2 + y^2 = 1$.

(f) In M1, the line with equation $2x - 8y + 3 = 0$ has slope 4.

(g) In M8, the Missing-Strip Incidence Plane, the lines with equation $y = x$ and $y = 3x - 4$ are parallel.

(h) M10, the Moulton Incidence Plane, is an affine plane.

(i) In M10, the line through $(-1, -1)$ and $(1, 1)$ contains $(0, 0)$.

(j) In M10, the line through $(-1, 1)$ and $(1, -1)$ contains $(0, 0)$.

5.6 Read "The Heroic Age in Geometry," which is Chapter 24 of Carl B. Boyer's excellent book *A History of Mathematics* (Wiley, 1968).

● **5.7** Show that M9 is an affine plane.

● **5.8** Why is M2 not isomorphic to any of the other fourteen models in Section 5.2?

5.9 Check that M2 and M3 are affine planes.

● **5.10** Show that M15 is a projective plane.

5.11 Show that M11 and M12 are hyperbolic planes.

5.12 For information on Dedekind, Poincaré, and Cantor read the last three chapters of E. T. Bell's classic *Men of Mathematics*.

5.13 In M8 find two lines l_1 and l_2 and a point P off each such that through P there are exactly two lines parallel to both l_1 and l_2.

5.14 For each of M5 and M11, find two lines l_1 and l_2 and a point P off each such that through P there is exactly one line parallel to both l_1 and l_2.

5.15 Show that M4 is not isomorphic to either M5 or M11.

5.16 In M8 find two lines l_1 and l_2 and a point P off each such that through P there are exactly three lines parallel to both l_1 and l_2.

5.17 For each of M5 and M11, find two lines l_1 and l_2 and a point P off each such that through P there are exactly two lines parallel to both l_1 and l_2.

*● **5.18** Find a model isomorphic to M1 such that the set of points is the set of real numbers.

***5.19** Read "A Simple Non-Desarguesian Plane Geometry" by F. R. Moulton in *Transactions of the American Mathematical Society* Vol. 3 (1902), pp. 192–195.

***5.20** Show that M1 and M9 are isomorphic. Show that M11, M12, and M13 are isomorphic.

***5.21** Show that all the isomorphisms between any two of the fifteen models are given by the previous exercise.

GRAFFITI

The truth is that other systems of geometry are possible, yet after all, these other systems are not spaces but other methods of space measurements. There is one space only, though we may conceive of many different manifolds, which are contrivances or ideal constructions invented for the purpose of determining space.

Paul Carus

Think of the image of the world in a convex mirror. . . . A well-made convex mirror of moderate aperture represents the objects in front of it as apparently solid and in fixed positions behind its surface. But the images of the distant horizon and of the sun in the sky lie behind the mirror at a limited distance, equal to its focal length. Between these and the surface of the mirror are found the images of all the other objects before it, but the images are diminished and flattened in proportion to the distance of their objects from the mirror. . . . Yet every straight line or plane in the outer world is represented by a straight line or plane in the image. The image of a man measuring with a rule a straight line from the mirror, would contract more and more the farther he went, but with his shrunken rule the man in the image would count out exactly the same number of centimeters as the real man. And, in general, all geometrical measurements of lines and angles made with regularly varying images of real instruments would yield exactly the same results as in the outer world, all lines of sight in the mirror would be represented by straight lines of sight in the mirror. In short, I do not see how men in the mirror are to discover that their bodies are not rigid solids and their experiences good examples of the correctness of Euclidean axioms. But if they could look out upon our world as we look into theirs without overstepping the boundary, they must declare it to be a picture in a spherical mirror, and would speak of us just as we speak of them; and if two inhabitants of the different worlds could communicate with one another, neither, as far as I can see, would be able to convince the other that he had the true, the other the distorted, relation. Indeed I cannot see that such a question would have any meaning at all, so long as mechanical considerations are not mixed up with it.

Helmholtz

CHAPTER 6

Incidence Axiom and Ruler Postulate

6.1 OUR OBJECTIVES

Our goal in this text is to learn something about the foundations of Euclidean geometry. We shall accomplish this by studying non-Euclidean geometry! Although this may strike you as strange at first, there are two good reasons for this approach. The principal reason is that you know too much about Euclidean geometry. It really is more difficult to study something that is very familiar because it is hard to keep in mind the distinction between the mathematical system that has been developed at any given time and what you feel has to be true. Of course, the second reason for this approach is to learn something about non-Euclidean geometry itself. The celebrated man-in-the-street has heard about non-Euclidean geometry, and every student of mathematics should know something about the subject.

Our aim is to develop that geometry that is very like the Euclidean plane except that the usual parallel postulate fails. The axioms we add to our system will be motivated by what we think the Euclidean plane should be but restricting ourselves to avoiding a parallel postulate for as long as is reasonably possible.

It is reasonable to ask why we shall be limiting ourselves to consideration of planes. Why not study systems motivated by our idea of Euclidean three-space? It turns out not to make much difference. The deep problems that arise involve consideration of only one plane at a time anyway. So to make matters easier we consider only planes in the beginning. Later, the extension from a plane to three-space is surprisingly easy.

Before starting the development of the axiom systems that are the topic of this book, we emphasize that it is the *formation* of a system that should have most of our attention in the beginning. As the system grows, our attention will be diverted more and more to the theory itself.

Excluding exercises, the *theory* consists of those paragraphs that are headed **Undefined terms, Axiom, DEFINITION, Theorem, Corollary,** or *Proof.* Everything else should be considered *discussion.* In the discussion we talk *about* the theory. Your life will be much happier if you keep in mind the distinction between the theory itself and the discussion about the theory. To aid you in doing this, the end of a proof is marked ■. The exercises add to the theory and to the discussion of the theory.

Italics in the discussion are used either for emphasis or to call attention to the fact that we are using words in an informal way. Definitions that occur in the theory are always in bold-face italic.

6.2 AXIOM 1: THE INCIDENCE AXIOM

We announce the setting for our axiom system by declaring our *preliminary assumptions* to be language, logic, set theory, and the real numbers.

The theory begins:

Undefined terms: $\mathscr{P}, \mathscr{L}, d, m.$

Axiom 1 *Incidence Axiom*

  a $\mathscr{P}$ and $\mathscr{L}$ are sets; an element of $\mathscr{L}$ is a subset of $\mathscr{P}$.
  b   If P and Q are distinct elements of $\mathscr{P}$, then there is a unique element of $\mathscr{L}$ that contains both P and Q.
  c   There exist three elements of $\mathscr{P}$ not all in any element of $\mathscr{L}$.

We are going to call the elements of $\mathscr{P}$ *points* and the elements of $\mathscr{L}$ *lines.* By (a) of the Incidence Axiom, we are taking the point of view that a line *is* a set of points. Thus, we automatically have an incidence relation for points and lines given by set membership. Because of (b), the Incidence Axiom might be called the *Straightedge Axiom.* We need (c) to get our *plane* off the ground, as without this there might be no points or lines at all or there might be just exactly one line.

DEFINITION 6.1 An element of $\mathscr{P}$ is called a **point;** an element of $\mathscr{L}$ is called a **line.** If point P is in line l, then we say that P is **on** l, l is **on** P, l **passes through** P, or that P and l are **incident. Off** means not on. If P is a point in each of two or more sets, then the sets **intersect** at P. We say that line l is **parallel** to line m and write $l \parallel m$ if either l and m do not intersect or $l = m$. A set S of points is **collinear** if S is a subset of a line. Two or more sets of points are **collinear** if their union is collinear. If two or more lines intersect at one point, then the lines are said to be **concurrent.** "Two points determine a line" means (b) of the Incidence Axiom. The unique line determined by distinct points P and Q is $\overleftrightarrow{PQ}$.

It is a good habit to read the symbol "$\overleftrightarrow{PQ}$," just defined, as "*line P Q*" since we are reserving the symbol "PQ" for something else. We are ready to prove our first theorem.

Theorem 6.2 If R and S are distinct points on $\overleftrightarrow{PQ}$, then $\overleftrightarrow{RS} = \overleftrightarrow{PQ}$. In particular, $\overleftrightarrow{QP} = \overleftrightarrow{PQ}$.

Proof Corollary of (b) in the Incidence Axiom. ■

Theorem 6.3 If l is a line, then $l \parallel l$. If l and m are lines, then $l \parallel$ m implies m $\parallel l$.

Proof The statements follow immediately from the definition of parallel lines. ■

Note that parallelism is a reflexive, symmetric relation on $\mathscr{L}$. We do *not* know that parallelism is an equivalence relation on the lines as we have no way of proving that parallelism is transitive.

Theorem 6.4 Two lines intersect in at most one point. Two non-parallel lines intersect in exactly one point. There exist three lines not all on one point.

Proof Two distinct lines cannot intersect in two distinct points by Theorem 6.2. If two lines are not parallel, then their intersection is not empty and, hence, must contain exactly one point. Requirements (b) and (c) of the Incidence Axiom imply the existence of three nonconcurrent lines. ■

Our first three theorems are necessarily simple and deal only with incidence. This must be so as we have only one axiom and that deals solely with incidence. We cannot infer the existence of non-parallel lines from Theorem 6.4. That theorem just says that *if* there are two nonparallel lines then they intersect in a unique point. Even though a parallel axiom would deal only with incidence, we intentionally do not state such an axiom. Recall our aim stated in Section 6.1.

At any given time our axiom system is called Σ. In the discussion "a model of Σ" means any interpretation of the axiom system as we have developed it up to that time. Thus the meaning of "Σ" and "a model of Σ" changes as we progress. This not unlike your own name which may stay the same even though you yourself change as time passes.

6.3 AXIOM 2: THE RULER POSTULATE

Letting $\mathscr{P} = \{A, B, C\}$ and $\mathscr{L} = \{\{A, B\}, \{A, C\}, \{B, C\}, \varnothing\}$, we have an uninteresting model of Σ. We *want* a line to have some points on it—lots of them! Any *respectable* line ought to suggest Figure 6.1, where there is a one-to-one correspondence between the points on the line and the real numbers. So for every line l there should be a bijection from l onto **R** which assigns a real number to every point on l. If point P is associated with real number p and point Q is associated with real number q, then the distance from P to Q should be $|q - p|$. Loosely speaking, a line is something like the edge of a *long ruler!* But what is *distance?* We don't have a *distance* yet! This is where the undefined term d enters the picture; d will give us *distance.* Our second axiom declares d to be a mapping that assigns to each ordered pair (P, Q) of points some real number PQ. Further, the mapping d determines one-to-one correspondences between the points on any particular line and the real numbers. The axiom is an *attempt* to make precise the idea conveyed by Figure 6.1.

Axiom 2 *Ruler Postulate* $d : \mathscr{P} \times \mathscr{P} \to \mathbf{R}$, $d : (P, Q) \mapsto PQ$ is a mapping such that for each line l there exists a bijection $f : l \to \mathbf{R}$, $f : P \mapsto f(P)$ where

$$PQ = |f(Q) - f(P)|$$

for all points P and Q on l.

DEFINITION 6.5 Mapping d is the **distance function,** and PQ is

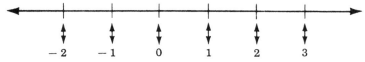

FIGURE 6.1

the ***distance*** from point P to point Q. If for line l, bijection $f : l \rightarrow \mathbf{R}$ is such that $PQ = |f(Q) - f(P)|$ for all points P and Q on l, then f is a ***coordinate system*** for l and $f(P)$ is a ***coordinate*** for P with respect to l and f.

You should spend some time *thinking* about what the Ruler Postulate says and what it does *not* say. Certainly the distance from P to Q ought to be a positive real number unless $P = Q$. Also, the distance from P to Q ought to be equal to the distance from Q to P.

Theorem 6.6 If P and Q are points, then

(D_1) $PQ \geqq 0$.

(D_2) $PQ = 0$ iff $P = Q$.

(D_3) $PQ = QP$.

Proof Exercise 6.1. ∎

Another property often associated with distance is the famous *triangle inequality* for points P, Q, and R:

(D_4) $PQ + QR \geqq PR$.

This is one of the things that the Ruler Postulate does *not* say. Reread the Ruler Postulate. Nothing prevents us from thinking of a model of Σ where distance is measured along some lines in *inches* while distance is measured along all other lines in *feet*. (Take $k = 12$ in Exercise 6.5.) With this in mind, it is not surprising that the triangle inequality is false for some models of Σ. Subsequent axioms will have to make demands on d so that d *behaves nicely*.

The Ruler Postulate requires that every line have a coordinate system determined by the distance function d. A coordinate system for a given line is not unique, however. Our next theorem says that we can *slide* the ruler along the line or we can *turn* the ruler around.

Theorem 6.7 If f is a coordinate system for line l, then g and h are coordinate systems for line l when for all points P on l, $g(P) = f(P) + a$ and $h(P) = -f(P)$ where a is any fixed real number.

Proof Clearly g and h are bijections from l into **R** since f is. Also, $|g(Q) - g(P)| = |f(Q) - f(P) + (a - a)| = |f(Q) - f(P)| = PQ$ and $|h(Q) - h(P)| = |-(f(Q) - f(P))| = PQ$. ■

Theorem 6.8 *Ruler Placement Theorem* Let P and Q be two points on line l, then l has a coordinate system f such that $f(P) = 0$ and $f(Q) > 0$.

Proof By the Ruler Postulate line l has some coordinate system g. So there exist real numbers a and b such that $g(P) = a$, $g(Q) = b$, and $a \neq b$. By the previous theorem, h is also a coordinate system for l where $h(X) = g(X) - a$ for every point X on l. So $h(P) = 0$ and $h(Q) = b - a$. If $b > a$, let $f = h$; if $a > b$, let $f = -h$. In either case, f is a coordinate system for l, $f(P) = 0$, and $f(Q) = |b - a| > 0$. ■

Although our preconceived concepts motivate us in formulating the axioms, we know nothing more about points, lines, and distance than what the axioms and theorems tell us. The undefined term m will not surface until Chapter 14.

6.4 EXERCISES

Henceforth the introductory phrases "Prove" or "For any model of Σ prove" are to be understood where they are lacking.

6.1 Theorem 6.6.

● **6.2** For each line in the Cartesian plane, find a coordinate system for that line.

● **6.3** The Ruler Postulate is independent of the Incidence Axiom.

6.4 If $k > 0$ and $d_2 : \mathscr{P} \times \mathscr{P} \to$ **R** is defined by $d_2(P, Q) = kPQ$, then d_2 also satisfies the Ruler Postulate.

6.5 Let l be a fixed line and assume $k > 0$. If $d_3 : \mathscr{P} \times \mathscr{P} \to$ **R** is defined by $d_3(P, Q) = kPQ$ when P and Q are on l and $d_3(P, Q) = PQ$ otherwise, then d_3 satisfies the Ruler Postulate.

● **6.6** The triangle inequality is not valid for every model of Σ.

- **6.7** True or False?

    (a) PQ is a number.

    (b) If lines l and m intersect, then they intersect in a unique point.

    (c) Two intersecting lines determine a point.

    (d) A line is parallel to itself.

    (e) A line is the shortest distance between two points.

    (f) There are an infinite number of lines.

    (g) Every line has three points. In fact, every line has an infinite number of points.

    (h) Parallelism is transitive for each model of Σ.

    (i) Parallelism is an equivalence relation for each model of Σ.

    (j) Every real number is a coordinate for point P.

6.8 There does not exist a d such that $(M2, d)$ is a model of Σ, where M2 is the Rational Cartesian Incidence Plane.

6.9 Which of the models in Section 5.2 satisfy the Incidence Axiom?

6.10 If S is any nonempty set and $d_4 : S \times S \to \mathbf{R}$ is defined by $d_4(P, Q) = 0$ when $P = Q$ and $d_4(P, Q) = 1$ when $P \neq Q$, then d_4 satisfies the properties D_1, D_2, D_3, and D_4.

6.11 The Ruler Postulate does not follow from the Incidence Axiom and the existence of a mapping $d : \mathscr{P} \times \mathscr{P} \to \mathbf{R}$ satisfying D_1 through D_4.

6.12 Let S be any nonempty set. Suppose $k > 0$ and $d_5 : S \times S \to \mathbf{R}$ satisfies D_1 through D_4. If $d_6 : S \times S \to \mathbf{R}$ is defined by $d_6(P, Q) = 0$ when $P = Q$ and $d_6(P, Q) = k + d_5(P, Q)$ when $P \neq Q$, then d_6 satisfies D_1 through D_4.

6.13 Mapping f is a bijection from the set of all reals between 0 and positive number a onto the set of all reals where $f(x) = \ln(x/(a-x))$.

*● **6.14** Although "$PQ \in \overleftrightarrow{PQ}$" is usually absurd, find a model of Σ where it is not.

*● **6.15** For which of the models in Section 5.2 does there exist a d satisfying the Ruler Postulate?

***6.16** If f is a coordinate system for a line, find all coordinate systems for that line.

GRAFFITI

Every teacher certainly should know something of
non-euclidean geometry. *Thus, it forms one of the few parts of
mathematics which, at least in scattered catch-words, is talked
about in wide circles, so that any teacher may be asked about it at
any moment. . . . Imagine a teacher of physics who is unable to
say anything about Röntgen rays, or about radium. A teacher of
mathematics who could give no answer to questions about non-
euclidean geometry would not make a better impression.
On the other hand, I should like to advise emphatically
against bringing non-euclidean into* regular school instruction
*(i.e., beyond occasional suggestions, upon inquiry by interested
pupils), as enthusiasts are always recommending. Let us be
satisfied if the preceding advice is followed and if the pupils learn
to really understand* euclidean geometry. *After all, it is in order for
the teacher to know a little more than the average pupil.*

Klein

*The most suggestive and notable achievement of the last
century is the discovery of Non-Euclidean geometry.*

Hilbert

"Why," said the Dodo, *"the best way to explain it is to do it."*

Lewis Carroll

*Astronomy was thus the cradle of the natural sciences and the
starting point of geometrical theories. The stars themselves gave
rise to the concept of a "point"; triangles, quadrangles and other
geometrical figures appeared in the constellations; the circle was
realized by the disc of the sun and the moon. Thus in an essentially
intuitive fashion the elements of geometrical thinking came into
existence.*

Lanczos

CHAPTER 7

Betweenness

7.1 ORDERING THE POINTS ON A LINE

Looking at Figure 7.1 below, we say that the circle is to the *left* of the other curve. However, this describes *our* position with respect to the figure rather than the position of the circle with respect to the other curve. Although we shall not hestitate to use such words as *left, right, above,* and *below* in the discussion, we realize that such words have no place in our theory at this time. These words simply haven't been defined, for the very good reason that it is impossible to give any reasonable definitions.

Is point *B between* points *A* and *C* for the curves in Figure 7.1? Of course, any answer would have to depend on the meaning of the word *between*. Certainly, given three points on a line, one point ought to be *between* the other two. We shall give a definition so that this is the case. Only in grade school is it almost as foolish to define *obvious*

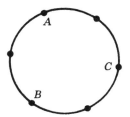 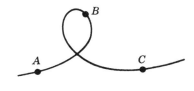

FIGURE 7.1

terms as it is to prove *obvious* theorems. Even Euclid overlooked the necessity of stating any axioms that would give him *betweenness;* he tacitly assumed the necessary properties as they were needed. It took a long time in history before it was realized that *between* is a technical word and that the betweenness properties for points must either follow from other axioms or else be explicitly stated as axioms. The initial work in developing axioms for betweenness was done by Moritz Pasch (1843–1930).

Point B ought to be *between* points A and C if for some long ruler (coordinate system) along $\overleftrightarrow{AC}$, the coordinate of B is *between* the coordinates of A and C. Note the two distinct uses of the word *between* in the last sentence. We know that real number b is between real numbers a and c iff either $a < b < c$ or $c < b < a$. In the theorems below, we shall be able to translate this known type of betweenness (for real numbers) into the new type of betweenness (for points). We shall define *between* in terms of distance, being motivated by Figure 7.2. The powerful Ruler Postulate will give us the desired results.

DEFINITION 7.1   We say point B is ***between*** points A and C and write $A-B-C$ if (1) A, B, C are three distinct points, (2) A, B, C are collinear, and (3) $AB + BC = AC$.

One of the conventions of mathematics is that any statement labeled *definition* is always assumed to be an if-and-only-if-statement even when not explicitly expressed as such. For example, the "if" in Definition 7.1 does carry the weight of "iff." So, if $A-B-C$, then statements (1), (2), and (3) in the definition must hold.

Theorem 7.2   If $A-B-C$, then $C-B-A$.

Proof   The hypothesis $A-B-C$ means that A, B, and C are three distinct collinear points and $AB + BC = AC$. Hence, for the conclusion, we have only to observe that $CB + BA = BC + AB = AB + BC = AC = CA$. So $C-B-A$ by definition.   ■

Is the method of proof in the last theorem obvious? Another proof might be: A and C are symmetric in Definition 7.1 because $PQ = QP$ for any points P and Q and $x + y = y + x$ for any real numbers x and y. In any case, we had to use Definition 7.1. This was the only recourse as all we knew about *between* was what Definition 7.1 told us. Now we know two things, the definition and one theorem.

FIGURE 7.2

Theorem 7.3 Suppose a line has coordinate system f and contains points A, B, and C. If $f(B)$ is between $f(A)$ and $f(C)$, then $A - B - C$.

Proof If $f(C) < f(B) < f(A)$, then $AC = |f(C) - f(A)| = f(A) - f(C) = f(A) - f(B) + f(B) - f(C) = |f(A) - f(B)| + |f(B) - f(C)| = AB + BC$. If $f(A) < f(B) < f(C)$, then $AC = AB + BC$ by interchanging "A" and "C" in the previous sentence. Hence, if $f(B)$ is between $f(A)$ and $f(C)$, then $AB + BC = AC$. A, B, C are collinear by hypothesis. A, B, C are distinct since f is a one-to-one mapping and $f(A), f(B), f(C)$ are distinct. ∎

Theorem 7.4 If $A - B - C$, then neither $A - C - B$ nor $C - A - B$.

Proof Suppose $A - B - C$ and $A - C - B$. From $AC = AB + BC$ and $AB = AC + CB$, we have $AB = AC + CB = (AB + BC) + CB = AB + 2BC$. So $BC = 0$ and $B = C$, contradicting $A - B - C$.

Suppose $A - B - C$ and $C - A - B$. From $AC = AB + BC$ and $CB = CA + AB$, we have $BC = CB = CA + AB = (AB + BC) + AB = BC + 2AB$. So $AB = 0$ and $A = B$, contradicting $A - B - C$. ∎

Theorem 7.5 If $A - B - C$, then $f(B)$ is between $f(A)$ and $f(C)$ for every coordinate system f of the line containing A, B, C.

Proof Let f be any coordinate system of the line containing the three points A, B, C. Exactly one of the three numbers $f(A)$, $f(B)$, $f(C)$ is between the other two. If $f(C)$ is between the other two numbers, then $A - C - B$; if $f(A)$ is between the other two numbers, then $C - A - B$. However, since $A - C - B$ and $C - A - B$ are each inconsistent with the hypothesis $A - B - C$ (Theorem 7.4), we must have $f(B)$ between $f(A)$ and $f(C)$. ∎

By the middle of the nineteenth century, analysts were facing the problem of giving a precise answer to the question "What *is* a real number?" Solutions to this problem given in 1872 by Cantor and Dedekind were motivated by the idea that there is an order preserving, one-to-one correspondence between the points on a *line* and the real numbers. Here, *line* is to be understood as the intuitive concept of a Euclidean line. In particular, there should be a distinct real number for each distinct point on the line. That the correspondence be order preserving requires that betweenness for one system corresponds to betweenness for the other. After having served as a motivation for the definition, the geometry is then entirely excluded from the formal definition of the real numbers. (For example, see the definition of a Dedekind cut in Section 3.2). With the definition of the real numbers and their natural order in hand, one then turns around and defines a *Cartesian line* so that there is an order-preserving, one-to-one correspondence between the points of the line and the real numbers. We define a *Euclidean line* to be isomorphic to a Cartesian line. Thus, Descartes' arithmetization of Euclidean geometry could not be completed until 1872.

That one intuitive idea can motivate the formal definition of a second which, in turn, is used to define the first is not uncommon in mathematics. (You probably defined area in calculus by the definite integral, whose definition was motivated by the idea of area in the first place.) The statement that there is an order-preserving, one-to-one correspondence between the points on a Euclidean line and the real numbers is known as the *Cantor–Dedekind Axiom*. Of course, for the Cantor–Dedekind Axiom to make sense, one has to know about the real numbers in the first place or else know exactly what a Euclidean line is. In the axiomatic development of our geometry, we have assumed knowledge of the field of real numbers. With the combination of the Ruler Postulate and our definition of betweenness for points, we can prove the Cantor–Dedekind Axiom.

Theorem 7.6 *Cantor–Dedekind Axiom* There is an order-preserving, one-to-one correspondence between the set of points on a line and the set of real numbers.

Proof Let l be a line with coordinate system f. Since f is a bijection from l onto the reals, f defines a one-to-one correspondence. Let A, B, C be points on l. Then B is between A and C iff $f(B)$ is between $f(A)$ and $f(C)$, (Theorems 7.3 and 7.5). Thus f is order preserving. ■

So a line in our geometry is isomorphic to a Euclidean line. The remaining theorems in this section are a consequence of that fact. However, it is important to note that just because every line in our plane is isomorphic to a Euclidean line, there is absolutely no justification in jumping to the conclusion that our plane is necessarily a Euclidean plane.

Theorem 7.7 For any three points on a line, exactly one is between the other two. If point P is on $\overleftrightarrow{AB}$, then exactly one of the following holds: $P-A-B$, $P=A$, $A-P-B$, $P=B$, or $A-B-P$.

Proof Exercise 7.1. ■

Theorem 7.8 If A and C are two points, then there exist points B and D such that $A-B-C$ and $A-C-D$.

Proof By the Ruler Placement Theorem (Theorem 6.8), $\overleftrightarrow{AC}$ has a coordinate system f such that $f(A) = 0$ and $f(C) = c > 0$. By the Ruler Postulate, there exist points B and D such that $f(B) = c/2$ and $f(D) =$

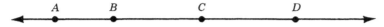

FIGURE 7.3

2c. Since $0 < c/2 < c$ and $0 < c < 2c$, we have $A-B-C$ and $A-C-D$ (Theorem 7.3). ■

Theorem 7.9 If $A-B-C$ and $A-B-D$, then $C=D$, $B-C-D$, or $B-D-C$.

Proof A, B, C are distinct and collinear; A, B, D are distinct and collinear. Hence A, B, C, D are on $\overleftrightarrow{AB}$. By Ruler Placement Theorem, $\overleftrightarrow{AB}$ has a coordinate system f such that $f(A) = 0$ and $f(B) > 0$. Since $A-B-C$ and $A-B-D$, we have $f(A) < f(B) < f(C)$ and $f(A) < f(B) < f(D)$. So either $f(C) = f(D)$, $f(B) < f(C) < f(D)$, or $f(B) < f(D) < f(C)$. The conclusion follows. ■

DEFINITION 7.10 $A-B-C-D$ iff $A-B-C$, $A-B-D$, $A-C-D$, and $B-C-D$.

Theorem 7.11 If $A-B-C$ and $B-C-D$, then $A-B-C-D$.

Proof Exercise 7.2. ■

Theorem 7.12 Any four collinear points can be named A, B, C, D such that $A-B-C-D$.

Proof Suppose four points on line l have coordinates w, x, y, z with respect to coordinate system f for l where $w < x < y < z$. Since f is a bijection, the original four points are A, B, C, D where $f(A) = w$, $f(B) = x$, $f(C) = y$, and $f(D) = z$. Since $w < x < y$, $x < y < z$, and f is order preserving, $A-B-C$ and $B-C-D$. So $A-B-C-D$. ■

7.2 TAXICAB GEOMETRY

Let's go back for a closer look at the definition of "point B is between points A and C." If $A = B$, then $AB + BC = 0 + AC = AC$. Since we don't really want "A is between A and C," it is reasonable to require that A, B, C be distinct for $A-B-C$. The idea behind the definition was using distance in the requirement

(※) $AC = AB + BC$.

But why did we also require that A, B, C be collinear?

For discussion only, we make the following definition: point B is *star-between* points A and C iff A, B, C are three distinct points such that $AC = AB + BC$. Also, for convenience of notation, $A \ast B \ast C$ iff point B is star-between points A and C. Clearly $A - B - C$ implies $A \ast B \ast C$. Does $A \ast B \ast C$ imply $A - B - C$? If the answer were "Yes," then we would have done nothing terribly wrong. It's just that there would be a redundancy in the definition of *between*. The geometry that we are about to describe will show that the answer is "No."

In the Real Cartesian Incidence Plane M1, we shall assume below that $P = (x_1, y_1)$ and $Q = (x_2, y_2)$. We all know that M1 satisfies the Ruler Postulate where

$$d(P, Q) = \sqrt{(x_2 - x_1)^2 + (y_2 - y_1)^2}.$$

However, Euclidean distance function d is not the only distance function that satisfies the Ruler Postulate. Consider $t : \mathscr{P} \times \mathscr{P} \to \mathbf{R}$ defined by

$$t(P, Q) = |x_2 - x_1| + |y_2 - y_1|.$$

M1 together with t is the *Taxicab Geometry*. Taxicab Geometry has the practical application of being the *real* geometry involved in getting from point P to point Q in a city where the streets are parallel, the avenues are parallel, and the streets are perpendicular to the avenues. See Figure 7.4.

If $x_1 = x_2$ or $y_1 = y_2$ (P and Q are on the same street or on the same avenue), then $t(P, Q) = d(P, Q)$; otherwise, $t(P, Q) \neq d(P, Q)$. See Figure 7.5. Assume $x_1 \neq x_2$ and $y_1 \neq y_2$, and let $A = (x_2, y_1)$. Then P, A, Q are three distinct points that are not collinear. We do not have $P - A - Q$. Nevertheless $P \ast A \ast Q$ as

$$t(P, Q) = |x_2 - x_1| + |y_2 - y_1| = d(P, A) + d(A, Q) = t(P, A) + t(A, Q).$$

So $P \ast A \ast Q$ does not imply $P - A - Q$! Anyone who has walked in such

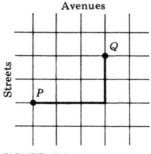

Avenues

Streets

FIGURE 7.4

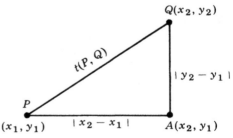

$Q(x_2, y_2)$

$t(P, Q)$

$|y_2 - y_1|$

P
(x_1, y_1)

$|x_2 - x_1|$

$A(x_2, y_1)$

FIGURE 7.5

a city will know that the points that are *star-between* P and Q when $x_1 \neq x_1$ and $y_1 \neq y_2$ consists of all the points, except P and Q, that are either on or else inside a particular rectangle with opposite vertices P and Q. See Figure 7.6. It follows that the points that are *between* two points P and Q in Taxicab Geometry are exactly those points that are between P and Q in the Cartesian plane.

We have yet to show that t satisfies the Ruler Postulate. Lines parallel to the axes (streets and avenues) have the usual Euclidean coordinate systems. Suppose l is a line through P and Q and that l is not parallel to an axis. So $x_1 \neq x_2$ and $y_1 \neq y_2$. If line l has slope m, then $m = (y_2 - y_1)/(x_2 - x_1)$. It is a simple exercise in algebra to show that

$$t(P, Q) = \frac{1 + |m|}{\sqrt{1 + m^2}} d(P, Q).$$

This equation tells us that *Taxicab distance* along any line is some positive constant multiple of *Euclidean distance* along the same line. The *Taxicab ruler* for any line is a scaled *Euclidean ruler;* the Taxicab rulers for different lines may have different scales! Every line has a coordinate system with respect to t. Taxicab Geometry is a model of Σ.

Does Taxicab Geometry satisfy the triangle inequality? If P and Q are not on the same street or avenue, then there are many ways of walking from P to Q traversing only the distance $t(P, Q)$. However, the triangle inequality does hold. Let $R = (x_3, y_3)$ and recall that $|a + b| \leq |a| + |b|$ for real numbers a and b. Then

$$\begin{aligned}
t(P, R) &= |x_3 - x_1| + |y_3 - y_1| \\
&= |(x_3 - x_2) + (x_2 - x_1)| + |(y_3 - y_2) + (y_2 - y_1)| \\
&\leq |x_3 - x_2| + |x_2 - x_1| + |y_3 - y_2| + |y_2 - y_1| \\
&= (|x_2 - x_1| + |y_2 - y_1|) + (|x_3 - x_2| + |y_3 - y_2|) \\
&= t(P, Q) + t(Q, R).
\end{aligned}$$

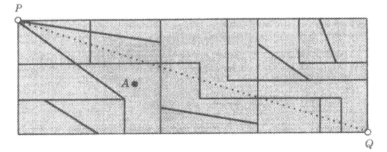

FIGURE 7.6

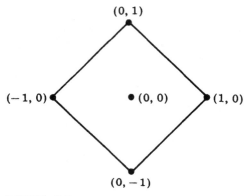

FIGURE 7.7

Since we have shown $t(P, Q) + t(Q, R) \geqq t(P, R)$ for arbitrary points P, Q, and R, we see that the triangle inequality does hold in Taxicab Geometry.

Everyone knows a *circle* is the locus (i.e., set) of all points equidistant from a fixed point. With $R = (x, y)$ and $O = (0, 0)$, the *unit circle* is the set of all points R such that the distance from O to R is 1. So the unit circle has the equation $1 = |x| + |y|$ in Taxicab Geometry. Strangely enough, the *unit circle* is a *square!* See Figure 7.7. To confirm this result, consider the equation in only one quadrant at a time: all (x, y) in the first quadrant such that $1 = x + y$, all (x, y) in the second quadrant such that $1 = -x + y$, all (x, y) in the third quadrant such that $1 = -x - y$, all (x, y) .in the fourth quadrant such that $1 = x - y$.

Everybody *knows* the locus of all points equidistant from two points P and Q is a line. If P and Q are on the same street or on the

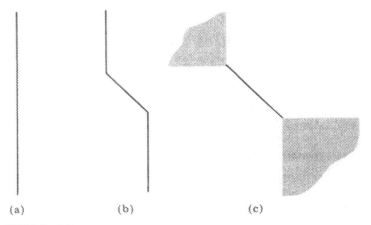

(a)   (b)   (c)

FIGURE   7.8

same avenue, this is true. In general the locus has equation $|x-x_1|+|y-y_1|=|x-x_2|+|y-y_2|$. The locus of all points equidistant from $(3, 2)$ and $(7, 4)$ is *not* a line but looks like Figure 7.8b. Even more surprising is the fact that the locus of all points equidistant from $(3, 2)$ and $(7, 6)$ looks like Figure 7.8c. This last set of points is certainly not a line in M1! Perhaps not everything that everybody knows is always true.

Taxicab Geometry is only one plane with *weird rulers*. Starting with M1, let d be Euclidean distance function and μ be any mapping from the set of lines in M1 into the set of positive real numbers. For distinct points P and Q, define $w(P, P) = 0$ and $w(P, Q) = \mu(\overleftrightarrow{PQ})d(P, Q)$. Then M1 together with w is a model of Σ.

7.3 EXERCISES

7.1 Theorem 7.7.

7.2 Theorem 7.11.

7.3 Give a coordinate system for the line with equation $y = mx + b$ in Taxicab Geometry.

● **7.4** Find the set of all points equidistant from $(2, 3)$ and $(4, 7)$ in Taxicab Geometry.

● **7.5** (M9, d), the Cubic Incidence Plane together with Euclidean distance, is not a model of Σ.

7.6 If A and B are points, give a reasonable definition of "the midpoint of A and B" and prove it exists.

● **7.7** True or False?

(a) $A-B-C$ iff $C-B-A$.

(b) $A-B-C$ iff $AC = AB + BC$.

(c) If $A-B-C$ and $B-C-A$, then $C-A-B$.

(d) In Taxicab Geometry, every circle is a square.

(e) In Taxicab Geometry, every square is a circle.

(f) $A-B-D$ and $A-C-D$ only if $A-B-C-D$.

(g) $A-B-D$ and $B-C-D$ only if $A-B-C-D$.

(h) $A-B-C$ and $A-C-D$ only if $A-B-C-D$.

(i) $A-B-C$ and $A-B-D$ only if $A-B-C-D$.

(j) $A-C-D$ and $B-C-D$ only if $A-B-C-D$.

7.8   Find the set of all points equidistant from $(2, 7)$ and $(6, 3)$ in Taxicab Geometry.

7.9   If line l makes an acute angle of measure θ with the x-axis, then for any points P and Q on l in the Taxicab Geometry we have $t(P, Q) = (\cos \theta + \sin \theta)\, d(P, Q)$.

7.10   To see how Taxicab Geometry might be used to enrich a high school geometry course, read "Taxicab Geometry—a Non-Euclidean Geometry of Lattice Points" by D. R. Byrkit in *The Mathematics Teacher* Vol. 64 (1971), pp. 418–422.

7.11   M1 together with r is a model of Σ where

$$r(P, Q) = \max \{|x_2 - x_1|,\ |y_2 - y_1|\}$$

gives the distance from point P to point Q when $P = (x_1, y_1)$ and $Q = (x_2, y_2)$.

7.12   For a comparison of the geometry of Exercise 7.11 and Taxicab Geometry read "A Geometric Duality for Two Metrics for the Coordinate Plane" by F. Rhodes in *The Mathematical Gazette* Vol. 54 (1970), pp. 19–23.

7.13   M1 together with h is a model of Σ where

$$h(P, Q) = \sqrt{(x_2 - x_1 + y_2 - y_1)^2 + (y_2 - y_1)^2}$$

gives the distance from point P to point Q when $P = (x_1, y_1)$ and $Q = (x_2, y_2)$.

7.14   There exist exactly two ways of renaming four points on a line A, B, C, D such that $A-B-C-D$.

*7.15   What is the locus of all points equidistant from two given points in Taxicab Geometry? Generalize Exercises 7.4 and 7.8.

GRAFFITI

Mathematics is an obscure field, an abstruse science, complicated and exact; yet so many have attained perfection in it that we might conclude almost anyone who seriously applied himself would achieve a measure of success.

Cicero

. . . *we must first base such words as "between" upon clear concepts, a thing which is quite feasible but which I have not seen done.*

Gauss

There is no rigorous definition of rigor.

So far as the theories of mathematics are about reality, they are not certain; so far as they are certain, they are not about reality.

Einstein

Reductio ad absurdum, *which Euclid loved so much, is one of a mathematician's finest weapons. It is a far finer gambit than any chess gambit: a chess player may offer the sacrifice of a pawn or even a piece, but a mathematician offers the game.*

Hardy

CHAPTER 8

Segments, Rays, and Convex Sets

8.1 SEGMENTS AND RAYS

Our definition of *segment* and *ray* will be motivated by Figure 8.1. It seems reasonable to say that a *segment* with *endpoints A* and *B* should contain all the points between *A* and *B*. Should *A* and *B* be included? This is initially a matter of *choice!* The reader of a textbook is sometimes unaware that the author has made several rather arbitrary decisions about his definitions. The author can decide whether he wants the endpoints included in a *segment* or not; he makes the decision for himself and his reader. Once that decision is made, the defined word must be used consistently. We are stuck with the *choice*.

Two thousand years of usage of the word *line* for today's word *curve* required the phrase *straight line* to distinguish what we now call a line from other curves. This usage has been almost completely abandoned. If the phrase *straight line* is so ingrained in your thoughts that you can hardly say "line" instead of "straight line," then at least be aware that you are using an old fashioned term that means no more and no less than *line*.

FIGURE 8.1

It is interesting that Euclid used *straight line* for what we would call a segment. This partially explains the usage of the phrase "a straight line is the shortest distance between two points." Although this phrase is known to almost every person on the street, it is still an anachronism. In modern usage of technical terminology the phrase is absurd since neither a line nor a segment is a distance. By universal usage, the whole phrase has come to have a meaning in the common language, although that meaning is not discerned from the individual words. We are powerless to keep this phrase off the streets, but we should keep it out of our geometry.

Although we'll have twenty-one definitions and theorems in this chapter, the first seventeen of these merely say that *segments* and *rays* behave as they *should*. It is hopeless to try to memorize all of our theorems and proofs. If you insist on memorizing something as a security blanket, then memorize the definitions. You do have to *know* the definitions to understand what you're talking about.

Let's say a few words about *rigor*, a word that strikes fear into the hearts of many undergraduates. The author would advise against cluttering up your proofs with trivial reasons, as they tend to obscure the principal ideas. For example, in " . . . So A and B are distinct points. $\overleftrightarrow{AB}$ exists because two points determine a line. Hence, $\overleftrightarrow{AB}$ intersects . . . ," the second sentence is really unnecessary to anyone who has been following the development of our theory. It is by no means easy for the beginning student to distinguish the unnecessary. The only way to learn how to write a proof is to write proofs and then have someone criticize your efforts. Avoid meaningless phrases such as "extend line to C" or "draw $\overleftrightarrow{AB}$." Another common error is not introducing points and lines in a proof before using them, even though the points and lines are indicated in your figure. Figures may accompany a proof as a mnemonic device to aid in reading the proof; they are *not* actually a part of the proof. Your proofs should look very much like those in the text, excluding those parenthetical references to previous theorems by number inserted to aid the reader. An excellent test is to read your proof aloud and ask yourself if it would be a convincing argument to everyone in the class.

DEFINITION 8.1 Let A and B be two distinct points. Then, $\overline{AB} = \{A, B\} \cup \{P | A - P - B\}$ and $\overrightarrow{AB} = \{P | P \in \overleftrightarrow{AB}$ but not $P - A - B\}$. $\overline{AB}$ is a *segment* with *endpoints* A and B. $\overrightarrow{AB}$ is a *ray* with *vertex* A. If T is

FIGURE 8.2

an arbitrary set of points containing point P, then P is **on** T and T **passes through** P.

Theorem 8.2 Given two points A and B,

(a) $\overline{AB} = \overline{BA}$,

(b) $\overrightarrow{AB} \neq \overrightarrow{BA}$,

(c) $\overrightarrow{AB} = \overline{AB} \cup \{P | A-B-P\}$,

(d) $\overline{AB} \subsetneq \overrightarrow{AB} \subsetneq \overleftrightarrow{AB}$.

Proof (a) Since $A-P-B$ iff $B-P-A$, the definition of $\overline{AB}$ is symmetric in "A" and "B." (b) There exists P such that $P-A-B$ (Theorem 7.8). If $P-A-B$, then P is not in $\overrightarrow{AB}$ but is in $\overrightarrow{BA}$. (c) If P is on $\overleftrightarrow{AB}$, then exactly one of the following holds: $P-A-B$, $P=A$, $A-P-B$, $P=B$, or $A-B-P$. So

$$\overrightarrow{AB} = \{A, B\} \cup \{P | A-P-B\} \cup \{P | A-B-P\} = \overline{AB} \cup \{P | A-B-P\}.$$

(d) The containments follow from (c). The existence of points C and D such that $D-A-B-C$ (Theorem 7.8) shows that the containments are proper. ∎

For emphasis we repeat the notation:

AB is a real number.

$\overleftrightarrow{AB}$ is a line.

$\overrightarrow{AB}$ is a ray.

$\overline{AB}$ is a segment.

Thus $\overleftrightarrow{AB}$, $\overrightarrow{AB}$, and $\overline{AB}$ are sets of points—in fact, three *different* sets of points. AB is not a set of points. The four symbols are not interchangeable!

Theorem 8.3 $\overline{AB} = \overrightarrow{AB} \cap \overrightarrow{BA}$; $\overleftrightarrow{AB} = \overrightarrow{AB} \cup \overrightarrow{BA}$.

Proof Exercise 8.1. ∎

Theorem 8.4 Let A and B be distinct points. Then $\overline{AB} = \overline{CD}$ iff $\{A, B\} = \{C, D\}$.

Proof If $\{A, B\} = \{C, D\}$, then either $A=C$ and $B=D$ or else $A=D$ and $B=C$. In either case, we have $\overline{AB} = \overline{CD}$ by definition of a segment.

Now suppose $\overline{AB} = \overline{CD}$, $A \neq C$, and $A \neq D$. Then, since A is on $\overline{CD}$ as A is on $\overline{AB}$, we must have $C-A-D$. Also, since C and D are distinct points on $\overline{AB}$ different from A, at least one of C or D is between A and B. So either $B-C-A$ or $B-D-A$. If $B-C-A$ and $C-A-D$, then $B-C-A-D$; if $B-D-A$ and $C-A-D$, then $B-D-A-C$ (Theorem 7.11). In either case, B is not in $\overline{CD}$, contradicting $\overline{AB} = \overline{CD}$. Hence $A = C$ or $A = D$. By symmetry B must be either C or D. Since $A \neq B$, we have $\{A, B\} = \{C, D\}$. ■

DEFINITION 8.5 If for $\overline{AB}$ and $\overline{CD}$ we have $AB = CD$, then $\overline{AB} \simeq \overline{CD}$. $\overline{AB}$ is *congruent* to $\overline{CD}$ iff $\overline{AB} \simeq \overline{CD}$. The *length* of $\overline{AB}$ is AB.

Theorem 8.6 Congruence of segments is an equivalence relation on the set of all segments.

Proof (R): $\overline{AB} \simeq \overline{AB}$, since $AB = AB$. (S): $\overline{AB} \simeq \overline{CD}$ implies $\overline{CD} \simeq \overline{AB}$, since $AB = CD$ implies $CD = AB$. (T): $\overline{AB} \simeq \overline{CD}$ and $\overline{CD} \simeq \overline{EF}$ implies $\overline{AB} \simeq \overline{EF}$, since $AB = CD$ and $CD = EF$ implies $AB = EF$. Thus congruence is an equivalence relation on the set of segments because equality is an equivalence relation on the real numbers. ■

Theorem 8.7 *Ray-Coordinatization Theorem* Given $\overrightarrow{VA}$ there is a unique coordinate system f for $\overleftrightarrow{VA}$ such that $f(V) = 0$ and $\overrightarrow{VA} = \{P|f(P) \geq 0\}$.

Proof By the Ruler Placement Theorem, $\overleftrightarrow{VA}$ has a coordinate system f such that $f(V) = 0$ and $f(A) > 0$. Since $0 = f(V) < f(A)$, for any point P on $\overleftrightarrow{VA}$ the following are equivalent: (i) P is off $\overrightarrow{VA}$, (ii) $P-V-A$, (iii) $f(V)$ is between $f(P)$ and $f(A)$, (iv) $f(P) < f(V) < f(A)$, (v) $f(P) < 0$. Hence $\overrightarrow{VA} = \{P|f(P) \geq 0\}$. Suppose g is another coordinate system for $\overleftrightarrow{VA}$ such that $g(V) = 0$ and $\overrightarrow{VA} = \{P|g(P) \geq 0\}$. Then for any point Q in $\overleftrightarrow{VA}$, we have $|g(Q)| = |g(Q) - g(V)| = VQ = |f(Q) - f(V)| = |f(Q)|$. If Q is on $\overrightarrow{VA}$, then $g(Q) = f(Q) \geq 0$. If Q is off $\overrightarrow{VA}$, we have $g(Q) < 0$, $f(Q) < 0$, and $|g(Q)| = |f(Q)|$. So $g(Q) = f(Q)$ for all points Q on $\overleftrightarrow{VA}$. Thus $g = f$ and f is unique. ■

Theorem 8.8 *Segment-Construction Theorem* Given $\overline{AB}$ and $\overrightarrow{VC}$, there exists a unique point D in $\overrightarrow{VC}$ such that $\overline{AB} \simeq \overline{VD}$.

Proof By the Ray-Coordinatization Theorem $\overleftrightarrow{VC}$ has a coordinate system f such that $\overrightarrow{VC} = \{P|f(P) \geq 0\}$ and $f(V) = 0$. For any point D in $\overrightarrow{VC}$, we have $f(D) = VD$. However, there is a unique point D in $\overrightarrow{VC}$

such that $f(D) = AB$. So there is a unique point D in $\vec{VC}$ such that $AB = VD$. ∎

There is a one-to-one correspondence between the set of rationals and the points on the x-axis of M2, the Rational Cartesian Incidence Plane. Since the rationals cannot be put in one-to-one correspondence with the reals, there does not exist any distance function d such that (M2, d) is a model of Σ because the Ruler Postulate can never be satisfied. However, for discussion, we define *distance d* on M2 by the usual Cartesian formula. We also apply the definitions in the theory to the geometry (M2, d). This provides an example of a geometry where Theorem 8.8 fails. Let $A = V = (0, 0)$, $B = (1, 1)$, and $C = (1, 0)$. Then $AB = \sqrt{2}$, but there is no point D on $\vec{VC}$ such that $VD = AB$ since $(0, \sqrt{2})$ is not a point in M2.

Theorem 8.9 *Segment-Addition Theorem* If $A-B-C$, $D-E-F$, $\overline{AB} \simeq \overline{DE}$, and $\overline{BC} \simeq \overline{EF}$, then $\overline{AC} \simeq \overline{DF}$.

Proof From the hypothesis we have $AB + BC = AC$, $DE + EF = DF$, $AB = DE$, and $BC = EF$. Thus $AC = AB + BC = DE + EF = DF$, as required. ∎

Theorem 8.10 *Segment-Subtraction Theorem* If $A-B-C$, $D-E-F$, $\overline{AB} \simeq \overline{DE}$, and $\overline{AC} \simeq \overline{DF}$, then $\overline{BC} \simeq \overline{EF}$.

Proof Exercise 8.2. ∎

Theorem 8.11 Point B is on $\vec{VA}$ and $B \neq V$ iff $\vec{VB} = \vec{VA}$.

Proof By the Ray-Coordinatization Theorem, $\overleftrightarrow{VA}$ has a unique coordinate system f such that $f(V) = 0$ and $\vec{VA} = \{P | f(P) \geq 0\}$. Suppose point B is on $\vec{VA}$ and $B \neq V$. Then, $f(B) > 0$ and $\vec{VB} = \{P | f(P) \geq 0\}$ by definition of $\vec{VB}$. Hence $\vec{VB} = \vec{VA}$. Conversely, B on $\vec{VB}$ and $\vec{VB} = \vec{VA}$ implies B on $\vec{VA}$ and $B \neq V$. ∎

Theorem 8.12 If $\vec{VA} = \vec{WB}$, then $V = W$.

Proof Suppose $\vec{VA} = \vec{WB}$ and $V \neq W$. Since W is on $\vec{VA}$ and V is on

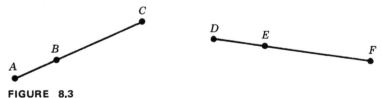

FIGURE 8.3

$\overrightarrow{WB}$, we apply the last theorem twice to get $\overrightarrow{VW} = \overrightarrow{VA} = \overrightarrow{WB} = \overrightarrow{WV}$, a contradiction (Theorem 8.2b). ∎

Corollary 8.13 Every ray has a unique vertex.

8.2 CONVEX SETS

DEFINITION 8.14 If $A-M-B$ and $AM=MB$, then M is a *mid-point* of $\overline{AB}$; point M is a *midpoint of A and B* if $A=M=B$ or M is a midpoint of $\overline{AB}$. If $A-V-B$, then $\overrightarrow{VA}$ is an *opposite ray* of $\overrightarrow{VB}$

Theorem 8.15 *Midpoint Theorem* Every segment has a unique midpoint. If A and B are points, then there exist unique points M and N such that M is a midpoint of A and B while B is a midpoint of A and N.

Proof If $A=B$, then clearly $M=N=A$. Suppose $A \neq B$. Let f be a coordinate system for $\overleftrightarrow{AB}$ such that $f(A)=0$ and $f(B)=b>0$. M is a midpoint of A and B iff $f(M)=AM=MB=AB-AM=b-f(M)$; B is a midpoint of A and N iff $b=AB=BN=AN-AB=f(N)-b$. Our theorem follows from the existence of unique points M and N such that $f(M)=\frac{1}{2}b$ and $f(N)=2b$. ∎

Theorem 8.16 Every ray has a unique opposite ray. If $\overrightarrow{VA}$ is an opposite ray of $\overrightarrow{VB}$, then $\overrightarrow{VB}$ is an opposite ray of $\overrightarrow{VA}$.

Proof Exercise 8.3. ∎

Theorem 8.17 If P and Q are points on $\overleftrightarrow{AB}$ such that $AP=AQ$ and $BP=BQ$, then $P=Q$.

Proof Let $\overleftrightarrow{AB}$ have coordinate system f such that $f(A)=0$ and $f(B)=b>0$. From $|f(P)|=AP=AQ=|f(Q)|$, we have $f(P)=\pm f(Q)$. If $f(Q)=f(P)$, then $P=Q$ and we are done. Suppose $f(Q)=-f(P)$. From $|f(P)-b|=BP=BQ=|f(Q)-b|$, we have $f(P)-b=\pm(f(Q)-b)$. In case $f(P)-b=+(f(Q)-b)$, we have $f(P)=f(Q)=-f(P)$. Then $f(P)=f(Q)=0$ and $P=Q=A$. In the other case, $f(P)-b=-(f(Q)-b)=f(P)+b$. Here we have the contradiction $b=0$. Therefore, $P=Q$ in all possible cases. ∎

DEFINITION 8.18 int $(\overline{AB})$, the *interior* of $\overline{AB}$, is $\overline{AB} \setminus \{A, B\}$; int $(\overrightarrow{VA})$, the *interior* of $\overrightarrow{VA}$, is $\overrightarrow{VA} \setminus \{V\}$. A *half line* is the interior

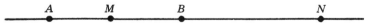

$$A \qquad M \qquad B \qquad\qquad N$$

FIGURE 8.4

FIGURE 8.5

of a ray. If T is a set of points such that $\overline{AB}$ is a subset of T for every two points A and B in T, then T is a **convex set.**

None of the regions in Figure 8.5 is a convex set in Euclidean geometry. To prove that a set T is not a convex set we must find *three* points A, B, C such that A and B are in T, C is not in T, and $A-C-B$. To prove that a set T is a convex set it is sufficient to show that Q is in T whenever P and R are in T and $P-Q-R$.

Theorem 8.19 The intersection of two (or more) convex sets is a convex set.

Proof If A and B are distinct points in the intersection, then A and B are distinct points in each of the convex sets. Since $\overline{AB}$ is a subset of each convex set by definition, $\overline{AB}$ is a subset of the intersection. ∎

Theorem 8.20 If A and B are distinct points, then each of $\varnothing$, $\{A\}$, $\overline{AB}$, int $(\overline{AB})$, $\overrightarrow{AB}$, int $(\overrightarrow{AB})$, $\overleftrightarrow{AB}$, and $\mathscr{P}$ is a convex set.

Proof The set $\mathscr{P}$ of all points is a convex set since any segment is a set of points. If P and Q are distinct points of $\overleftrightarrow{AB}$, then $\overline{PQ} \subset \overleftrightarrow{PQ} = \overleftrightarrow{AB}$. So $\overleftrightarrow{AB}$ is a convex set. Let f be a coordinate system for $\overleftrightarrow{AB}$ such that $f(A) = 0$ and $f(B) > 0$. Then $\overrightarrow{AB} = \{P | f(P) \geqq 0\}$ and int $(\overrightarrow{AB}) = \{P | f(P) > 0\}$. If P and R are distinct points of $\overrightarrow{AB}$ and $P-Q-R$, then $f(Q)$ is positive. Therefore, Q is in both $\overrightarrow{AB}$ and int $(\overrightarrow{AB})$. So $\overrightarrow{AB}$ and int $(\overrightarrow{AB})$ are convex sets. Applying the previous theorem, we see that $\overline{AB}$ and int $(\overline{AB})$ are convex sets as $\overrightarrow{AB} \cap \overrightarrow{BA} = \overline{AB}$ and int $(\overrightarrow{AB}) \cap$ int $(\overrightarrow{BA}) = (\overrightarrow{AB} \cap \overrightarrow{BA}) \setminus \{A, B\} = $ int $(\overline{AB})$. Both $\varnothing$ and $\{A\}$ are convex sets since neither contains two points. ∎

Theorem 8.21 *Line-Separation Theorem* For every point V in line l there exist convex sets H_1 and H_2 such that (i) $l \setminus \{V\} = H_1 \cup H_2$ and (ii) if $P \in H_1$, $Q \in H_2$, and $P \neq Q$, then $\overline{PQ} \cap \{V\} \neq \varnothing$.

Proof Let $A-V-B$ with $l=\overleftrightarrow{AB}$. Then H_1 and H_2 clearly satisfy the requirements where $H_1=\text{int}\,(\overrightarrow{VA})$ and $H_2=\text{int}\,(\overrightarrow{VB})$. ∎

To get the Moulton Incidence Plane M10, we took the Cartesian Incidence Plane but *bent* the Cartesian lines having positive slope. If we *bend* the rulers for these lines also, then we have a model of Σ. More precisely, (M10, s) is a model of Σ where s is the distance function given by Euclidean arclength along Moulton lines. Let $A=(-2, -1)$, $B=(2, 2)$, $C=(0, \frac{1}{2})$, and $W=(0, 1)$. See Figure 8.6. The Moulton line through A and B contains W but not C; the Euclidean line through A and B contains C but not W. Then $s(A, B)=d(A, W)+d(W, B)=2\sqrt{2}+\sqrt{5}$ and $s(A, C)+s(C, B)=d(A, C)+d(C, B)=d(A, B)=5$. In (M10, s) the triangle inequality fails as $s(A, C)+s(C, B)<s(A, B)$.

A segment in (M10, s) is a Euclidean segment unless the segment is the union of two noncollinear Euclidean segments each having an endpoint on the y-axis. The region *above* the Moulton line through the points A and B in Figure 8.6 is certainly not a convex set in Euclidean geometry. However, this region *is* a convex set in (M10, s). To see that there is nothing special about this particular line, we generalize the situation and suppose l is the line with all points (x, y) such that $y=mx+b$ if $x\leqq 0$ and $y=\frac{1}{2}mx+b$ if $x\geqq 0$. The line contains $(0, b)$. The line in Figure 8.6 is the special case $m=b=1$. Let P and Q be any two points *above* line l. If P and Q are not on *opposite sides* of the y-axis, it is easy to see that all points between P and Q are

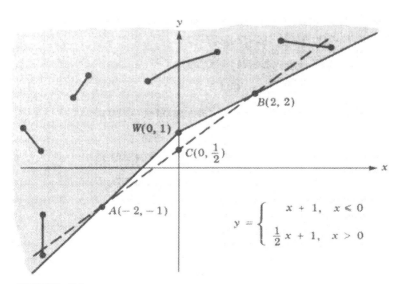

FIGURE 8.6

above l. Assume $P = (x_1, y_1)$ and $Q = (x_2, y_2)$ where $x_1 < 0 < x_2$. The Moulton line through P and Q passes through the point $(0, k)$ on the y-axis where $k = (x_2 y_1 - 2x_1 y_2)/(x_2 - 2x_1)$. Since P and Q are *above* l, we must have $y_1 > mx_1 + b$ and $y_2 > \frac{1}{2}mx_2 + b$. So $x_2 y_1 > x_2(mx_1 + b)$ and $(-2x_1)(y_2) > (-2x_1)(\frac{1}{2}mx_2 + b)$. Then $x_2 y_1 - 2x_1 y_2 > (x_2 - 2x_1)b$. Since $x_2 - 2x_1$ is positive, we have $k > b$. Therefore, the segment intersects the y-axis at a point *above* l, and every point between P and Q is *above* l. Thus by checking the definition of a convex set, we have shown that the region *above* l is a convex set in $(M10, s)$. Replacing "*above*" by "*below*" and ">" by "<" in the argument shows that the region *below* such a line l is a convex set in $(M10, s)$.

8.3 EXERCISES

8.1 Theorem 8.3.

8.2 Theorem 8.10.

8.3 Theorem 8.16.

• **8.4** In the Line-Separation Theorem (Theorem 8.21), the sets H_1 and H_2 are unique except for order.

• **8.5** Find a set of points in the Cartesian Incidence Plane that is a convex set in Euclidean geometry but not in $(M10, s)$.

• **8.6** True or False?

(a) $\overline{AB} = \overline{CD}$ only if $A = C$ or $A = D$.

(b) If $AB = CD$, then $A = C$ or $A = D$.

(c) If $\overline{AB} \simeq \overline{CD}$, then $\{A, B\} = \{C, D\}$.

(d) $\overline{AB} = \overline{CD}$ only if $\overline{AB} \simeq \overline{CD}$.

(e) The definition of "segment" depends on distance.

(f) The endpoints of a segment are unique.

(g) If M is the midpoint of A and B, then $A - M - B$.

(h) A point on $\overleftrightarrow{AB}$ is uniquely determined by its distance from A and from B.

(i) If A and B are points, then $\{A, B\}$ is a convex set.

(j) The union of two convex sets is a convex set.

8.7 Let l be a line in $(M10, s)$ with equation $x = a$. The region to the *right* of l and the region to the *left* of l are convex sets in (M10, s).

● **8.8**   What are the segments and rays in Taxicab Geometry?

● **8.9**   Find a counterexample to $\overline{AB} = \{P | AP + PB = AB\}$ in Taxicab Geometry.

8.10   Equation $\overrightarrow{VC} = \{P | PC = |VC - VP|\}$ holds in (M1, d) but not in Taxicab Geometry.

● **8.11**   Equation $AM = MB = \frac{1}{2}AB$ does not imply that M is the midpoint of A and B for a model of Σ.

8.12   Read "What is a convex set?" by V. Klee in *The American Mathematical Monthly* Vol. 78 (1971), pp. 616–631.

*__*8.13__   In (M10, s) what is the unit circle? What is the equation for the set of all points of distance 2 from $(-1, 0)$?

*__*8.14__   In (M10, s) find all points P such that $AP + PB = AB$ where $A = (-2, -1)$ and $B = (2, 2)$.

*__*8.15__   In (M10, s) find all points M such that $AM = MB = \frac{1}{2}AB$ when $A = (-2, -1)$ and $B = (2, 2)$.

GRAFFITI

The Greeks made Space the subject-matter of a science of supreme simplicity and certainty. Out of it grew, in the mind of classical antiquity, the idea of pure science. Geometry became one of the most powerful expressions of that sovereignty of the intellect that inspired the thought of those times. At a later epoch, when the intellectual despotism of the Church, which had been maintained through the Middle Ages, had crumbled, and a wave of scepticism threatened to sweep away all that had seemed most fixed, those who believed in Truth clung to Geometry as to a rock, and it was the highest ideal of every scientist to carry on his science "more geometrico."

Weyl

I should rejoice to see . . . Euclid honourably shelved or buried "deeper than did ever plummet sound" out of the schoolboys' reach; morphology introduced into the elements of algebra; projection, correlation, and motion accepted as aids to geometry; the mind of the student quickened and elevated and his faith awakened by early initiation into the ruling ideas of polarity, continuity, infinity, and familiarization with the doctrines of the imaginary and inconceivable.

> **Sylvester**

The critical mathematician has abandoned the search for truth. He no longer flatters himself that his propositions are or can be known to him or to any other human being to be true; and he contents himself with aiming at the correct, or the consistent. The distinction is not annulled nor even blurred by the reflection that consistency contains immanently a kind of truth. He is not absolutely certain, but he believes profoundly that it is possible to find various sets of a few propositions each such that the propositions of each set are compatible, that the propositions of each set imply other propositions, and that the latter can be deduced from the former with certainty. That is to say, he believes that there are systems of coherent or consistent propositions, and he regards it his business to discover such systems. Any such system is a branch of mathematics.

> **Keyser**

The essence of mathematics lies in its freedom.

> **Cantor**

Mathematicians are like Frenchmen: whatever you say to them they translate into their own language and forthwith it is something entirely different.

> **Goethe**

CHAPTER 9

Angles and Triangles

9.1 ANGLES AND TRIANGLES

Our idea of an *angle* is simply a set of points which is the union of two noncollinear rays with the same vertex. See Figure 9.1. Note that this is not saying the same thing as the union of any two rays $\overrightarrow{VA}$ and $\overrightarrow{VB}$. If $V-A-B$, the union is just a ray; if $A-V-B$, the union is a line. We *choose* to have our *angles* distinct from rays and lines. Otherwise when making statements, we would always have to keep making exceptions for these cases. They would be more bother than they would be worth. Perhaps you are shocked that this eliminates the so-called *straight angle* from the class of *angles*. Well, what is a *straight angle* anyway? If it is just a line, who needs it? Should it be a flag — an ordered pair (P, l) with point P on line l? If so, then it is certainly different from a set of points. Perhaps it should be an *angle* of 180 or 540 *degrees* or, maybe, a special kind of *rotation* or a special equivalence class of *rotations*. Fortunately, because life is simpler for us,

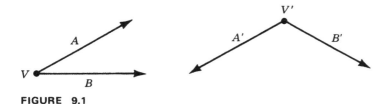

FIGURE 9.1

these ideas are not presently available as we have no such things as *degrees* or *rotations*. We do confess that perhaps it would be nice to use "elementary angle" where we shall use "angle" to allow for a more sophisticated use of the latter term later. We abstain from doing so mainly because we are too lazy to keep using the longer term.

DEFINITION 9.1 If A, V, B are three noncollinear points, then $\angle AVB = \overrightarrow{VA} \cup \overrightarrow{VB}$. $\angle AVB$ is an **angle** having **vertex** V and **sides** $\overrightarrow{VA}$ and $\overrightarrow{VB}$.

Theorem 9.2 If A, B, C are three noncollinear points, then

$$\angle ABC = \angle CBA \neq \angle ACB.$$

Proof Since $\overrightarrow{BA} \cup \overrightarrow{BC} = \overrightarrow{BC} \cup \overrightarrow{BA}$, we have $\angle ABC = \angle CBA$. Let D be such that $C-D-A$. Then we have $\overleftrightarrow{DA} = \overleftrightarrow{AC} = \overleftrightarrow{DC}$, $\overleftrightarrow{BA} \neq \overleftrightarrow{AC}$, and $\overleftrightarrow{AC} \neq \overleftrightarrow{BC}$. But D in $\overrightarrow{BA}$ implies $\overleftrightarrow{AB} = \overleftrightarrow{AD} = \overleftrightarrow{AC}$, and D in $\overrightarrow{BC}$ implies $\overleftrightarrow{BC} = \overleftrightarrow{DC} = \overleftrightarrow{AC}$. Hence, since D is in neither $\overrightarrow{BA}$ nor $\overrightarrow{BC}$, point D is not in $\angle ABC$. However, since D is in $\overrightarrow{CA}$, point D is in $\angle ACB$. Therefore, $\angle ABC \neq \angle ACB$. ■

Theorem 9.3 Given $\angle AVB$, point C in int $(\overrightarrow{VA})$, and point D in int $(\overrightarrow{VB})$, then $\angle AVB = \angle CVD$.

Proof Since $\overrightarrow{VC} = \overrightarrow{VA}$ and $\overrightarrow{VD} = \overrightarrow{VB}$ by hypothesis (Theorem 8.11) and V, C, D are not collinear as V, A, B are not collinear, the result follows from the definition of an angle. ■

Theorem 9.4 If $\angle AVB = \angle CVD$, then either $\overrightarrow{VA} = \overrightarrow{VC}$ or $\overrightarrow{VA} = \overrightarrow{VD}$.

Proof Since A is in $\angle AVB$, point A must be in $\overrightarrow{VC}$ or $\overrightarrow{VD}$. Since $A \neq V$, point A must be in either int $(\overrightarrow{VC})$ or int $(\overrightarrow{VD})$. Hence either $\overrightarrow{VA} = \overrightarrow{VC}$ or $\overrightarrow{VA} = \overrightarrow{VD}$ (Theorem 8.11). ■

Theorem 9.5 If $\angle AVB = \angle AWB$, then $V = W$.

Proof We know V is in $\angle AWB$. Assume V is in int $(\overrightarrow{WA})$. Then V is off $\overrightarrow{WB}$ and there exists D such that $V-D-B$. So D is in $\angle AVB$. Since V, D, B are collinear and V, W, A are collinear, we must have D off both $\overrightarrow{WA}$ and $\overrightarrow{WB}$ as otherwise A, V, B are collinear. Hence D is off $\angle AWB$ but on $\angle AVB$, contradicting the hypothesis. Our assumption must be false; point V is not in int $(\overrightarrow{WA})$. By symmetry, point V

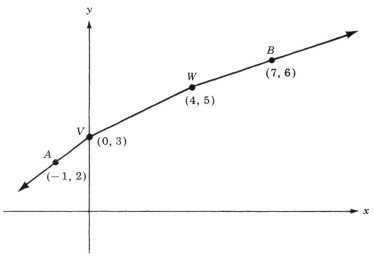

FIGURE 9.2

is not in int $(\overrightarrow{WB})$. Therefore, since V is in $\angle AWB$, there is only one remaining possibility $V = W$. ■

Theorem 9.6 If $\angle AVB = \angle CWD$, then $V = W$ and either $\overrightarrow{VA} = \overrightarrow{VC}$ or $\overrightarrow{VA} = \overrightarrow{VD}$.

Proof Not all points of $\angle CWD$ are on $\overrightarrow{VA}$ nor are they all on $\overrightarrow{VB}$. So there exist point P in int $(\overrightarrow{WC})$ and point Q in int $(\overrightarrow{WD})$ such that either P is in int $(\overrightarrow{VA})$ and Q is in int $(\overrightarrow{VB})$ or else P is in int $(\overrightarrow{VB})$ and Q is in int $(\overrightarrow{VA})$. In either case, (by Theorem 9.3) we have $\angle PVQ = \angle AVB = \angle CWD = \angle PWQ$. Therefore, (by Theorem 9.5) we have $V = W$. So $\angle AVB = \angle CVD$ and (by Theorem 9.4) either $\overrightarrow{VA} = \overrightarrow{VC}$ or $\overrightarrow{VA} = \overrightarrow{VD}$. ■

Even though it may seem intuitively *obvious* that an angle has a unique vertex, we still had to *prove* the fact for every model of Σ. In Figure 9.2, representing (M10, s), point V is on the side $\overrightarrow{WA}$ of $\angle AWB$ and not a vertex of $\angle AWB$. In this figure $\angle AVB$ is not illustrated and "$\angle AVW$" is meaningless as A, V, W are collinear.

DEFINITION 9.7 Given $\angle AVB$, $A - V - A'$, and $B - V - B'$, then $\angle AVB$ and $\angle A'VB'$ are **vertical angles**. Also, $\angle AVB$ and $\angle A'VB$ are a **linear pair** of angles.

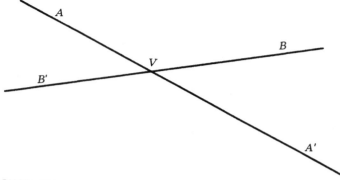

FIGURE 9.3

Theorem 9.8 Given $\angle AVB$, if $\overrightarrow{VA'}$ is the opposite ray of $\overrightarrow{VA}$ and $\overrightarrow{VB'}$ is the opposite ray of $\overrightarrow{VB}$, then

(a) $\angle AVB$ and $\angle A'VB'$ are vertical angles,

(b) $\angle AVB'$ and $\angle BVA'$ are vertical angles,

(c) $\angle AVB$ and $\angle BVA'$ are a linear pair,

(d) $\angle AVB$ and $\angle AVB'$ are a linear pair,

(e) $\angle BVA'$ and $\angle A'VB'$ are a linear pair,

(f) $\angle A'VB'$ and $\angle B'VA$ are a linear pair.

Proof Exercise 9.1. ∎

How shall we *choose* to define a triangle? A few of the reasonable possibilities that are open to us are suggested by Figure 9.4. We make the following *choice*.

DEFINITION 9.9 If A, B, C are three noncollinear points, then $\triangle ABC = \overline{AB} \cup \overline{BC} \cup \overline{CA}$. We say $\triangle ABC$ is a *triangle* with *vertices* A, B, and C and with *sides* $\overline{AB}, \overline{BC}$, and $\overline{AC}$. The angles $\angle BAC, \angle ABC$,

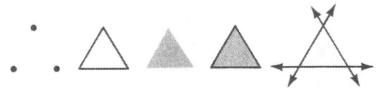

FIGURE 9.4

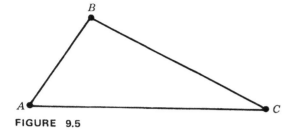

FIGURE 9.5

and $\angle ACB$ are **angles of** $\triangle ABC$ and are called $\angle A$, $\angle B$, and $\angle C$, respectively, when no confusion is likely.

Theorem 9.10 If A, B, C are three noncollinear points, then

$$\triangle ABC = \triangle CBA = \triangle ACB \qquad \text{and} \qquad \overline{AB} = \triangle ABC \cap \overleftrightarrow{AB}.$$

Proof The first statement follows immediately from the definition of a triangle. For the second part

$$
\begin{aligned}
\triangle ABC \cap \overleftrightarrow{AB} &= (\overline{AB} \cup \overline{BC} \cup \overline{CA}) \cap \overleftrightarrow{AB} \\
&= (\overline{AB} \cap \overleftrightarrow{AB}) \cup (\overline{BC} \cap \overleftrightarrow{AB}) \cup (\overline{CA} \cap \overleftrightarrow{AB}) \\
&= \overline{AB} \cup \{B\} \cup \{A\} \\
&= \overline{AB}. \quad \blacksquare
\end{aligned}
$$

Theorem 9.11 If $\triangle ABC = \triangle DEF$, then $\{A, B, C\} = \{D, E, F\}$.

Proof Assume D is neither A, B, nor C. If we reach a contradiction, then the theorem follows from the symmetry stated in the first part of the previous theorem. Since there are certainly four points on $\overleftrightarrow{DE}$, at least two of these four points must be on the same side $\overline{AB}$, $\overline{BC}$, or $\overline{CA}$ of $\triangle ABC$. By the symmetry stated in the first part of the previous theorem, we may say that these two points are on $\overline{AB}$. Then $\overleftrightarrow{DE} = \overleftrightarrow{AB}$. So $\overline{DE} = \triangle DEF \cap \overleftrightarrow{DE} = \triangle ABC \cap \overleftrightarrow{AB} = \overline{AB}$. Hence $\{D, E\} = \{A, B\}$ (Theorem 8.4). Therefore, $D = A$ or $D = B$, contradicting our assumption and proving the theorem. $\quad \blacksquare$

Corollary 9.12 The three vertices, the three sides, and the three angles of a triangle are unique.

If Figure 9.6 represents the Moulton plane with Euclidean arc length along lines defining distance, then it is nonsense to talk about

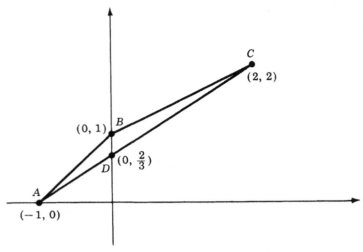

FIGURE 9.6

$\triangle ABC$ as B is on $\overset{\leftrightarrow}{AC}$. In this figure, since D is off $\overset{\leftrightarrow}{AC}$, D is a vertex of $\triangle ACD$ and B is on a side of $\triangle ACD$.

Surely a line cannot intersect a triangle in each of its three sides and not pass through a vertex. In other words (see Figure 9.7), does $\triangle ABC$, $A-D-B$, $B-E-C$, and $A-G-C$ imply that $D-G-E$ is impossible? It would be a fine exercise to try to prove this; in making the attempt, you would review all the previous definitions and theorems. Any proof obtained should be examined with the utmost scrutiny.

9.2 MORE MODELS

We want to talk about the *interior* of an angle and the *interior* of a triangle. What we have in mind is Figure 9.8. If we can define the *interior* of an angle, then it will be easy to define the *interior* of a triangle as the intersection of the *interiors* of its angles. Consider defining the interior of $\angle AVB$ to be the intersection of the *side* of $\overset{\leftrightarrow}{VA}$ that contains B with the *side* of $\overset{\leftrightarrow}{VB}$ that contains A. That seems reasonable. However, it only *seems* reasonable! What is a *side* of a line? Any fool knows that a line divides a plane into two sides. However, we are not basing our geometry on what any fool knows. We must have a definition in terms of the theory we have already developed or else introduce a new axiom.

One solution to our problem is to either introduce as an axiom

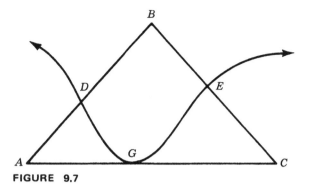

FIGURE 9.7

or prove PSP, the *Plane-Separation Postulate*, which says that for any line l there exist convex sets H_1 and H_2 (these will be the desired *half-planes* or *sides* of line l) such that every point off l is on H_1 or H_2 and such that point P in H_1 and point Q in H_2 with $Q \neq P$ implies $\overline{PQ}$ intersects l (so intuitively you can't *go* from one side to the other without *crossing* the line). Requiring that H_1 and H_2 be convex is some attempt to assure that lines are what we think of as *straight*. Now we are not defining "straight," but we do admit to some intuitive motivation behind the axioms. Given PSP the *interior* of an angle can be defined as the intersection of certain halfplanes of the lines containing the sides of the angle. Then, in turn, *Crossbar* makes sense, where Crossbar is the harmless looking statement, "If point P is in the interior of $\angle AVB$, then $\overrightarrow{VP}$ intersects $\overline{AB}$." See Figure 9.9. Beware of what fools know! Of course, a good question is whether it is necessary to introduce PSP as an axiom. Possibly PSP is already a theorem of Σ that we just haven't proved.

Returning to Figure 9.8 and our idea of the interior of an angle, Crossbar suggests another attack. Although Crossbar is meaningless without a definition of "interior," we can turn things around and use

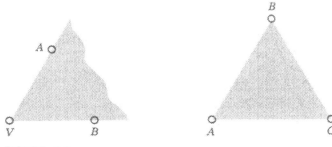

FIGURE 9.8

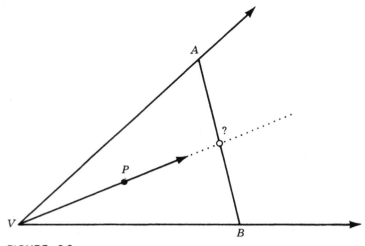

FIGURE 9.9

the idea of Crossbar to define *shade*. (We are cautiously keeping "interior" in reserve.) The *shade* of $\angle AVB$ is the set of all points excluding V that are on any one of the rays $\overrightarrow{VP}$ where P is in int $(\overline{AB})$. Draw a little picture to see that this *seems* reasonable. Perhaps you are suspicious; rightfully so. Let's examine the so-called definition of *shade* more closely. Considering Figure 9.10, if $\angle AVB = \angle AVD$, then certainly the shade of $\angle AVB$ is equal to the shade of $\angle AVD$. Is this the case? In other words, is "shade" *well-defined?* If our term is not well-defined, then it is less than useless! The question is whether

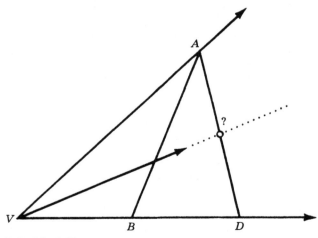

FIGURE 9.10

every ray with vertex V which intersects int $\overline{(AB)}$ also intersects int $\overline{(AD)}$ and conversely. Put in other words, if line $\overleftrightarrow{VE}$ intersects a side of $\triangle ABD$ not at a vertex, then does $\overleftrightarrow{VE}$ intersect another side of $\triangle ABD$? What else can $\overleftrightarrow{VE}$ do? Not being able to imagine any other possibility is hardly a proof. We have arrived at another statement which either must be proved or adopted as an axiom: If a line intersects a side of a triangle not at a vertex, then the line intersects another side of the triangle. This statement is known as *Pasch's Postulate*.

It is time we turned to some models of Σ to clear the air. The Missing-Strip Incidence Plane M8 obviously satisfies the Incidence Axiom. We can define a distance function e on M8 such that (M8, e) is a model of Σ. The easy way to define e is to use Euclidean distance d but so that a ruler has a *blank space* inserted where it crosses the *missing strip*. The details are left for Exercise 9.9, but the idea should be clear from the examples in Figure 9.11 where $e(A, G) = d(A, G)$ but $e(V, A) = d(V, E) + d(R, A)$ and $e(V, B) = d(V, B) - 1$. We can use the real number $d(R, A)$ even though R is not a point in M8.

Figure 9.11 points out that there is something shady about our so-called definition of shade. Point G is in the shade of $\angle AVB$ but G is not in the shade of $\angle AVD$ even though $\angle AVB = \angle AVD$. Since the angles are equal but have different shades, we do *not* have a definition for "shade" that is independent of the notation for the angle. Since "shade" is not well-defined, it must be relinquished to the trash heap.

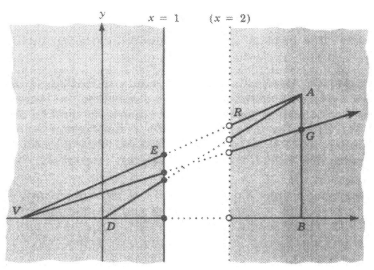

FIGURE 9.11

Pasch's Postulate fails in (M8, e) because $\overleftrightarrow{VG}$ intersects $\triangle ABD$ only at G where $A-G-B$. Although it is somewhat difficult to actually prove, *sides* H_1 and H_2 of $\overleftrightarrow{VB}$ cannot be defined such that PSP holds.

Model (M8, e) shows that both PSP and Pasch's Postulate are independent in Σ, which now contains only the Incidence Axiom and the Ruler Postulate. Recall that our aim is to develop that geometry which is like the Euclidean plane but without any parallel postulate. It is clear that another axiom is needed. We certainly want PSP and Pasch's Postulate to hold in our geometry. Should they both be listed as axioms? Later we shall see that this would be unnecessary as each follows from the other under our present two axioms.

Shade of $\angle AVB$ had to be discarded because its supposed definition was not independent of the particular notation for the angle. The *idea* can be saved, however, as follows.

DEFINITION 9.13 Point P is in the **ray-interior** of $\angle AVB$ if there exist C in int $(\overrightarrow{VA})$, D in int $(\overrightarrow{VB})$, and E in int $(\overrightarrow{CD})$ such that P is in int $(\overrightarrow{VE})$. If P is in the ray-interior of $\angle AVB$, then $\overrightarrow{VP}$ is an **interior ray** of $\angle AVB$. Point P is in the **inside** or **segment-interior** of $\angle AVB$ if there exist C in int $(\overrightarrow{VA})$ and D in int $(\overrightarrow{VB})$ such that P is in int $(\overrightarrow{CD})$.

The ray-interior and inside of an angle are certainly well-defined. Is the ray-interior of an angle different from the inside of the angle? We shall soon give a model of Σ where the ray-interior is not contained in the segment-interior of an angle. However, the following theorem is a trivial result of the definition.

Theorem 9.14 The inside of $\angle AVB$ is contained in the ray-interior of $\angle AVB$.

Model (M4, d), the Space Incidence Plane with Euclidean distance d, is a model of Σ. Perhaps it has not occurred to you that our two axioms do not restrict models to what we *usually* think of as a *plane*. In (M4, d) the ray-interior and inside of an angle are equal. Pasch's Postulate and PSP fail for (M4, d), but the following modified form of Crossbar does hold: If P is a point in the ray-interior of $\angle AVB$, then $\overrightarrow{VP}$ intersects $\overline{AB}$.

The mapping f from $\{x \mid x \in \mathbf{R}, \ 0 < x < s\}$ onto $\mathbf{R}$ defined by $f(x) = \frac{1}{2}\ln [x/(s-x)]$ is a bijection for each positive real number s. See Figure 9.12. That f is onto follows from $f(e^{2t}s/(1+e^{2t})) = t$ for t in $\mathbf{R}$. The bijection f also has the property of being order preserving, since $0 < x_1 < x_2 < s$ implies $f(x_1) < f(x_2)$. Thus there is an order-preserving one-to-one correspondence between the set of points in the interior of

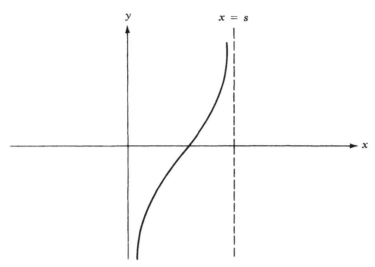

FIGURE 9.12

any Euclidean segment and **R**. Therefore, for each line l in the Cayley –
Klein Incidence Plane M13, there is a bijection f_l from l onto **R**. For
example, if $l = \overleftrightarrow{PQ}$ in Figure 9.13, then

$$f_l(P) = \tfrac{1}{2}\ln \frac{d(P, T)}{d(S, T) - d(P, T)} = \tfrac{1}{2}\ln \frac{d(P, T)}{d(P, S)}$$

defines such a bijection. Interchanging the Cartesian points S and T,
which are not in the model, would only change the sign of $f_l(P)$. For

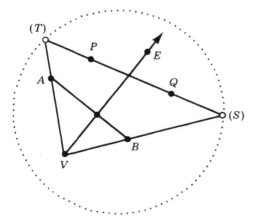

FIGURE 9.13

each line l in M13, we suppose f_l is defined this way. Then define distance function h on M13 by $h(P, P) = 0$ and

$$h(P, Q) = \left| f_l(Q) - f_l(P) \right| = \frac{1}{2} \left| \ln \frac{d(P,S)\ d(Q,T)}{d(P,T)\ d(Q,S)} \right|$$

where $l = \overleftrightarrow{PQ}$ with S and T the Cartesian points on the unit circle determined by l. So f_l is a coordinate system for l. Although $h(P, Q)$ can be expressed in terms of only the coordinates of P and Q, we shall postpone this calculation until it is needed in Chapter 23. For the present we observe that (M13, h) is a model of Σ having the property that $A - B - C$ in (M13, h) only if $A - B - C$ in (M1, d). In Figure 9.13 we see that E is in the ray interior of $\angle AVB$ but not in the inside of $\angle AVB$. Hence the ray-interior of an angle is not necessarily contained in the inside of the angle in a model of Σ.

The methods of the last paragraph can be generalized. If set $\mathscr{P}$ of points and set $\mathscr{L}$ of lines satisfy the Incidence Axiom and if for each line l in $\mathscr{L}$ there exists some fixed bijection $f_l: l \to \mathbf{R}$, then there exists a distance function d such that $(\mathscr{P}, \mathscr{L}, d)$ is a model of Σ. All we have to do to obtain such a model is to define $d(P, Q)$ for points P and Q on line m by $d(P, Q) = |f_m(Q) - f_m(P)|$.

We shall now look at a model which emphasizes that in Σ the definitions of *triangle, angle, ray, segment,* and *between* all rely on *distance* as given by the Ruler Postulate. We start sanely enough with M1, the Real Cartesian Incidence Plane. So the points and lines are the familiar ones. We shall distort the usual Euclidean distance d however. Define bijections h and k from $\mathbf{R}$ into $\mathbf{R}$ by $h(x) = k(x) = x$ if x is not an integer; but if x is an integer then $h(x) = x + 2$ and $k(x) = -x$. Next, define distance d' on M1 by

$$d'((x_1, y), (x_2, y)) = |h(x_2) - h(x_1)| \qquad \text{for all } y,$$

$$d'((x, y_1), (x, y_2)) = |k(y_2) - k(y_1)| \qquad \text{for all } x,$$

$$d'((x_1, y_1), (x_2, y_2)) = d((x_1, y_1), (x_2, y_2)) \text{ if } x_1 \neq x_2 \text{ and } y_1 \neq y_2.$$

So we have the usual distance between two points that are not on lines parallel to one of the axes but rather peculiar distances otherwise. Lines parallel to the x-axis have a coordinate system f where $f((x, y)) = h(x)$. Lines parallel to the y-axis have a coordinate system f where $f((x, y)) = k(y)$. Lines intersecting both axes have the usual Euclidean coordinate systems. Therefore, (M1, d') is a model of Σ.

What makes (M1, d') interesting is that the weird distance function d' gives a betweenness for points that is essentially different from the usual Euclidean case (M1, d). For example, by the *definition* of *between* it follows that ($5/2$, 0) is between (0, 0) and (1, 0). We have not *moved* any of the points! The points and lines are in their usual places! It is just that the betweenness relation for points in (M1, d') is so very different from Euclidean betweenness. Since segments and rays are defined in terms of betweenness, the segments and rays in (M1, d') are rather bizarre. Picking out the endpoints of the nine segments and the vertices of the nine rays pictured in Figure 9.14 will test your knowledge of the definitions. Don't tell your roommate that Figure 9.15 depicts an angle while Figure 9.16 depicts two triangles, because he just may move out on you. All we know about such things as angles and triangles is what the axioms, definitions, and theorems tell us. If nothing else, this model demonstrates that the theorems we have proved so far may not be quite as trivial as they might appear. The model is called the *weird plane*.

Check back to Figure 7.2 and Definition 7.1. Probably most of us *see* a lot more in the figure than the definition actually states. Figure 7.2 might illustrate that A is between B and C in such weird planes as the one just considered. In order for our usual ideas of ruler and betweenness to be consolidated into the theory, the Ruler Postulate must be accompanied by another axiom such as PSP or Pasch's Pos-

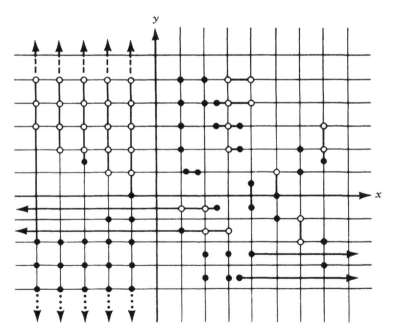

FIGURE 9.14

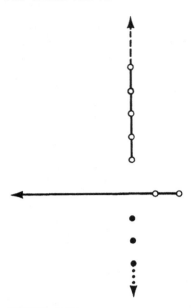

FIGURE 9.15

tulate. We shall add a new axiom to our system in Chapter 12, which can be read next without loss of continuity.

We shall usually abbreviate such a phrase as "if $\triangle ABC$ exists" to "if $\triangle ABC$." Obviously, $\triangle ABC$ exists iff A, B, and C are three non-collinear points. Thus we frequently use the concise phrase "if $\triangle ABC$" in place of the longer phrase "if A, B, and C are three noncollinear points," even though the concept of noncollinearity is more primitive than that of a triangle.

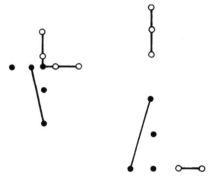

FIGURE 9.16

9.3 EXERCISES

9.1 Theorem 9.8.

● **9.2** In the weird plane sketch the triangle with vertices $(0, \frac{1}{2})$, $(\frac{1}{2}, \frac{1}{2})$, $(\frac{1}{2}, -2)$, the triangle with vertices $(6, -1)$, $(6, -3)$, $(5, -3)$, and $\angle AVB$ where $A = (0, -2)$, $V = (0, -1)$, and $B = (\frac{1}{2}, -1)$.

9.3 What are the endpoints of the segments and the vertices of the rays pictured in Figure 9.14?

● **9.4** Give an example where a line intersects the three sides of a triangle but does not pass through a vertex.

● **9.5** What is the ray-interior of the angle in Exercise 9.2? The inside of this angle is not the ray-interior of the angle.

● **9.6** True or False?

    (a)   A line intersects an angle in at most two points.

    (b) $\triangle ABC = \triangle ACB = \triangle BAC = \triangle BCA = \triangle CAB = \triangle CBA$.

    (c) $\overrightarrow{AB} = \overleftrightarrow{AB} \cap \angle ABC$.

    (d) $\overline{AB} = \overleftrightarrow{AB} \cap \triangle ABC$.

    (e)   A triangle is the union of three angles.

    (f)   If $\triangle ABC$, then $\overleftrightarrow{AB} \cup \overleftrightarrow{BC} \cup \overleftrightarrow{CA} = \angle A \cup \angle B \cup \angle C$.

    (g)   If $\triangle ABC$, then $\overleftrightarrow{AB} \cup \overleftrightarrow{BC} \cup \overleftrightarrow{CA}$ is the union of twelve angles.

    (h)   If $\triangle ABC$, then $\angle A \cap \angle B = \overline{AB}$.

    (i) $\triangle ABC = (\angle A \cap \angle B) \cup (\angle B \cap \angle C) \cup (\angle C \cap \angle A)$.

    (j) $\triangle ABC = (\angle A \cup \angle B) \cap (\angle B \cup \angle C) \cap (\angle C \cup \angle A)$.

9.7 Pasch's Postulate does not hold in $(M1, d')$.

9.8 A line off the vertex of an angle intersects the angle in at most two points.

9.9 Give formulas for distance function e in $(M8, e)$.

9.10 Give an example where $\triangle ABC$, $A - D - B$, $B - E - C$, and $\overleftrightarrow{AE} \parallel \overleftrightarrow{DC}$.

9.11 Find a set of points that is convex in $(M8, e)$ but not in $(M1, d')$.

9.12 If $\overrightarrow{VB}$ is the opposite ray of $\overrightarrow{VA}$, then int $(\overrightarrow{VB}) = \overleftrightarrow{VA} \setminus \overrightarrow{VA}$.

*__9.13__ What is the inside of the angle in Exercise 9.2?

*9.14   PSP fails for both (M8, e) and (M1, d').

*9.15   Find distance functions satisfying the Ruler Postulate for M5, M6, and M7.

GRAFFITI

As lightning clears the air of impalpable vapours, so an incisive paradox frees the human intelligence from the lethargic influence of latent and unsuspected assumptions. Paradox is the slayer of Prejudice.

Sylvester

"When I use a word," Humpty-Dumpty said, "it means just what I choose it to mean — neither more nor less."

Lewis Carroll

Geometrical axioms are neither synthetic a priori *conclusions nor experimental facts. They are conventions: our choice, amongst all possible conventions, is guided by experimental facts; but it remains free, and is only limited by the necessity of avoiding all contradiction. . . . In other words, axioms of geometry are only definitions in disguise.*

That being so what ought one to think of this question: Is the Euclidean Geometry true?

The question is nonsense. One might as well ask whether the metric system is true and the old measures false; whether Cartesian co-ordinates are true and polar co-ordinates false.

Poincaré

Some persons have contended that mathematics ought to be taught by making the illustrations obvious to the senses. Nothing can be more absurd or injurious: it ought to be our never-ceasing effort to make people think, not feel.

Coleridge

CHAPTER 10

The Golden Age of Greek Mathematics

10.1 ALEXANDRIA

The Academy opened its gates at Athens in the year 387 B.C. The marvelous list of accomplishments that we call the Greek Miracle was still growing. This was four centuries after Homer had recited his poems and two centuries after Thales had introduced deduction as a means of obtaining truth. The Golden Age of Pericles, which saw high achievement in art and literature, had recently passed, and ancient Greece would soon be threatened by Macedonia. The century-old Pythagorean idea that the earth was round (spherical) was an obvious fact, but there was much discussion about which heavenly bodies revolved about what. Theatetus was laying the foundation for the study of irrationals that would later appear in Book X of Euclid's *Elements*. Theatetus was also the first to determine that there are exactly five regular solids, which are usually named after his student, colleague, friend, and founder of the Academy, Plato.

The Academy was the first institution of higher learning. Above its gateway was the admonishment, "Let no one ignorant of geometry enter here." Plato did not produce any significant mathematics, but he did produce significant mathematicians. Of these, Eudoxus (circa 408–355 B.C.) ranks as one of the greatest mathematicians of all time. He is also called the father of scientific astronomy. From the so-called Archimedes' axiom, Eudoxus developed the geometric aspect of the integral calculus, known as the method of exhaustion. Of even more significance, Eudoxus solved the great quandary of the Pythagoreans by developing a theory of proportion that included the irrationals. This theory, which leads directly to the work of Dedekind, is preserved in Book V of Euclid's *Elements.*

Menaechmus, a student of Eudoxus, introduced the conics (the parabola, the ellipse, the hyperbola) into mathematics. As a tutor to the young man who would become Alexander the Great, Menaechmus informed his student that there are no royal roads to geometry. Alexander was, in turn, also the student of, the friend of, and the protector of another member of the Academy, Aristotle. Like Plato (circa 430–349 B.C.), Aristotle (384–322 B.C.) had a high regard for mathematics, although not a mathematician himself. Aristotle's work influenced mathematics, as it did all branches of learning. From the numerous mathematical examples in Aristotle's writing, the extent of pre-Euclidean mathematics can be gleaned. Aristotle's student Eudemus wrote a *History of Geometry,* but it is unfortunately lost.

Philip II of Macedonia conquered Greece, and his son Alexander III conquered most of the world. Setting out to Hellenize the world, Alexander ended up trying to harmonize it. Men still strive toward Alexander's dream of a united world. Alexander became king of Macedonia at the age of twenty in 336 B.C. and died of malaria at the age of thirty-three at Babylon in 323 B.C. His death, together with the death of Aristotle the next year, marks the end of the Hellenic Age of Greek Civilization. The next period is called the Hellenistic Age or the Alexandrian Age.

With Alexander dead, his empire fell apart and Egypt went to one of Alexander's leading generals and best friends since childhood, Ptolemy, son of Lagos. (In fact, he may also have been a half-brother since his mother had been a concubine of Philip II.) Declaring himself King Ptolemy I, called Soter (savior), in 306 B.C. he established the Ptolemaic dynasty. Ptolemy I was succeeded by Ptolemy II, called Philadelphus (sister-loving), and then by Ptolemy III, called Evergetes (benefactor), who died in 221 B.C. When there is doubt about the identity of a particular Ptolemaic king, we shall simply call him King Ptolemy. The fifth century B.C. had seen the zenith of Greek literature, the fourth witnessed the flowering of philosophy, and, under the direction of these three kings, the third century would see the Golden Age of Greek Mathematics and Science.

Before departing for Persia, the ancient enemy of the Greeks, Alexander in Egypt in the winter of 332–331 B.C. established *Alexandria-near-Egypt.* Of the seventeen or more cities named after Alexander, this is *the* Alexandria and the capital of the Ptolemaic dynasty. Cooled by sea breezes and located between Lake Mariout and the Mediterranean Sea, the city was directly connected by canal to the Canopic branch of the Nile and from the Nile to the Red Sea by canal. By both land and sea, Alexandria was at the crossroads of Asia, Africa, and Europe and soon became the most important city of the world. The architect of this truly cosmopolitan city was Dinocrates, builder of the Temple of Diana at Ephesus. All the buildings were made of stone, and most public buildings were faced with marble. The two principal avenues were illuminated at night by oil lamps. Alexandria was created as one of the most beautiful cities the world has ever known.

At the center of Alexandria stood the Soma, containing the body of Alexander. The highest spot in the city was occupied by the Serapium, which contained the Temple of Serapis. (Later, some would say that the magnificance of the Serapium was excelled only by the Capitol at Rome.) Serapis was a composite figure of Greek and Egyptian gods constructed by a committee at the direction of King Ptolemy with the political aim of providing the population with a common cult. This committee was more than successful. The cult of Serapis and his consort Isis spread through the Mediterranean world and was one of the last to hold out against Christianity. Six hundred years later a memorial to the Roman Emperor Diocletian would be built near the Serapium. This eighty-four foot column is (incorrectly) called Pompey's Pillar and is the only structure from ancient Alexandria that still stands in place.

The most famous of Alexandria's buildings is the Pharos, one of the Seven Wonders of the Ancient World. With its gleaming white marble united with molten lead, the lighthouse was about five hundred feet tall, perhaps even taller. Left unattended in later years, the lantern and statue of Poseidon that topped the Pharos fell about 700 A.D. (This was a thousand years after the clever architect Sostratos had inscribed his name in the stone and then covered it with plaster bearing the royal inscription of King Ptolemy.) The circular third story with its helical staircases and the octagonal second story fell as a result of an earthquake about 1100. Finally, the square first story, which could have housed three hundred attendants, was destroyed by another earthquake during the fourteenth century. The Pharos was also a fort. Since its demise, several forts have been built and destroyed on the location.

The Museum and the Library at Alexandria were organized under the direction of Ptolemy I by Demetrius of Phaleron. Because of these institutions, Alexandria became the intellectual capital of the world from the third century B.C. until the Middle Ages. At Demetrius' instigation, Ptolemy II had the *Holy Scriptures* of the Jews translated into Greek. This version, which is called the *Septuagint,* became the major source for the *Old Testament* since the original sources were lost. The name comes from the fact that some seventy Jewish scholars were called to Alexandria to make the translation. According to legend, the scholars were segregated into separate cells to make separate translations but all translations were identical, proving they were inspired by God. The Alexandrian Jews produced an enormous religious literature during the last two centuries B.C., including *Ecclesiastes,* part of the *Psalms,* and most of the *Apocrypha.*

The third century B.C. was the Silver Age of Greek literature. After its foundation by Demetrius, the Library was directed by Zenodotus of Phesus. We owe to him the editing and preservation of Homer's works. Theocritus of Syracuse, the only Alexandrian poet still read for pleasure, founded pastoral or idyllic poetry. The elegy and epigram were established as major forms for poetic expression. Poetry was written about everyday things. Romantic love between men and women, darts, hearts, etc., were introduced into literature. On the avant-garde side, poems were written in the shape of a bird, and whole books were written without the letter "s." The systematic study of grammar was introduced. Apollonius of Rhodes was a director of the Library. This Apollonius is the author of the *Argonauts,* the familiar tale about Jason, Medea, and the Golden Fleece. Perhaps the best known director of the Library is Eratosthenes of Cyrene. He was an intellectual giant who excelled in every field. His calculation of the diameter of the earth was off by only fifty miles! In mathematics classes today, he is best known for his *sieve* for isolating prime numbers.

The Library housed the largest collection of books assembled before the printing press, as well as being the center of research and production in the humanities. It probably never exceeded a million rolls. (A roll was ten inches by an average of thirty-five feet.) The Library grew by direct purchases, such as that of Aristotle's library. Also, all travelers to Alexandria had to surrender their books and, if they were not in the Library, they would be kept with copies on cheap papyrus given to the owners. Fearing the growth of the library at Pergamum, the export of papyrus from Egypt was made illegal. King Ptolemy borrowed the original manuscripts of Aeschylus, Sophocles, and Euripides from libraries at Athens to have copies made and deposited a certain sum to guarantee their return. Upon obtaining the works, he considered them more valuable than his large deposit and returned the copies. The Library was soon so large that most of the collection spilled over to the "daughter Library" in the Serapium.

The direction of the Museum was determined by Straton of Lampsacos, whom Ptolemy I called to tutor Ptolemy II. Straton was a physicist, and so the Museum became a court institution for the investigation of mathematics and science. (The living part of any museum today that deserves the name is its research staff; a proper museum is a collection of people and not a collection of things.) Aristarchus of Samos, one of Straton's students, calculated that the sun is larger than the earth and proposed that the sun, not the earth is the center of the universe. Herophilus of Chalcedon, the founder of anatomy, discovered the circulation of the blood. By dissecting (perhaps live) subjects at the Museum, he determined that the brain is the seat of intelligence, contrary to Aristotle's idea that it was the heart. Mathematics, geography, astronomy, and medicine all grew to maturity at the Museum within a short time. The third century B.C. was the greatest period of scientific growth that civilization has known. The Golden Age of Greek Mathematics was a part of this and was dominated by three giants, Euclid, Archimedes, and Apollonius.

Euclid was probably educated at the Academy. He was in Alexandria about 300 B.C. Other than his works, the only other things we know about Euclid are two stories, both of which may be apocryphal. One is that he assured King Ptolemy that there were no royal roads to geometry. (Shades of Menaechmus and Alexander.) The other is that in response to the beginning student's eternal question, "What shall I gain from learning these things?" Euclid directed that the student be given a small coin. Although there is no record that Euclid was at the Museum/Library, it is difficult to imagine otherwise. Euclid wrote many books on several subjects. His *Optics* is still extant; the *Conics* itself is lost but is preserved in the first three books of Apollonius' *Conics;* and the *Elements of Music* is completely lost. Euclid would have to be listed as a first class mathematician, even without the work that is synonymous with his name and with geometry itself, the *Elements*.

There was no other book like the *Elements* before. The *Elements* placed mathematics on an axiomatic basis! Euclid's method was so successful that the works of his predecessors were discarded (unfortunately, for the historian). The *Elements* does not contain all of mathematics that was known at the time; for example, the conics are not treated. The *Elements* is not only about geometry; several of its Books deal with irrational numbers and arithmetic (i.e., the theory of numbers). Euclid's *Elements* became the standard against which all mathematical and scientific writing was compared for two thousand years. There will never be another book like the *Elements*.

The second mathematical giant of the Golden Age is Archimedes of Syracuse (287–212 B.C.). After studying at Alexandria, he returned to Sicily to live under King Hieron II and his son Gelon II. He was on intimate terms with these Greek kings, if not related to them. The very short *Sand-reckoner,* which is a part of every mathematician's baggage, is addressed to King Gelon. All of Archimedes' works are short. *The Method,* which is addressed to Eratosthenes at Alexandria and was lost until 1906, explains how he derived his theorems. The historian W. W. R. Ball claims that Archimedes deliberately misstated some of his results to ensnare poachers. Everyone knows the "Eureka story," the many stories about his mechanical inventions, and that Archimedes was slain when Syracuse fell to the Romans. Archimedes attached no importance to his mechanical inventions but prided himself on his mathematics. As he had directed, his tomb was decorated only with a cylinder (with ends) enscribed about a sphere with his discovery that the ratio of the surface areas of the solids and the ratio of the volumes are both 3/2. (Cicero later found and repaired the tomb, but its location is now lost.) The theorems on the tomb were results that appear in Archimedes' greatest masterpiece, *On the Sphere and Cylinder.*

The third mathematical giant of the Golden Age is Apollonius of Perga. Apollonius, who was twenty-five years younger than Archimedes, studied in Alexandria and stayed there. His *Tangencies* discusses what is known as the Problem of Apollonius: Given three things, each of which is either a point, a line, or a circle, determine the circles tangent to the three. Apollonius' reputation as "the Great Geometer" rests on a single book, the *Conics*. The first seven of its eight Books are extant. From the dedication of the last five Books of the *Conics* to King Attalus I of Pergamum, we might infer that Apollonius and the new king, Ptolemy IV, were not on the best of terms. The *Conics* is the last Greek mathematical work that is an unqualified masterpiece.

The end of the third century B.C. saw the end of the Golden Age of Greek Mathematics. After the death of Ptolemy III in 221 B.C., the fortunes of the Museum/Library waxed and waned, but mostly waned. The vicissitudes of peace and turmoil within the city of Alexandria (and everywhere else) did not provide the environment for creation. Of the mathematicians from the next three centuries, only one made a significant contribution, and he was a mathematician by necessity. In order to do his work, Hipparchus of Nicaea (circa 180–125 B.C.), one of the great astronomers, founded trigonometry.

After Ptolemy III, the Ptolemaic dynasty degenerated. However, the last member of the dynasty to rule was certainly an exception. Brilliant, daring, ruthless, unscrupulous, totally charming, and equally ambitious, she was Cleopatra VII. This pure Macedonian Greek queen was the first of the Ptolemaic dynasty to speak the language of the Egyptians. She had many accomplishments. Even after her plan to rule the world with Julius Caesar came to an abrupt end on 15 March 44 B.C., Rome feared Cleopatra—and with good reason. The story of her second plan, this time with Mark Antony, belongs to poetry. She committed suicide in 30 B.C.

Cleopatra plays a role in two chapters of the history of the Library. The first incident may be best known from Shaw's *Caesar and Cleopatra.* "Horror unspeakable! . . .Oh, worse than the death of ten thousand men! Loss irreparable to mankind! . . .The fire has spread from your ships. The first of the seven wonders of the world perishes. The library at Alexandria is in flames." A few of the 700,000 rolls may have been destroyed at the harbor by the fire set by Caesar, but these would have been those that Cleopatra had given to Caesar as a gift (investment). The marble library itself could hardly have been "in flames," and its contents did not perish in 47 B.C. In the next episode, the Library is compensated by the addition of 200,000 rolls in 41 B.C. For it was then that Antony packed up the second most prominent library in the world, the library at Pergamum, and sent it off to Cleopatra as a token of his affection.

With the end of the Ptolemaic dynasty and the creation of the Roman Empire by Augustus, Rome became the center of art and literature but would never excel the glory that was Greece. The roman contribution to mathematics and science was nil. The advances in these fields still came from Alexandria.

Menelaus of Alexandria made advances in geometry and spherical trigonometry at the end of the first century. Galen, the supreme authority of medicine, lived in the second century as did Claudius Ptolemy of Alexandria, who is *the* Ptolemy and is not related to the earlier rulers. The work started by Hipparchus in the second century B.C. culminated in Ptolemy's *Mathematical Synthesis*. This Greek work is better known by its Arabic title *Almagest* (The Greatest). The *Almagest* and his *Geography* became the standard textbooks in their fields for at least fourteen centuries. Ptolemy's modest estimate of the size of the earth encouraged Columbus to undertake his voyage. In Ptolemy's defense, it should be added that the so-called Ptolemaic theory fit the available data better than the so-called Copernican theory. Hard as it is to believe, the author of the *Almagest* is the same Ptolemy that wrote the famous *Tetrabiblos* on astrology.

The Silver Age of Greek Mathematics is the century from A.D. 250 to 350. Although his dates are uncertain, we may place Heron of Alexandria at the beginning of this revival. Heron wrote on almost every area of mathematics and physics, including a commentary on the *Elements.* Nothing seems to have come of his invention of the steam engine. Diophantus of Alexandria is more important to the history of mathematics. In his *Arithmetica* we see the beginnings of of modern algebra. Today Diophantine analysis is a branch of the theory of numbers. The Silver Age ends with the last giant of Greek mathematics, the geometer Pappus of Alexandria. Pappus made several significant contributions to mathematics in his *Synagoge,* which is usually called the *Collection*. (See Section 34.1 for one of them.) Only a fragment of his commentaries on the *Elements* and the *Almagest* are extant. In the history of geometry, Pappus (circa 320 A.D.) is followed by Descartes (circa 1637). In fact, Descartes invented analytic geometry in trying to solve a problem posed in the *Collection*.

Theon of Alexandria (circa 365 A.D.) edited the *Elements* and wrote a commentary on the *Almagest*. These were probably discussed less than Philostratus' century-old biography of Apollonius of Tyana (circa 30 A.D.), which describes Apollonius' divine birth, miracles, and ascension into heaven. Alexandria had turned to philosophy and theology. Much of the theological debate centered on the nature of Jesus of Nazareth. The great schisms among the Christian sects and the power politics that established Christian doctrine do not concern us here. The struggle between the Christians and the pagans does. The patriarchs of Alexandria became the real rulers of the city. Having the opportunity in 391 A.D., the patriarch roused the mob to destroy the contents of the Serapium, which housed the daughter branch of the Library. Certainly the god Serapis and all his trappings were destroyed. Exactly how much of the Library was destroyed is not known — opinions vary from only a few of the books to almost all of the books.

Hypatia of Alexandria, the daughter of Theon, was a mathematician and philosopher of some fame. She wrote commentaries on the works of Apollonius, Ptolemy, and Diaphantus. Hypatia was an ardent devotee of pagan learning and culture. The Christian persecution of the pagans and of the Jews that resulted in the destruction of a part of the Library in 391 A.D. was still going on. For her defense of things Greek, Hypatia was literally torn limb from limb by the fanatic Christian mob in 415 A.D. To Hypatia goes the honor of being the first known woman mathematician, one of the first martyrs of science, and the last mathematician to lend glory to the Museum.

After studying at Alexandria, Proclus Lycius (A.D. 410–485) went to Athens and became the greatest director of the Academy in the last century of its existence. Proclus was more of a philosopher than a mathematician. However, his *Commentary on Book I of the Elements* is invaluable since it contains a large proportion of all the available information on the history of pre-Euclidean geometry. (A new English translation of this interesting commentary is available; see Exercise 11.5.) The traditional date of the "fall" of Rome is 476 A.D. Proclus was writing about this time and had before him all the great mathematical works from Eudemus' *History of Geometry* to Pappus' *Commentary on the Elements*. The last director of the Academy was Damascus. His student Simplicius of Cilicia wrote a commentary on Book I of the *Elements*. Simplicius' work is valuable because it preserves the work of others, including much of Aristotle and some of Eudemus.

In 529 A.D., Justinian closed the Academy because its pagan and (supposedly) perverse learning was a threat to Christianity. If the fall of Rome in 476 marks the beginning of the Middle Ages and some part of the Middle Ages is to be called the Dark Ages, then the year 529 is a best choice for the beginning of the Dark Ages. Even the Dark Ages were not completely black, of course. Also in 529, St. Benedict founded the monastery at Monte Cassino (later destroyed by Allied bombings in 1944). An unsung hero of this time is King Chosroes of Persia. When Justinian closed the schools, Damascus and Simplicius were among those from the Academy who escaped to Baghdad. King Chosroes extracted from Justinian that those members of the "Academy-in-Exile" who wished to return be exempt from the laws against pagan subjects of the empire.

The Byzantine rule of Alexandria was interrupted by the Persians for the decade beginning with 618 A.D. Little is known about the Persian Interlude except that it was a peaceful period, unusual for Alexandria since the days of Ptolemy III. Also, the Library was used for research, mostly in theology and medicine.

The second city of the Empire fell not with a bang but with a whimper. Amidst the struggles among the bishops for power, among the successors to the emperor for power, and between the two, Alexandria simply surrendered to a small army of Arabs led by Amr ibn al As in 641 A.D. The terms of the surrender provided for an eleven-month armistice, during which time anyone could leave Alexandria by sea with movable property. On these generous terms, what gems in the form of precious stones and in the form of priceless manuscripts left the city? What fragments of a thousand years of Greek civilization were evacuated? Amr sent a letter to the Caliph Omar asking what to do with the Library. Omar answered that if the books contain what is in the *Koran* then they are superfluous and may be destroyed, but if they contain things contrary to the *Koran* then they are pernicious and should be destroyed. The books were burned. According to one legend, the Library furnished fuel for the four thousand baths in the city for a period of six months. With its intellectual lighthouse gone, the earth became flat and the Dark Ages of Western civilization would last six more centuries.

10.2 EXERCISES

10.1 Place the names of the following ten mathematicians in chronological order: Apollonius, Archimedes, Euclid, Eudoxus, Hypatia, Pappus, Proclus, Ptolemy, Pythagoras, Thales.

● **10.2** There exist exactly five regular (convex) solids.

● **10.3** True or False?

(a) The Library at Alexandria was one of the Seven Wonders of the World.

(b) The scholars at the Museum were paid by the rulers of Alexandria.

(c) Everyone at the Museum knew the earth was round.

(d) Although mice, mold, and termites must have taken their toll, the three recorded incidents of destruction of the books at the Library were due to military operations, religious fanaticism, and simple ignorance.

(e) The Elements placed mathematics on axiomatic basis.

(f) Aristotle was a tutor to Alexander the Great.

(g) Euclid's *Elements* is a summary of all geometry that was known by 300 B.C.

(h) The Academy, which lasted a millenium, was founded by Euclid.

(i) Euclid alone looked on beauty bare.

(j) The Golden Age of Greek Mathematics was approximately from 300 to 200 B.C.

10.4 What is the method in Archimedes' *The Method?*

10.5 For *the* history of Greek mathematics, read the two volume *A History of Greek Mathematics* by T. L. Heath (Oxford, 1921). For *the* history of Greek science, read the two-volume Norton paperback *A History of Science* by George Sarton (Harvard, 1952). However, if you can take only small doses of history, then the short paperback *Ancient Science and Modern Civilization* by Sarton (U. Nebraska, 1954) is for you.

10.6 If you like your history in story form, read the two volume *Alexandria, The Golden City* by H. T. Davis (Principia of Illinois, 1957), *Hypatia* by Charles Kingsley (in Everyman's Library), or *Cleopatra's Children* by Alice Desmond (Dodd, Mead, 1971).

10.7 For a history of Alexandria or of the Library read, respectively, *The Golden Age of Alexandria* by John Marlowe (Gollancz, 1971) or *The Alexandrian Library* by E. A. Parsons (Elsevier, 1952).

CHAPTER 11

Euclid's Elements

11.1 THE ELEMENTS

The trivium, consisting of grammar, logic, and rhetoric, was added to the quadrivium to form *the seven arts* of the medieval curriculum. The quadrivium consists of the *mathemata* or *subjects of study* that go back at least to Archytas of Taras, one of the last of the Pythagoreans. Of course, the word *mathematics* no longer means all *learning*, as it once did. Of the seven liberal arts, the original four that were deemed worthy of study were the mathemata called arithmetica, harmonica, geometria, and astrologia. Today the quadrivium is usually listed as arithmetic, music, geometry and astronomy. This is somewhat misleading and especially so for Americans who don't realize that the subjects they call arithmetic and theory of numbers go by the opposite names in the rest of the world. An American student would better understand the content of the quadrivium if it were listed as theory of numbers, mathematical theory of music, geometry, and mathematical astronomy.

Euclid wrote on each of the mathemata of the quadrivium. His *Elements of Music* is completely lost, but the *Phaenomena* on spherical geometry for astronomy is extant. Needless to say, Euclid wrote a book on arithmetic and geometry that is called the *Elements*. No other secular book has circulated more widely over the world or has been more edited or studied. Very little development of the mathematical method took place after Euclid until modern times! It is only in our own century that the *Elements* has been universally replaced as the high school geometry textbook.

It is always fun for little men to point out the flaws of a work that superseded all of its predecessors and stood as a standard for two millennia. Although the *Elements* does have flaws, it has most of the necessary virtues. We shall give below a small amount of commentary in which some of Euclid's flaws are indicated. We do so, however, humbled by the knowledge that it is unlikely a book having the significance of the *Elements* will ever be written again.

We shall now look at the contents of the first Book of the *Elements*. The theory from Book I that is quoted below is taken from *The Thirteen Books of Euclid's Elements* by Thomas L. Heath with the kind permission of the publisher, the Cambridge University Press. Heath is *the* source in English on the *Elements.* Heath's three volume book of text and commentary is reprinted in paperback by Dover.

As do most of the thirteen books, Book I begins with Definitions, which is just a list without any discussion. The first items in the list are not actually definitions but are descriptions to let the reader, who is just beginning the study of geometry, know in what sense the words are being used. It will be seen that Euclid uses the word *line* as we would use the word *curve,* Euclid's *straight line* would be our *segment,* Euclid's *circle* would be our *disc,* and Euclid's *triangle* would be our *triangular region* (our triangle together with its interior). (Look at Figure 9.4 again.) While *rhombus* is defined, the word is not used in the *Elements.* On the other hand, some technical words that are used are not defined at all. For example, Euclid would suppose that everyone knew the *circumference of a circle* is the line bounding a circle. (We would say that a circle is the curve bounding a disc. However, even Euclid occasionally confused his terms *circle* and *circumference of a circle.*) These are just a few of the things you might look for in reading the Definitions of Book I.

Euclid's Definitions of Book I   1. A *point* is that which has no part. 2. A *line* is breadthless length. 3. The extremities of a line are points. 4. A *straight line* is a line which lies evenly with the points on itself. 5. A *surface* is that which has length and breadth only. 6. The extremities of a surface are lines. 7. A *plane surface* is a surface which lies evenly with the straight lines on itself. 8. A *plane angle* is the inclination to one another of two lines in a plane which meet one another and do not lie in a straight line. 9. And when the lines containing the angle are straight, the angle is called *rectilineal.* 10. When a straight line set up on a straight line makes the adjacent angles equal to one another, each of the equal angles is *right,* and the straight line standing on the other is called a *perpendicular* to that on which it stands. 11. An *obtuse angle* is an angle greater than a right angle. 12. An *acute angle* is an angle less than a right angle. 13. A *boundary* is that which is an extremity of anything. 14. A *figure* is that which is contained by any boundary or boundaries. 15. A *circle* is a plane figure contained by one line such that all the straight lines falling upon it from one point among those lying within the figure are equal to one another. 16. And the point is called the *centre* of the circle. 17. A *diameter* of the circle is any straight line drawn through the centre and terminated in both directions by the circumference of the circle, and such a straight line also bisects the circle. 18. A *semicircle* is the figure contained by the diameter and the circumference cut off by it. And the centre of the semicircle is the same as that of the circle. 19. *Rectilineal* figures are those which are contained by straight lines, *trilateral* figures being those contained by three, *quadrilateral* those contained by four, and *multilateral* those contained by more than four straight lines. 20. Of trilateral figures, an *equilateral triangle* is that which has its three sides equal, an *isosceles triangle* that which has two of its sides alone equal, and a *scalene triangle* that which has its three sides unequal. 21. Further, of trilateral figures, a *right-angled triangle* is that which has a right angle, an *obtuse-angled triangle* that which has an obtuse angle, and an *acute-angled triangle* that which has its three angles acute. 22. Of quadrilateral figures, a *square* is that which is both equilateral and right-angled; an *oblong* that which is right-angled but not equilateral; a *rhombus* that which is equilateral but not right-angled; and a *rhomboid* that which has its opposite sides and angles equal to one another but is neither equilateral nor right-angled. And let quadrilaterals other than these be called *trapezia*. 23. *Parallel* straight lines are straight lines which, being in the same plane and being produced indefinitely in both directions, do not meet one another in either direction.

The Postulates and the Common Notions (or Axioms) appear next in Book I. In reading these you might well wonder what most of these items actually mean. As the philosopher Schopenhauer has pointed out, it is surprising that the fourth axiom was not the subject of attack rather than that stroke of genius, the fifth postulate.

Euclid's Postulates Let the following be postulated: 1. To draw a straight line from any point to any point. 2. To produce a finite straight line continuously in a straight line. 3. To describe a circle with any centre and distance. 4. That all right angles are equal to one another. 5. That, if a straight line falling on two straight lines make the interior angles on the same side less than two right angles, the two straight lines, if produced indefinitely, meet on that side on which are the angles less than the two right angles.

Euclid's Common Notions 1. Things which are equal to the same thing are also equal to one another. 2. If equals be added to equals, the wholes are equal. 3. If equals be subtracted from equals, the remainders are equal. 4. Things which coincide with one another are equal to one another. 5. The whole is greater than the part.

The forty-eight propositions of Book I follow the Common Notions. The existence of points is tacitly assumed. The existence of straight lines and circles is then assured by the postulates. Beyond this, the existence of any entity must be proved. Euclid knew very well that the definition of a thing does not imply its existence. The first three propositions are *problems* rather than *theorems*. Following the statement of a problem, Euclid proves the existence of some desired entity. For example, we would state the first proposition as follows: Given $\overline{AB}$, there exists an equilateral triangle with side $\overline{AB}$. Euclid argues that if point C is the intersection of the circumference of the circle with center A and radius $\overline{AB}$ and of the circumference of the circle with center B and radius $\overline{AB}$ then $\triangle ABC$ is equilateral. This is fine. However, Euclid asserts the existence of such a point C. This is a major flaw! There is nothing in the postulates to verify this assertion. Here and in some other cases, it seems clear that Euclid resorts to arguing from a figure. Euclid's postulates and axioms are not adequate to describe what we call Euclidean geometry.

At the end of some of the statements of the propositions from Book I that are listed below, there is a parenthetical note indicating where that proposition occurs in our development. Even with our powerful axioms, the proof of the existence of the point C for the first proposition is by no means trivial. In fact, Euclid's Proposition I.1 is among the last of his propositions that we shall be able to prove. Clearly, Euclid should have had some additional postulates.

In arguing that $\triangle ABC$ in the proof of his first proposition is equilateral, Euclid would say that $\overline{AC}$ and $\overline{AB}$ are *equal* by his fifteenth definition. Euclid's use of the word *equal* is not a flaw. Of course, it would be wrong for *us* to say that $\overline{AC}$ and $\overline{AB}$ are equal. (We would say $\overline{AC}$ and $\overline{AB}$ are *congruent*.) However, Euclid uses the one word *equal* for several different relations. It is not Euclid's fault that the meaning of the word in modern mathematics is restricted to *is exactly the same as*. It is the burden of the modern reader to distinguish the various meanings from the context, just as it is the burden of the reader to distinguish the several meanings of our word *side* from its context. As is always the case in reading mathematics, in order to make proper sense of what is being read, the reader must keep in mind the definitions and use of terminology.

The attentive reader of the *Elements* has the additional burden of compensating for the gaps left by Euclid. The only postulate that explicitly deals with the relation of betweenness is the second. Otherwise, it did not occur to Euclid that such an obvious idea needed any explanation. We infer from the Definitions that Euclid tacitly assumes a relation on the set of straight lines that he denotes by the word *equal*. However, given two straight lines which are not radii of the same circle, how can we tell whether they are equal or not? (In our terminology, the problem is to determine when two segments are congruent.)

We would state Euclid's second proposition as follows: Given $\overline{BC}$ and point A, there exists a point L such that $\overline{AL} \simeq \overline{BC}$. From this result, *we* can derive an answer to the question of determining when two straight lines are equal. The second proposition has another application. To say that a proof of the existence of some entity is *constructive* means that the proof tells you how to find the entity, which is very different from knowing only that the entity exists. Euclid's solutions to his problems are constructive existence proofs, which give rise to the ruler and compass game of Euclidean constructions (Section 34.3). In regard to this, the second proposition extends the use of the compass. By Euclid's proof (see our Theorem 21.2), to *describe* a circle we need to be given only its center and a *straight line* that is *equal* to a radius, rather than its center and a radius as the third postulate requires.

Euclid's Propositions I.1 through I.4   1. On a given finite straight line to construct an equilateral triangle. (Theorem 21.1) 2. To place at a given point (as an extremity) a straight line equal to a given straight line. (Theorem 21.2) 3. Given two unequal straight lines, to cut off from the greater a straight line equal to the less. (Theorem 8.8) 4. If two triangles have the two sides equal to two sides respectively, and have the angles contained by the equal straight lines equal, they will also have the base equal to the base, the triangle will be equal to the triangle, and the remaining angles will be equal to the remaining angles respectively, namely those which the equal sides subtend. (Axiom 5)

Proposition I.4 is Euclid's first *theorem* and does not depend on the three *problems* that precede it. The proposition is called the Side-Angle-Side Theorem for obvious reasons. There might be a question about the meaning of the statement. In particular, when are two plane angles equal? We infer, from the Definitions that Euclid tacitly assumes, an undefined relation on the set of plane angles that he denotes by the word *equal*. (It is not until after Proposition I.8, which is called the Side-Side-Side Theorem, that *we* have a criterion for determining whether two angles are equal or not. In essence, Proposition I.8 defines the meaning of Euclid's word *equal* as a relation on the set of plane angles.) Of course, rather than using Euclid's *equal*, we would use *is congruent to*, which is how the symbol $\simeq$ is to be read below in all cases. Euclid's Proposition I.4 is usually replaced by the following statement, which is called SAS: Given $\triangle ABC$ and $\triangle DEF$, if $\overline{AB} \simeq \overline{DE}$, $\angle A \simeq \angle D$, and $\overline{AC} \simeq \overline{DF}$, then $\angle B \simeq \angle E$, $\overline{BC} \simeq \overline{EF}$, and $\angle C \simeq \angle F$. The statement SAS covers the content of the proposition except for the part that says "the triangles will be equal to the triangles." In what will be our terminology, the hypothesis of SAS must also imply that the *interiors* of $\triangle ABC$ and $\triangle DEF$ are *congruent* (Exercise 19.2).

Euclid's proof of Proposition I.4 illustrates one more gap in his system. Euclid tacitly assumes that a given line has two *sides*. As in our development in the other chapters, some axioms such as PSP and Pasch's Postulate are necessary to rule out such weird planes as we considered at the end of Section 9.2. This necessity was not realized until the nineteenth century. Euclid's proof is based on the idea of superimposing one triangle onto another and seeing that they fit. That Euclid realized he was on shaky ground is indicated by his reluctance to use this method. The method is called *superposition* and is supposedly valid by the fourth common notion. Actually, some sort of converse of the fourth common notion would be more relevant. The idea of superposition is a good one. The difficulty is in making the idea precise. One way of overcoming this difficulty is to assume SAS as a postulate. In Chapters 16 and 17, we shall discuss this as well as an alternative method.

> **Euclid's Propositions I.5 through I.7**   5. In isosceles triangles the angles at the base are equal to one another, and if the equal straight lines be produced further, the angles under the base will be equal to one another. (Theorem 17.5) 6. If in a triangle two angles be equal to one another, the sides which subtend the equal angles will also be equal to one another. (Theorem 18.7) 7. Given two straight lines constructed on a straight line (from its extremities) and meeting in a point, there cannot be constructed on the same straight line (from its extremities), and on the same side of it, two other straight lines meeting in another point and equal to the former two respectively, namely each to that which has the same extremity with it. (Exercise 17.8)

We would state Proposition I.7 as follows: Given $\triangle ABC$, if point P is on the same side of $\overset{\leftrightarrow}{AB}$ as C, $\overline{AP} \simeq \overline{AC}$, and $\overline{BP} \simeq \overline{BC}$, then $P = C$. This is called the Hinge Theorem, as it is an abstraction of the idea that hinging three rods together gives a rigid configuration. Closely related to this is the *Hinge Axiom:* Given $\triangle DEF$ and $\overline{AB} \simeq \overline{DE}$, there exists a unique point C on a given side of $\overset{\leftrightarrow}{AB}$ such that $\overline{AC} \simeq \overline{DF}$ and $\overline{BC} \simeq \overline{EF}$.

We claim (without proof) that the following system is a categorical axiom system for Euclidean plane geometry. We take

(a)   Our theory in Chapters 6 through 9.

(b)   Pasch's Postulate as an axiom.

(c)   The following definition, which is based on Euclid's Proposition I.8: $\angle AVB$ *is congruent to* $\angle CWD$ if there exist segments $\overline{VE}$ on $\overrightarrow{VA}$, $\overline{VF}$ on $\overrightarrow{VB}$, $\overline{WG}$ on $\overrightarrow{WC}$, and $\overline{WH}$ on $\overrightarrow{WD}$ such that $\overline{VE} \simeq \overline{WG}$, $\overline{VF} \simeq \overline{WH}$, and $\overline{EF} \simeq \overline{GH}$.

(d)   The Hinge Axiom as an axiom.

(e)   SAS as an axiom.

(f)   And Euclid's Parallel Postulate as an axiom.

Our approach in succeeding chapters will be somewhat different.

Euclid's Propositions I.8 through I.28   8. If two triangles have the two sides equal to two sides respectively, and have also the base equal to the base, they will also have the angles equal which are contained by the equal straight lines. (Theorem 17.14) 9. To bisect a given rectilineal angle. (Theorem 14.8) 10. To bisect a given finite straight line. (Theorem 8.15) 11. To draw a straight line at right angles to a given straight line from a given point on it. (Theorem 14.18) 12. To a given infinite straight line, from a given point which is not on it, to draw a perpendicular straight line. (Theorem 18.1) 13. If a straight line set up on a straight line makes angles, it will make either two right angles or angles equal to two right angles. (Theorem 14.9) 14. If with any straight line, and at a point on it, two straight lines not lying on the same side make the adjacent angles equal to two right angles, the two straight lines will be in a straight line with one another. (Theorem 14.10) 15. If two straight lines cut one another, they make the vertical angles equal to one another. (Theorem 14.11) 16. In any triangle, if one of the sides be produced, the exterior angle is greater than either of the interior and opposite angles. (Theorem 17.9) 17. In any triangle two angles taken together in any manner are less than two right angles. (Exercise 17.5) 18. In any triangle the greater side subtends the greater angle. (Theorem 18.10) 19. In any triangle the greater angle is subtended by the greater side. (Theorem 18.11) 20. In any triangle two sides taken together in any manner are greater than the remaining one. (Theorem 18.12) 21. If on one of the sides of a triangle, from its extremities, there be constructed two straight lines meeting within the triangle, the straight lines so constructed will be less than the remaining two sides of the triangle, but will contain a greater angle. (Theorem 18.16) 22. Out of three straight lines, which are equal to three given straight lines, to construct a triangle: thus it is necessary that two of the straight lines taken together in any manner should be greater than the remaining one. (Theorem 20.15) 23. On a given straight line and at a point on it to construct a rectilineal angle equal to a given rectilineal angle. (Theorem 14.3) 24. If two triangles have the two sides equal to two sides respectively, but have the one of the angles contained by the equal straight lines greater than the other, they will also have the base greater than the base. (Theorem 18.18) 25. If two triangles have the two sides equal to two sides respectively, but have the base greater than the base, they will also have the one of the angles contained by the equal straight lines greater than the other. (Theorem 18.19) 26. If two triangles have the two angles equal to two angles respectively, and one side equal to one side, namely, either the side adjoining the equal angles, or that subtending one of the equal angles, they will also have the remaining sides equal to the remaining sides and the remaining angle to the remaining angle. (Theorems 17.12 and 17.13) 27. If a straight line falling on two straight lines make the alternate angles equal to one another, the straight lines will be parallel to one another. (Corollary 21.6) 28. If a straight line falling on two straight lines make the exterior angle equal to the interior and opposite angle on the same side, or the interior angles on the same side equal to two right angles, the straight lines will be parallel to one another. (Corollary 21.7)

The part of Book I that is independent of the fifth postulate is called *absolute geometry*. Euclid's absolute geometry ends with his Proposition I.28. The controversial postulate on parallels is used for the first time in the proof of Proposition I.29.

Euclid's Propositions I.29 through I.48 29. A straight line falling on parallel straight lines makes the alternate angles equal to one another, the exterior angle equal to the interior and opposite angle, and the interior angles on the same side equal to two right angles. 30. Straight lines parallel to the same straight line are also parallel to one another. 31. Through a given point to draw a straight line parallel to a given straight line. 32. In any triangle, if one of the sides be produced, the exterior angle is equal to the two interior and opposite angles, and the three interior angles of the triangle are equal to two right angles. 33. The straight lines joining equal and parallel straight lines (at the extremities which are) in the same directions (respectively) are themselves also equal and parallel. 34. In parallelogrammic areas the opposite sides and angles are equal to one another, and the diameter bisects the areas. 35. Parallelograms which are on the same base and in the same parallels are equal to one another. 36. Parallelograms which are on equal bases and in the same parallels are equal to one another. 37. Triangles which are on the same base and in the same parallels are equal to one another. 38. Triangles which are on equal bases and in the same parallels are equal to one another. 39. Equal triangles which are on the same base and on the same side are also in the same parallels. 40. Equal triangles which are on equal bases and on the same side are also in the same parallels. 41. If a parallelogram have the same base with a triangle and be in the same parallels, the parallelogram is double of the triangle. 42. To construct, in a given rectilineal angle, a parallelogram equal to a given triangle. 43. In any parallelogram the complements of the parallelograms about the diameter are equal to one another. 44. To a given straight line to apply, in a given rectilineal angle, a parallelogram equal to a given triangle. 45. To construct, in a given rectilineal angle, a parallelogram equal to a given rectilineal figure. 46. On a given straight line to describe a square. 47. In right-angled triangles the square on the side subtending the right angle is equal to the squares on the sides containing the right angle. 48. If in a triangle the square on one of the sides be equal to the squares on the remaining two sides of the triangle, the angle contained by the remaining two sides of the triangle is right.

Thus Book I ends with the Pythagorean theorem and its converse. (It's the converse that is sometimes more useful.) The *equal* in Proposition I.47 is the same *equal* that is introduced without warning by Euclid in Proposition I.35. This *equal* is not our *equal,* nor is it our *congruent* either! In high school this *equal* is usually replaced by *equal in area,* which is an exaggerated oversimplification of Euclid's usage. (Many high school students can prove the Pythagorean formula $c^2 = a^2 + b^2$, but few of their teachers can prove Euclid's Proposition I.47.) Propositions I.35 through I.48 deal with the idea of *piecewise congruence.* Loosely speaking, regions R and S are *piecewise congruent* if each can be cut up into n regions R_i and S_i, respectively, such that R_i is congruent to S_i for $i = 1, 2, \ldots, n$. (For a more precise formulation of the idea and to see how one can extend the idea of area of a triangle to area of a polygonal region in Euclidean geometry, see Section 33.2, which can be read now.) If you think about all this for a moment, you will have to admit that Euclid's Book I is still a *tour de force!*

11.2 EXERCISES

11.1 Euclid's Proposition I.1 does not hold on a sphere.

11.2 List the different meanings of the word *equal* in Book I of the *Elements.* (Don't forget the only meaning that is allowed in modern mathematics.)

11.3 What is the content of Euclid's *Porisms?*

11.4 Which of Euclid's proofs would be followed by "Q.E.F." and which by "Q.E.D."?

11.5 Read Proclus, *A Commentary on the First Book of Euclid's Elements* by G. R. Morrow (Princeton, 1970).

GRAFFITI

*As to writing another book on geometry, the middle ages
would as soon have thought of composing another New Testament.*

De Morgan

*It would be foolish to give credit to Euclid for pangeometrical
conceptions; the idea of geometry different from the common-sense
one never occurred to his mind. Yet, when he stated the fifth
postulate, he stood at the parting of the ways. His subconscious
prescience is astounding. There is nothing comparable to it in the
whole history of science.*

Sarton

*As to the need of improvement there can be no question whilst
the reign of Euclid continues. My own idea of a useful course is to
begin with arithmetic, and then not Euclid but algebra. Next, not
Euclid, but practical geometry, solid as well as plane; not
demonstration, but to make acquaintance. Then not Euclid, but
elementary vectors, conjoined with algebra, and applied to geometry.
Addition first; then the scalar product. Elementary calculus should
go on simultaneously, and come into the vector algebraic geometry
after a bit. Euclid might be an extra course for learned men, like
Homer. But Euclid for children is barbarous.*

Heaviside

*Euclid avoids it [the treatment of the infinite]; in modern
mathematics it is systematically introduced, for only then is
generality obtained.*

Cayley

*The science of figures is most glorious and beautiful. But how
inaptly it has received the name geometry!*

Frischlinus

CHAPTER 12

Pasch's Postulate and Plane Separation Postulate

12.1 AXIOM 3: PSP

Neither the Plane-Separation Postulate (PSP) nor Pasch's Postulate (PASCH) have been formally introduced into our theory. Neither can be a theorem for Σ at this time. Which one shall we take as a new axiom? It turns out not to make any difference! We shall first show that PASCH implies PSP. The proof is quite long and may even be omitted without loss of continuity because we shall actually take PSP as our third axiom. Then PASCH later turns up as a theorem.

DEFINITION 12.1 *Pasch's Postulate* or PASCH: If a line intersects a triangle not at a vertex, then the line intersects two sides of the triangle. *Plane-Separation Postulate* or PSP: For every line l there exist convex sets H_1 and H_2 whose union is the set of all points off l and such that if P and Q are two points with P in H_1 and Q in H_2 then $\overline{PQ}$ intersects l.

You should recognize that the following three statements are equivalent to PASCH: (1) If a line intersects the interior of a side of a triangle, then the line intersects another side of the triangle. (2) If a line intersects a triangle, then the line intersects two sides of the triangle. (3) If a line does not intersect either of two sides of a triangle, then the line does not intersect the third side of the triangle. Of course, a line may intersect all three sides of a triangle. For example, given

$\triangle ABC$ and $B-E-C$, then $\overleftrightarrow{AE}$ intersects all three sides of $\triangle ABC$ as does $\overleftrightarrow{AB}$.

Theorem 12.2 If PASCH, then PSP.

Proof Let l be any line. Let L be a point on l. Let A and B be points off l such that $A-L-B$. Let H_1 be the set consisting of A and all points P such that l does not intersect $\overline{AP}$. Let H_2 be the set consisting of B and all points Q such that l does not intersect $\overline{BQ}$.

To show H_1 is convex, assume P and R are distinct points in H_1 and $P-S-R$. We must show S is in H_1. If A, P, R are not distinct, then S is in H_1 by definition of H_1. If A, P, R are distinct and collinear, then S is in $\overline{AP}$ or $\overline{AR}$. In this case S is in H_1 by definition of H_1 as $\overline{AS}$ is contained in $\overline{AP}$ or $\overline{AR}$. If $\triangle APR$, then l cannot intersect $\overline{PR}$ by PASCH applied to $\triangle APR$. Then, since $\overline{PS}$ is contained in $\overline{PR}$, line l cannot intersect $\overline{AS}$ by PASCH applied to $\triangle APS$. So S is in H_1 in all possible cases. H_1 is convex. Likewise, to show H_2 is convex replace "H_1" by "H_2" and "A" by "B" in the argument.

Before proving every point off l is in H_1 or H_2, we shall show that l cannot intersect all three sides of $\triangle ABV$ when V is a point off both l and $\overleftrightarrow{AB}$. Assuming otherwise, we suppose $A-M-V$ and $B-N-V$ with M and N on l. Since L, M, N are distinct and collinear, one of the points must be between the other two. If $L-M-N$, then $\overleftrightarrow{AV}$ intersects $\triangle BLN$ only at M which is an interior point of side $\overline{LN}$. If $L-N-M$, then $\overleftrightarrow{BV}$ intersects $\triangle ALM$ only at N which is an interior point of side $\overline{LM}$. If $M-L-N$, then $\overleftrightarrow{AB}$ intersects $\triangle VMN$ only at L which is an interior point of side $\overline{MN}$. Since each of the possibilities of one of L, M, N being between the other two contradicts PASCH, line l cannot intersect the interior of each side of $\triangle ABV$.

If point T is in int $(\overrightarrow{LA})$, then T is in H_1 and off H_2; if point T is in int $(\overrightarrow{LB})$, then T is in H_2 and off H_1. If point T is off l and off $\overleftrightarrow{AB}$, then l cannot intersect both $\overline{AT}$ and $\overline{BT}$ since l cannot intersect the interior of each side of $\triangle ABT$. Thus T is in H_1 or H_2. Hence $\mathscr{P} \setminus l = H_1 \cup H_2$.

Finally, suppose P is a point in H_1, Q is a point in H_2, and $P \neq Q$. If P is in $\overrightarrow{LA}$ and Q is in $\overrightarrow{LB}$, then l intersects $\overline{PQ}$ at L. If P is in $\overrightarrow{LA}$ and Q is off $\overrightarrow{LB}$, then l intersects $\overline{PQ}$ by PASCH applied to $\triangle BPQ$; if P is off $\overrightarrow{LA}$ and Q is in $\overrightarrow{LB}$, then l intersects $\overline{PQ}$ by PASCH applied to $\triangle APQ$. If P and Q are both off $\overleftrightarrow{AB}$, then l intersects int $(\overline{PB})$ by PASCH

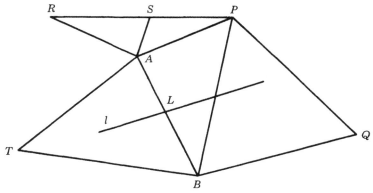

FIGURE 12.1

applied to $\triangle ABP$. Thus l intersects $\overline{PQ}$ by $PASCH$ applied to $\triangle BPQ$. Therefore l intersects $\overline{PQ}$ in all cases. So H_1 and H_2 satisfy PSP. ■

Axiom 3 *PSP* $\forall l \in \mathscr{L}$ $\exists$ convex sets H_1 and H_2 $\ni$

 1 $\mathscr{P} \setminus l = H_1 \cup H_2$,
 2 $P \in H_1, Q \in H_2, P \neq Q \Rightarrow \overline{PQ} \cap l \neq \varnothing$.

DEFINITION 12.3 The sets H_1 and H_2 in Axiom 3 are **halfplanes** of line l, and l is an **edge** of each halfplane. A halfplane of $\overleftrightarrow{AB}$ is a **halfplane** of $\overrightarrow{AB}$ and a **halfplane** of $\overline{AB}$.

We now set out to prove that these halfplanes have the properties we think a *side* of a line should have.

Theorem 12.4 If $\overline{AB} \cap l \neq \varnothing$ where A and B are two points off line l, then A and B are not in the same halfplane of l.

Proof Assume A and B are two points in halfplane H of line l. Then $\overline{AB}$ is a subset of H since H is convex. Because a halfplane of l does not contain a point of l by PSP, we then have $\overline{AB} \cap l = \varnothing$, contradicting the hypothesis. Thus both A and B cannot be in H. ■

Theorem 12.5 If H_1 and H_2 are a pair of halfplanes of line l as given in PSP, then $H_1 \neq \varnothing$, $H_2 \neq \varnothing$, but $H_1 \cap H_2 = \varnothing$.

Proof Since $H_1 \cup H_2 = H_2 \cup H_1$ and $\overline{PQ} = \overline{QP}$, we see that PSP is symmetric in "H_1" and "H_2." So, if point A is off line l, we may assume A

is in H_1 without loss of generality. Let L be any point on l and B be any point such that $A - L - B$. Then B is in H_2 (Theorem 12.4), proving neither H_1 nor H_2 is empty. Assume A is also in H_2. Since A and B are distinct points in convex set H_2, then L must be in H_2, a contradiction. Hence no point is in both H_1 and H_2. ■

Theorem 12.6   If H_1 and H_2 are a pair of halfplanes of line l as given in PSP, then $\mathscr{P}$ is the union of the three mutually disjoint sets H_1, H_2, and l. Also, $H_2 = \mathscr{P} \setminus (H_1 \cup l) = (\mathscr{P} \setminus l) \setminus H_1$.

Proof   Follows directly from $\mathscr{P} \setminus l = H_1 \cup H_2$ and Theorem 12.5. ■

Theorem 12.7   The halfplanes of a line are unique (except for order). If A is any point off line l, then the halfplanes of l are

$$\{P | P \in \mathscr{P} \setminus l \text{ and } \overline{AP} \cap l \neq \varnothing\},$$

$$\{A\} \cup \{Q | Q \in \mathscr{P} \setminus l \text{ and } \overline{AQ} \cap l = \varnothing\}.$$

Proof   Let H_1 and H_2 be a pair of halfplanes of line l as given in PSP. Assume H_1' and H_2' are another such pair. We may suppose point A is in $H_1 \cap H_1'$ (Theorem 12.5). If A and P are two points off line l, then A and P are in different halfplanes of l iff $\overline{AP}$ intersects l (PSP and Theorem 12.4). Thus H_2 and H_2' each consists of exactly those points P in $\mathscr{P} \setminus l$ such that $\overline{AP}$ intersects l. With $H_2 = H_2'$, it follows that $H_1 = H_1'$ as each of H_1 and H_1' must be $(\mathscr{P} \setminus l) \setminus H_2$, (Theorem 12.6). ■

As a result of our first theorems, we see that we could have assumed the stronger axiom: If l is a line then there exists a unique pair H_1 and H_2 of disjoint, nonempty convex sets such that (1) $\mathscr{P} \setminus l = H_1 \cup H_2$ and (2) $P \in H_1$, $Q \in H_2 \Rightarrow \overline{PQ} \cap l \neq \varnothing$.

The exception with respect to order in Theorem 12.7 is expected since there is no way to distinguish between halfplanes H_1 and H_2 of line l without reference to some point off l. Theorem 12.7 states that a line has unique halfplanes. We must show that conversely a halfplane determines a unique line.

Theorem 12.8   No two lines have the same halfplanes; the edge of a halfplane is unique.

Proof   Let line l have the two halfplanes H and $H_2(l)$; let line m have the two halfplanes H and $H_2(m)$. (Although $l \cup H_2(l) = m \cup H_2(m)$, we need to show $l = m$ to prove the theorem.) Assume there exists a point A on $l \setminus m$. Since A is off H and m, point A is in $H_2(m)$. Let B be any point of H. Then m intersects $\overline{AB}$ at some point M by PSP. Since

$A-M-B$ and A is on l, points B and M are in the same halfplane of l (Theorem 12.7). So M is in H, a contradiction since M cannot be both a point of m and in a halfplane of m. Hence $l \setminus m = \varnothing$ and $l = m$. ∎

DEFINITION 12.9 A halfplane of $\overleftrightarrow{AB}$ is a *side* of $\overleftrightarrow{AB}$, a *side* of $\overrightarrow{AB}$, and a *side* of $\overline{AB}$. Each of the two halfplanes of a line is the *opposite* side of the other. Points P and Q are *on opposite sides* of line l if P is in one halfplane of l and Q is in the other halfplane of l.

Theorem 12.10 Let A and B be points on opposite sides of line l. Then A and B are not on the same side of l. If B and C are points on opposite sides of l, then A and C are on the same side of l. If B and D are points on the same side of l, then A and D are on opposite sides of l.

Proof Trivial. ∎

We have followed the popular convention of having two words "halfplane" and "side" with one meaning. A good argument against this is that "side" already had two different meanings. So now "side" has three meanings! There is little likelihood of confusion however. In fact, these meanings are so much a part of our culture that, unless it were pointed out, most students would not notice that "side" is used in two different ways in the next proof.

Theorem 12.11 If line l contains no vertex of $\triangle ABC$, then l cannot intersect all three sides of the triangle.

Proof Assume line l contains no vertex of $\triangle ABC$, but l intersects each side of $\triangle ABC$. Then A and B are on opposite sides of l, and B and C are on opposite sides of l. Hence A and C are on the same side of l, contradicting the assumption that l intersects the side $\overline{AC}$ of $\triangle ABC$. ∎

Theorem 12.12 PASCH.

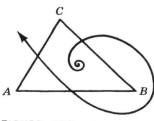 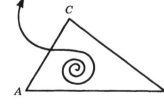

FIGURE 12.2

Proof Exercise 12.1. ■

The shortest proof of Theorem 12.12 is very much like the proof of Theorem 12.11. The assertion that PSP *and* PASCH *are equivalent under our first two axioms* has now been demonstrated.

Theorem 12.13 *Peano's Postulate* If $\triangle ABC, B-C-D$, and $A-E-C$, then there exists point F on $\overleftrightarrow{DE}$ such that $A-F-B$.

Proof Since D is off $\triangle ABC$ and off $\overleftrightarrow{AE}$, line $\overleftrightarrow{DE}$ exists and does not pass through A. Also $\overleftrightarrow{DE}$ cannot intersect $\overline{BC}$ because E is off $\overleftrightarrow{BC}$. Then $\overleftrightarrow{DE}$ must intersect the interior of $\overline{AB}$ in some point F by PASCH. ■

Theorem 12.14 If $\triangle ABC$, $B-C-D$, and $A-F-B$, then there exists E on $\overleftrightarrow{DF}$ such that $A-E-C$ and $D-E-F$.

Proof The proof that $\overleftrightarrow{DF}$ is a line containing E such that $A-E-C$ is quite similar to our proof of Peano's Postulate (Exercise 12.2). We must show $D-E-F$. Points D, E, F are clearly distinct and collinear. Since B and D are on opposite sides of $\overleftrightarrow{AC}$ and B and F are on the same side of $\overleftrightarrow{AC}$, it follows that D and F are on opposite sides of $\overleftrightarrow{AC}$. Then, since E is on $\overleftrightarrow{AC}$, we have $D-E-F$. ■

Theorem 12.15 If $\triangle ABC$, then every point lies on a line that intersects the triangle at two points.

Proof Let P be an arbitrary point. Let Q be any point different from P such that $A-Q-B$. If $\overleftrightarrow{PQ}=\overleftrightarrow{AB}$, the result is trivial. Otherwise the result follows from PASCH. ■

The next theorem is included only because it is an interesting

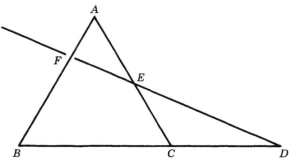

FIGURE 12:3

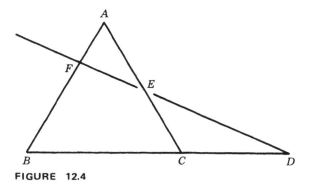

FIGURE 12.4

curiosity. First conjectured by J. J. Sylvester in 1893, the statement remained a conjecture for the Euclidean plane until proved by T. Gallai forty years later.

Theorem 12.16 *Sylvester's Theorem* If n points are not all collinear, then there exists a line containing exactly two of the points.

Proof Let P_1, P_2, P_3 be three of the given n points such that P_1 is off $\overleftrightarrow{P_2 P_3}$. Let Q be any point on $\overleftrightarrow{P_2 P_3}$ such that $\overrightarrow{P_1 Q}$ contains none of the given points except P_1. Let R be the unique point in int $(\overrightarrow{P_1 Q})$ such that R is on a line l through two of the given points but no line containing two of the given points intersects int $(\overline{P_1 R})$. Possibly $R = Q$; in any case, R is not one of the given points. If l contains exactly two of the given points, we are done. Otherwise we may assume l contains three of the given points P_4, P_5, P_6 such that $R - P_4 - P_5$ and P_6 is off $\overline{RP_5}$. If one of the given points is in int $(\overline{P_1 P_5})$, then the line through this point and P_6 intersects int $(\overline{P_1 R})$; if one of the given points is in $\overrightarrow{P_1 P_5} \setminus \overline{P_1 P_5}$, then the line through this point and P_4 intersects int $(\overline{P_1 R})$ (Theorem 12.13 and 12.14). Since no line containing two of the given points intersects int $(\overline{P_1 R})$, we have that $\overleftrightarrow{P_1 P_5}$ is a line containing exactly two of the given points. ■

12.2 PASCH, PEANO, PIERI, AND HILBERT

From Proclus we learn Pappus (circa 300) had added some axioms to those of Euclid: (1) All parts of the plane and of the line coincide with each other; (2) a point divides a line; (3) a line divides a plane; (4) a plane divides a solid; and (5) for any segment there is another segment whose length is greater and one whose length is less. These great insights were dismissed by Proclus as unnecessary with the claim that they follow from Euclid's definitions. Proclus shared the human problem of distinguishing between the obvious and that which only appears to be obvious.

By the sixteenth century Euclid's deductive system of geometry had been reduced to the *science of space* based on *self-evident truths*. The following two centuries saw the axiomatic method applied to social, political, and philosophical theories. Mathematics, strangely enough, was a Johnny-come-lately in appreciating its potential. The axiomatic method, which would be so fruitful in mathematics, blossomed anew in geometry in the nineteenth century. This was out of necessity. Mathematicians were beginning to hear about non-Euclidean geometry. It was time for geometers to put their house in order. Algebraists and analysts were busy doing the same thing.

Moritz Pasch (1843–1930) stated that pure geometry must be formal in the strict sense that everything necessary to deduce the theorems must be found in the axioms. His *Vorlesungen über neuere Geometrie* (1882) is the first rigorous axiomatic development of Euclidean geometry. For the first time *betweenness* for points is treated axiomatically. To realize the necessity of doing so is a great intellectual achievement. (We obtained our *betweenness* from the Ruler Postulate.) Pasch gives a complete set of axioms for the real projective plane described in Section 4.2. He requires that the theorems be independent of any figure and, further, that the theorems be independent of any *particular meaning* assigned to the technical terms in the axioms. However, Pasch regarded his axioms, which he called *nuclear propositions,* as truths that could be verified by observation.

Giuseppe Peano (1858–1932) brought the substance of Pasch's work across the Alps with two important differences. First, Pasch's empiricism is gone. In *I principii di geometria, logicamente esposti* (1889) Peano begins with an arbitrary set of *points,* emphasizing that "point" is undefined and may be thought of as any entity. The modern view of axiomatic geometry is firmly established; no longer are axioms necessarily limited to experience. The second important difference is unfortunate. The geometry is expressed as a symbolic calculus. Peano invented many of the symbols, and some are still in use today. However, this made the work difficult to follow. Hence this important book was not popular, even ridiculed by Poincaré.

Soon many books on the foundations of geometry appeared. In 1891 Giuseppe Veronese (1854–1917) produced the first geometries where Archimedes' axiom fails. This means that there are two segments of *length* t and b, respectively, such that $nt < b$ for every positive integer n. In 1892 Gino Fano (1871–1952) and, four years later, the American Eliakim H. Moore (1862–1950) produced the first finite geometries (Section 4.2). The non-Archimedean geometries and the finite geometries are contrary to all intuition regarding *space*. That geometry is only the *science of space* is no longer tenable.

Mario Pieri (1860–1913) wrote *Della geometria elementare come sistema ipotetico deduttivo* in 1899. This axiomatic development of the Euclidean plane was not widely accepted at the time and is not as well known now as it might be. Pieri reduces the number of undefined terms down to two, *point* and *motion*. The motions are mappings on the set of points that satisfy certain axioms. (Think of things like translations and rotations.) For example, one axiom states that if there is a nonidentity motion which fixes three points A, B, C, then every motion which fixes A and B must also fix C. In this case A, B, C are *defined* to be *collinear*. Then a *line* is *defined* by saying, in our notation, that $\overleftrightarrow{AB}$ consists of A, B, and all points C such that A, B, C are collinear. This beautiful treatment is very modern in flavor and will be described in more detail in Section 15.2. In succeeding years both Pieri and Peano gave developments using only *point* and *distance* as undefined terms. For example, Pieri's axioms involve a ternary relation I on the points such that $I(A, B, C)$ iff the distance from A to B equals the distance from A to C. Also both Pieri and Alessandro Padoa (1868–1937) gave developments using only *point* and *congruent* as undefined terms. Congruent is a relation on pairs of points and was considered preferable to using motion because a motion is a relation on infinite sets of points.

The idea of motion is of fundamental importance in geometry. (The automorphisms of any mathematical system are always of fundamental importance in the study of that system.) It turns out that the motions for the Euclidean plane and for the non-Euclidean planes can be expressed as products of *reflections* in lines. Based on ideas from *Neue Begründung der ebenen Geometrie* (1907) by Johannes Hjelmslev (1873–1950), an axiomatic development of plane geometry can be given where the number of undefined terms is just one, *reflection*. Since this is one less than Pieri's two, the game of one-downmanship comes to an end. Chances are that you cannot think how to define a *point* without knowing any more about the system. This problem can be solved once we have studied motions. In this century there have been hundreds of new axiomatic systems for plane geometry. They continue to appear as plane geometry is still one of the most fascinating games known to mankind.

The capstone to the nineteenth century efforts of axiomatizing geometry is the ninety-two page *Grundlagen der Geometrie* (1899) by David Hilbert (1862–1943). Most of the mathematical systems then known had been axiomatized in the last half of the century. Contrary to what is often reported, Hilbert's *Foundations of Geometry* is not the first systematic development of Euclidean geometry free of intuition. The credit for this must go to Pasch, regardless of his restriction of the axioms to those motivated by experience. Certainly Peano and Fano had previously specified that *points* were an arbitrary set of elements. Peano and Padoa had already demonstrated the idea of independence of axioms by constructing counterexamples using different interpretations of the undefined terms. In the *Grundlagen,* Pasch is cited in a footnote with reference to PASCH but Peano is never even mentioned. Disputes over priorities concerning the ideas in this book continue to this day (see Exercise 12.6).

Besides the axiomatic development, which is described fully in Section 15.1, Hilbert's *Grundlagen der Geometrie* discusses theory of proportion, plane area, and geometric constructions. Also the significance of the theorems of Desargues and Pappus is thoroughly examined, thus consummating the union of algebra and geometry begun by Descartes. The first example of a non-Desarguesian plane is replaced in later editions by the simpler Moulton Incidence Plane M10 of Section 5.2. With other alterations and the addition of appendices and supplements, the eleventh edition (1972) as edited by P. Bernays is twice as long as the original. The tenth edition has been translated into a second English edition (Open Count, 1971).

The first International Congress of Mathematicians, held in Zurich in 1897, was followed by the second in Paris in 1900. At the Paris congress Hilbert made his famous speech on the future problems of mathematics. The philosophers held their first international congress in Paris in 1900. Knowledge of Hilbert's *Grundlagen der Geometrie* was soon widespread. This book, more than any other, has been the basis for the modern view of geometry and influencing mathematics' turn toward axiomatics in the twentieth century. *Grundlagen der Geometrie* has been the third most influential book in geometry, even more influential than those describing non-Euclidean geometry for the first time. A book called *Elements* must rank first. The contemporary view of axiomatic systems for geometry is typified by a remark made by Hilbert in 1891 but not published until 1935: "*One must be able to say at all times — instead of points, straight lines, and planes — tables, chairs, and beer mugs.*" Geometry is happily freed of the infamous "*A point is that which has no part.*"

12.3 EXERCISES

12.1 Theorem 12.12.

12.2 Supply the missing part of the proof of Theorem 12.14.

- **12.3** In Peano's Postulate as stated in Theorem 12.13, we have $D-E-F$.

12.4 Our first two axioms and Peano's Postulate together do not imply PSP.

- **12.5** True or False?

(a) There exist nonempty convex sets H_1 and H_2 such that for every line l a point off l is in H_1 or H_2.

(b) There exist nonempty convex sets H_1 and H_2 such that for every line l if P and Q are two points with P in H_1 and Q in H_2, then $\overline{PQ}$ intersects l.

(c) There exist nonempty convex sets H_1 and H_2 such that for every line l a point off l is in H_1 or H_2 and, if P and Q are two points with P in H_1 and Q in H_2, then $\overline{PQ}$ intersects l.

(d) Although PSP and PASCH are both dependent on the first two axioms for their meaning, both PSP and PASCH are independent of the first two axioms.

(e) If A, D, V are three points on one line and V, B, C are three points on another line, then $\overleftrightarrow{AB}$ intersects $\overleftrightarrow{CD}$.

(f) $\triangle ABC, A-C'-B, B-A'-C, C-B'-A \Rightarrow \triangle A'B'C'$.

(g) If a line intersects a triangle, then the line intersects two sides of the triangle.

(h) If a line does not intersect two sides of a triangle, then the line contains a vertex of the triangle.

(i) A line can intersect a triangle at three points.

(j) A halfplane can have two edges.

12.6 Read "The Origins of Modern Axiomatics: Pasch to Peano" by H. C. Kennedy in *The American Mathematical Monthly* Vol. 79 (1972), pp. 133–136.

- **12.7** (M10, s), the Moulton Incidence Plane together with distance function s given by Euclidean arclength along Moulton lines, is a model of Σ.

12.8 (M2, d), the Rational Cartesian Incidence Plane together with Euclidean distance, is not a model of Σ but does satisfy PSP.

- **12.9** Does (M2, d) also satisfy PASCH?

12.10 In the proof of Sylvester's Theorem, possibly $\{P_2, P_3\}$ and $\{P_4, P_5, P_6\}$ are not disjoint.

12.11 The union of a halfplane and its edge is a convex set.

12.12 If n is a positive integer, then a set of $n + 1$ points is not a convex set.

12.13 Do our first two axioms together with Theorem 12.11 imply PSP?

12.14 Let l be a line. Find at least three pairs of convex sets H_1 and H_2 such that every point off l is in H_1 or H_2, such that if P and Q are two points with P in H_1 and Q in H_2 then $\overline{PQ}$ intersects l, and such that neither H_1 nor H_2 is a halfplane.

● **12.15** It is not possible to arrange any finite number of points so that a line through every two of them shall pass through a third, unless all the points lie on one line.

****12.16** Is there a d such that (M5, d) is a model of Σ? Is there a d such that (M6, d) is a model of Σ? Is there a d such that (M7, d) is a model of Σ?

*● **12.17** If $\triangle ABC$ and line l intersects $\overleftrightarrow{AB}$, then l intersects $\overleftrightarrow{BC}$ or $\overleftrightarrow{AC}$?

****12.18** Under our first two axioms, Peano's Postulate and Theorem 12.15 together are equivalent to our Axiom 3.

****12.19** Give properties of $h : \mathscr{L} \to 2^{\mathscr{P}}$ such that the existence of h is equivalent to Axiom 3, assuming our first two axioms.

GRAFFITI

"Would you tell me please, which way I ought to go from here?"
"That depends a good deal on where you want to get to," said the Cat.

Lewis Carroll

The axioms of geometry are — according to my way of thinking — not arbitrary, but sensible, statements, which are, in general, induced by space perception and are determined as to their precise content by expediency.

Klein

I once knew an otherwise excellent teacher who compelled his students to perform all their demonstrations with incorrect *figures, on the theory that it was the* logical *connection of the concepts, not the figure, that was essential.*

Mach

It is true that a mathematician, who is not somewhat of a poet, will never be a perfect mathematician.

Weierstrass

But in the present century, thanks in good part to the influence of Hilbert, we have come to see that the unproved postulates with which we start are purely arbitrary. They must *be consistent, they* had better *lead to something interesting.*

Coolidge

The quote from Kant that begins Hilbert's Foundations of Geometry:

All human knowledge thus begins with intuitions, proceeds thence to concepts, and ends with ideas.

CHAPTER 13

Crossbar and Quadrilaterals

13.1 MORE INCIDENCE THEOREMS

Before thinking about adding any new axioms to our system, we shall prove several more incidence theorems that follow from the three axioms we already have. Among these is the very useful little theorem mentioned in Section 9.2 that is known as Crossbar. Crossbar can be stated as a theorem only after a formal definition of the *interior* of an angle has been given. We begin by extending the definition of *in* and *on* so that our theory encompasses such phrases as "$\triangle ABC$ is on a side of line *l*" and "$\overline{CD}$ is in the ray-interior of $\angle AVB$." Also, for example, we shall be able to talk about two segments being on opposite sides of line *l*.

DEFINITION 13.1 If S and T are nonempty sets of points and $S \subset T$, then S is *on* T or S is *in* T.

Theorem 13.2 Let each of A, B, C, D be either a point or a nonempty set of points. Let A and B be on opposite sides of line *l*. Then A and B are not on the same side of *l*. If B and C are on opposite sides of *l*, then A and C are on the same side of *l*. If B and D are on the same side of *l*, then A and D are on opposite sides of *l*.

Proof Trivial (Definition 12.9, Definition 13.1, and Theorem 12.10). ■

Theorem 13.3 If S is a nonempty convex set which does not intersect line l, then S is on one side of l.

Proof Let A be a point of S on side H_1 of line l. Assume there exists a point B of S on the opposite side of H_1. Then $\overline{AB}$ intersects l by PSP. So, since S is a convex set (Definition 8.18), we have the contradiction that S intersects l. Hence all the points of S are on H_1. ■

Corollary 13.4 If x is a line, ray, or segment which does not intersect line l, then x is on one side of l. If line l intersects $\overleftrightarrow{AC}$ only at point V such that $A-V-C$, then int $(\overrightarrow{VA})$ and int $(\overline{VA})$ are on the same side of l as is A but int $(\overrightarrow{VA})$ and int $(\overrightarrow{VC})$ are on opposite sides of l.

DEFINITION 13.5 The *interior* of $\angle AVB$ is the intersection of the side of $\overleftrightarrow{VA}$ that contains B and the side of $\overleftrightarrow{VB}$ that contains A; int $(\angle AVB)$ is the interior of $\angle AVB$.

The interior of $\angle AVB$ is illustrated in Figure 13.1. That the interior of an angle is well-defined follows from Corollary 13.4. From the definition and the corollary preceding it, we have several immediate results that are lumped together as the next theorem.

Theorem 13.6 Point P is in int $(\angle AVB)$ iff points A and P are on the same side of $\overleftrightarrow{VB}$ and points B and P are on the same side of $\overleftrightarrow{VA}$. Given $\angle AVB$, if $A-P-B$, then P is in int $(\angle AVB)$. Given $\triangle ABC$, then int $(\overline{AB})$ is on int $(\angle ACB)$.

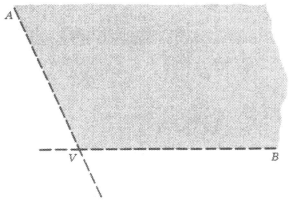

FIGURE 13.1

Theorem 13.7 If P is in int $(\angle AVB)$ and $\overrightarrow{AP}$ intersects $\overrightarrow{VB}$ at D, then $A-P-D$.

Proof Either $A-P-D$ or $A-D-P$. Assume $A-D-P$. Then A and P are on opposite sides of $\overleftrightarrow{VD}$. Since $\overleftrightarrow{VD}=\overleftrightarrow{VB}$, P is not on the side of $\overleftrightarrow{VB}$ that contains A. Thus P is not in int $(\angle AVB)$, a contradiction. So $A-P-D$. ∎

Theorem 13.8 *Crossbar* If point P is in int $(\angle AVB)$, then $\overrightarrow{VP}$ intersects int $(\overline{AB})$.

Proof Let C be any point such that $B-V-C$. (See Figure 13.2.) Since P and B are on the same side of $\overleftrightarrow{VA}$ and since B and C are on opposite sides of $\overleftrightarrow{VA}$, then P and C are on opposite sides of $\overleftrightarrow{VA}$. So int $(\overrightarrow{VP})$ and int $(\overline{AC})$ are on opposite sides of $\overleftrightarrow{VA}$. Thus $\overrightarrow{VP}$ cannot intersect $\overline{AC}$. Then, since int $(\overline{AC})$ is on the same side of $\overleftrightarrow{VC}$ as P, we know $\overleftrightarrow{VP}$ cannot intersect $\overline{AC}$. So $\overleftrightarrow{VP}$ intersects int $(\overline{AB})$ at some point Q by PASCH applied to $\triangle ABC$. Finally, since P and Q are on the same side of $\overleftrightarrow{VB}$, point Q must be on $\overrightarrow{VP}$. ∎

Theorem 13.9 Given $\angle AVB$, if $\overrightarrow{VP}$ intersects int $(\overline{AB})$, then P is in int $(\angle AVB)$.

Proof Let $\overrightarrow{VP}$ intersect int $(\overline{AB})$ at Q. (Possibly $P=Q$.) Then A and Q are on the same side of $\overleftrightarrow{VB}$, and Q and P are on the same side of $\overleftrightarrow{VB}$. Hence P is on the side of $\overleftrightarrow{VB}$ that contains A. Likewise, since P, Q, B are on the same side of $\overleftrightarrow{VA}$, P is on the side of $\overleftrightarrow{VA}$ that contains B. Therefore P is in int $(\angle AVB)$. ∎

Theorem 13.10 If points B and P are on the same side of $\overleftrightarrow{VA}$, then P

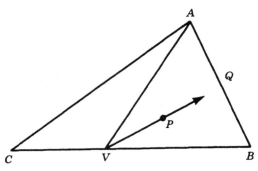

FIGURE 13.2

is in int $(\angle AVB)$ iff points A and B are on opposite sides of $\overleftrightarrow{VP}$. If $A - V - C$, then point P is in int $(\angle AVB)$ iff point B is in int $(\angle CVP)$. If points B and P are on the same side of $\overleftrightarrow{VA}$, then either $\overrightarrow{VB} = \overrightarrow{VP}, P$ is in int $(\angle AVB)$, or B is in int $(\angle AVP)$.

Proof If point P is in int $(\angle AVB)$, then $\overrightarrow{VP}$ intersects int $(\overline{AB})$ by Crossbar. Hence A and B are on opposite sides of $\overleftrightarrow{VP}$. Conversely, suppose that B and P are on the same side of $\overleftrightarrow{VA}$ and that A and B are on opposite sides of $\overleftrightarrow{VP}$. Then $\overleftrightarrow{VP}$ intersects int $(\overline{AB})$ by PSP. Since int $(\overline{AB})$ and P are on the same side of $\overleftrightarrow{VA}$, then $\overrightarrow{VP}$ must intersect int $(\overline{AB})$. So P is in int $(\angle AVB)$ by the preceding theorem. We have now proved the first statement in the theorem. Suppose $A - V - C$. Since $\overleftrightarrow{VA} = \overleftrightarrow{VC}$ and since points B and C are on the same side of $\overleftrightarrow{VP}$ iff points A and B are on opposite sides of $\overleftrightarrow{VP}$, it follows that the second statement is only a restatement of the first (Theorem 13.6). If points B and P are on the same side of $\overleftrightarrow{VA}$, $A - V - C$, and $\overrightarrow{VB} \neq \overrightarrow{VP}$, then P is in either int $(\angle AVB)$ or int $(\angle CVB)$. However, P in int $(\angle CVB)$ implies B is in int $(\angle AVP)$ by the second statement in the theorem. Therefore the third statement follows from the second. ∎

Theorem 13.11 If $\angle AVB = \angle CVD$ and $\overrightarrow{VE}$ intersects int $(\overline{CD})$, then $\overrightarrow{VE}$ intersects int $(\overline{AB})$.

Proof Since $\overrightarrow{VE}$ intersects int $(\overline{CD})$, then E is in int $(\angle CVD)$, (Theorem 13.9). Thus E is in int $(\angle AVB)$, and $\overrightarrow{VE}$ intersects int $(\overline{AB})$ by Crossbar. ∎

Theorem 13.12 Given $\angle AVB$, the following are equivalent:

(a) Point P is in int $(\angle AVB)$.

(b) $\overrightarrow{VP}$ intersects int $(\overline{AB})$.

(c) Point P is in the ray-interior of $\angle AVB$.

(d) $\overrightarrow{VP}$ is an interior ray of $\angle AVB$.

Proof That (a) implies (b) follows from Crossbar; (b) implies (c) by Definition 9.13; and (c) implies (a) by Theorem 13.9. So the first three statements are equivalent. Since (c) is equivalent to (d) by Definition 9.13, all four statements are equivalent. ∎

Corollary 13.13 The ray-interior of $\angle AVB$ is the interior of $\angle AVB$.

Theorem 13.11 confirms that in the presence of our third axiom *shade* from Section 9.2 is well-defined. However, there is no reason to introduce *shade* now as it is the same thing as *ray-interior*. In fact, we can do without *ray-interior* as well in the light of Corollary 13.13. Can we also dispose of the *inside* of an angle? Is the inside of an angle just the interior of the angle? No. (M13, h), the Cayley–Klein Incidence Plane with distance h of Section 9.2 (see Figure 9.13) is still a model of Σ as PASCH and PSP hold (Exercise 13.3). We know the inside of an angle is contained in the interior of the angle (Theorem 9.14), but the converse does not necessarily hold.

We leave as an exercise the proof of the following statement which is famous as a tacit assumption made by Euclid in proving his Proposition I.16.

Theorem 13.14 If $\triangle ABC$, $B-C-D$, $A-E-C$, and $B-E-F$, then F is in int $(\angle ACD)$.

Proof Exercise 13.1. ■

DEFINITION 13.15 The *interior* of $\triangle ABC$ or int $(\triangle ABC)$ is the intersection of three sets: (1) the side of $\overleftrightarrow{AB}$ containing C, (2) the side of $\overleftrightarrow{BC}$ containing A, and (3) the side of $\overleftrightarrow{AC}$ containing B. For those who like a lot of concise notation, if line l is off point A, define $H_A(l)$ to be the side of l containing A and int $(\triangle ABC) = H_A(\overleftrightarrow{BC}) \cap H_B(\overleftrightarrow{CA}) \cap H_C(\overleftrightarrow{AB})$. The *exterior* of an angle is the set of all points that are neither on the angle nor on the interior of the angle; the *exterior* of a triangle is the set of all points that are neither on the triangle nor on the interior of the triangle. The *outside* of an angle is the set of all points that are neither on the angle nor on the inside of the angle.

Theorem 13.16 int $(\angle AVB)$ and int $(\triangle ABC)$ are convex sets.

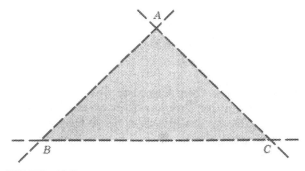

FIGURE 13.3

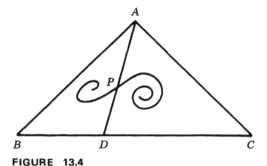

A

P

B   D   C

FIGURE   13.4

Proof   Each is defined as an intersection of convex sets and must be a convex set (Theorem 8.19). ■

Theorem   13.17   Given $\triangle ABC$,   then   int $(\triangle ABC)$ = int $(\angle A)$ ∩ int $(\angle B)$ = int $(\angle A)$ ∩ int $(\angle B)$ ∩ int $(\angle C)$.

Proof   Exercise 13.2. ■

Theorem 13.18   *Line – Triangle Theorem*   If a line intersects the interior of a triangle, then the line intersects the triangle exactly twice.

Proof   Let point P be on line l and in int $(\triangle ABC)$. (See Figure 13.4). So P is in int $(\angle A)$ and in int $(\angle B)$. By Crossbar $\overrightarrow{AP}$ intersects int $(\overline{BC})$ at some point D. Then $A-P-D$ (Theorem 13.7). If $l = \overleftrightarrow{AD}$, then l intersects $\triangle ABC$ at least twice. Suppose $l \neq \overleftrightarrow{AD}$. So A and D are both off l. Then l intersects $\overline{AB}$ or $\overline{BD}$ by PASCH applied to $\triangle ABD$, and l intersects $\overline{AC}$ or $\overline{DC}$ by PASCH applied to $\triangle ACD$. Thus, in any case, l intersects $\triangle ABC$ at least twice. A line that intersects a triangle in three points must contain a side of the triangle (Theorem 12.11) and does not intersect the interior of the triangle. Therefore l intersects $\triangle ABC$ exactly twice. ■

13.2   QUADRILATERALS

At a quick glance the next definition looks horrendous. Actually we are only introducing the common terms regarding a *quadrilateral.* See Figure 13.5.

DEFINITION   13.19   Let $\square ABCD = \overline{AB} \cup \overline{BC} \cup \overline{CD} \cup \overline{DA}$ if A, B, C, D are four points such that no three are collinear and such that no two of int $(\overline{AB})$, int $(\overline{BC})$, int $(\overline{CD})$, and int $(\overline{DA})$ intersect

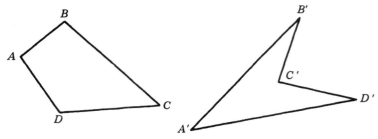

FIGURE 13.5

each other. Then □*ABCD* is a *quadrilateral* with *vertices* A, B, C, D; *sides* $\overline{AB}, \overline{BC}, \overline{CD}, \overline{DA}$; and *diagonals* $\overline{AC}$ and $\overline{BD}$. Also, $\angle DAB$, $\angle ABC, \angle BCD, \angle CDA$ are *angles* of □*ABCD*, and they may be denoted by $\angle A, \angle B, \angle C, \angle D$, respectively. The endpoints of a diagonal are *opposite* vertices. If two sides intersect each other, they are *adjacent* sides; otherwise two sides are *opposite.* If the intersection of two angles contains a side, then the angles are *adjacent;* otherwise two angles are *opposite.*

Theorem 13.20 Given □*ABCD*, then □*ABCD*=□*DCBA* and □*ABCD*=□*BCDA*=□*CDAB*=□*DABC*. If □*ABCD* and □*ABDC* both exist, then they are not equal.

Proof The first part follows from the symmetry of Definition 13.19. The second part follows from the fact that int $(\overline{AC})$ does not intersect □*ABCD* but is contained in □*ABDC*. ∎

Theorem 13.21 The four vertices, the four sides, the two diagonals, and the four angles of a quadrilateral are unique.

Proof Suppose □*ABCD*=□*A'B'C'D'*. Since $\overline{A'B'}$ has at least five points, we may assume two of them are on $\overline{AB}$ by symmetry (Theorem 13.20). Then A' and B' are on $\overline{AB}$. Neither A' nor B' can be in int $(\overline{AB})$ as that would leave A or B off □*A'B'C'D'*. So $\{A', B'\} = \{A, B\}$. We may assume $A' = A$ and $B' = B$ by symmetry. (We have now exhausted all the symmetry of Theorem 13.20.) Now neither C' nor D' can be in either int $(\overline{BC})$ or int $(\overline{DA})$ as this would leave C or D off □*A'B'C'D'*. Neither C' nor D' can be in int $(\overline{CD})$ as this would leave C or D off □*A'B'C'D'*. Hence $\{C', D'\} = \{C, D\}$. But $C' = D$ and $D' = C$ is impossible as □*ABCD* ≠ □*ABDC*. Hence $C' = C$ and $D' = D$. Hence □*ABCD* and □*A'B'C'D'* have the same vertices and the same sides. The rest of the theorem then follows immediately. ∎

DEFINITION 13.22 A *convex quadrilateral* is a quadrilateral with the property that each side of the quadrilateral is on a halfplane of the opposite side of the quadrilateral.

In Figure 13.5 □$A'B'C'D'$ is not a convex quadrilateral in the Euclidean plane. Note that "convex" has two totally different meanings in mathematics. In Figure 13.5, □$ABCD$ is a convex quadrilateral in the Euclidean plane but is certainly not a convex set. While we are talking about language, why do you suppose "side" was not used in place of "halfplane" in Definition 13.22?

Theorem 13.23 A quadrilateral is a convex quadrilateral iff the vertex of each angle of the quadrilateral is in the interior of its opposite angle.

Proof Given □$ABCD$, if A is in int ($\angle BCD$), then A and B are on the same halfplane of $\overleftrightarrow{CD}$. Since a halfplane is a convex set, then $\overline{AB}$ is on a side of $\overleftrightarrow{CD}$. Likewise, if B is in int ($\angle CDA$), then $\overline{BC}$ is on a side of $\overline{DA}$; if C is in int ($\angle DAB$), then $\overline{CD}$ is on a side of $\overline{AB}$; and if D is in int ($\angle ABC$), then $\overline{DA}$ is on a side of $\overline{BC}$. Hence (Definition 13.19), if the vertex of each angle of the quadrilateral is in the interior of its opposite angle, then the quadrilateral is a convex quadrilateral. Conversely, suppose □$ABCD$ is a convex quadrilateral. Because of the symmetry (Theorem 13.20), it is sufficient to show that C is in int ($\angle A$) in order to complete the proof. With C and D on the same side of $\overleftrightarrow{BA}$ and with C and B on the same side of $\overleftrightarrow{AD}$, it follows that C is in int ($\angle BAD$). ∎

Theorem 13.24 The diagonals of a convex quadrilateral intersect each other. Conversely, if the diagonals of a quadrilateral intersect each other, then the quadrilateral is a convex quadrilateral.

Proof Suppose □$ABCD$ is a convex quadrilateral. Then $\overrightarrow{AC}$ intersects $\overline{BD}$ at some point P such that $B-P-D$ by Crossbar. Also $\overrightarrow{BD}$ intersects $\overline{AC}$ at some point Q such that $A-Q-C$ by Crossbar. Since $\overleftrightarrow{AC} \neq \overleftrightarrow{BD}$ but each of P and Q is on both $\overleftrightarrow{AC}$ and $\overleftrightarrow{BD}$, then $P=Q$. So P is on both $\overline{BD}$ and $\overline{AC}$. Hence the diagonals intersect at P.

Conversely, suppose □$ABCD$ is such that $\overline{AC}$ and $\overline{BD}$ intersect at V. It follows that $A-V-C$, $B-V-D$, and $\overleftrightarrow{AC} \neq \overleftrightarrow{BD}$. Since V is an interior point of $\angle BCD$ (Theorem 13.6) and $\overrightarrow{CV}=\overrightarrow{CA}$, then A is in int ($\angle BCD$). By exactly the same reasoning, it is shown that B is in int ($\angle CDA$), C is in int ($\angle DAB$), and D is in int ($\angle ABC$). Thus □$ABCD$ is a convex quadrilateral (Theorem 13.23). ∎

13.3 EXERCISES

● **13.1** Theorem 13.14.

13.2 Theorem 13.17.

● **13.3** (M13, h) of Section 9.2 is a model of Σ.

● **13.4** Fano's Axiom: Given $\square ABCD$ with point E on $\overleftrightarrow{AB}$ and $\overleftrightarrow{CD}$, point F on $\overleftrightarrow{AD}$ and $\overleftrightarrow{BC}$, and point G on $\overleftrightarrow{AC}$ and $\overleftrightarrow{BD}$, then E, F, G are not collinear.

● **13.5** True or False?

(a) Two lines are parallel iff each is on a side of the other.

(b) If point C is in int $(\angle AVB)$, then $\overrightarrow{VA} \cup \overrightarrow{VC}$ is an angle.

(c) Each side of a triangle except for its endpoints is in the interior of the opposite angle.

(d) The interior of a triangle is the intersection of the interiors of any two of its angles.

(e) The exterior of an angle may be a convex set.

(f) The interior of an angle is on the inside of the angle.

(g) If C is in int $(\angle BAD)$, then $\square ABCD$ is a convex quadrilateral.

(h) If C and D are points on opposite sides of $\overline{AB}$, then $\square ABCD$ does not exist.

(i) If $\overline{AB}$ is on a side of $\overline{CD}$ and $\overline{BC}$ is on a side of $\overline{AD}$, then $\square ABCD$ is a convex quadrilateral.

(j) A convex quadrilateral is a convex set.

13.6 If int $(\angle ABC) = $ int $(\angle DEF)$, then $\angle ABC = \angle DEF$.

13.7 Let $I = \{P | \exists D \ni B-D-C$ and $A-P-D\}$, given $\triangle ABC$. Show that I is independent of any permutation of the letters "A," "B," and "C."

13.8 If a line l does not intersect $\triangle ABC$, then $\triangle ABC$ is on a side of l and int $(\triangle ABC)$ is on the same side.

13.9 The union of an angle and its interior is a convex set; the union of a triangle and its interior is a convex set.

13.10   The inside of an angle is a convex set.

13.11   The set of all points that are interior but not inside an angle is a convex set.

13.12   If A and C are on the same side of $\overleftrightarrow{VB}$ and B and C are on opposite sides of $\overleftrightarrow{VA}$, then A and B are on the same side of $\overleftrightarrow{VC}$.

13.13   Given four points such that no three are collinear, then the four points are either the vertices of exactly one convex quadrilateral or else the vertices of exactly three quadrilaterals none of which is a convex quadrilateral.

13.14   What is a *quadrangle?* What is a *trilateral?* What is a *tetragon?*

13.15   If l and m are parallel lines, then a halfplane of l is contained in a halfplane of m.

13.16   If a line intersects the interior of an angle, does the line intersect the angle?

*13.17   Give a reasonable definition of the *interior* of a quadrilateral. Is the interior of $\square ABCD$ a convex set iff $\square ABCD$ is a convex quadrilateral?

*13.18   Does the inside of an angle contain a ray?

GRAFFITI

Euclid always contemplates a straight line as drawn between two definite points, and is very careful to mention when it is to be produced beyond this segment. He never thinks of the line as an entity given once for all as a whole. This careful definition and limitation, so as to exclude an infinity not immediately apparent to the senses, was very characteristic of the Greeks in all their many activities. It is enshrined in the difference between Greek architecture and Gothic architecture, and between Greek religion and modern religion. The spire of a Gothic cathedral and the importance of the unbounded straight line in modern Geometry are both emblematic of the transformation of the modern world.

Whitehead

The nineteenth century which prides itself upon the invention of steam and evolution, might have derived a more legitimate title to fame from the discovery of pure mathematics.

Russell

The moving power of mathematical invention is not reasoning but imagination.

De Morgan

What Vesalius was to Galen, what Copernicus was to Ptolemy, that was Lobatchewsky to Euclid. There is, indeed, a somewhat instructive parallel between the last two cases. Copernicus and Lobatchewsky were both of Slavic origin. Each of them has brought about a revolution in scientific ideas so great that it can only be compared with that wrought by the other. And the reason of the transcendent importance of these two changes is that they are changes in the conception of the Cosmos. . . . And in virtue of these two revolutions the idea of the Universe, the Macrocosm, the All, as subject of human knowledge, and therefore of human interest, has fallen to pieces.

Clifford

The chair of "mathematics" [held by Galileo in 1592] then covered the teaching of geometry, astronomy, military engineering, and fortification.

Santillana

Angling may be said to be so like the mathematics, that it can never be fully learnt.

Walton

CHAPTER 14

Measuring Angles and the Protractor Postulate

14.1 AXIOM 4: THE PROTRACTOR POSTULATE

In 1733 a geometry book by a Jesuit priest named Saccheri appeared. Although the book caused some stir at the time, soon it was almost completely forgotten. We shall learn a great deal about this book later. In 1832 there appeared a geometry text containing a short appendix written by the son of the author of the text. This appendix has been described by G. B. Halsted as ". . . the most extraordinary two dozen pages in the whole history of thought!" The author of the *Appendix* was John Bolyai, a name that will live as long as any advanced form of human civilization exists. It is Bolyai and Lobachevsky that are recognized as the cofounders of non-Euclidean geometry. More on Bolyai and Lobachevsky later. We now jump one more century to 1932. That year saw the publication of *A Set of Postulates for Plane Geometry Based on Scale and Protractor* written by George David Birkhoff. Although nobody would suggest this paper is as important as Bolyai's *Appendix*, the paper is significant to the reader of this book. Our approach to axiomatizing plane geometry is based on Birkhoff's axiom system as given in that paper.

    G. D. Birkhoff (1884–1944) was an American mathematician noted for his work in differential equations, dynamics, and relativity. His 1932 paper can be traced back to his *The Origin, Nature, and Influence of Relativity* (Macmillan, 1925), written for the layman. In this book we find "The facts concerning geometry in the plane can be

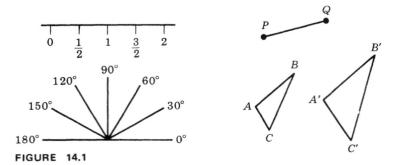

FIGURE 14.1

taken to repose upon the following four assumptions: I. Measurement of distance in a line can be made by means of the ruler; II. Measurement of the angle between lines can be made by means of the protractor; III. One and only one straight line contains two given points; IV. The plane is alike and even similar to itself in all its parts." Each of the assumptions was accompanied by a figure to help explain the meaning. See Figure 14.1. It would be unfair to stop and criticize these assumptions from a mathematical point of view, since the book was aimed at the general reader. You may wonder about the lack of some sort of parallel postulate. However, as you will learn later, there is no such lack. The parallel axiom is hidden, but it is there.

In 1929 Birkhoff teamed up with educator Ralph Beatley to write an article for *The Teaching of Geometry,* the Fifth Yearbook of the National Council of Teachers of Mathematics. The article, *A New Approach to Elementary Geometry,* expanded only slightly on the assumptions above and gave some advice to high school teachers. The article is still worth reading (Exercise 14.9). The two Harvard professors experimented with using this approach to teach geometry. The result was the text *Basic Geometry* (Scott Foresman, 1941; Chelsea, 1959). The book deserved to be more popular than it was; it was too revolutionary in its outlook. The approach has been growing in popularity, popularized today by such eminent mathematicians as Edwin Moise and such influential organization as the School Mathematics Study Group.

In this text we are undertaking the axiomatic development of Euclidean geometry and the non-Euclidean geometry of Bolyai and Lobachevsky. We are following G. D. Birkhoff's idea of using axioms that are motivated by the ruler and the protractor. The *ruler* for any line is given to us by the distance function d in the Ruler Postulate. Given d, the theorems involving order (i.e., betweenness) on a line follow directly from the properties of the real numbers. In analogous fashion we are about to postulate an angle measuring function m. The idea is that we can *place* a *protractor* in a halfplane with edge

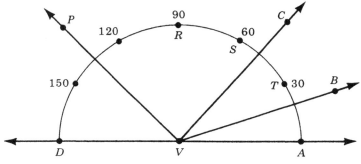

FIGURE 14.2

$\overleftrightarrow{VA}$ so that there is a *one-to-one correspondence* between the set of real numbers between 0 and 180 and the set of all rays $\overrightarrow{VP}$ with P in the halfplane. Besides being in a one-to-one correspondence the numbers should *increase as we go around*. Although the last remark is hardly precise (Around what? In fact, what is *around?*), there is an idea there that should be made precise in our axiom. Can you think how this might be done? In Figure 14.2, which illustrates the idea of a protractor, the numbers increase *uniformly*. Here is a third requirement that might be made precise and then be incorporated into our axiom.

In the notation above we would have $\overrightarrow{VP}$ in Figure 14.2 correspond to the real *number* $m\angle AVP$, the *measure* of $\angle AVP$. In the same figure, $m\angle AVB + m\angle BVC$ should be equal to $m\angle AVC$, and $m\angle AVP + m\angle PVD$ is probably 180. "Why 180?" Because 180 is half of·360. "Then why 360?" Why not?! This snide answer will have to do because the truth is that nobody knows for sure why 360. The number goes back at least as far as ancient Babylonia. (Perhaps a year had three hundred and sixty days.) Actually, 120 would do just as well for elementary geometry. (Then one could construct with straightedge and compass an angle of unit measure.) You can probably think of a good reason to use 1 in place of 360. In the author's mind, 4 would be a nice replacement for 360, as then an angle of unit measure would be a *right angle*. The whole point is that it doesn't make much difference which positive real number is picked for $m\angle AVP + m\angle PVD$ in Figure 14.2. Only when one starts to use calculus is it seen that another best choice is 2π in place of 360 — for reasons that have nothing to do with the geometry itself but rather to make formulas and calculations easier. Since we shall be doing some calculations in later chapters, we shall be consistent and use this so-called *radian measurement* in the theory throughout the book. If you want a definition of the number π independent of geometry, you may take Leibniz' formula:

$\pi/4 = 1 - 1/3 + 1/5 - 1/7 + 1/9 - 1/11 + \cdots$

We are ready for the statement of our next axiom, which determines some properties of the undefined term m.

Axiom 4 *Protractor Postulate* m is a mapping from the set of all angles into $\{x \mid x \in \mathbf{R}, 0 < x < \pi\}$ such that

a if $\overrightarrow{VA}$ is a ray on the edge of halfplane H, then for every r such that $0 < r < \pi$ there is exactly one ray $\overrightarrow{VP}$ with P in H such that $m\angle AVP = r$;

b if B is a point in the interior of $\angle AVC$, then $m\angle AVB + m\angle BVC = m\angle AVC$.

You should stop and examine the Protractor Postulate in detail. As well as deciding what the axiom says, you should think about what it does not say. How close does Axiom 4 come to incorporating all that you *see* when you look at a protractor?

DEFINITION 14.1 Mapping m is called the ***angle measure function***. The ***measure*** of $\angle AVB$ is $m\angle AVB$. If an angle has measure $k\pi$, then the angle is said to be ***of*** 180k ***degrees***. $\angle AVB \simeq \angle CWD$ iff $m\angle AVB = m\angle CWD$, in which case we say that $\angle AVB$ is ***congruent*** to $\angle CWD$.

Since people are generally reluctant to give up something with four thousand years seniority, we have included *degrees* in the formal definitions by popular demand. However, we insist that neither $\angle AVB$

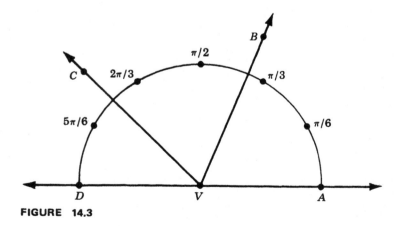

FIGURE 14.3

nor $m\angle AVB$ is ever *equal* to, say, sixty degrees. Remember $\angle AVB$ is a *set of points* and $m\angle AVB$ is a *number*.

As congruence of segments is defined in terms of length given by d, so congruence of angles is defined in terms of measure given by m. Congruence of angles must be an equivalence relation since equality of real numbers is an equivalence relation.

Theorem 14.2 Congruence of angles is an equivalence relation on the set of all angles.

Proof (R): $\angle AVB \simeq \angle AVB$, since $m\angle AVB = m\angle AVB$. (S): $\angle AVB \simeq \angle CWD$ implies $\angle CWD \simeq \angle AVB$, since $m\angle AVB = m\angle CWD$ implies $m\angle CWD = m\angle AVB$. (T): $\angle AVB \simeq \angle CWD$ and $\angle CWD \simeq \angle EXF$ implies $\angle AVB \simeq \angle EXF$, since $m\angle AVB = m\angle CWD$ and $m\angle CWD = m\angle EXF$ implies $m\angle AVB = m\angle EXF$. ■

Theorem 14.3 *Angle-Construction Theorem* Given $\angle AVB$, $\overrightarrow{WC}$, and halfplane H of $\overleftrightarrow{WC}$, then there exists exactly one ray $\overrightarrow{WD}$ such that D is in H and $\angle AVB \simeq \angle CWD$.

Proof By the Protractor Postulate there is exactly one ray $\overrightarrow{WD}$ with D in H such that $m\angle CWD = m\angle AVB$. ■

Combining the Segment-Construction Theorem with the Angle-Construction Theorem, we have a result that will often be useful:

Corollary 14.4 *Angle-Segment-Construction Theorem* If $0 < a < \pi$, $0 < r$, and H is a halfplane of $\overleftrightarrow{AB}$, then there is a unique point C in H such that $m\angle ABC = a$ and $BC = r$.

Theorem 14.5 *Angle-Addition Theorem* Suppose C is a point in int $(\angle AVB)$, C' is a point in int $(\angle A'V'B')$, and $\angle AVC \simeq \angle A'V'C'$. Then $\angle AVB \simeq \angle A'V'B'$ iff $\angle CVB \simeq \angle C'V'B'$.

Proof Exercise 14.1. ■

Theorem 14.6 If B and C are two points on the same side of $\overleftrightarrow{VA}$ and $m\angle AVB < m\angle AVC$, then B is in int $(\angle AVC)$.

Proof Assume B is not in int $(\angle AVC)$. Since B is on the same side of $\overleftrightarrow{VA}$ as C, either B is on $\overrightarrow{VC}$ or else B and A are on opposite sides of $\overleftrightarrow{VC}$. That B is on $\overleftrightarrow{VC}$ is impossible as then $\overrightarrow{VB} = \overrightarrow{VC}$ and $m\angle AVB = m\angle AVC$. So B and A are on opposite sides of $\overleftrightarrow{VC}$, and C is in int $(\angle AVB)$, (Theorem 13.10). Then, by the Protractor Postulate, $m\angle AVC + m\angle CVB = m\angle AVB$. So $m\angle AVC < m\angle AVB$, contradicting the hypothesis that $m\angle AVB < m\angle AVC$. Therefore B is in int $(\angle AVC)$. ■

Should an *angle bisector* be a ray or a line? There seems to be a *choice* to be made. Why don't the other two terms in the next definition involve a choice?

DEFINITION 14.7 If $m\angle AVB = m\angle BVC$ and B is in int $(\angle AVC)$, then $\overrightarrow{VB}$ is an **angle bisector** of $\angle AVC$. If $m\angle AVB + m\angle CWD = \pi$, then $\angle AVB$ and $\angle CWD$ are **supplementary**; if $m\angle AVB + m\angle CWD = \pi/2$, then $\angle AVB$ and $\angle CWD$ are **complementary**.

Theorem 14.8 Every angle has a unique angle bisector.

Proof Follows from (a) of the Protractor Postulate. ■

By the *Rational Cartesian Plane* we mean (M2, d, m) where d and m are the restrictions of the usual Cartesian distance and angle measure to the Rational Cartesian Incidence Plane M2. Let $A = (1, 0)$, $V = (0, 0)$, and $B = (1, 1)$. In the Cartesian plane $\angle AVB$ has an angle bisector which is contained in the line with equation $y = (-1 + \sqrt{2})x$. Since there is no such line in M2, it follows that not every angle in the Rational Cartesian Plane has an angle bisector. Of course, (M2, d, m) satisfies neither the Ruler Postulate nor the Protractor Postulate.

Theorem 14.9 *Euclid's Proposition I.13* If two angles are a linear pair, then the two angles are supplementary.

Proof Let B be on halfplane H of $\overleftrightarrow{VC}$. Let $C - V - A$. Let $x = m\angle AVB$ and $y = m\angle BVC$. Our task is to prove that $x + y = \pi$. Assume $x + y < \pi$. Then there exists unique $\overrightarrow{VD}$ with D on H such that $m\angle AVD = x + y$. Since $m\angle AVB = x < x + y$, we have B in int $(\angle AVD)$, (Theorem 14.6). Then D is in int $(\angle CVB)$, (Theorem 13.10). Thus, by the Protractor

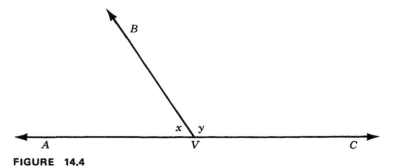

FIGURE 14.4

Postulate, it follows that $x + m\angle BVD = x + y$ and $m\angle CVD + m\angle DVB = y$. So $m\angle CVD = 0$, a contradiction. Now assume $x + y > \pi$. Then there exists a unique $\overrightarrow{VE}$ with E on the same side of $\overleftrightarrow{VB}$ as C such that $m\angle BVE = \pi - x$. Since $\pi - x < y = m\angle BVC$, we have E is in int $(\angle BVC)$. So (Theorem 13.10), we have B is in int $(\angle AVE)$. Thus $m\angle AVE = (x) + (\pi - x) = \pi$, a contradiction. Therefore $x + y = \pi$. ∎

Theorem 14.10 *Euclid's Proposition I.14* If points A and C are on opposite sides of $\overleftrightarrow{VB}$, then $m\angle AVB + m\angle BVC = \pi$ implies $A - V - C$.

Proof Exercise 14.2. ∎

Theorem 14.11 *Euclid's Proposition I.15* Vertical angles are congruent. Also, if $A - V - C$ and points B and D are on opposite sides of $\overleftrightarrow{AC}$, then $\angle AVB \cong \angle CVD$ implies $B - V - D$.

Proof The first statement follows from the fact that $x + y = \pi$ and $x + z = \pi$ implies $y = z$. The second statement follows from the same reasoning as the proof of Theorem 14.10. ∎

DEFINITION 14.12 If the measure of an angle is $\pi/2$, the angle is *right*. If the measure of an angle is less than $\pi/2$, the angle is *acute;* if the measure of an angle is greater than $\pi/2$, the angle is *obtuse.*

Theorem 14.13 If two congruent angles are a linear pair, then each of the angles is a right angle.

Proof Follows directly from the fact that $x + y = \pi$ and $x = y$ implies $x = y = \pi/2$. ∎

Theorem 14.14 *Four-Angle Theorem* If $A' - V - A$, $B' - V - B$, and $\angle AVB$ is a right angle, then each of $\angle AVB'$, $\angle A'VB$, and $\angle A'VB'$ is a right angle.

Proof Follows directly from three applications of the fact that $x + y = \pi$ and $x = \pi/2$ implies $y = \pi/2$. ∎

Theorem 14.15 If $m\angle AVB + m\angle BVC = m\angle AVC$, then B is in int $(\angle AVC)$.

Proof If B and C are on the same side of $\overleftrightarrow{VA}$, then we are done (Theorem 14.6). Assume B and C are on opposite sides of $\overleftrightarrow{VA}$. Let $A' - V - A$. Now A and C cannot be on the same side of $\overleftrightarrow{VB}$ as then (by Theorem 13.10) point A would be in int $(\angle BVC)$, contradicting the hypothesis $m\angle BVC < m\angle AVC$. Therefore, we must suppose A' and C are on the same side of $\overleftrightarrow{VB}$. So A' is in int $(\angle BVC)$. Then $m\angle BVC = m\angle BVA' + m\angle A'VC$. Hence, since $m\angle AVB + m\angle BVA' = \pi$ (Theorem 14.9), we have $m\angle AVC = m\angle AVB + m\angle BVC = m\angle AVB + m\angle BVA' + m\angle A'VC = \pi + m\angle A'VC$. So $m\angle AVC > \pi$, a contradiction. ∎

DEFINITION 14.16 If l and m are two lines whose union contains a right angle, then we write $l \perp m$ and say that l is **perpendicular** to m. We agree that the following are equivalent:

$$\overleftrightarrow{AB} \perp \overleftrightarrow{CD}, \quad \overleftrightarrow{AB} \perp \overrightarrow{CD}, \quad \overleftrightarrow{AB} \perp \overline{CD},$$

$$\overrightarrow{AB} \perp \overleftrightarrow{CD}, \quad \overrightarrow{AB} \perp \overrightarrow{CD}, \quad \overrightarrow{AB} \perp \overline{CD},$$

$$\overline{AB} \perp \overleftrightarrow{CD}, \quad \overline{AB} \perp \overrightarrow{CD}, \quad \overline{AB} \perp \overline{CD}.$$

Theorem 14.17 If a is a segment, ray, or line and b is a segment, ray, or line, then $a \perp b$ implies $b \perp a$.

Proof Let a be on $\overleftrightarrow{AB}$ and b be on $\overleftrightarrow{CD}$. Then $a \perp b$ iff $\overleftrightarrow{AB} \perp \overleftrightarrow{CD}$. However, by the symmetry in the first part of the definition above, $\overleftrightarrow{AB} \perp \overleftrightarrow{CD}$ iff $\overleftrightarrow{CD} \perp \overleftrightarrow{AB}$. Hence $a \perp b$ iff $b \perp a$. ∎

Theorem 14.18 If P is a point on line l, then there exists a unique line through P that is perpendicular to l.

Proof Let $l = \overleftrightarrow{PA}$. Let H be a halfplane of l. By the Protractor Postulate, there exists a unique ray $\overrightarrow{PB}$ with B in H such that $m\angle APB = \pi/2$. So $\overleftrightarrow{PB}$ is a line through P and perpendicular to l. Suppose m is a line through P and perpendicular to l. Since m intersects l and is distinct from l, there is a point C that is on both m and H. Since $m \perp l$, then $\angle APC$ is a right angle by the Four-Angle Theorem. So $m\angle APC = \pi/2$. By the uniqueness of $\overrightarrow{PB}$ above, we have $\overrightarrow{PC} = \overrightarrow{PB}$. Therefore $m = \overleftrightarrow{PB}$, and $\overleftrightarrow{PB}$ is the unique line through P and perpendicular to l. ∎

If to Taxicab Geometry of Section 7.2 we add the *usual* Cartesian angle measure function m, then we have that (M1, t, m) is a model of Σ. Since betweenness in (M13, h) of Section 9.2 corresponds to Euclidean betweenness in the Cartesian plane, by taking m to be the *usual* angle measure function restricted to M13, we see that (M13, h, m) is a model of Σ.

The Cartesian plane (M1, d, m) with the *usual* distance function d and angle measure function m is a model of Σ. If a proposition is false for any model of Σ, then the proposition is necessarily not a theorem of Σ. However, it must be emphasized that, since the Cartesian plane is only one model of Σ, a proposition which holds for this model might not be a theorem of Σ.

The Moulton Incidence Plane with distance function s given by Euclidean arclength along the Moulton lines satisfies the Incidence Axiom and the Ruler Postulate. The argument in Section 8.2 shows that (M10, s) also satisfies PSP. To find m_2 such that (M10, s, m_2) is a model of Σ, we only need an angle measure function m_2 for (M10, s) that satisfies the Protractor Postulate. Since some angles in (M10, s) are lines in the Cartesian plane, it is clear that the Cartesian angle function m will not work. However, the only time that m and the Protractor Postulate are incompatible is when V, as it appears in Axiom 4, is *on* the y-axis. Thus, given $\angle AVB$ in (M10, s) with V *off* the y-axis and assuming without loss of generality (Theorem 9.3) that A and B are on the same side of the y-axis as V, we define $m_2\angle AVB$ for $\angle AVB$ in (M10, s) to be $m\angle AVB$ for $\angle AVB$ in the Cartesian plane. That m_2 is well-defined for all angles $\angle AVB$ in (M10, s) with V off the y-axis follows from the fact that each halfplane of the y-axis in (M10, s) is indistinguishable from a halfplane of the y-axis in (M1, d). It also follows that m_2 satisfies the Protractor Postulate as long as V, as it appears in Axiom 4, is *off* the y-axis. There remains the task of defining $m_2\angle AVB$ when V is *on* the y-axis. This is accomplished by using the construct that M10 is obtained from M1 by *bending* that part of certain lines that is on the *right* side of the y-axis. The idea is to have used m to measure the angles *before* doing the bending. See Figure 14.5, where the Cartesian line $\overleftrightarrow{P'Q}$

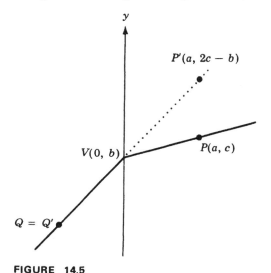

FIGURE 14.5

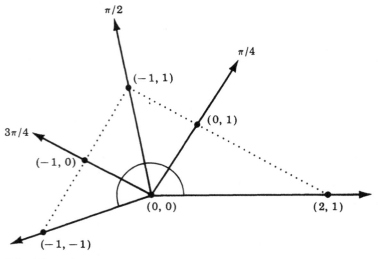

FIGURE 14.6

was *bent* to form the Moulton line $\overleftrightarrow{PQ}$. For any number b and point P with $P = (x, y)$, define P_b to be $(x, 2y - b)$ if $x > 0$ and $y > b$ but $P_b = P$ otherwise. Then, for $V = (0, b)$, define $m_2 \angle AVB$ to be $m \angle A_b V B_b$. Since m satisfies the Protractor Postulate for the Cartesian plane, it follows that m_2 satisfies the Protractor Postulate for (M10, s, m_2). Figure 14.6 illustrates a *protractor* with initial ray $\overrightarrow{VA}$ on the halfplane containing $(-1, 1)$ when $V = (0, 0)$ and $A = (2, 1)$.

In forming the definition of m_2 we used the construct that M10 is obtained from M1 by *bending* that part of some lines which is on the *right* side of the y-axis. We might just as well have taken the view that the *bending* was done on the *left* side of the y-axis. Then, by a construction analogous to the one giving m_2, we would obtain a different angle measure function m_3 such that (M10, s, m_3) is a model of Σ. By the way, although it may seem unorthodox, we can consider that M1 is formed by *bending* some of the lines in M10!

Looking at Figure 14.7, we observe some peculiar properties of the plane (M10, s, m_2). First note that V is on $\overleftrightarrow{AB}$ and on $\overleftrightarrow{CD}$. Since $m_2 \angle AVC = \pi/2$, we conclude that $\angle AVC$, $\angle AVD$, $\angle BVC$, and $\angle BVD$ are all right angles. (Perhaps the Four-Angle Theorem (Theorem 14.14) is not really trivial.) In particular, $\overleftrightarrow{DV} \perp \overleftrightarrow{AB}$. If $H = (\tfrac{4}{5}, \tfrac{2}{5})$, then H is on $\overleftrightarrow{AB}$, $H \neq V$, and $\overleftrightarrow{DH} \perp \overleftrightarrow{AB}$. So $\overleftrightarrow{DV}$ and $\overleftrightarrow{DH}$ are *two* distinct lines through D that are perpendicular to $\overleftrightarrow{AB}$. However, if $G = (-1, \tfrac{3}{2})$, then there is no line through G that is perpendicular to $\overleftrightarrow{AB}$. Shocking!

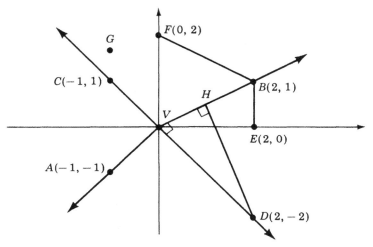

FIGURE 14.7

We might notice that $\triangle VDH$ has two right angles. Excessive, to say the least! Also, looking at $\triangle BVE$ and $\triangle BVF$, we note that Euclid's Proposition I.4, the *Side-Angle-Side Theorem*, fails for this plane: $\overline{BV} \simeq \overline{BV}$, $\angle BVE \simeq \angle BVF$, and $\overline{VE} \simeq \overline{VF}$ but $\angle VBE \neq \angle VBF$, $\overline{BE} \neq \overline{BF}$, and $\angle BEV \neq \angle BFV$. Totally unacceptable!

Before stating Axiom 4, we had in mind three requirements for any respectable protractor. (See the fourth paragraph of this chapter.) Condition (a) of the Protractor Postulate gives the correspondences we were after, and condition (b) makes precise the requirement that the numbers on a protractor "increase as we go around." However, we have not incorporated into the axiom the third requirement, that for any protractor these numbers should increase "uniformly." (For example, in Figure 14.2 we expect that $RS = ST$ since $\angle RVS$ and $\angle SVT$ have the same measure. Cf. Exercise 14.7.) You might say our fourth axiom should have a third condition. Rather than calling such a condition part (c) of Axiom 4, the author has decided to label the condition Axiom 5. The numbering of axioms is an arbitrary game anyway. For example, condition (b) of the Protractor Postulate could just as well have been listed as a separate axiom. So we are in somewhat the same situation as we were after introducing the Ruler Postulate. Either PSP or PASCH was required to complete the full intent of our motivation behind the Ruler Postulate. Now, we need another axiom to complete the full intent of our motivation behind the Protractor Postulate. As you may have guessed, Euclid's Side-Angle-Side Theorem is itself one possibility for this axiom. Another important possibility is the subject of Chapter 16.

14.2 PECULIAR PROTRACTORS

In this section, the development of our axiom system is examined a little more thoroughly. There is an essential difference between the present situation and the situation after the introduction of the Ruler Postulate. The Ruler Postulate was independent of the previous axiom. We shall show below that the Protractor Postulate is *not* independent of our first three axioms. Axiom 4 is only a prologue to Axiom 5. Our fourth and fifth axioms could be replaced by axioms that do not directly involve the real numbers. (Some suggestions appear in Section 15.1.) With these axioms, the proof that there is an angle measure function m satisfying our fourth and fifth axioms is tedious and not very informative. The only penalty for having stronger axioms than necessary is that there is more work in checking that a given model satisfies all the axioms. In our case, this is outweighed by having the angle measure function handed to us. As was the Ruler Postulate, the Protractor Postulate is somewhat of a shortcut. Accepting the Protractor Postulate is certainly in complete accord with our aim as expressed at the beginning of Section 6.1.

Suppose $(\mathscr{P}, \mathscr{L}, d)$ is any model that satisfies our first three axioms. We shall construct an angle measure function m such that $(\mathscr{P}, \mathscr{L}, d, m)$ is a model of Σ. In fact, let $\angle AVT$ be any angle in the model; we shall construct m such that $\angle AVT$ is a right angle. If this seems impossible, remember that all we know about right angles or anything else in our theory is what the axioms, definitions, and theorems tell us. A right angle is simply an angle whose measure is $\pi/2$. The following definition will make the discussion more concise.

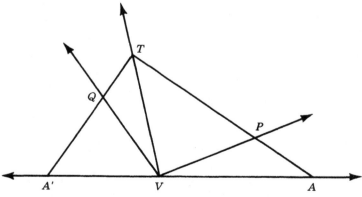

FIGURE 14.8

We say *m determines a protractor on H with initial ray* $\overrightarrow{VA}$ if $\overrightarrow{VA}$ is a ray on the edge of halfplane H and if m is a mapping on the angles such that

(a) for every r such that $0 < r < \pi$ there is exactly one ray $\overrightarrow{VP}$ with P in H such that $m\angle AVP = r$;

(b) if B and C are points on H with B in int $(\angle AVC)$, then $m\angle AVB + m\angle BVC = m\angle AVC$.

In any model satisfying our first three axioms, suppose $A - V - A'$ and point T is in halfplane H of $\overleftrightarrow{VA}$. If $A - P - T$, define $m\angle AVP$ to be $(\pi/2)(AP/AT)$; define $m\angle AVT$ to be $\pi/2$; and, if $T - Q - A'$, define $m\angle AVQ$ to be $(\pi/2) + (\pi/2)(TQ/TA')$. See Figure 14.8. By the Ruler Postulate, it follows that m has property (a) above. If B and C are points on H with B in int $(\angle AVC)$, then define $m\angle BVC$ to be $m\angle AVC - m\angle AVB$. Hence, m determines a protractor on H with initial ray $\overrightarrow{VA}$. So far we've had a lot of leeway. If our efforts to construct m are successful, it is clear that there are infinitely many different angle measuring functions that can be imposed on any model satisfying our first three axioms to form a model of Σ. We shall see that the measure of all angles with vertex V is now determined. In other words, one *protractor* with initial ray $\overrightarrow{VA}$ determines all the protractors with initial ray having vertex V.

Now, for point P in H, define $m\angle A'VP$ to be $\pi - m\angle AVP$. We claim that m determines a protractor on H with initial ray $\overrightarrow{VA'}$. Property (a) is obvious. Let B and C be on H with B in int $(\angle A'VC)$. Then C is in int $(\angle AVB)$ by Theorem 13.10. See Figure 14.9. Then property (b) follows from the calculation

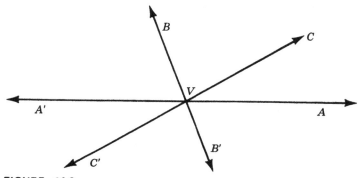

FIGURE 14.9

$$m\angle A'VB + m\angle BVC = (\pi - m\angle AVB) + (m\angle AVB - m\angle AVC)$$

$$= \pi - m\angle AVC$$

$$= m\angle A'VC.$$

Since m is to satisfy the Protractor Postulate, the definition of $m\angle A'VP$ above is forced by the fact that the angles of a linear pair must be supplementary (Theorem 14.9).

Let H' be the side of $\overleftrightarrow{VA}$ opposite H. For any points B and C in H but not collinear with V, let $B'-V-B$ and $C'-V-C$. Since vertical angles must be congruent (Theorem 14.11), we are forced to make the following definitions. Define $m\angle AVB'$ to be $m\angle A'VB$; define $m\angle A'VB'$ to be $m\angle AVB$; and define $m\angle B'VC'$ to be $m\angle BVC$. Then m determines a protractor on H' with initial ray $\overrightarrow{VA}$, and m determines a protractor on H' with initial ray $\overrightarrow{VA'}$.

So $m\angle PVQ$ has been defined for $\angle PVQ$ except when P and Q are on opposite sides of $\overleftrightarrow{VA}$, in which case exactly one of A or A' is in int $(\angle PVQ)$. If A is in int $(\angle PVQ)$, define $m\angle PVQ$ to be $m\angle PVA + m\angle AVQ$; if A' is in int $(\angle PVQ)$, define $m\angle PVQ$ to be $m\angle PVA' + m\angle A'VQ$. This definition is forced on us by property (b) of Axiom 4. It now takes very little calculation to show that, if P is any point except V and if K is either halfplane of $\overleftrightarrow{VP}$, then m determines a protractor on K with initial ray $\overrightarrow{VP}$.

The game we have just played with point V can be played with any other point in the model. (To avoid having the definition of m depend on any infinite number of choices, we could set up a list of rules to follow so that no more choices would be necessary.) Thus, we may suppose that if V and P are any two points and if K is either halfplane of $\overleftrightarrow{VP}$ then m is defined so that m determines a protractor on K with initial ray $\overrightarrow{VP}$. It follows that $(\mathscr{P}, \mathscr{L}, d, m)$ satisfies the Protractor Postulate. Hence, the Protractor Postulate is not independent of our first three axioms.

To end this section, we discuss one more question. Why not consider the possibility of an angle measure function that has arbitrarily large values? After all, the distance function has arbitrarily large values. Thus, we might examine what we shall call the *Big-Protractor Postulate* or BPP, which is identical with Axiom 4 except that "$0 < r < \pi$" is replaced by "$0 < r$." However, as we shall see, the Big-Protractor Postulate is inconsistent with our first three axioms. For this discussion we are assuming only our first three axioms and BPP. Let $\angle BVA$ and $\angle BVC$ be a linear pair. Let $m\angle AVB = x$ and $m\angle BVC = y$. By (a) of BPP there exists point D on the same side of $\overleftrightarrow{VA}$ as B such that $m\angle AVD = x + y$. Since $A - V - C$ we have D is off $\overleftrightarrow{VC}$. Also (Theorem 14.6), B is in int ($\angle AVD$). "But, but, but Theorem 14.6 comes *after* the assumption of the Protractor Postulate in our theory," you protest. Good point! However, checking back, we see that the proof of Theorem 14.6 is valid under BPP. Continuing, by (b) of BPP we have $m\angle AVB + m\angle BVD = x + y$. Since $x + y = m\angle AVB + m\angle BVC$, we have $m\angle BVD = m\angle BVC$. Hence, $\overrightarrow{VD} = \overrightarrow{VC}$ by (a) of BPP. So D is on $\overrightarrow{VC}$, and D is off $\overleftrightarrow{VC}$. The axiom system is inconsistent. Exit BPP.

14.3 EXERCISES

14.1 Theorem 14.5.

14.2 Theorem 14.10.

14.3 One angle of a linear pair is acute iff the other angle is obtuse.

• **14.4** The mapping on the points of (M1, d) that sends any point P to P' where $P' = (\sqrt{3}x, y)$ if $P = (x, y)$ is a collineation that preserves betweenness but is not an isomorphism. (M1, d, m') is a model of Σ where m' is given by defining $m'\angle ABC$ to be $m\angle A'B'C'$.

• **14.5** For (M1, d, m') of Exercise 14.4: Euclid's Proposition I.4 fails, through any point there is a unique line perpendicular to any given line, and the sum of the measures of the three angles of any triangle is π.

• **14.6** True or False?

(a) $180 = \pi$.

(b) $0 < m\angle AVB < 180$.

(c) If $m\angle AVB < m\angle AVC$, then B is in int ($\angle AVC$).

(d) If $m\angle AVC = m\angle CVB$, then $\overrightarrow{VC}$ is the angle bisector of $\angle AVB$.

(e) If two angles are supplementary, then the two angles are a linear pair.

(f) The union of two right angles is a subset of the union of two lines.

(g) If the union of two lines contains a right angle, then the union contains three right angles.

(h) Perpendicularity is a reflexive relation on lines.

(i) Perpendicularity is a symmetric relation on lines.

(j)   Perpendicularity is a transitive relation on lines.

● **14.7**   For the plane in Exercise 14.4, sketch the protractor on half-plane H with initial ray $\overrightarrow{VA}$ where (i) $V = (0, 0), A = (1, 1)$, and H contains $(-1, 1)$, and (ii) $V = (0, 0), A = (-1, -1)$, and H contains $(-1, 1)$.

14.8   Does the Rational Cartesian Plane satisfy the Segment-Construction Theorem or the Angle-Construction Theorem?

14.9   Read "A New Approach to Elementary Geometry" in *The Teaching of Geometry*, the Fifth Yearbook of the National Council of Teachers of Mathematics, or as reprinted in volume three of *George David Birkhoff, Collected Mathematical Papers* (American Mathematical Society, 1950; Dover, 1968).

14.10   If $\overrightarrow{VA} \perp \overrightarrow{VB}$, then points A and C are on opposite sides of $\overrightarrow{VB}$ iff $\angle AVC$ is obtuse or $A-V-C$.

14.11   Those who have not heard about an American legislature passing a bill $67-0$ in favor of making π rational should read "The Modern Circle-Squarers," Chapter 17 of *A History of π* by P. Beckmann (Golem Press, 1970).

14.12   The mapping that sends (x, y) to (x, y^3) is a collineation from M1 onto M9. Define d' and m' such that $(M9, d', m')$ is isomorphic to the Cartesian plane.

14.13   Let l be the line through $(0, 0)$ and $(2, 1)$ in $(M10, s, m_2)$. What are the loci of all points P on zero, one, or two perpendiculars to l, respectively?

14.14   What is the smallest measure of an angle in $(M10, s, m_2)$ that is a line in $(M1, d, m)$?

14.15   Let $A = (-1, -1), B = (0, 0)$, and $C = (1, 1)$. Find the measure of each angle of $\triangle ABC$ in $(M10, s, m_2)$.

14.16 $(M10, s)$ is not isomorphic to $(M1, d)$.

14.17   Is BPP consistent with our first two axioms?

14.18   Read "What is an Angle?" by H. Zassenhaus in *The American Mathematical Monthly* Vol. 61 (1954), pp. 369–378.

14.19   Give a reasonable definition of a *right triangle*.

14.20   Give a reasonable definition for the statement "$\overrightarrow{VB}$ is *between* $\overrightarrow{VA}$ and $\overrightarrow{VC}$." What properties does this new betweenness have?

14.21   Give some reasonable definitions for the statement, "Line l is *between* lines m and n."

*14.22   Read "A Set of Postulates for Plane Geometry Based on Scale and Protractor" by Birkhoff in *Annals of Mathematics* Vol. 33 (1932), pp. 329 – 345 or as reprinted in the same volume of Birkhoff's collected works mentioned in Exercise 14.9.

*14.23   Read "Metric Postulates for Plane Geometry" by S. MacLane in *The American Mathematical Monthly* Vol. 66 (1959), pp. 543 – 555.

*14.24   Suppose m and m' both satisfy Axiom 4. If $m' = gm$, what can be said about the function g?

*14.25   In Euclidean geometry an angle of three degrees can be constructed with straight-edge and compass but not an angle of two degrees.

CHAPTER 15

Alternative Axiom Systems

15.1 HILBERT'S AXIOMS

This chapter is a digression. We pause in the development of our axiom system to take a quick look at some other axiom systems for geometry. It is intended that this chapter be read rather quickly and not studied in detail at this time. Most imperative, the reader should understand that none of the postulates, definitions, notation, or theorems is to be assumed in the other chapters!

The first rigorous axiom system for Euclidean geometry is due to Pasch in 1882, when the *order* of points on a line was axiomatized for the first time. Following the tradition of Pasch and Peano, Oswald Veblen (1880–1960) gave a system based on the two undefined terms *point* and *order* in "A system of axioms for geometry" in *Transactions of the American Mathematical Society* Vol. 5 (1904), pp. 343–384. Veblen's revised system of 1917 may be found in the reference given in Exercise 15.1. Many different axiom systems for geometry have been and are still being invented. References for eight of these may be found in Exercise 15.1 through 15.8. This chapter focuses on two other systems, one due to Hilbert and one due to Pieri.

The most famous of the axiom systems for Euclidean geometry is due to David Hilbert (1862–1943). Based on lectures at the University of Göttingen during the winter semester of 1898–99, the first edition of *Grundlagen der Geometrie* was presented as a memorial address published in connection with the unveiling of the Gauss–

Weber monument at Göttingen in June 1899. The material below is based on the original 1899 edition.

The undefined terms in Hilbert's system denote three sets and five relations. The three sets are $\mathscr{P}$, $\mathscr{L}$, and $\mathscr{E}$. The elements of these sets are called *points, lines,* and *planes,* respectively. A relation between $\mathscr{P}$ and $\mathscr{L}$ is denoted by the word *on.* A relation between $\mathscr{P}$ and $\mathscr{E}$ is also denoted by the word *on.* The first of five sets of axioms deals with these two undefined relations.

I. Hilbert's *Axioms of Incidence:* (1) Any two points are on at least one line. (2) Any two points are on at most one line. (3) Any three points not all on one line are on at least one plane. (4) Any three points not all on one line are on at most óne plane. (5) If two points on a line are on a plane, then every point on the line is on the plane. (6) If a point is on each of two planes, then there is another point on each of the two planes. (7) There are at least two points on each line; there are at least three points on each plane; and there are four points not all on one plane [and not all on one line].

Remark. Hilbert's axiom I.6 assures that the geometry is at most three-dimensional. Axiom I.7, including our bracketed addition, assures that the geometry is at least three-dimensional. Without our addition, there is no assurance that any planes exist at all since the axioms would be satisfied if all points were on one line. An incidence relation between lines and planes can then be defined in the obvious way. The usual phrases denoting incidence are now assumed. It follows that a line and a point off the line determine a plane.

Hilbert's second set of axioms, which he noted was first studied in detail by Pasch, concerns the third undefined relation. This is a ternary relation on the set of points and is denoted by the word *between.* Hilbert explicitly states that the points in a triple satisfying the relation are collinear and tacitly assumes they are also distinct. We have augmented the original axiom II.1. Further, we have omitted the original II.4, which is our Theorem 7.12 with Definition 7.10, since E. H. Moore proved the postulate to be not independent in 1902. Actually, the axioms given below can be further weakened. For example, it is easy to prove the existence of point B in II.2 in the presence of II.5.

II. Hilbert's *Axioms of Order:* (1) If point B is between points A and C, then A, B, C are three collinear points and B is between C and A. (2) If A and C are two points, then there is at least one point B that is between A and C and there is at least one point D such that C is between A and D. (3) Among any three points on a line, exactly one is between the other two. (5) Let A, B, C be three points not on a line. Let l be a line off A, B, C but in the plane containing A, B, C. Let l contain a point that is between A and B. Then l contains either a point between A and C or a point between B and C.

Remark. In order to list the axioms in the second set together, we have stated II.5 in terms of betweenness rather than in terms of "a point of a segment," as does Hilbert. Hilbert's definitions of "segment" and "a point of a segment" are unfortunate. The set of points of a segment is occasionally confused with the segment itself, although the meaning is always clear. (Anyone who reads the English translations should be aware that some ambiguities have been translated into inconsistencies.) All the definitions needed for the statement of the remaining axioms can now be given. An equivalent definition of "ray" is given below; otherwise, the definitions are due to Hilbert. Among the theorems that then follow are PSP with respect to any plane and the three-dimensional analog of PSP. We shall suppose *half-plane* has been defined in a fashion analogous to that of our Theorem 12.2.

Definition. Let A, B, C be three points not all on one line. The set $\{A, B\}$ is a *segment,* which is denoted by either AB or BA. If point P is between points A and B, then P is a *point of the segment AB*. The set $\{AB, BC, CA\}$ is a *triangle,* which is denoted by ABC. The set of all points P such that A is between B and P is a *ray* with *vertex* A. If h and k are two rays with vertex C on different lines, then the set $\{h, k\}$ is an *angle* with *vertex* C; further, if A is on h and B is on k, then the angle is denoted by $\angle ACB$.

III. Hilbert's *Axiom of Parallels:* If point P is off line l, then there exists exactly one line in the plane containing P and l that does not intersect l.

Remark. The fourth set of axioms is concerned with the remaining two undefined relations. Each is denoted by *is congruent to*. The first is a relation between segments, and the second is a relation between angles.

IV. Hilbert's *Axioms of Congruence:* (1) Given segment AB and a ray with vertex A', there exists one and only one point B' on the ray such that segment AB is congruent to segment $A'B'$; every segment is congruent to itself. (2) If segment AB is congruent to segment $A'B'$ and to segment $A''B''$, then $A'B'$ is congruent to $A''B''$. (3) If point B is between points A and C, point B' is between points A' and C', AB is congruent to $A'B'$, and BC is congruent to $B'C'$, then AC is congruent to $A'C'$. (4) Given angle $\{h, k\}$, ray h', and a halfplane H of the line containing h', then there exists one and only one ray k' on H such that angle $\{h', k'\}$ is congruent to angle $\{h, k\}$; every angle is congruent to itself. (5) If angle $\{h, k\}$ is congruent to angle $\{h', k'\}$ and to angle $\{h'', k''\}$, then $\{h', k'\}$ is congruent to $\{h'', k''\}$. (6) Given triangles ABC and $A'B'C'$ such that AB is congruent to $A'B'$, $\angle BAC$ is congruent to $\angle B'A'C'$, and AC is congruent to $A'C'$, then $\angle ABC$ is congruent to $\angle A'B'C'$.

Remark. Hilbert's fifth and final set of axioms originally contained only Archimedes' axiom. This assures that there are not *too many* points on a line. However, Archimedes' axiom alone is not sufficient to give a categorical axiom system, as Hilbert pointed out. In order to assure that there are *enough* points on a line to have a categorical axiom system with Cartesian three-space as a model, one more axiom is required. Hilbert's own *completeness axiom,* added in other editions as V.2, takes the somewhat awkward form of requiring that it be impossible to properly extend the sets and relations satisfying the other axioms so that all the other axioms still hold.

V. Hilbert's *Axiom of Continuity:* (1) If points A_1 is between points A and B, then there exist points $A_2, A_3, \ldots, A_n$ such that (i) A_k is between A_{k-1} and A_{k+1} for $k = 1, 2, \ldots, n-1$ with $A_0 = A$, (ii) segment $A_k A_{k+1}$ is congruent to segment AA_1 for $k = 1, 2, \ldots, n-1$, and (iii) point B is between A and A_n.

15.2 PIERI'S POSTULATES

Mario Pieri (1860–1913), a student of Peano, gave several axiomatic systems for Euclidean geometry. The first appeared in 1896. The paper we are going to consider is dated April 1899 and was approved at the 14 May 1899 meeting of the Royal Academy of Science at Torino. The full title is "Della geometria elementare come sistema impotetico deduttivo, Monografia del punto e del moto." The paper appears in *Memorie della Reale Academia delle Scienze di Torino* Vol. 49 (1899), pp. 173–222. The idea is to start with two undefined terms: *point* and *motion.* The motions are the mappings that are motivated by physical motions in space, considering only the initial and final positions. (Mappings motivated by reflections in planes are not rigid motions and are not included among Pieri's motions.) The twenty postulates stated below are as Pieri gave them, using his alternative form for Postulate 17. The definitions are essentially the original definitions. However, the notation does not follow that used by Pieri. The theorems stated below are only a selection of those stated and proved by Pieri. Pieri's system now follows.

Undefined terms. $\mathscr{P}$, $\mathscr{M}$.

Postulate 1. $\mathscr{P}$ and $\mathscr{M}$ are sets.

Definition. The elements of $\mathscr{P}$ are called *points*, and the elements of $\mathscr{M}$ are called *motions*. A *figure* is a set of points.

Postulate 2. There exists at least one point.

Postulate 3. If P is a point, then there exists a point different from P.

Postulate 4. A motion is a bijection on the set of points.

Postulate 5. If μ is a motion, then μ^{-1} is a motion.

Postulate 6. If μ and ν are motions, then $\nu\mu$ is a motion.

Postulate 7. If A and B are two points, then there exists a nonidentity motion fixing both A and B.

Postulate 8. If a nonidentity motion fixes three points A, B, C, then every motion that fixes both A and B also fixes C.

Remark. Pieri's fifth and sixth postulates together state that the motions form a group under composition if there exist any motions at all. The seventh postulate assures $\mathscr{M}$ is not empty and the geometry is at least two dimensional. The eighth postulate assures the geometry is at most three-dimensional.

Definition. Points A, B, C are *collinear* if there exists a nonidentity motion that fixes each of A, B, C. If A and B are two points, then the *line* $\overleftrightarrow{AB}$ is the set of all points collinear with A and B. The words *on*, *through*, etc., have their usual meaning. If A, B, C are noncollinear points, then the union of the three sets (i) the set of all points collinear with A and a point of $\overleftrightarrow{BC}$, (ii) the set of all points collinear with B and a point of $\overleftrightarrow{AC}$, and (iii) the set of all points collinear with C and a point of $\overleftrightarrow{AB}$ is a *plane*. If A and B are points, then the *sphere* B_A with *center* A is the set of all points P for which there is a motion fixing A and sending P to B.

Theorem. Two points determine a line. There exist three noncollinear points. The motions form a group under composition. A motion preserves the set of lines and the set of planes.

Postulate 9. If A, B, C are noncollinear points and D is a point on $\overleftrightarrow{BC}$ other than B, then a plane through A, B, D exists and is a subset of a plane through A, B, C.

Theorem. Three noncollinear points determine a plane. If a line intersects a plane at two distinct points, then the line is a subset of the plane. Point A is on B_A iff $A = B$, in which case $B_A = \{A\}$. If point C is on B_A, then $B_A = C_A$. A motion preserves the set of spheres. A motion that fixes point A fixes every sphere with center A. If A and B are two points and C is a point such that C_A and C_B intersect only at C, then A, B, C are collinear.

Postulate 10. If A and B are distinct points, then there exists a motion fixing A and sending B to a point on $\overleftrightarrow{AB}$ different from B.

Postulate 11. If A and B are distinct points and μ and ν are motions fixing A and sending B to a point on $\overleftrightarrow{AB}$ different from B, then $\mu B = \nu B$.

Postulate 12. If A and B are distinct points, then there exists a motion sending A to B and fixing some point of $\overleftrightarrow{AB}$.

Definition. If $\overleftrightarrow{AB}$ intersects B_A at a point C different from B, then A is a *midpoint* of B and C. Point A is the *midpoint* of A and A.

Theorem. Let A and B be distinct points. Then a midpoint of A and B exists, is unique, and is the midpoint of B and A. If motion μ fixes A and sends B to point B' on $\overleftrightarrow{AB}$, then $\mu B' = B$. If point M is on $\overleftrightarrow{AB}$ and motion μ fixes M and interchanges A and B, then any motion interchanging A and B must fix M. If μ is a motion sending A to B and fixing some point M on $\overleftrightarrow{AB}$, then M is the midpoint of A and B. The midpoint of A and B is the unique point on $\overleftrightarrow{AB}$ that is the center of a sphere through both A and B. A motion preserves midpoints. There exists a motion sending A to B, fixing $\overleftrightarrow{AB}$, but fixing no point of $\overleftrightarrow{AB}$; if ν is such a motion, then B is the midpoint of A and νB while A is the midpoint of B and $\nu^{-1}A$.

Remark. Postulates 10, 11, and 12 deal with motions that fix a line. Postulates 10 through 14 are motivated by the revolutions that fix a plane.

Postulate 13. If A, B, C are noncollinear points, then there exists a motion fixing A and B but sending C to a point different from C on the plane through A, B, C.

Postulate 14. If A, B, C are noncollinear points and D and E are points on the plane through A, B, C that are different from C but common to C_A and C_B, then $D = E$.

Definition. A *circle* with *center* A is the intersection of a plane through A and a sphere with center A. If l and m are intersecting lines and there exists a motion that fixes l pointwise and fixes m but not pointwise, then we say l is *perpendicular* to m and write $l \perp m$.

Theorem. Let A, B, C be noncollinear points on plane z. There are exactly two points on z that are both on C_A and C_B. A line intersects a sphere in at most two points. If D is a point on z and if μ and ν are nonidentity motions fixing A, B, and z, then $\mu D = \nu D$ and $\mu^2 D = D$. If a motion fixes both A and B and sends C to a point on $\overleftrightarrow{AC}$ different from C, then this point is on C_B. If D is a point on C_B and A is the midpoint of C and D, then $\overleftrightarrow{AB} \perp \overleftrightarrow{AC}$. If $\overleftrightarrow{AB} \perp \overleftrightarrow{AC}$, then A_B intersects $\overleftrightarrow{AC}$ only at A. If point D is on C_A, C_B, and z and point E is on $\overleftrightarrow{AB}$, then D is on C_E. If line l is perpendicular to line m, then m is perpendicular to l. If point P is off line l, then there exists a unique point Q on l such that $\overleftrightarrow{PQ} \perp l$. If motion ρ fixes A, sends B to a point on $\overleftrightarrow{AB}$ different from B, and sends C to a point on $\overleftrightarrow{AC}$ different from C, then A is the midpoint of P and ρP for every point P on z; further, such a motion ρ exists. A motion preserves perpendicularity.

Postulate 15. If points A, B, C are not collinear, then there exists a point off the plane through A, B, C.

Postulate 16. If points A, B, C, D are not in the same plane, then there exists a motion fixing A and B and sending D to a point on the plane through A, B, C.

Definition. Two points B and C are *equidistant* from point A if both are on a sphere with center A.

Theorem. Let A, B, C be noncollinear points on plane z. There exists a motion sending A to B, sending B to a point of $\overleftrightarrow{AB}$ other than B, and sending C to a point on z. There exists a point Q on z such that $\overleftrightarrow{AQ} \perp \overleftrightarrow{AB}$. The set of all points on z that are equidistant from A and B is a line perpendicular to $\overleftrightarrow{AB}$ at the midpoint of A and B. If A, B, E are three collinear points and D is an arbitrary point, then any point on two of the spheres D_A, D_B, D_E is also on the third. If D is on z, then D_A, D_B, D_C intersect only at D. A line through a point off z that is perpendicular to both $\overleftrightarrow{AB}$ and $\overleftrightarrow{AC}$ is perpendicular to every line on z that passes through A.

Definition. The *polar sphere* of two points A and B is the sphere through A and B with center the midpoint of A and B. A point is in the *interior* of a sphere if the point is the midpoint of two points on the sphere. Point P is *between* two points A and B if P is on $\overleftrightarrow{AB}$ and in the interior of the polar sphere of A and B. If A and B are two points, then *segment* $\overline{AB}$ is the union of $\{A, B\}$ and the set of all points between A and B while *ray* $\overrightarrow{AB}$ is the union of $\overline{AB}$ and all points P such that B is between A and P.

Remark. By Postulate 15 the geometry is at least three-dimensional. Postulate 16 is motivated by the rotation of a plane onto another plane about a common line. The next three postulates deal with betweenness.

Postulate 17. If A, B, C, D are four collinear points, then D cannot be in exactly one of the segments $\overline{AB}$, $\overline{AC}$, $\overline{BC}$.

Postulate 18. If point C is between points A and B, then no point can be both between A and C and between B and C.

Postulate 19. Given noncollinear points A, B, C on plane z, if line l is on z and contains a point between A and B, then l contains a point between A and C or a point between B and C provided l is off A, B, C.

Definition. If A, B, C are noncollinear points, then *angle* $\angle BAC$ is the union of all rays $\overrightarrow{AP}$ with P on $\overline{BC}$ and *triangle* $\triangle ABC$ is the union of all segments $\overline{AP}$ with P on $\overline{BC}$. Figures h and k are *congruent* if there is a motion μ such that $\mu h = k$. Given $\overline{AB}$ and $\overline{CD}$, we say $\overline{AB}$ is *shorter* than $\overline{CD}$ and write $\overline{AB} < \overline{CD}$ if there is a motion sending A to C and sending B to a point between C and D.

Remark. Among the terms Pieri defines next is *halfplane*. PSP is proved. Indeed, at this point it is possible to prove all of the theorems in the first and third books of Euclid that are independent of Euclid's Parallel Postulate, with one exception! Euclid's Proposition I.22 cannot be proved without a new axiom. Pieri's final axiom is an axiom of continuity, which is taken from Peano's *Principii di Geometria.*

Postulate 20. If A and B are distinct points and k is a figure containing a point between A and B, then there exists a point X which is either equal to B or between A and B such that no point of k is between X and B and such that if point Y is between A and X then there exists a point of k which is either equal to X or between Y and X.

Remark. Finally, Pieri proves Archimedes' axiom and then Euclid's Proposition I.22. Pieri's twenty postulates describe three-dimensional *absolute geometry,* so called because Euclid's Parallel Postulate is independent of the twenty postulates. Although it is not obvious, if we add Euclid's Parallel Postulate to Pieri's system, we then obtain a categorical axiom system for three-dimensional Euclidean geometry.

15.3 EXERCISES

● **15.1** For Veblen's axiom system based on the undefined terms *point* and *order* see Section 29 of Volume II of *Projective Geometry* by O. Veblen and J. W. Young (Ginn, 1946).

15.2 For an axiom system formed by selecting axioms from those systems by Hilbert and Veblen, see Chapter 5 of *The Foundations of Geometry* by G. de B. Robinson (University of Toronto Press, 1959).

15.3 For E. V. Huntington's axiom system based on the undefined terms *sphere* and *inclusion* see Appendix I of the book in Exercise 15.2.

15.4 For H. G. Forder's axioms see Chapter XIII of his *The Foundations of Euclidean Geometry* (Dover, 1958).

15.5 For Pieri's axiom system based on the undefined terms *point* and *congruence* (a relation on the set of ordered pairs of points) see Chapter XIV of the book in Exercise 15.4.

15.6 For an axiom system based on the undefined terms *point* and *distance,* see Chapter 5 of *A Modern View of Geometry* by L. M. Blumenthal (Freeman, 1961).

15.7 For a totally different approach, see pages 7 through 24 of *Linear Algebra and Geometry* by J. Dieudonné (Houghton Mifflin, 1969).

15.8 If you can read just a little German, begin to read *Aufbau der Geometrie aus dem Spiegelungsbegriff* by F. Bachmann (Springer-Verlag, 1959).

15.9 Which of Pieri's postulates excludes the reflections in planes from the set of motions?

15.10 In Pieri's axiom system, two points determine a line.

15.11 Hilbert's axiom V.2 is independent of his other axioms but is inconsistent without axiom V.1.

GRAFFITI

A Model of the Elliptic Plane ($\mathscr{P}$, $\mathscr{L}$, d, m):

   A **point** *is a Euclidean line through* $(0,0,0)$ *in* $\mathbf{R}^3$. *Set $\mathscr{P}$ is the set of all points.*

   For each Euclidean plane through $(0,0,0)$ *in* $\mathbf{R}^3$, *the set of all (elliptic) points (euclidean) on the plane is a* **line**. *Set $\mathscr{L}$ is the set of all lines.*

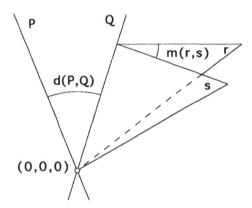

   Distance function $d : \mathscr{P} \times \mathscr{P} \to \mathbf{R}$ *is such that* $d(P, Q)$ *is the Euclidean angle measure between the Euclidean lines P and Q.*

   Angle measure $m : \mathscr{L} \times \mathscr{L} \to \mathbf{R}$ *is inherited as the Euclidean dihedral angle measure between Euclidean planes in* $\mathbf{R}^3$.

CHAPTER 16

Mirrors

16.1 RULERS AND PROTRACTORS

Figuratively, we can slide a ruler along its line and turn the ruler around. This is the idea behind the Ruler-Placement Theorem (Theorem 6.8). Figuratively, we can slide a protractor around a point and turn the protractor over. We don't have a Protractor-Placement Theorem because this idea is already contained in the Protractor Postulate. (We can start our protractor with any ray $\overrightarrow{VA}$ and use either halfplane of $\overleftrightarrow{VA}$.) Our intuition may be galloping ahead of our theory. Nothing has been said about picking up a ruler for line l and putting the ruler down on line m if $m \neq l$; nothing has been said about picking up a protractor with initial ray $\overrightarrow{VA}$ and putting the protractor down on $\overrightarrow{WB}$ if $W \neq V$. Nothing has been said about such things for a very good reason. The Ruler Postulate concerns itself with only one line at a time; the Protractor Postulate concerns itself with only one point at a time.

Suppose $(\mathcal{P}, \mathcal{L}, d, m)$ is a model of Σ. Let l be any line in this model. Let f be a coordinate system for l. Let h be any bijection on the reals which is strictly increasing. So h is one-to-one, onto, and $h(x_1) < h(x_2)$ when $x_1 < x_2$. Define distance d' on $(\mathcal{P}, \mathcal{L})$ by $d'(P, Q) = d(P, Q)$ unless P and Q are distinct points on l in which case $d'(P, Q) = |h(f(Q)) - h(f(P))|$. Then $(\mathcal{P}, \mathcal{L}, d')$ satisfies the Ruler Postulate. Lines different from l have their old coordinate systems, and l now has coordinate system hf. Since h is strictly increasing, we have $A-B-C$ in $(\mathcal{P}, \mathcal{L}, d')$ iff $A-B-C$ in $(\mathcal{P}, \mathcal{L}, d)$. Thus the segments, rays, angles, and triangles of one plane coincide with those of the other. It follows that $(\mathcal{P}, \mathcal{L}, d', m)$ is a model of Σ.

To illustrate the last construction, we start with the Cartesian plane $(\mathrm{M1}, d, m)$. Let l be the x-axis. Then f is a coordinate system for l where $f(x, 0) = x$. Let h be defined on the reals by $h(x) = x^3$. So $hf(x, 0) = x^3$. If $P = (x_1, y_1)$ and $Q = (x_2, y_2)$, then define $d'(P, Q) = |x_2^3 - x_1^3|$ if $y_1 = y_2 = 0$ and $d'(P, Q) = PQ$ otherwise. Then $(\mathrm{M1}, d')$ satisfies the Ruler Postulate and $(\mathrm{M1}, d', m)$ is a model of Σ. For a less spectacular example take $h(x) = 2x$ instead of $h(x) = x^3$. Then we could think of the x-axis as just having a different *scale* than the other lines.

Recalling Taxicab Geometry (Section 7.2), we see that this whole idea of obtaining a model of Σ from a given model of Σ by defining a new distance function determined by altering coordinate systems for the lines is not new to us. It is clear that our axiom system is not strong enough to relate distance on two distinct lines.

The same sort of game can be played with the protractors as with the rulers. Suppose $(\mathcal{P}, \mathcal{L}, d, m)$ is a model of Σ. Let $\overrightarrow{VA}$ be some ray in the model. Let H be a halfplane of $\overleftrightarrow{VA}$. Let g be any real function such that $g(0) = 0$, $g(\pi) = \pi$, and $g(x_1) < g(x_2)$ when $0 < x_1 < x_2 < \pi$. For example, let $g(x) = x^2/\pi$, $g(x) = x^3/\pi^2$, or $g(x) = \pi \sin \frac{1}{2}x$. We shall find a new angle measure function m' by altering the measure of those angles with vertex V. Let $m' \angle CWD = m \angle CWD$ when $W \neq V$. Define $m' \angle AVB = g(m \angle AVB)$ if B is in H. By the nature of g we have a new protractor on H with initial ray $\overrightarrow{VA}$. There remains to define m' so

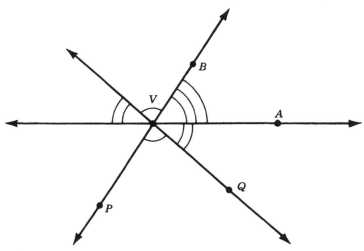

FIGURE 16.1

that we always have a protractor on a side of $\overrightarrow{VP}$ with initial ray $\overrightarrow{VP}$. We know from Chapter 14 that this can be done by defining $m'\angle PVQ$ under the assumption of (b) in the Protractor Postulate and the assumption that the angles in a linear pair are supplementary. It follows that $(\mathscr{P}, \mathscr{L}, d, m')$ is a model of Σ. We see that our axiom system is not strong enough to prevent such tinkering with an angle measure function.

16.2 MIRROR AND SAS

One conclusion to be drawn from the remarks made so far is that there is no provision for *symmetry* in our axiom system. We shall next consider an axiom to overcome this lack of symmetry. Put briefly we want every line to be a *mirror*. How can we make this idea precise? In Figure 16.2 point P' is the *mirror image* of P in line m. Every line m should determine a mapping on the points. This mapping should be a permutation on the points that sends lines to lines. In other words, the mapping should be a collineation. In Figure 16.2 we should have $AB = A'B'$ and $m\angle ABC = m\angle A'B'C'$. So the collineation should preserve distance and angle measure. A point on m should be its own image, but the image of a point off m and the point should be on opposite sides of m. Anything else? One might think to require that P' be defined such that m is the perpendicular bisection of $\overline{PP'}$ in Figure 16.2. The difficulty is that given point P off line m we have no theorem telling us that there exists a unique perpendicular from P to m. In fact, the model $(\text{M10}, s, m_2)$ from Section 14.1 shows that there can be no such theorem. This difficulty will melt away as a consequence of the requirements we have already listed. We have enough properties to make precise the idea that every line is a *mirror*.

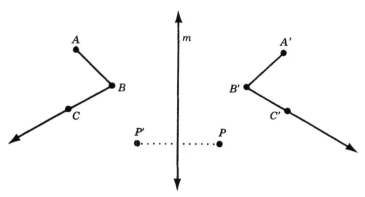

FIGURE 16.2

DEFINITION 16.1 *Mirror Axiom* or MIRROR: For every line m there exists a collineation that preserves distance and angle measure, fixes m pointwise, and interchanges the halfplanes of m.

We shall see that, if accepted as our next axiom, MIRROR would allow us to make precise the idea of picking up a triangle and putting it down without deformation. This idea, known as *superposition,* has been a problem in the foundations of geometry from the beginning. Schopenhauer wrote that Euclid's Common Notion 4, which supposedly allowed superposition, was more to be questioned than the parallel postulate. Euclid obviously had deep reservations about the method as he used it only twice, once for the fundamental Proposition I.4, the Side-Angle-Side Theorem, and then again for Proposition I.8, the Side-Side-Side Theorem. Since the use of superposition in Euclid's Proposition I.8 can be avoided (by Philo's proof of our Theorem 17.13), our attention is necessarily focused on Euclid's Side-Angle-Side Theorem.

DEFINITION 16.2 *Side-Angle-Side* or SAS: Given $\triangle ABC$ and $\triangle A'B'C'$, if $\overline{AB} \simeq \overline{A'B'}$, $\angle A \simeq \angle A'$, and $\overline{AC} \simeq \overline{A'C'}$, then $\angle B \simeq \angle B'$, $\overline{BC} \simeq \overline{B'C'}$, and $\angle C \simeq \angle C'$.

In some model of Σ suppose A, B, C, A', B', C' are six distinct points such that $\triangle ABC$ and $\triangle A'B'C'$ satisfy the hypothesis of SAS as stated in Definition 16.2. Then we can tinker with distance along $\overleftrightarrow{BC}$ and angle measure for angles with vertex B and for angles with vertex C to produce a model of Σ having two triangles that satisfy the hypothesis of SAS but none of the conclusions. Yet, given our aim (Section 6.1), SAS is certainly a reasonable statement. Indeed, SAS is such an *obvious fact* that the proposition is often taken as an axiom.

We now have two desirable propositions, MIRROR and SAS, that are not theorems of our axiom system. Each has the very important property of making requirements on distance d and angle measure m together. We might say that each would make d and m *behave* and *cooperate.* If you suspect that we are in the same relative position as we were when deciding between PASCH and PSP, you are correct.

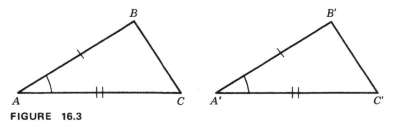

FIGURE 16.3

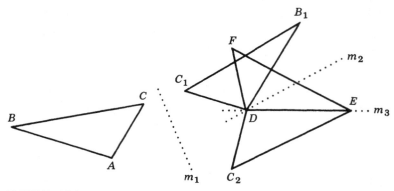

FIGURE 16.4

Under our four axioms, MIRROR and SAS are equivalent! In this chapter we shall show that MIRROR implies SAS. In the next chapter we shall take SAS as our fifth axiom. Of course, after taking SAS as an axiom, we shall not be able to use MIRROR until it is later proved as a theorem. In a more elementary approach both MIRROR and SAS could be taken as axioms.

Although the proof that MIRROR implies SAS is rather long, the idea behind the proof is simple. Consider Figure 16.4. Suppose $AB = DE$, $m\angle BAC = m\angle EDF$, and $AC = DF$. How can we *move* $\triangle ABC$ onto $\triangle DEF$ to show the remaining parts of the two triangles fit? This is accomplished by three mappings, placing one vertex at a time. First get A to D by using m_1 as a mirror. This sends B to B_1 and C to C_1. Then, using m_2 as a mirror, get B_1 to E leaving D fixed. This sends C_1 to C_2. Finally get C_2 to F by using m_3 as a mirror, leaving D and E fixed. Now $\triangle ABC$ has been moved on top of $\triangle DEF$ and all the parts fit. Easy enough! The first part of the proof shows that mirrors *move* triangles and how to find the desired mirrors. In the second part we carry out the moving just as we have described. Of course we don't actually *move* any triangle. What we do is consider a triangle and its image under a mapping. The colorful language employed in talking about mappings should not be interpreted too literally.

Theorem 16.3 If MIRROR, then SAS.

Proof For a given line m, a collineation given by MIRROR for that line will be called a *mirror map* in m. Before getting to the main part of the proof, we make three preliminary observations.

Suppose $P\text{–}Q\text{–}R$ and P, Q, R have images P', Q', R' under a mirror map. Then P', Q', R' are distinct and collinear because a mirror map is a collineation. Since a mirror map preserves distance, we have $P'R' = PR = PQ + QR = P'Q' + Q'R'$. So $P'\text{–}Q'\text{–}R'$. Hence a mirror map preserves betweenness. We observe that any product of mirror maps preserves betweenness and, hence, must preserve segments, rays, angles, and triangles as well as distance and angle measure.

Our second observation is that if m is perpendicular to $\overline{ST}$ at the midpoint of $\overline{ST}$, then T is the image of S under any mirror map in m. To prove this, suppose $\overleftrightarrow{MN}$ is perpendicular to $\overline{ST}$ at M, the midpoint of $\overline{ST}$. (See Figure 16.5.) Let S' be the image of S under the mirror map in $\overleftrightarrow{MN}$. Then S' and T are on the same side of $\overleftrightarrow{MN}$. Since the mapping fixes M and N and preserves right angles, we must have S' on $\overrightarrow{MT}$. Since $MS' = MS = MT$, we have $S' = T$, as desired.

If line m contains angle bisector $\overrightarrow{VI}$ of $\angle GVH$ and $VG = VH$, then H is the image of G under the mirror map in m. This third observation follows from the Angle-Segment-Construction Theorem as H is on the opposite side of $\overleftrightarrow{VI}$ from G and $m\angle IVH = \frac{1}{2}m\angle GVH$.

Now suppose $\triangle ABC$ and $\triangle DEF$ are any triangles such that $AB = DE$, $m\angle A = m\angle D$, and $AC = DF$. To prove our theorem we need to show $m\angle B = m\angle E$, $BC = EF$, and $m\angle C = m\angle F$.

If $A = D$, then let $\sigma_1 = \iota$, the identity map on the set of all points. If $A \neq D$, then let σ_1 be a mirror map in the unique line perpendicular to $\overline{AD}$ at the midpoint of $\overline{AD}$. By our second observation, $D = \sigma_1 A$ in in either case. Let $B_1 = \sigma_1 B$ and $C_1 = \sigma_1 C$.

Since σ_1 preserves distance, we have $DB_1 = AB = DE$. So, if B_1 is on $\overrightarrow{DE}$, then $B_1 = E$. In this case let $\sigma_2 = \iota$. Also, if B_1 is on the opposite ray of $\overrightarrow{DE}$, then D is the midpoint of $\overline{B_1E}$. In this case let σ_2 be the mirror map in the line perpendicular to $\overline{B_1E}$ at D. If $\angle B_1DE$, then let σ_2 be the mirror map in the line containing the angle bisector of $\angle B_1DE$. By our second and third observations, $D = \sigma_2 D$ and $E = \sigma_2 B_1$. Let $C_2 = \sigma_2 C_1 = \sigma_2\sigma_1 C$.

Now $D = \sigma_2 D = \sigma_2\sigma_1 A$, $E = \sigma_2 B_1 = \sigma_2\sigma_1 B$, and $C_2 = \sigma_2 C_1 = \sigma_2\sigma_1 C$. Since $\sigma_2\sigma_1$ preserves distance and angle measure, it follows that $DC_2 = AC = DF$ and $m\angle C_2DE = m\angle CAB = m\angle FDE$. So, if C_2 is on the same side of $\overleftrightarrow{DE}$ as F, then $C_2 = F$. In this case let $\sigma_3 = \iota$. Also, if C_2 is on the

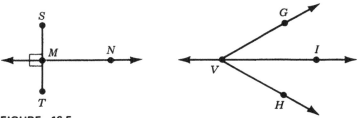

FIGURE 16.5

opposite side of $\overleftrightarrow{DE}$ as F, then $\overleftrightarrow{DE}$ contains the angle bisector of $\angle C_2DF$ or $\overleftrightarrow{DE} \perp \overleftrightarrow{C_2F}$. In this case let σ_3 be the mirror map in $\overleftrightarrow{DE}$. In either case $D = \sigma_3 D$, $E = \sigma_3 E$, and $F = \sigma_3 C_2$.

We now have $D = \sigma_3 D = \sigma_3 \sigma_2 \sigma_1 A$, $E = \sigma_3 E = \sigma_3 \sigma_2 \sigma_1 B$, and $F = \sigma_3 C_2 = \sigma_3 \sigma_2 \sigma_1 C$. Finally, since $\sigma_3 \sigma_2 \sigma_1$ preserves distance and angle measure, it follows that $m\angle DEF = m\angle ABC$, $EF = BC$, and $m\angle DFE = m\angle ACB$, as desired. ■

SAS brings to mind the so-called *congruence theorems* such as ASA, the *Angle-Side-Angle Theorem*, which you probably encountered in high school. Suppose we write down the eight possible three letter *words* using only "A" and "S." Each of these *words* suggests a proposition. We can cross SAS off our list as that has already been defined. We can cross off AAS because the AAS proposition is the same thing as the SAA proposition. Likewise SSA is a repetition of ASS. The definition of each of the five remaining *words* is left as an exercise. All that has to be done is to imitate Definition 16.2 in each case. We are not claiming that all these propositions will turn out to be theorems.

DEFINITION 16.4 ASA, SAA, ASS, SSS, AAA: (Exercise 16.1).

Our next theorem shows that ASA implies SAS. It is also true (Exercise 16.2) that SAS implies ASA. Hence ASA is equivalent to SAS and, we claim, is equivalent to MIRROR. Although this result is of some interest, for our purpose the importance of the theorem is in its proof. If we don't have mirror maps to move triangles about, how do we get at the idea of superposition? The solution is elementary. We simply *build a copy* of a given triangle where we want it. For example, in the next proof we build $\triangle ABE$ as a copy of $\triangle A'B'C'$. This method of superposition will be used frequently. Note that the statement of the next theorem is actually of the form: If (if p then q), then (if r then s). To prove such a proposition assume both (r) and (if p then q) in order to deduce (s).

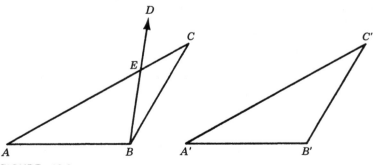

FIGURE 16.6

Theorem 16.5 If ASA, then SAS.

Proof Let $\triangle ABC$ and $\triangle A'B'C'$ be such that $AB = A'B'$, $m\angle A = m\angle A'$, and $AC = A'C'$. Assume $m\angle B \neq m\angle B'$. Without loss of generality we may suppose $m\angle B' < m\angle B$. Let D be a point on the same side of $\overleftrightarrow{AB}$ as C such that $m\angle ABD = m\angle A'B'C'$. Then D is in int $(\angle ABC)$, (by Theorem 14.6). So $\overrightarrow{BD}$ intersects int $(\overline{AC})$ by Crossbar, say at E. Hence $A - E - C$. Applying ASA to $\triangle ABE$ and $\triangle A'B'C'$, we obtain $AE = A'C'$. Since $A'C' = AC$, we now have $AE = AC$ and $A - E - C$, a contradiction. Hence $m\angle B = m\angle B'$. Then, applying ASA to $\triangle ABC$ and $\triangle A'B'C'$, we also have $BC = B'C'$ and $m\angle C = m\angle C'$. Therefore SAS, as desired. ■

We have previously remarked that our Incidence Axiom could just as well be called the Straightedge Axiom. Our third axiom, PSP, could be called the Scissors Axiom since the idea is that a line *cuts* the points off a line into two halfplanes. If we were to take MIRROR instead of its equivalent SAS as our fifth axiom, then the first five axioms would be (1) STRAIGHTEDGE, (2) RULER, (3) SCISSORS, (4) PROTRACTOR, and (5) MIRROR.

16.3 EXERCISES

In doing the exercises below it must be remembered that we now have only four axioms. We shall not assume SAS until the next chapter. Of course we cannot use MIRROR (or any *theorem* remembered from high school) until we have actually proved it.

16.1 Complete Definition 16.4.

● **16.2** SAS implies ASA.

16.3 ASS is false for every model of Σ.

16.4 Prove the following postulate of Hilbert implies SAS: Given $\triangle ABC$ and $\triangle A'B'C'$, if $\overline{AB} \simeq \overline{A'B'}$, $\angle A \simeq \angle A'$, and $\overline{AC} \simeq \overline{A'C'}$, then $\angle B \simeq \angle B'$.

● **16.5** There does not exist a mirror map for the line with equation $y = 2x$ in the Taxicab Plane (M1, t, m) where m is the Cartesian angle measure function.

● **16.6** True or False?

(a) If P and Q are distinct points, then there exists a line l perpendicular to $\overline{PQ}$ at the midpoint of $\overline{PQ}$.

(b)   If point P is off line l, then there exists a line perpendicular to l through P.

(c)   SAS is independent of our four axioms.

(d)   SAS holds for some model of Σ.

(e)   SAS holds for every model of Σ.

(f)   MIRROR implies ASA.

(g)   ASA implies SAS.

(h)   AAA.

(i)   SAS implies AAA.

(j)   (If (if p, then q), then (if r, then s)) iff (if (r) and (if p, then q), then (s)).

16.7   Do there exist any lines in the Taxicab Plane of Exercise 16.5 for which a mirror map does exist?

● **16.8**   Paper-Pencil-Pin Problem: Starting with $\triangle ABC$ on a piece of paper, reproduce Figure 16.4 using only a pencil to draw segments and a pin to locate points. No other tools allowed.

16.9   Relate the following shopping list to our study: Flag, Ruler, Knife, Protractor, Mirror.

● **16.10**   Is a collineation of a model of Σ necessarily distance preserving? A distance preserving map on the points of a model of Σ is not necessarily a collineation.

16.11   Give a model of Σ where AAA fails.

16.12   Give a model of Σ such that the sum of the measures of the three angles of some triangle is $\pi/6$ and of some other triangle is 2π.

16.13   Use Taxicab Geometry to show that each of the propositions AAA, ASA, SAA, SAS, and SSS is independent of our first four axioms.

***16.14**   Does SAA imply SAS?
Does SSS imply SAS?
Does AAA imply SAS?

***16.15**   Can "angle measure" be replaced by "perpendicularity" in Definition 16.1 and still have Theorem 16.3?

GRAFFITI

Bolyai, when in garrison with cavalry officers, was provoked by thirteen of them and accepted all their challenges on condition that he be permitted after each duel to play a bit on his violin. He came out victor from his thirteen duels, leaving his thirteen adversaries on the square.

Halsted

In most sciences one generation tears down what another has built and what one has established another undoes. In Mathematics alone each generation builds a new story to the old structure.

Hankel

The advancement and perfection of mathematics are intimately connected with the prosperity of the State.

Napoleon

To Thales . . . the primary question was not What do we know, *but* How do we know it.

Aristotle

"Then you should say what you mean," the March Hare went on.
"I do," Alice hastily replied; "at least — at least I mean what I say — that's the same thing, you know."
"Not the same thing a bit!" said the Hatter. "Why you might as well say that 'I see what I eat' is the same thing as 'I eat what I see'!"
"You might just as well say," added the March Hare, "that 'I like what I get' is the same thing as 'I get what I like'!"

Lewis Carroll

CHAPTER 17

Congruence and the Penultimate Postulate

17.1 CONGRUENCE FOR TRIANGLES

It is time to launch into a discussion of *congruence* for triangles. This turns out to be not as trivial as one might suppose. The standard conventions are somewhat misleading. There are many comments to make about the next two definitions.

DEFINITION 17.1 Suppose $\triangle ABC$ and $\triangle DEF$ exist. Then $\triangle ABC \simeq \triangle DEF$ if all six of the following relations hold:

$$\overline{AB} \simeq \overline{DE}, \qquad \overline{BC} \simeq \overline{EF}, \qquad \overline{AC} \simeq \overline{DF},$$
$$\angle A \simeq \angle D, \qquad \angle B \simeq \angle E, \qquad \angle C \simeq \angle F.$$

DEFINITION 17.2 We say $\triangle ABC$ and $\triangle DEF$ are **congruent** if at least one of the following six statements holds:

$$\triangle ABC \simeq \triangle DEF, \qquad \triangle BCA \simeq \triangle DEF, \qquad \triangle CAB \simeq \triangle DEF,$$
$$\triangle CBA \simeq \triangle DEF, \qquad \triangle ACB \simeq \triangle DEF, \qquad \triangle BAC \simeq \triangle DEF.$$

Notice that the two definitions are indeed different. Referring to Figure 17.1, where the represented sides and angles that look congruent are assumed to be congruent, "$\triangle ABC \simeq \triangle DEF$" is a false

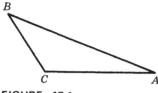 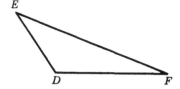

FIGURE 17.1

statement because $\overline{AB}$ is not congruent to $\overline{DE}$. However $\triangle ABC$ and $\triangle DEF$ are congruent because $\triangle ABC \simeq \triangle FED$.

You may have to write out some of the six statements in Definition 17.2 (using Definition 17.1) to convince yourself that these six statements are different. Doing this will give you a better idea of what each definition says and what it does not say. You should also convince yourself that it is these six statements in Definition 17.2 that we *want*, as the triangles represented in Figure 17.1 are still *congruent* no matter how the letters "A," "B," and "C" are permuted in the figure.

You may have noticed that *congruence* (Definition 17.2) is defined in terms of congruence. This is not a cyclic definition, however, as it is the previously defined congruence of segments and the previously defined congruence of angles that are used to define a new kind of congruence. This could even be avoided by writing "$AB = DE$" for "$\overline{AB} \simeq \overline{DE}$," "$m\angle A = m\angle D$" for "$\angle A \simeq \angle D$," etc. Although using the word *congruent* in several contexts doesn't help an already confusing situation, it really shouldn't be a bother.

There is no short notation to say that $\triangle ABC$ and $\triangle DEF$ are congruent; as in Definition 17.2 we must write out the words. There is little need for any symbolism because the congruence of Definition 17.2 is of such a general nature that it is hardly ever used. The idea of Definition 17.2 might be ignored in elementary texts. What is used— and used often—is Definition 17.1.

In Definition 17.1 you are not told how to read the statement "$\triangle ABC \simeq \triangle DEF$." This causes no problem at all in the written language. What happens when one wants to use the spoken language? We read "$\overline{AB} \simeq \overline{DE}$" aloud as "Segment A B is congruent to segment D E." If you have not been told otherwise, chances are that you are reading the statement in question aloud as "Triangle A B C is congruent to triangle D E F." Indeed everyone does just that, including the author. Then one has to agree that the two statements "$\triangle ABC$ is congruent to $\triangle DEF$" and "$\triangle ABC$ and $\triangle DEF$ are congruent" mean different things. There is nothing wrong with that. (Compare "Joe is opposed to Jim" with "Joe and Jim are opposed.") Since we shall have a generalized definition of *congruent* which applies to arbitrary sets of points and encompasses our present three uses (Definitions 8.5, 14.1, and 17.2), we shall not *formally* attach the word *congruent* to Definition 17.1.

The problem with "$\triangle ABC \simeq \triangle DEF$" is not how to read it but that it can be misleading in other ways. From $\triangle CAB \simeq \triangle DEF$ and the obvious fact that $\triangle ABC = \triangle CAB$, we cannot conclude by simple substitution that $\triangle ABC \simeq \triangle DEF$. (This substitution would be somewhat analogous to concluding that $72 + 63 = 672 + 111$ from $783 = 672 + 111$ and $2 + 6 = 8$.) Hence "$\triangle ABC \simeq \triangle DEF$" should be viewed as one symbol. Although Definition 17.1 looks like it defines an equivalence relation on the set of triangles, the fact is that the definition does not even define a relation on the set of triangles! The definition not only involves triangles but also involves some ordering of the vertices of these triangles.

Perhaps in a more advanced text an author would introduce *ordered triangles*, say denoted by $\overrightarrow{\triangle} ABC$, which could be defined as an ordered 4-tuple $(\triangle ABC, A, B, C)$. Then $\overrightarrow{\triangle} ABC \neq \overrightarrow{\triangle} CBA$ in general. The six relations in Definition 17.1 would then be used to define $\overrightarrow{\triangle} ABC \simeq \overrightarrow{\triangle} DEF$, giving an equivalence relation on the set of all ordered triangles. This, after all, is really the whole idea behind Definition 17.1. The symbol "$\triangle ABC \simeq \triangle DEF$" would then be free for use in Definition 17.2 if desired. However, this author has chosen to stick with the standard notation and conventions.

Theorem 17.3 The following are equivalent: $\triangle ABC \simeq \triangle DEF$, $\triangle BCA \simeq \triangle EFD$, $\triangle CBA \simeq \triangle FED$, and $\triangle DEF \simeq \triangle ABC$.

Proof Follows directly from Definition 17.1. ■

The previous discussion shows that Theorem 17.3 is not completely trivial. The next theorem has the form it has because $\simeq$ is *not* a relation on the set of all triangles.

Theorem 17.4 Congruence of triangles is an equivalence relation on the set of triangles. Further,

(a) $\triangle ABC \simeq \triangle ABC$;

(b) if $\triangle ABC \simeq \triangle DEF$, then $\triangle DEF \simeq \triangle ABC$;

(c) if $\triangle ABC \simeq \triangle DEF$ and $\triangle DEF \simeq \triangle GHI$, then $\triangle ABC \simeq \triangle GHI$.

Proof The second part follows from Definition 17.1 and the fact that congruence of segments and congruence of angles are each equivalence relations. The first part then follows from the second part by Definition 17.2. ∎

17.2 AXIOM 5: SAS

Recall Taxicab Geometry (M1, t) from Section 7.2. From the coincidence of betweenness for points in Taxicab Geometry with betweenness for points in the Cartesian plane follows the coincidence of segments, halfplanes, rays, angles, and triangles in the two planes. Therefore, where m agrees with the angle measure function of the Cartesian plane, (M1, t, m) is a model of Σ. This model satisfies the triangle inequality, which is not yet one of our theorems. Whatever its interest or use for practical applications, this geometry should be excluded by an axiom system developed with our aim (see Section 6.1). In (M1, t, m) there are equilateral right triangles and right triangles where the length of the hypotenuse equals the sum of the lengths of the other two sides. In Figure 17.2, $\triangle ABC$ and $\triangle DEF$ are two triangles in (M1, t, m) such that $\overline{AB} \simeq \overline{DE}$, $\angle A \simeq \angle D$, and $\overline{AC} \simeq \overline{DF}$, but $\triangle ABC \neq \triangle DEF$ because $\overline{BC} \neq \overline{EF}$. Hence (M1, t, m) does not satisfy Euclid's Proposition I.4, the Side-Angle-Side Theorem. If two sides and the included angle of one triangle are respectively congruent to two sides and the included angle of another triangle, then certainly the two triangles are congruent. (See Figure 17.3.) This proposition makes strong requirements on both the distance function and the angle measure function together. We shall use Euclid's I.4 to formulate our next to the last axiom, Axiom 5.

> **Axiom 5** SAS Given $\triangle ABC$ and $\triangle DEF$,
> if $\overline{AB} \simeq \overline{DE}$, $\angle A \simeq \angle D$, and $\overline{AC} \simeq \overline{DF}$,
> then $\triangle BAC \cong \triangle EDF$.

With the new axiom at hand, it is only fitting that the next theo-

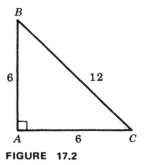
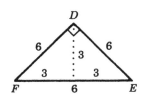

FIGURE 17.2

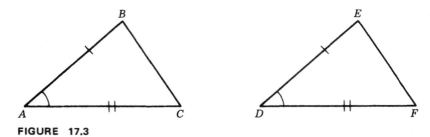

FIGURE 17.3

rem be Euclid's Proposition I.5, the famous Pons Asinorum or Bridge of Asses.

Theorem 17.5 *Pons Asinorum* Given $\triangle ABC$, if $\overline{AB} \simeq \overline{AC}$, then $\angle B \simeq \angle C$.

Proof *(Pappus)* Since $\overline{AB} \simeq \overline{AC}$, $\angle A \simeq \angle A$, and $\overline{AC} \simeq \overline{AB}$, it follows that $\triangle BAC \simeq \triangle CAB$ by SAS. Hence $\angle B \simeq \angle C$. ∎

According to Proclus this beautiful proof is due to Pappus. The proof is based on the fact that nowhere in the statement of SAS in Definition 17.1 is it required that the triangles be distinct. Those readers who dropped out before the Pons Asinorum have been labeled. To those still with us, *Congratulations!*

A hundred years ago Pappus' proof of the Pons Asinorum might have been stated in terms of picking up the triangle, turning it over, and replacing it on top of itself. For the wrong reasons, such a proof would not have been accepted at Oxford in the classes taught by the minor mathematician Charles Lutwidge Dodgson (1832–1897). Dodgson, better known to the world as Lewis Carroll, author of *Alice in Wonderland,* would have pointed out that since there is only one triangle in the first place, when you pick it up it is no longer there to be put back on itself. Remember that triangles were then considered to be actual physical entities. Also, given two points A and B, one had to *draw* $\overline{AB}$ or otherwise $\overline{AB}$ somehow didn't seem to exist. Dodgson's *Euclid and his Modern Rivals* (1879), which has been reissued recently by Dover, is a long defense for using only Euclid's *Elements* as a beginning textbook and contains a sophisticated discussion of the parallel postulate. In particular Dodgson exposes those geometers who use the word *direction* without definition or axiom. He also wrote *A New Theory of Parallels* (1888), which proposes a substitute for the parallel postulate of Euclid. In this work Dodgson makes tacit assumptions about *area*, without defining the word.

Theorem 17.6 Given $\triangle ABC$ and $A-C-D$, then $m\angle BCD > m\angle B$.

Proof Let E be the midpoint of $\overline{BC}$. By the Midpoint Theorem (Theorem 8.15) there exists a unique point F such that $A-E-F$ and $AE = EF$. So $\overline{AE} \simeq \overline{FE}$, $\angle AEB \simeq \angle FEC$ since vertical angles are congruent, and $\overline{EB} \simeq \overline{EC}$. Thus $\triangle AEB \simeq \triangle FEC$ by SAS. So $\angle ABE \simeq \angle FCE$. Therefore $m\angle B = m\angle BCF$.

Since $A-C-D$, points A and D are on opposite sides of $\overleftrightarrow{BC}$. Since $A-E-F$ and $B-E-C$, points A and F are on opposite sides of $\overleftrightarrow{BC}$. So D and F are on the same side of $\overleftrightarrow{BC}$. Also B, E, and F are on the same side of $\overleftrightarrow{CD}$. Hence F is in the interior of $\angle BCD$. Then, by the Protractor Postulate, $m\angle BCD > m\angle BCF$. This last result, together with $m\angle B = m\angle BCF$, gives $m\angle BCD > m\angle B$. ∎

DEFINITION 17.7 If $\triangle ABC$ and $A-C-D$, then $\angle BCD$ is an **exterior angle** of $\triangle ABC$ with **remote interior angles** $\angle A$ and $\angle B$. If $\triangle ABC$, then $\angle A$ is the angle **opposite** $\overline{BC}$ and $\overline{BC}$ is the side **opposite** $\angle A$. If two sides of a triangle are congruent, then the triangle is **isosceles.** The angles opposite the congruent sides of an isosceles triangle are **base angles.** A triangle that is not isosceles is **scalene.** If all three sides of a triangle are congruent, then the triangle is **equilateral;** if all three angles of a triangle are congruent, then the triangle is **equiangular.**

There should be no surprises in the definition above. However, note that in Figure 17.4, if $AB \neq BC$ but $AB = AC$, then $\angle B$ is a base angle of $\triangle ABC$ but $\angle A$ is not a base angle of $\triangle ABC$. The position of the figure is immaterial. We can restate the Pons Asinorum: *The base angles of an isosceles triangle are congruent.* The next definition introduces two relations on the set of angles by using angle measure in the obvious way. (In writing be careful to distinguish between "$\angle$" and "$<$.")

DEFINITION 17.8 We say $\angle ABC$ is **larger** than $\angle DEF$ and write $\angle ABC > \angle DEF$ if $m\angle ABC > m\angle DEF$; we say $\angle DEF$ is **smaller**

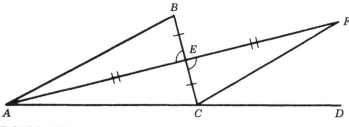

FIGURE 17.4

than $\angle ABC$ and write $\angle DEF < \angle ABC$ if $m\angle DEF < m\angle ABC$.

Theorem 17.9 *Euclid's Proposition I.16* An exterior angle of a triangle is larger than either of its remote interior angles.

Proof Suppose $\triangle ABC$ and $A-C-D$. We must prove $\angle BCD$ is larger than $\angle A$ and larger than $\angle B$. Let F be such that $B-C-F$. Then, by the previous theorem, we immediately have $\angle BCD > \angle B$ and also $\angle ACF > \angle A$. Since $\angle ACF$ and $\angle BCD$ are vertical angles, it follows that $\angle BCD > \angle A$. ∎

The last theorem is the first of the three theorems on exterior angles which we shall eventually obtain. The importance of this key theorem cannot be overemphasized.

Theorem 17.10 The base angles of an isosceles triangle are acute.

Proof Suppose $\triangle ABC$ with $AB = BC$ and $A-C-D$. By the Pons Asinorum $m\angle A = m\angle C$. By the previous theorem $m\angle BCD > m\angle A$. So $m\angle BCD > m\angle C$. But $\angle BCD$ and $\angle C$ are a linear pair. Hence $\angle BCD$ is obtuse and $\angle C$ is acute. Therefore $\angle A$ and $\angle C$ are acute. ∎

17.3 CONGRUENCE THEOREMS

The SAS axiom suggests other possible propositions which may or may not be theorems of our axiom system. Anyone who has passed over the Pons Asinorum can figure out the derivation of the names ASA, SAA, ASS, SSS, and AAA and what each stands for. Each of the names is read by pronouncing its three letters in order. Axiom 5 can be restated: If $\overline{AC} \simeq \overline{DF}$, $\angle ACB \simeq \angle DFE$, and $\overline{CB} \simeq \overline{FE}$, then $\triangle ACB \simeq \triangle DFE$. The best restatement of SAS: Given $\triangle ABC$ and $\triangle DEF$, if $\overline{AB} \simeq \overline{DE}$, $\angle B \simeq \angle E$, and $\overline{BC} \simeq \overline{EF}$, then $\triangle ABC \simeq \triangle DEF$. These restatements follow a certain convention. For example, "$\triangle BAC \simeq \triangle EDF$ by SAS" indicates that the congruence follows from $\overline{BA} \simeq \overline{ED}$, $\angle A \simeq \angle D$, and $\overline{AC} \simeq \overline{DF}$ and not from some other combination of side-angle-side. Likewise, "$\triangle CAB \simeq \triangle FED$ by ASA" will tell us that the congruence follows from $\angle C \simeq \angle F$, $\overline{CA} \simeq \overline{FE}$, and $\angle A \simeq \angle E$. In this way the notation is extremely effective in giving maximum information. To test your understanding of this convention see if you can write down the hypotheses of ASA, SAA, and ASS when the conclusion is $\triangle ABC \simeq \triangle DEF$. Then check with the following definition.

DEFINITION 17.11 ASA: Given $\triangle ABC$ and $\triangle DEF$, if $\angle A \simeq \angle D$, $\overline{AB} \simeq \overline{DE}$, and $\angle B \simeq \angle E$, then $\triangle ABC \simeq \triangle DEF$.

SAA: Given $\triangle ABC$ and $\triangle DEF$, if $\overline{AB} \simeq \overline{DE}$, $\angle B \simeq \angle E$, and $\angle C \simeq \angle F$, then $\triangle ABC \simeq \triangle DEF$.

ASS: Given $\triangle ABC$ and $\triangle DEF$, if $\angle A \simeq \angle D$, $\overline{AB} \simeq \overline{DE}$, and $\overline{BC} \simeq \overline{EF}$, then $\triangle ABC \simeq \triangle DEF$.

SSS: Given $\triangle ABC$ and $\triangle DEF$, if $\overline{AB} \simeq \overline{DE}$, $\overline{BC} \simeq \overline{EF}$, and $\overline{CA} \simeq \overline{FD}$, then $\triangle ABC \simeq \triangle DEF$.

AAA: Given $\triangle ABC$ and $\triangle DEF$, if $\angle A \simeq \angle D$, $\angle B \simeq \angle E$, and $\angle C \simeq \angle F$, then $\triangle ABC \simeq \triangle DEF$.

Theorem 17.12 ASA.

Proof Suppose $\triangle ABC$ and $\triangle DEF$ are such that $\angle A \simeq \angle D$, $\overline{AB} \simeq \overline{DE}$, and $\angle B \simeq \angle E$. If $BC \neq EF$, then we may assume without loss of generality that $BC > EF$. Then there exists C' such that $B-C'-C$ and $BC' = EF$. So $\triangle ABC' \simeq \triangle DEF$ by SAS. Hence $m\angle BAC' = m\angle EDF$. But, since C' is in int $(\angle BAC)$, $m\angle BAC' < m\angle BAC$. Then $m\angle D < m\angle A$, contradicting the hypothesis $\angle A \simeq \angle D$. So $BC = EF$. Therefore $\overline{BC} \simeq \overline{EF}$, and $\triangle ABC \simeq \triangle DEF$ by SAS. ■

Theorem 17.13 SAA.

Proof Suppose $\triangle ABC$ and $\triangle DEF$ are such that $\overline{AB} \simeq \overline{DE}$, $\angle B \simeq \angle E$, and $\angle C \simeq \angle F$. If $BC \neq EF$, then we may assume without loss of generality that $BC > EF$. Then there exists C' such that $B-C'-C$ and $BC' = EF$. So $\triangle ABC' \simeq \triangle DEF$ by SAS. Hence $m\angle AC'B = m\angle DFE$. Since $m\angle DFE = m\angle ACB$ by hypothesis, we have $m\angle AC'B = m\angle ACB$. However, looking at $\triangle ACC'$, we see that $\angle AC'B$ is an exterior angle with remote interior angle $\angle ACC'$. Hence $m\angle AC'B > m\angle ACC'$. Then $m\angle ACB > m\angle ACC'$, a contradiction. So $BC = EF$. Therefore $\overline{BC} \simeq \overline{EF}$, and $\triangle ABC \simeq \triangle DEF$ by SAS. ■

Euclid combined ASA and SAA into Proposition I.26. Our proofs follow those of Euclid. The idea is to make $\triangle ABC'$ a *copy* of $\triangle DEF$ so

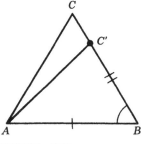

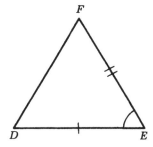

FIGURE 17.5

that the copy can be compared with $\triangle ABC$. Euclid's Proposition I.8 is SSS. In proving SSS, we shall not follow Euclid but rather follow a proof which, according to Proclus, is due to Philo. Again the idea is to make $\triangle ABC'$ a copy of $\triangle DEF$. (In Figure 17.6 $\triangle DEF$ is not shown.) Showing that $\triangle ABC'$ is also a copy of $\triangle ABC$ implies that $\triangle ABC$ and $\triangle DEF$ are copies of each other.

Theorem 17.14 SSS.

Proof *(Philo)* Suppose $\triangle ABC$ and $\triangle DEF$ are such that $\overline{AB} \simeq \overline{DE}$, $\overline{BC} \simeq \overline{EF}$, and $\overline{CA} \simeq \overline{FD}$. By the Angle-Segment-Construction Theorem, let C' be the unique point such that C and C' are on opposite sides of $\overleftrightarrow{AB}$, $\angle ABC' \simeq \angle DEF$, and $\overline{BC'} \simeq \overline{EF}$. Then $\triangle ABC' \simeq \triangle DEF$ by SAS. Suppose we can show $\angle ACB \simeq \angle AC'B$. Then $\triangle ACB \simeq \triangle AC'B$ by SAS. So $\triangle ABC \simeq \triangle ABC'$ (Theorem 17.3). Then, since $\triangle ABC' \simeq \triangle DEF$, we would have $\triangle ABC \simeq \triangle DEF$ (Theorem 17.4) as desired. Therefore to complete the proof we need only show that $\angle ACB \simeq \angle AC'B$.

Since C and C' are on opposite sides of $\overleftrightarrow{AB}$, let point H be the intersection of $\overline{CC'}$ with $\overleftrightarrow{AB}$. So $C-H-C'$ with H on $\overleftrightarrow{AB}$. The remainder of the proof depends on the order of $A, B,$ and H. If $H = A$, then $\angle ACB \simeq \angle AC'B$ follows from application of the Pons Asinorum to $\triangle CBC'$.

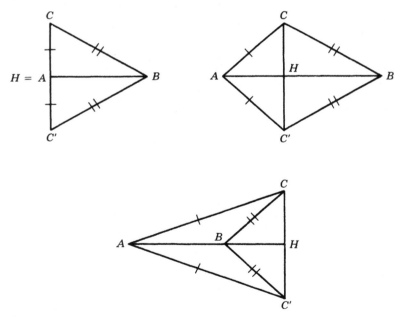

FIGURE 17.6

If $H = B$, then $\angle ACB \simeq \angle AC'B$ follows from application of the Pons Asinorum to $\triangle CAC'$. Now suppose H is distinct from A and from B. Applications of the Pons Asinorum to $\triangle CAC'$ and $\triangle CBC'$ give $\angle HCA \simeq \angle HC'A$ and $\angle HCB \simeq \angle HC'B$. If $A-H-B$, then $\angle ACB \simeq \angle AC'B$ by the Angle-Addition Theorem (Theorem 14.5) as H is in int $(\angle ACB)$ and int $(\angle AC'B)$. There remain only the cases $H-A-B$ and $A-B-H$ to consider. However, as in the case $A-H-B$, again $\angle ACB \simeq \angle AC'B$ follows from the Angle-Addition Theorem (Exercise 17.1). ∎

Three of the five propositions of Definition 17.11 have been proved. Neither ASS nor AAA can be a theorem at this time since each fails for the Euclidean plane. However, it should not be assumed that each is false for every model of Σ.

17.4 EXERCISES

17.1 Give the details proving the last sentence in the proof of Theorem 17.14.

17.2 Theorem 17.3 implies there are twelve ways to express the fact that $\triangle ABC$ is congruent to $\triangle DEF$. List them.

● **17.3** An equilateral triangle is equiangular.

● **17.4** State and prove the converse of the Pons Asinorum.

17.5 Prove Euclid's Proposition I.17: Given $\triangle ABC$, $m\angle A + m\angle B < \pi$.

● **17.6** True or False?

   (a) The first four axioms and the triangle inequality imply SAS.

   (b) SAS is independent of our first four axioms.

   (c) $\overline{AB} \simeq \overline{DE}$ iff $\overline{BA} \simeq \overline{DE}$.

   (d) $\triangle ABC \simeq \triangle DEF$ iff $\triangle BAC \simeq \triangle DEF$.

   (e) $\triangle ABC \simeq \triangle BCA$.

   (f) The converse of SAS is ASA.

   (g) An equilateral triangle is an isosceles triangle.

   (h) $\angle A < \angle B$ iff $\angle B > \angle A$.

   (i) If neither $\angle A > \angle B$ nor $\angle B > \angle A$, then $\angle A = \angle B$.

   (j) "Penultimate" means "next to the last."

• **17.7** Does the strict triangle inequality (if $\triangle ABC$, then $AB + BC > AC$) imply SAS under our first four axioms?

17.8 Prove Euclid's Proposition I.7 using SSS.

• **17.9** If P is a point off line l, then there is a unique line through P that is perpendicular to l.

17.10 Given $\triangle ABC$, $m\angle A + m\angle B + m\angle C < 3\pi/2$.

17.11 Assume the statement of SAS requires the triangles be distinct. Then prove SAS as we have it.

17.12 Read "The Mathematical Manuscripts of Lewis Carroll" by W. Weaver in the *Proceedings of the American Philosophical Society* Vol. 98 (1954), pp. 377 – 381.

• **17.13** Give a geometric interpretation of the model (M1, p, m) where m is Cartesian angle measure function and p is the distance function defined by having the distance from (x_1, y_1) to (x_2, y_2) be $\sqrt{(x_2 - x_1)^2 + 5(y_2 - y_1)^2}$. Use this model to prove the independence of SAS.

***17.14** Use (M13, h, m), where m is the Cartesian angle measure function restricted to (M13, h) of Section 9.2, to prove the independence of SAS.

***17.15** Under our first four axioms is SAS equivalent to the following property? If $\angle ABC$ and $\angle DEF$ are right angles, $\overline{AB} \simeq \overline{DE}$, and $\overline{BC} \simeq \overline{EF}$, then $\triangle ABC \simeq \triangle DEF$.

GRAFFITI

Newton had so remarkable a talent for mathematics that Euclid's Geometry seemed to him "a trifling book," and he wondered that any man should have taken the trouble to demonstrate propositions, the truth of which was so obvious to him at the first glance. But, on attempting to read the more abstruse geometry of Descartes, without having mastered the elements of the science, he was baffled, and was glad to come back again to his Euclid.

James Parton

It is the glory of geometry that from so few principles, fetched from without, it is able to accomplish so much.

Newton

God ever geometrizes.

Plato

The proof of self-evident propositions may seem to the uninitiated, a somewhat frivolous occupation. To this we might reply that it is often by no means self-evident that one obvious proposition follows from another obvious proposition; so that we are really discovering new truths when we prove what is evident by a method which is not evident. But a more interesting retort is, that since people have tried to prove obvious propositions, they have found that many of them are false. Self-evidence is often a mere will-o'-the-wisp, which is sure to lead us astray if we take it as our guide.

Russell

We already know . . . that the behavior of measuring-rods and clocks is influenced by . . . the distribution of matter. This in itself is sufficient to exclude the exact validity of Euclidean geometry in our universe.

Einstein

CHAPTER 18

Perpendiculars and Inequalities

18.1 A THEOREM ON PARALLELS

There are many theorems to prove, here and in the next several chapters, in order to fully understand the implications of adding the powerful SAS to our axiom system. We begin with a theorem motivated by Euclid's Proposition I.11 and Proposition I.12.

Theorem 18.1   Given point P and line l, there is a unique line through P that is perpendicular to l.

Proof   We know the theorem holds when P is on l (Theorem 14.18). Suppose P is off l and $l = \overleftrightarrow{AB}$. By the Angle-Segment-Construction

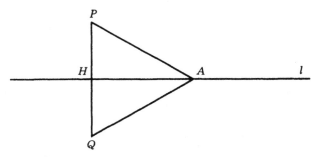

FIGURE 18.1

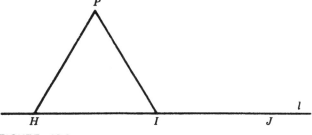

FIGURE 18.2

Theorem there is a unique point Q such that P and Q are on opposite sides of l, $\angle BAP \cong \angle BAQ$, and $\overline{AP} \cong \overline{AQ}$. If $B-A-C$, then $\angle CAP \cong \angle CAQ$. Let $\overline{PQ}$ intersect l at point H. If $H = A$, then $\angle BHP$ and $\angle BHQ$ are a linear pair of congruent angles and $\overline{PH} \perp l$ (Theorem 14.13). If $H \neq A$, then $\triangle HAP \cong \triangle HAQ$ by SAS. Hence $\angle PHA$ is congruent to $\angle QHA$. Since these two angles are a linear pair, each must be a right angle. Thus $\overline{PH} \perp l$. In any case, we have $\overleftrightarrow{PH}$ is perpendicular to l, establishing the existence of the desired line. For the uniqueness assume $\overleftrightarrow{PI}$ is also perpendicular to l with I distinct from H and on l. Let $H-I-J$. Then $\triangle HIP$ has exterior angle $\angle PIJ$ congruent to its remote interior angle $\angle PHI$, a contradiction (Theorem 17.9). Therefore, $\overleftrightarrow{PH}$ is the unique line through P that is perpendicular to l. ∎

If you have the feeling that you have seen the existence part of the last proof before, you are correct. It is almost a repetition of Philo's proof of SSS.

Theorem 18.2 If two lines are perpendicular to the same line, then the two lines are parallel.

Proof Exercise 18.1. ∎

Theorem 18.3 If P is a point off line l, then there is a line through P that is parallel to l.

Proof Let h be the line through P that is perpendicular to l, and let m be the line through P that is perpendicular to h (Theorem 18.1). Then m and l are parallel (Theorem 18.2) with m through P. ∎

There is an inevitable sequence from Philo's proof of SSS to Theorem 18.3. *This last theorem should be a surprise!* Following our aim to develop that geometry that is very much like Euclidean geometry but avoiding a parallel postulate, we still have obtained a theorem that excludes the possibility of no parallel lines (Axiom 2' of Section 4.2). If we are eventually to have a geometry that is different from the Euclidean plane we had better analyze our present position very carefully. It is by no means obvious that we have not gone too far already! It seems quite possible that the parallel line in Theorem 18.3 is always unique and that our five axioms form a categorical system. The only reason you have to expect that this is not the case is that the author has promised you non-Euclidean geometry.

We shall now give the obvious definition of a *perpendicular bisector* and then prove a fundamental theorem. In this theorem, as in every other theorem where it occurs the word *locus* may be replaced by the word *set*.

DEFINITION 18.4 If P and Q are distinct points, then the **perpendicular bisector** of $\overline{PQ}$ or the **perpendicular bisector** of P and Q is the line perpendicular to $\overline{PQ}$ that passes through the midpoint of $\overline{PQ}$. (Existence and uniqueness are consequences of the Protractor Postulate and the Four-Angle Theorem.)

Theorem 18.5 The locus of all points equidistant from distinct points P and Q is the perpendicular bisector of $\overline{PQ}$.

Proof As for every locus theorem there are two things to prove. Here we must show (1) every point equidistant from P and Q is on the perpendicular bisector of $\overline{PQ}$ and, conversely, (2) every point on the perpendicular bisector of $\overline{PQ}$ is equidistant from P and Q. Let M be the midpoint of $\overline{PQ}$. We first show (1). The only point on $\overleftrightarrow{PQ}$ that is equidistant from P and Q is M (Theorem 8.17), which is certainly on the perpendicular bisector of $\overline{PQ}$. Suppose point A is off $\overleftrightarrow{PQ}$ and is equidistant from P and Q. Then $\triangle AMP \simeq \triangle AMQ$ by SSS. So $\angle AMP \simeq \angle AMQ$. Hence $\angle AMP$ and $\angle AMQ$ are right angles since they are a linear pair of congruent angles. Therefore A is on the perpendicular bisector of $\overline{PQ}$. Now we need to prove (2). The only point common to the perpendicular bisector and $\overleftrightarrow{PQ}$ is M which is certainly equidistant from P and Q. Suppose B is off $\overleftrightarrow{PQ}$ and is on the perpendicular bisector of $\overline{PQ}$. Then $\angle BMP \simeq \angle BMQ$ as each of the angles is a right angle. So $\triangle BMP \simeq \triangle BMQ$ by SAS. Therefore $BP = BQ$, as desired. ∎

Theorem 18.6 If point P is off line l, then there exists a unique point Q such that l is the perpendicular bisector of $\overline{PQ}$.

Proof Any such point Q must be on a line through P and perpendicular to l. There is a unique line m satisfying these two conditions

(Theorem 18.1). Let l and m intersect at M. There is a unique point Q on m such that M is the midpoint of P and Q by the Midpoint Theorem. ■

A way to remember the distinction between the Pons Asinorum and its converse is to remember that Pappus' proof of the Pons Asinorum uses the *axiom* SAS. The next theorem is the converse of the Pons Asinorum, Euclid's Proposition I.6, which is proved by imitating Pappus' proof of the Pons Asinorum but using the *theorem* ASA.

Theorem 18.7 Given $\triangle ABC$, if $\angle B \simeq \angle C$, then $\overline{AB} \simeq \overline{AC}$.

Proof $\triangle BCA \simeq \triangle CBA$ by ASA. So $\overline{AB} \simeq \overline{AC}$. ■

Theorem 18.8 A triangle is equilateral iff the triangle is equiangular.

Proof That an equilateral triangle is equiangular follows directly from the Pons Asinorum. That an equiangular triangle is equilateral follows directly from the converse of the Pons Asinorum. ■

18.2 INEQUALITIES

We already have an important inequality for triangles, namely Theorem 17.9 which states that an exterior angle of a triangle is larger than either of its remote interior angles. After a definition analogous to Definition 17.8, we shall prove Euclid's Proposition I.18 and Proposition I.19. For the proofs of these two theorems we follow Euclid.

DEFINITION 18.9 We say $\overline{AB}$ is *longer* than $\overline{CD}$ and write $\overline{AB} > \overline{CD}$ if $AB > CD$; we say $\overline{AB}$ is *shorter* than $\overline{CD}$ and write $\overline{AB} < \overline{CD}$ if $AB < CD$.

Theorem 18.10 If two sides of a triangle are not congruent, then the angle opposite the longer side is larger than the angle opposite the shorter side.

Proof Given $\triangle ABC$ with $\overline{AC} > \overline{AB}$, we wish to show $\angle B > \angle C$. There exists a point D such that $A - D - C$ and $\overline{AD} \simeq \overline{AB}$ by the Segment-Construction Theorem. Then $\angle BDA > \angle C$ because $\angle BDA$ is an exterior angle of $\triangle BDC$ with remote interior angle $\angle C$. Also $\angle DBA \simeq \angle BDA$ by application of the Pons Asinorum to $\triangle BAD$. So $\angle DBA > \angle C$. Since D is in int $(\angle B)$, $\angle B > \angle DBA$. Hence $\angle B > \angle C$, as desired. ■

Theorem 18.11 If two angles of a triangle are not congruent, then the side opposite the larger angle is longer than the side opposite the smaller angle.

Proof Suppose $\triangle ABC$ with $\angle B > \angle C$. By the Pons Asinorum, if $\overline{AC} \simeq \overline{AB}$, then $\angle B \simeq \angle C$, a contradiction. By the previous theorem, if $\overline{AC} < \overline{AB}$, then $\angle B < \angle C$, a contradiction. The only possibility remaining is $\overline{AC} > \overline{AB}$, as desired. ∎

Theorem 18.12 In any triangle the sum of the lengths of any two sides is greater than the length of the third side.

Proof Given $\triangle ABC$ we wish to show $AB + BC > AC$. Let D be the point such that $A-B-D$ and $BD = BC$. Then $AD = AB + BD = AB + BC$. Also, since B is in int $(\angle ACD)$, $m\angle ACD > m\angle BCD$. Since $m\angle BCD = m\angle BDC$ by application of the Pons Asinorum to $\triangle BCD$, then $m\angle ACD > m\angle BDC$. Hence $AD > AC$ by application of the previous theorem to $\triangle ADC$. Therefore $AB + BC > AC$, as desired. ∎

Theorem 18.12, which is Euclid's Proposition I.20, is sometimes called the *triangle inequality*. We have been calling the first statement in the next theorem the *triangle inequality*. This discrepancy in conventions of nomenclature makes little difference now as they are equivalent. Our next theorem tells the full story.

Theorem 18.13 *Triangle Inequality* $AB + BC \geqq AC$ for any points A, B, C. Also, $AB + BC = AC$ iff point B is on $\overline{AC}$ or $A = B = C$. Further, $\triangle ABC$ iff $AB + BC > AC$, $AC + CB > AB$, and $BA + AC > BC$.

Proof If $B = A$ or $B = C$, then $AB + BC = AC$ trivially. If $A = C$ but

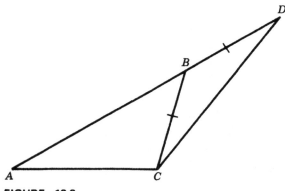

FIGURE 18.3

$A \neq B$, then $AB + BC > AC$ trivially. Now suppose A, B, C are distinct points. If $\triangle ABC$, then $AB + BC > AC$, $AC + CB > AB$, and $BA + AC > BC$ by the previous theorem. If $A - B - C$, then $AB + BC = AC$ by definition. If B is off $\overline{AC}$, then $\triangle ABC$, $B - A - C$, or $A - C - B$. For each of these cases, we have $AB + BC > AC$. Therefore $AB + BC \geq AC$ for any points A, B, C. Also, if A, B, C are distinct, then $AB + BC = AC$ iff $A - B - C$. Therefore, $AB + BC = AC$ iff B is on $\overline{AC}$ or $A = B = C$. If $AB + BC > AC$, $AC + CB > AB$, and $BA + AC > BC$, then A, B, C are distinct points such that none of $A - B - C$, $A - C - B$, or $B - A - C$ holds. Thus, in this case, we have $\triangle ABC$. ∎

Corollary 18.14 If $AR = BR = \frac{1}{2}AB$, then R is the midpoint of A and B. Further, if A and B are distinct points, then

$$\overline{AB} = \{P \mid AP + PB = AB\} \qquad \text{and} \qquad \overrightarrow{AB} = \{P \mid BP = |AP - AB|\}.$$

Proof Exercise 18.2. ∎

Corollary 18.15 *Polygon Inequality* If $P_1, P_2, \ldots, P_n$ are points, then $P_1 P_n \leq P_1 P_2 + P_2 P_3 + P_3 P_4 + \cdots + P_{n-1} P_n$ for $n \geq 2$.

Proof Exercise 18.3. ∎

Is there an m' such that (M10, s, m') is a model of Σ where (M10, s) is the Moulton Incidence Plane with distance function s given by Euclidean arc length along Moulton lines? The fact that we cannot think of any function that would work might only prove a lack of omniscience on our part. However, in this case, we can show that such a function actually does not exist. We need to find some property of (M10, s) that is necessarily false for a model of Σ. Because Triangle Inequality is a theorem, the existence of a triangle having one side whose length is greater than the sum of the lengths of the other two sides is such a property. ($AB > AC + CB$ for $\triangle ABC$ in Figure 8.6.) Therefore, there does not exist any m' such that (M10, s, m') is a model of Σ.

Theorem 18.16 *Euclid's Proposition I.21* If point D is in int $(\triangle ABC)$, then $BD + DC < BA + AC$ and $\angle BDC > \angle BAC$.

Proof *(Euclid)* By Crossbar let $\overrightarrow{BD}$ intersect $\overline{AC}$ at E. Then $A - E - C$ and $B - D - E$. (See Figure 18.4.) Applying the Triangle Inequality to $\triangle BAE$ and $\triangle DEC$, we obtain $BA + AE > BE$ and $DE + EC > DC$. Substituting $AE = AC - EC$ and $DE = BE - BD$ in the inequalities, we obtain $BA + AC > BE + EC$ and $BE + EC > BD + DC$. Hence $BA +$

$AC > BD + DC$, as desired. Since $\triangle DEC$ has exterior angle $\angle BDC$ with remote interior angle $\angle BEC$, we have $\angle BDC > \angle BEC$. Since $\triangle BAE$ has exterior angle $\angle BEC$ with remote interior angle $\angle BAC$, we have $\angle BEC > \angle BAC$. Therefore $\angle BDC > \angle BAC$. ∎

Theorem 18.17 If $\triangle ABC$, $A-D-B$, and $BC \cong AC$, then $CD < BC$.

Proof Since $BC \cong AC$, we have $m\angle CAB \cong m\angle CBD$ (Theorem 17.5 and 18.10). (See Figure 18.5.) Since $\triangle CDA$ has exterior angle $\angle CDB$ with remote interior angle $\angle CAD$, then $m\angle CDB > m\angle CAB$. Hence $m\angle CDB > m\angle CBD$. So $\overline{CB} > \overline{CD}$ (Theorem 18.11). Therefore $CD < BC$. ∎

Theorem 18.17 might be stated in words as follows. Any segment with one endpoint the vertex of an angle of a triangle and the other endpoint in the interior of the side opposite the angle is shorter than the longer of the other two sides of the triangle. Try to state the next theorem in your own words.

Theorem 18.18 *Euclid's Proposition I.24* Suppose $\triangle ABC$ and $\triangle DEF$ are such that $\overline{AB} \cong \overline{DE}$, $\overline{AC} \cong \overline{DF}$, and $\angle A > \angle D$, then $\overline{BC} > \overline{EF}$.

Proof Without loss of generality we may suppose $DE \cong DF$. Let G be the point such that G and E are on the same side of $\overleftrightarrow{DF}$, $m\angle FDG = m\angle CAB$, and $DG = AB$. Then $\triangle FDG \cong \triangle CAB$ by SAS. So $GF = BC$. Also $\angle DGE \cong \angle DEG$ by the Pons Asinorum.

Since E is in int $(\angle FDG)$, let $\overrightarrow{DE}$ intersect $\overline{FG}$ at H. Then $F-H-G$. Since $DG = DE \cong DF$, by the previous theorem we know $DH < DG = DE$. Hence $D-H-E$. (See Figure 18.6.) Now H and F are in int $(\angle DGE)$, implying $\angle FGE < \angle DGE$. Also, since H and D are in int $(\angle GEF)$, then $\angle DEG < \angle GEF$. Since $\angle DGE \cong \angle DEG$, we

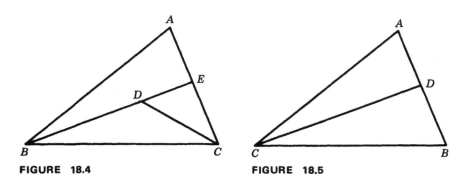

FIGURE 18.4   **FIGURE 18.5**

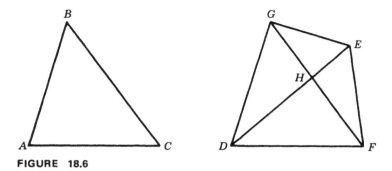

FIGURE 18.6

have $\angle FGE < \angle GEF$. Therefore (Theorem 18.11), $FE < GF$. Finally since $GF = BC$, then $FE < BC$ and $\overline{BC} > \overline{EF}$. ■

Theorem 18.19 *Euclid's Proposition I.25* Suppose $\triangle ABC$ and $\triangle DEF$ are such that $\overline{AB} \simeq \overline{DE}$, $\overline{AC} \simeq \overline{DF}$, and $\overline{BC} > \overline{EF}$, then $\angle A > \angle D$.

Proof We can't have $\angle A \simeq \angle D$ as then $\overline{BC} \simeq \overline{EF}$ by SAS. We can't have $\angle D > \angle A$ as then $\overline{EF} > \overline{BC}$ by the previous theorem. Hence $\angle A > \angle D$. ■

18.3 RIGHT TRIANGLES

Theorem 18.20 A triangle has at most one right angle. If a triangle has a right angle or an obtuse angle, then the other two angles of the triangle are acute.

Proof Let $\triangle ABC$ be such that $m\angle C \geqq \pi/2$. Let $B - C - D$. Then $m\angle ACD \leqq \pi/2$. Since $\angle A$ and $\angle B$ are the remote interior angles of

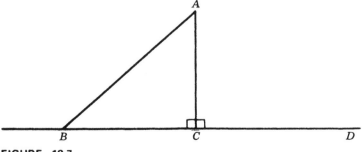

FIGURE 18.7

the exterior angle $\angle ACD$ of $\triangle ACB$, then $\angle A$ and $\angle B$ are acute (Theorem 17.9). ∎

DEFINITION 18.21 A triangle having a right angle (necessarily unique) is a **right triangle.** The side opposite the right angle of a right triangle is the **hypotenuse,** and a side opposite an acute angle of a right triangle is a **leg.** If perpendicular lines l and m intersect at F, and P is any point on l, then F is the **foot** of the perpendicular from P to m.

Theorem 18.22 The hypotenuse of a right triangle is longer than either leg. The shortest segment joining a point to a line is the perpendicular segment.

Proof Let point A be off line l. Let C be the foot of the perpendicular from A to l. Let B be any point on l other than C. Then $\triangle ABC$ is a right triangle with $\angle C$ a right angle. By the previous theorem $\angle A$ and $\angle B$ are acute. The theorem now follows from the fact that the longest side of $\triangle ABC$ must be opposite the largest angle (Theorem 18.11). ∎

Think of all the machinery that is required to prove the following *simple* theorem.

Theorem 18.23 Let F be the foot of the perpendicular from A to $\overleftrightarrow{BC}$. If $\overline{BC}$ is a longest side of $\triangle ABC$, then $B-F-C$.

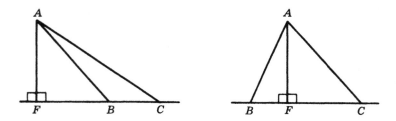

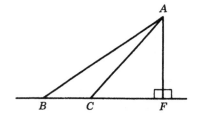

FIGURE 18.8

Proof   If either $F=B$ or $F=C$, then either $\angle B$ or $\angle C$ is a right angle and $\overline{BC}$ is not a longest side of $\triangle ABC$ by the previous theorem. Suppose F is off $\overline{BC}$. Then either $F-B-C$ or $B-C-F$. Using the previous theorem we have, in the first case, $BC < FC < AC$ and, in the second case, $BC < BF < AB$. So, in either case, $\overline{BC}$ is not a longest side of $\triangle ABC$. The theorem follows as we have proved its contrapositive. ∎

If the hypotenuse and an acute angle of one right triangle are respectively congruent to the hypotenuse and an acute angle of another right triangle, that the two triangles are congruent follows as a special case of SAA. If the two legs of one right triangle are respectively congruent to the two legs of another right triangle, that the two triangles are congruent follows as a special case of SAS.

Theorem 18.24   *Hypotenuse-Leg Theorem*   If the hypotenuse and a leg of one right triangle are respectively congruent to the hypotenuse and a leg of another right triangle, then the two triangles are congruent.

Proof   Suppose $\triangle ABC$ and $\triangle DEF$ are such that $m\angle C = m\angle F = \pi/2$, $AB = DE$ and $AC = DF$. Let B' be on $\overrightarrow{CB}$ such that $CB' = FE$. Then $\triangle ACB' \simeq \triangle DFE$ by SAS. So $AB' = DE = AB > AC$. Thus $C - B - B'$ and $C - B' - B$ are impossible (Theorem 18.17). It follows that $B' = B$ and $\triangle ACB \simeq \triangle DFE$. ∎

18.4   EXERCISES

18.1   Theorem 18.2.

18.2   Corollary 18.14.

18.3   Corollary 18.15.

• **18.4**   There does not exist any angle measure function m' such that $(M1, t, m')$, Taxicab Geometry together with m', is a model of Σ.

• **18.5**   True or False?

(a)   If $l \perp n$ and $n \perp m$, then $l \perp m$.

(b)   If point P is on line l, then there exists a unique line through P that is parallel to l.

(c)   If P is a point and l is a line, then there exists a line through P that is parallel to l.

(d)   If P is a point off line l, then there exists a line through P that is parallel to l.

(e)   The difference of the lengths of any two sides of a triangle is less than the length of the third side.

(f)   If $\overline{AB} > \overline{CD}$ and $\overline{CD} > \overline{EF}$, then $\overline{AB} > \overline{EF}$.

(g)   Exactly one of the following holds, given $\overline{AB}$ and $\overline{CD}$: $\overline{AB} > \overline{CD}$, $\overline{AB} = \overline{CD}$, $\overline{AB} < \overline{CD}$.

(h) $AB + BC > AC$ iff $\triangle ABC$.

(i) $AB + BC = AC$ iff $A - B - C$.

(j)   If each leg of one right triangle is congruent to a leg of another right triangle, then the triangles are congruent.

18.6   The converse of Theorem 18.23 is false.

18.7   A triangle is isosceles iff the perpendicular bisector of some side contains the angle bisector of some angle of the triangle.

18.8   If F is the foot of the perpendicular from A to $\overleftrightarrow{BC}$ and $B - C - F$, then $AB > AC$.

18.9   If F is the foot of the perpendicular from A to $\overleftrightarrow{BC}$ and $F \neq B$, then there exists a unique point D on $\overleftrightarrow{BC}$ such that $D \neq B$ but $AD = AB$.

● **18.10**   Given $\triangle ABC$ with $m\angle C = \pi/2$ and $B - E - C$, there exists D in int $(\triangle ABC)$ such that $BD + DE > BA + AC$.

18.11   If there exist two distinct lines through P that are parallel to line l, then there exist infinitely many such lines.

18.12   If $\angle A > \angle B$, $\angle A \simeq \angle C$, and $\angle B \simeq \angle D$, then $\angle A > \angle D$, $\angle C > \angle B$, and $\angle C > \angle D$.

18.13   Give reasonable definitions of a *median* and of an *altitude* of a triangle. Then prove that if any two of the following are collinear for a triangle then the triangle is isosceles: the perpendicular bisector of some side, the angle bisector of some angle of the triangle, the median to some side, the altitude to some side.

18.14   If there exist D, E, F such that E is in int $(\triangle ABC)$, $A - D - F - C$, and $DE + EF = AB + BC$, then there exists G in int $(\triangle ABC)$ such that $DG + GF > AB + BC$. Compare this with Theorem 18.16 and Exercise 18.15.

*18.15   Given $\triangle ABC$ in the Euclidean plane such that $AC > AB \geqq$

BC, find D, E, F such that E is in int $(\triangle ABC)$, $A-D-F-C$, and $DE + EF = AB + BC$.

*18.16 If you would like to read about geometries where there are lines l and m such that l is perpendicular to m but m is not perpendicular to l (i.e., perpendicularity of lines is not a symmetric relation), then see *The Geometry of Geodesics* by H. Busemann (Academic Press, 1955).

GRAFFITI

The study of "non-Euclidean" Geometry brings nothing to students but fatigue, vanity, arrogance, and imbecility.
. . . "Non-Euclidean" space is the false invention of demons, who gladly furnish the dark understandings of the "non-Euclideans" with false knowledge. . . . The "Non-Euclideans," like the ancient sophists, seem unaware that their understandings have become obscured by the promptings of the evil spirits.

Matthew Ryan (1905)

Gauss' motto:

Pauca sed matura.

If we consider him [Euclid] as meaning to be what his commentators have taken him to be, a model of the most unscrupulous formal rigour, we can deny that he has altogether succeeded, though we admit that he made the nearest approach.

De Morgan

CHAPTER 19

Reflections

19.1 INTRODUCING ISOMETRIES

Isometry literally means same-distance-measuring.

DEFINITION 19.1 An *isometry* is a mapping from the set of points into itself that preserves distance. The set of all isometries is $\mathcal{I}$. The identity mapping on the set of points is ι.

Let $\alpha:\mathcal{P} \to \mathcal{P}$, $\alpha:P \mapsto P'$. Then, by definition, α is an isometry iff $P'Q' = PQ$ for all points P and Q. Suppose α is any isometry. Just what kind of an animal is α? We shall discover several facts before stating any theorems. In the development, P' is always αP for any point P.

If $A' = B'$ for points A and B, then $0 = A'B' = AB$. Since $AB = 0$ implies $A = B$, we have $A' = B'$ implies $A = B$. Thus, by definition, α is a one-to-one mapping. In other words, for our first fact we have:

1 An isometry is an injection.

If $A - B - C$, then $AB + BC = AC$. Since $A'B' = AB$, $B'C' = BC$, and $A'C' = AC$, we have $A'B' + B'C' = A'C'$. Since A', B', C' are distinct by (1), then $A' - B' - C'$ by the Triangle Inequality. Conversely, if $A' - B' - C'$, then $A'B' + B'C' = A'C'$, $AB + BC = AC$, and $A - B - C$. Hence:

2 $A-B-C$ iff $A'-B'-C'$; an isometry preserves betweenness.

(To say α preserves betweenness means only that $A-B-C$ implies $A'-B'-C'$.)

The sum of any two of AB, BC, CA is greater than the third iff the sum of any two of $A'B'$, $B'C'$, $C'A'$ is greater than the third. Therefore, by the Triangle Inequality:

3 $\triangle ABC$ iff $\triangle A'B'C'$; an isometry preserves triangles.

Given any angle $\angle ABC$ we have $\triangle ABC$ and $\triangle A'B'C'$ by (3). Since $\triangle A'B'C' \simeq \triangle ABC$ by SSS, then $m\angle A'B'C' = m\angle ABC$. Therefore:

4 An isometry preserves both d and m.

An isometry preserves d by definition. That α preserves m means only that $m\angle A'B'C' = m\angle ABC$. This last equation should not be misunderstood; it does not mean what it might seem to mean. It is true that if $\angle ABC$, then $\angle A'B'C'$ and the two angles are congruent. However, the equation does not say that α preserves angles. Even though α preserves betweenness, it is not yet clear that α preserves rays or lines. (The function f given by $f(x) = e^x$ preserves betweenness on the real line but does not preserve rays on the real line.) The trouble is that we do not yet know that α is a surjection on the points. We remedy this situation next.

Suppose P is a point. We wish to show that P is the image of some point under α. Take any $\triangle ABC$. Since P cannot be on all three sides of $\triangle A'B'C'$, we may suppose $\triangle PA'B'$ without loss of generality. Let $m\angle B'A'P = a$ and $A'P = b$. By the Angle-Segment-Construction Theorem, there exist a unique pair of points Q_1 and Q_2, one on each side of $\overleftrightarrow{AB}$, such that $m\angle BAQ_1 = m\angle BAQ_2 = a$ and $AQ_1 = AQ_2 = b$. (See Figure 19.1.) Hence, by (4), $m\angle B'A'Q_1' = m\angle B'A'Q_2' = a$ and $A'Q_1' = A'Q_2' = b$. $Q_1' \neq Q_2'$ by (1). By the Angle-Segment-Construction Theorem it follows that either $P = Q_1'$ or $P = Q_2'$. In either case, P is the image of some point and α is onto. So an isometry is a surjection. This fact together with (1) can be summarized:

5 An isometry is a bijection on $\mathscr{P}$.

If l is a line, let $l' = \{P' | \text{point } P \text{ is on } l\}$. The points in l' are collinear by (2). If $l = \overleftrightarrow{AB}$, then every point on $\overleftrightarrow{A'B'}$ is on l' by (2) and (5). Also no point off $\overleftrightarrow{A'B'}$ is in l' by (3). Hence l' is a line. So α induces

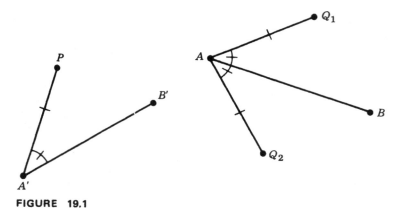

FIGURE 19.1

a mapping from $\mathscr{L}$ into $\mathscr{L}$. That this mapping is a surjection follows immediately from the fact that α is a surjection on the points. To see that this mapping is also an injection, assume $l' = m'$ where l and m are lines. There are two points P' and Q' on l'. Then P and Q are two points on both l and m. Hence $l = m$. So $l' = m'$ implies $l = m$, proving α induces an injection on the lines. The results of this paragraph can be summarized:

6 An isometry induces a bijection on $\mathscr{L}$.

Since a line in our geometry is always a set of points, that point P is on line l iff point P' is on line l' follows from the definition of l'. So α is a collineation. In other words, (5) and (6) together say that an isometry is an isomorphism of the incidence structure $(\mathscr{P}, \mathscr{L})$. This together with (4) means that an isometry is an isomorphism of the whole ball game $(\mathscr{P}, \mathscr{L}, d, m)$. (Everything in our geometry can be traced back to incidence, distance, and angle measure.) Because mathematicians call an isomorphism from a mathematical system onto itself an automorphism (*auto* means self), we may summarize (4), (5), and (6) by saying that an isometry is an automorphism of our geometry $(\mathscr{P}, \mathscr{L}, d, m)$.

Theorem 19.2 An isometry is an automorphism. Conversely, an automorphism is an isometry.

Proof Since an automorphism must preserve distance in particular, an automorphism is necessarily an isometry. That every isometry is also an automorphism follows from the long argument above. ∎

Theorem 19.3 The isometries form a group $(\mathscr{I}, \circ, \iota)$. The group of all isometries is the group of all automorphisms.

Proof We must verify that $(\mathscr{I}, \circ, \iota)$ satisfies the three axioms of a group (see Section 3.1). The second statement in the theorem then follows from the previous theorem.

  1. Composition is an associative binary operation on $\mathscr{I}$, the set of isometries: Since isometries are special cases of permutations on the set $\mathscr{P}$, composition of isometries is associative. We must show that $\beta\alpha$ is an isometry when β and α are isometries. (We write $\beta\alpha$ for $\beta\circ\alpha$.) Let $\alpha:P \mapsto P'$ and $\beta:P' \mapsto P''$ for any point P. Then $\beta\alpha:P \mapsto P''$ as $\beta\alpha(P) = \beta(\alpha P) = \beta(P') = P''$ by definition of composition. Now $\beta\alpha$ is an isometry as $P''Q'' = PQ$ for any points P and Q because $P''Q'' = P'Q'$, since β is an isometry, and $P'Q' = PQ$, since α is an isometry.

  2. ι is an isometry such that $\iota\alpha = \alpha = \alpha\iota$ for any isometry α: The identity permutation on $\mathscr{P}$ is trivially an isometry by definition.

  3. If α is an isometry, then there exists isometry α^{-1} such that $\alpha^{-1}\alpha = \iota = \alpha\alpha^{-1}$: Let α be an isometry and α^{-1} its inverse permutation. So $\alpha^{-1}(P) = P'$ iff $\alpha(P') = P$, for any point P. Since α is an isometry, $PQ = P'Q'$. So $P'Q' = PQ$ and α^{-1} is an isometry. ∎

19.2 REFLECTION IN A LINE

Although we know the isometries form a group, we know about the existence of only one – namely the trivial isometry ι. After all the work above, you have every right to expect that there are more.

DEFINITION 19.4 The **reflection** in line l is the mapping $\rho_l:\mathscr{P} \to \mathscr{P}$ such that $\rho_l P = P$ if point P is on l and $\rho_l P = P'$ where l is the perpendicular bisector of $\overline{PP'}$ if point P is off l. Line l is called the **center** of ρ_l.

The reflection in line l is unique and well defined (Theorem 18.6). In this book ρ_l is always the reflection in line l. The following definitions are standard.

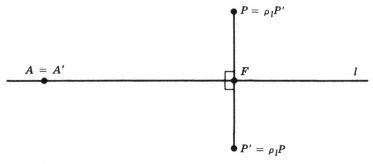

FIGURE 19.2

DEFINITION 19.5 Let σ be any mapping on $\mathscr{P}$. Let T be any set of points. Then σ **fixes** point P if $\sigma(P) = P$; σ **fixes** set T if $\sigma T = T$ where $\sigma T = \{\sigma P | P$ is a point in $T\}$; and σ fixes set T **pointwise** if $\sigma P = P$ for every point P in T. Further, σ is an **involution** if $\sigma \neq \iota$ but $\sigma^2 = \iota$ where σ^2 is $\sigma\sigma$.

Theorem 19.6 ρ_l is an involution that interchanges the halfplane of l. Further, ρ_l fixes line m pointwise iff $m = l$, and ρ_l fixes line m iff $m = l$ or $m \perp l$.

Proof From the definition of ρ_l it follows that ρ_l is an involution which fixes l pointwise, fixes every line perpendicular to l, and interchanges the halfplanes of l (Theorem 18.1 and Theorem 18.6). Now suppose ρ_l fixes line m with $m \neq l$. Let P be a point on m but off l. Since ρ_l fixes m, then P' must be on m where $P' = \rho_l P$. So $m = \overleftrightarrow{PP'}$, and $m \perp l$ by definition of ρ_l. ∎

Let l and m be lines. Then the second sentence of Theorem 19.6 can be restated in symbols by

(1) $\rho_l = \rho_m$ iff $l = m$;

(2) $\rho_l m = m$ iff $l = m$ or $l \perp m$.

We have yet to prove the fact that ρ_l is an isometry.

Theorem 19.7 A reflection is an involutary isometry.

Proof We already know a reflection is an involution. To show ρ_l is an isometry let $P' = \rho_l P$ and $Q' = \rho_l Q$ where P and Q are any two points. We need to prove $P'Q' = PQ$. There are four cases.

1. Suppose P and Q are both on l. Then $P'Q' = PQ$ since $P' = P$ and $Q' = Q$.

2. Suppose exactly one of P or Q is on l, say Q is on l. Then Q is equidistant from P and P' since Q is on the perpendicular bisector of $\overline{PP'}$ (Theorem 18.5).

3. Suppose $\overleftrightarrow{PQ}$ is perpendicular to l at L. Let f be any coordinate system for $\overleftrightarrow{PQ}$ such that $f(L) = 0$. If $f(P) = p$ and $f(Q) = q$, then $f(P') = -p$ and $f(Q') = -q$ by the definition of ρ_l. Hence, by the Ruler Postulate, $P'Q' = |q - p| = PQ$.

4. Suppose P and Q are both off l with $\overleftrightarrow{PQ}$ not perpendicular to l. Let l bisect $\overline{PP'}$ at M and $\overline{QQ'}$ at N. Then $M \neq N$ as $\overleftrightarrow{PQ}$ is not perpendicular to l. So $\triangle PMN \simeq \triangle P'MN$ by SAS. Then $PN = P'N$ and $m\angle PNM = m\angle P'NM$. It follows that $m\angle PNQ = m\angle P'NQ'$. (See

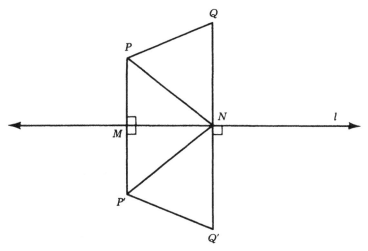

FIGURE 19.3

Figure 19.3 for the case P and Q on the same side of l and Figure 19.4 for the case P and Q on opposite sides of l.) Hence $\triangle PNQ \simeq \triangle P'NQ'$ by SAS. Therefore $PQ = P'Q'$. ∎

For obvious reasons we might call a reflection a *mirror map*. The previous theorem completes the proof of the assertion that MIRROR *and* SAS *are equivalent under our first four axioms.*

Theorem 19.8 If an isometry fixes two points on a line, then the

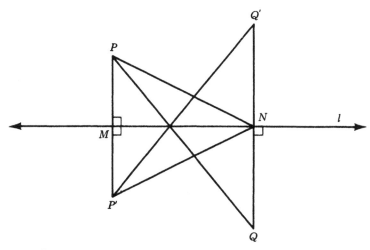

FIGURE 19.4

isometry fixes that line pointwise. If an isometry fixes three non-collinear points, then the isometry is ι.

Proof Suppose σ fixes two points A and B. So $\sigma\overleftrightarrow{AB} = \overleftrightarrow{AB}$ as $\sigma A = A$ and $\sigma B = B$. Let P be any point on $\overleftrightarrow{AB}$ and $\sigma P = Q$. Then Q is on $\overleftrightarrow{AB}$, $AP = AQ$, and $BP = BQ$. Hence $P = Q$ (Theorem 8.17). Therefore σ fixes every point on $\overleftrightarrow{AB}$.

Suppose σ fixes three noncollinear points A, B, C. Then σ fixes every point on $\triangle ABC$ by the first result. Let P be any point. Let Q be a point on the interior of a side of $\triangle ABC$ but different from P. Then $\overleftrightarrow{PQ}$ intersects $\triangle ABC$ at a point R different from Q. Since σ fixes Q and R, then σ fixes $\overleftrightarrow{QR}$ pointwise by the first result. So σ fixes P. Since σ fixes every point, σ must be the identity isometry. ■

For a picture of what is going on in the next proof look at Figure 16.4. The next proof imitates the proof of Theorem 16.3.

Theorem 19.9 If $\triangle ABC \simeq \triangle DEF$, then there exists a unique isometry σ such that $\sigma A = D$, $\sigma B = E$, and $\sigma C = F$.

Proof If $A = D$, let $\sigma_1 = \iota$; if $A \neq D$, let σ_1 be the reflection in the perpendicular bisector of $\overline{AD}$. In either case, $\sigma_1 A = D$. Let $\sigma_1 B = B_1$ and $\sigma_1 C = C_1$. If $B_1 \neq E$, then D is on the perpendicular bisector of $\overline{B_1 E}$ (Theorem 18.5) as $DE = AB = DB_1$ since σ_1 is an isometry. If $B_1 = E$, let $\sigma_2 = \iota$; if $B_1 \neq E$, let σ_2 be the reflection in the perpendicular bisector of $\overline{B_1 E}$. In either case, $\sigma_2 D = D$ and $\sigma_2 B_1 = E$. Let $\sigma_2 C_1 = C_2$. If $C_2 \neq F$, then D and E are on the perpendicular bisector of $\overline{C_2 F}$ (Theorem 18.5) as $DF = AC = DC_1 = DC_2$ and $EF = BC = B_1 C_1 = EC_2$ since σ_1 and σ_2 are isometries. If $C_2 = F$, let $\sigma_3 = \iota$; if $C_2 \neq F$, let σ_3 be the reflection in the perpendicular bisector of $\overline{C_2 F}$. In either case, $\sigma_3 D = D$, $\sigma_3 E = E$, and $\sigma_3 C_2 = F$. Letting $\sigma = \sigma_3 \sigma_2 \sigma_1$ we have $\sigma A = D$, $\sigma B = E$, and $\sigma C = F$. We have demonstrated the existence of the desired isometry. For the uniqueness suppose σ and τ are isometries sending A, B, C respectively to D, E, F. Then $\tau^{-1}\sigma A = \tau^{-1}D = A, \tau^{-1}\sigma B = \tau^{-1}E = B$, and $\tau^{-1}\sigma C = \tau^{-1}F = C$. So $\tau^{-1}\sigma = \iota$ by the previous theorem. Therefore, $\tau(\tau^{-1}\sigma) = \tau\iota$ and $\sigma = \tau$, proving the uniqueness of σ. ■

The proof of Theorem 19.9 is very important. (It is also quite lovely, and you should make every effort to understand it.) Our next result is a corollary of the proof of Theorem 19.9 more than a corollary of the theorem itself. Theorem 19.10 will be essential later when we look at the isometries in considerable detail.

Theorem 19.10   Every isometry is a product of at most three reflections. If an isometry fixes a point, then the isometry is either a reflection or a product of two reflections. An isometry fixes two points on a line iff the isometry is the reflection in that line or the identity.

Proof   Let σ be an isometry. Let A, B, C be three noncollinear points. Let $\sigma A = D$, $\sigma B = E$, and $\sigma C = F$. Then $\triangle ABC \simeq \triangle DEF$. Hence σ is a product of at most three reflections by the proof of Theorem 19.9. If σ fixes a point, let A be that point. In this case σ is a product of at most two reflections by the proof of Theorem 19.9 as then $\sigma = \sigma_3 \sigma_2 \sigma_1 = \sigma_3 \sigma_2$ since $\sigma_1 = \iota$. If σ fixes two points, let A and B be those points. Then $\sigma = \sigma_3$ in the proof of Theorem 19.9.   ∎

Theorem 19.11   If isometry σ interchanges two points A and B, then σ fixes the midpoint of A and B.

Proof   Let M be the midpoint of A and B. Let $M' = \sigma M$. Since $AM = BM = \frac{1}{2}AB$, we have $BM' = AM' = \frac{1}{2}AB$ and M' is the midpoint of A and B (Corollary 18.14). Hence $\sigma M = M$.   ∎

We are now in a position to see why it is quite natural that the one word *congruence* be used to describe three different relations, one on each of the set of segments, the set of angles, and the set of triangles.

Theorem 19.12   Two segments, two angles, or two triangles are congruent iff there is an isometry taking one onto the other.

Proof   Exercise 19.1.   ∎

DEFINITION 19.13   If T_1 and T_2 are sets of points, then T_1 and T_2 are **congruent** if there exists an isometry σ such that $\sigma T_1 = T_2$.

19.3   EXERCISES

● **19.1**   Theorem 19.12.

19.2   If $\triangle ABC \simeq \triangle DEF$, then int $(\triangle ABC)$ and int $(\triangle DEF)$ are congruent.

● **19.3**   Give a collineation of the Cartesian plane that preserves m but is not an isometry.

● **19.4**   How many isometries are there that map $\triangle ABC$ onto congruent triangle $\triangle DEF$ if each triangle is scalene, isosceles, or equilateral, respectively?

19.5 Congruence is an equivalence relation on the set of all subsets of points.

● **19.6** True or False?

    (a) Every isometry is a collineation.

    (b) Every collineation is an isometry.

    (c) An isometry preserves perpendicularity.

    (d) An isometry preserves parallelness.

    (e) If $\sigma \in \mathcal{I}$, $\sigma A = A'$, $\sigma B = B'$, then $AA' = BB'$.

    (f) $\rho_a = \rho_b$ iff $a = b$, where a and b are lines.

    (g) $\rho_a b = b$ iff $a = b$, where a and b are lines.

    (h) ι is a product of two reflections.

    (i) Isometry σ fixes three points iff $\sigma = \iota$.

    (j) A product of four reflections is an isometry.

● **19.7** The mapping of the Cartesian plane sending (x, y) to (x', y') is the reflection in the line with equation $Ax + By + C = 0$ iff

$$x' = x - (Ax + By + C)(2A/(A^2 + B^2)),$$
$$y' = y - (Ax + By + C)(2B/(A^2 + B^2)).$$

19.8 Three noncollinear points (and their images) completely determine an isometry.

19.9 Any two rays are congruent.

19.10 Congruence is an equivalence relation on the set of all convex quadrilaterals.

19.11 The Cartesian reflection in the line with equation $y = 2x$ is not an isometry in Taxicab Geometry.

*● **19.12** Find all collineations of the Cartesian plane.

*● **19.13** Find all isometries of the Cartesian plane.

*__19.14__ What are the reflections in Taxicab Geometry?

*__19.15__ What are the isometries in Taxicab Geometry?

GRAFFITI

There is no branch of mathematics, however abstract, which may not some day be applied to phenomena of the real world.

Lobachevsky

How can it be that mathematics, being after all a product of human thought independent of experience, is so admirably adapted to the objects of reality?

Einstein

[*Euclid's* Elements] *has been for nearly twenty-two centuries the encouragement and guide of that scientific thought which is one thing with the progress of man from a worse to a better state. The encouragement; for it contained a body of knowledge that was really known and could be relied on, and that moreover was growing in extent and application. For even at the time this book was written — shortly after the foundation of the Alexandrian Museum — Mathematics was no longer the merely ideal science of the Platonic school, but had started on her career of conquest over the whole world of Phenomena. The guide; for the aim of every scientific student of every subject was to bring his knowledge of that subject into a form as perfect as that which geometry had attained. Far up on the great mountain of Truth, which all the sciences hope to scale, the foremost of that sacred sisterhood was seen, beckoning for the rest to follow her.*

Clifford

The full impact of the Lobatchewskian method of challenging axioms has probably yet to be felt. It is no exaggeration to call Lobatchewsky the Copernicus of Geometry [as did Clifford], for geometry is only a part of the vaster domain which he renovated; it might even be just to designate him as a Copernicus of all thought.

E. T. Bell

CHAPTER 20

Circles

20.1 INTRODUCING CIRCLES

The first of the following definitions may be new to you. The second says, in the light of Theorem 18.22, that the distance from point P to line l is defined to be the shortest distance from P to l when P is off l. The remaining definitions should be no surprise. As usual, *radius* has two meanings—one meaning as a segment and another as a number. Likewise *the diameter* of a circle is the length of *a diameter*.

DEFINITION 20.1   Line l is a *line of symmetry* for set T of points if $\rho_l T = T$. The *distance from point P to line l* is the length of the perpendicular segment from P to l when P is off l and 0 when P is on l. If C is a point and r a positive number, then the locus of all points P such that $CP = r$ is a *circle* with *center C* and *radius r*. Let A and B be two points on circle $\mathscr{C}$ with center C and radius r. Then $\overline{CA}$ is a *radius* of $\mathscr{C}$ with *outer end A; $\overline{AB}$* is a *chord* of $\mathscr{C}$; if $A-C-B$, then $\overline{AB}$ is a *diameter* of $\mathscr{C}$; $2r$ is the *diameter* of $\mathscr{C}$; $\overleftrightarrow{AB}$ is a *secant* of $\mathscr{C}$; a line which intersects $\mathscr{C}$ in exactly one point is a *tangent* of $\mathscr{C}$; the locus of all points Q such that $CQ < r$ is the *interior of $\mathscr{C}$*, denoted by int $(\mathscr{C})$; and the locus of all points Q such that $CQ > r$ is the *exterior* of $\mathscr{C}$. Two or more circles having a common center are *concentric*.

Now we are going to prove some of the elementary theorems

about circles. Perhaps these theorems will be more exciting if you consider which of them are valid in Taxicab Geometry and which hold in Moulton Geometry. (Recall that in Taxicab Geometry a circle is a square!) In doing this you may be convinced that the theorems are not really trivial. In mathematics, as in life in general, there are many cases where a fact seems obvious only because in our ignorance we are unable to conceive alternative possibilities.

Theorem 20.2 Each of the center and the radius of a circle is unique. The perpendicular bisector of a chord of a circle passes through the center of the circle. The segment joining the center of a circle to the midpoint of a chord which is not a diameter is perpendicular to the chord, and the perpendicular from the center of a circle to a chord of the circle bisects the chord. Three points on a circle determine the circle. Two circles intersect in at most two points. A line intersects a circle in at most two points.

Proof Let P, Q, R be three distinct points on some circle. Such points exist since every line through a center of any circle contains exactly two points on the circle by the Segment-Construction Theorem. Let l and m be the perpendicular bisectors of $\overline{PQ}$ and $\overline{QR}$, respectively. Since any center of a circle containing P, Q, R must be equidistant from P, Q, and R, such a center must be on both l and m (Theorem 18.5). Now $l \neq m$ as otherwise Q and the midpoints of $\overline{PQ}$ and $\overline{QR}$ are the vertices of a triangle with two right angles, a contradiction. So l and m intersect in exactly one point, which must be the unique center of the unique circle containing the three noncollinear points P, Q, R. Surprisingly, each statement in the theorem follows from this argument. ∎

The preceding theorem says that three points *on a circle* determine the circle. It is not true that three points determine a circle as three collinear points do not. (Lines l and m would be parallel in the proof above.) The theorem also does not make the rash statement that three noncollinear points determine a circle!

Theorem 20.3 If P is a point on a circle with center C, then every point on the circle is the image of P under some reflection in a line through C. Conversely, any point which is the image of P under some reflection in a line through C is on the circle. Every line through the center of a circle is a line of symmetry for the circle. Conversely, every line of symmetry for a circle passes through the center of the circle.

Proof Let $\mathscr{C}$ be a circle with center C and containing point P. So $\mathscr{C}$

has radius CP. Clearly, $P = \rho_l P$ if $l = \overleftrightarrow{CP}$. Also, if Q is a point on $\mathscr{C}$ different from P, then $Q = \rho_l P$ where l is the perpendicular bisector of chord $\overline{PQ}$. Then l must pass through C (Theorem 20.2), proving the first statement in the theorem. If $R = \rho_l P$ where l is a line through C, then $CR = CP$ since $C = \rho_l C$ and ρ_l is an isometry. So R is on $\mathscr{C}$ by the definition of a circle, proving the second and third statements in the theorem. Finally, suppose l is any line of symmetry for $\mathscr{C}$. Let $Q = \rho_l P$ where P is any point on $\mathscr{C}$ but off l. Then $Q \neq P$ and l is the perpendicular bisector of $\overline{PQ}$. Since l is a line of symmetry for $\mathscr{C}$, then Q is on $\mathscr{C}$ and $\overline{PQ}$ is a chord of $\mathscr{C}$. So l passes through C by the previous theorem. ∎

Theorem 20.4 If a line is perpendicular to a radius of a circle at its outer end, then the line is a tangent of the circle. Conversely, every tangent of a circle is perpendicular to a radius of the circle containing the point of intersection. Every point of a circle lies on a unique tangent of the circle. A tangent of a circle contains no interior points of the circle.

Proof Let $\mathscr{C}$ be a circle with center C and containing point P. Let t be the perpendicular to $\overline{CP}$ at P. Suppose t intersects $\mathscr{C}$ at some other point Q with $Q \neq P$. Let $l = \overleftrightarrow{CP}$ and $R = \rho_l Q$. Since l is a line of symmetry for both t and $\mathscr{C}$ (Theorem 19.6, Theorem 20.3), then P, Q, R are three distinct points on both t and $\mathscr{C}$, a contradiction (Theorem 20.2). Hence t intersects $\mathscr{C}$ only at P, and t is a tangent. Now suppose s is a tangent of $\mathscr{C}$ containing point P. Let F be the foot of the perpendicular from C to s. (See Figure 20.2.) By the Midpoint Theorem there exists S such that F is the midpoint of P and S. If $P \neq F$, then $\triangle PFC \simeq \triangle SFC$ by SAS. Then $CP = CS$, and S is on $\mathscr{C}$, contradicting s is a tangent. Therefore, $F = P$ and $s = t$. The third statement in the theorem follows

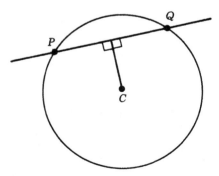

FIGURE 20.1

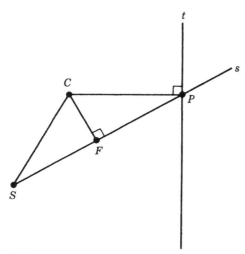

FIGURE 20.2

from the first two, and the last statement follows from the fact that radius CP is the shortest distance from C to t. ∎

Theorem 20.5 If $\overline{AB}$ is a chord of a circle with radius r, then $AB \leqq 2r$ and $AB = 2r$ iff $\overline{AB}$ is a diameter.

Proof Let $\overline{AB}$ be a chord of a circle with center C and radius r. By the Triangle Inequality, $AB \leqq AC + CB = 2r$ with equality iff $A-C-B$. ∎

Theorem 20.6 If t is a tangent of circle $\mathscr{C}$ with center C and σ is an isometry, then $\sigma\mathscr{C}$ is a circle with center σC and σt is a tangent of $\sigma\mathscr{C}$. Two circles are congruent iff they have the same radius.

Proof Exercise 20.1. (Of course *radius* in the statement of the theorem means a number and not a segment.) ∎

Theorem 20.7 In the same circle or congruent circles, any two congruent chords are equidistant from the center, and conversely.

Proof Let $\overline{AB}$ be a chord of a circle with center C and radius r; let $\overline{DE}$ be a chord of a circle with center F and radius r. Let G be the midpoint of $\overline{AB}$; let H be the midpoint of $\overline{DE}$. Now $CG = 0$ iff $AB = 2r$, and $FH = 0$ iff $DE = 2r$ (Theorem 20.5). So suppose $\triangle ABC$ and $\triangle DEF$. (See Figure 20.3.) Then $\triangle CGA$ and $\triangle FHD$ are right triangles with congruent hypotenuses. Hence $CG = FH$ iff $AG = DH$ (Theorem 18.24). Since $AG = DH$ iff $AB = DE$, we have $CG = FH$ iff $AB = DE$. as desired. ∎

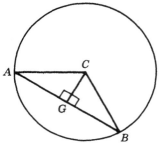 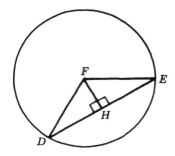

FIGURE 20.3

Theorem 20.8 The interior of a circle is a convex set.

Proof Let $\mathscr{C}$ be a circle with center C and radius r. Suppose A and B are distinct points in int $(\mathscr{C})$. If C is on $\overleftrightarrow{AB}$, then $\overline{AB}$ is in int $(\mathscr{C})$ since the interior of any diameter of $\mathscr{C}$ is in int $(\mathscr{C})$. Suppose $\triangle ABC$ and A–P–B. We need to show P is in int $(\mathscr{C})$. Now either $CP < CA$ or $CP < CB$ (Theorem 18.17). In either case, since $CA < r$ and $CB < r$, we have $CP < r$ and P is in int $(\mathscr{C})$. ∎

20.2 THE TWO-CIRCLE THEOREM

We now come to some theorems that might seem even more obvious than those we have already proved. Certainly, if a line intersects the interior of a circle, then the line intersects the circle. This result is actually quite deep. If we look at the Rational Cartesian Plane $(M2, d, m)$, where d and m are taken from the Cartesian plane, we see some peculiar things happening. The line with equation $y = x$ does contain the point $(0, 0)$ which is in the interior of the circle with equation $x^2 + y^2 = 1$. Yet this line does not intersect the circle! From this example we see that the existence of the intersection of a line and a circle or of two circles is not trivial. The proof of the next theorem uses the intermediate-value theorem from calculus. An alternative proof using the greatest lower bound (see Section 3.2) is indicated in Exercise 20.3.

Theorem 20.9 *Compass-Construction Axiom* If $\overrightarrow{PQ} \perp \overline{CP}$ and $CP < r$, then there exists a unique point T on $\overrightarrow{PQ}$ such that $CT = r$.

Proof Let f be a coordinate system for $\overleftrightarrow{PQ}$ such that $f(P) = 0$ and $f(Q) > 0$. Without loss of generality we may suppose $f(Q) = PQ = r$ (Theorem 8.11). Then $CQ > r$ as the hypotenuse of a right triangle is

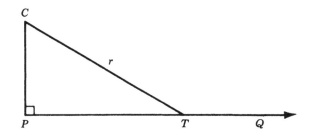

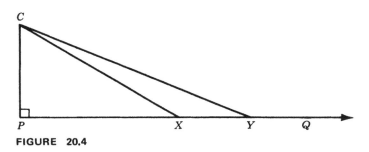

FIGURE 20.4

longer than either leg. For each real number x there is a unique point X on $\overleftrightarrow{PQ}$ such that $f(X) = x$; we then define $g(x) = CX$. So g is a function from the reals into the reals. For distinct points X and Y on $\overleftrightarrow{PQ}$ we have $|CX - CY| < XY$ by the Triangle Inequality. Then $|g(x) - g(y)| < |x - y|$ for all distinct real numbers x, y. So G is a continuous real function. Since $g(0) = CP < r$ and $g(r) = CQ > r$, there exists a real number t such that $0 < t < r$ and $g(t) = r$ by the intermediate-value theorem. Letting $f(T) = t$, we have $CT = g(t) = r$ with T in $\overline{PQ}$. The uniqueness of point T on $\overrightarrow{PQ}$ such that $CT = r$ follows from the fact that if A and B are two points on $\overrightarrow{PQ}$ such that $P - A - B$ then $CA < CB$ (Theorem 18.17). ∎

Theorem 20.10 If S is a point in the interior of a circle and Q is a point in the exterior of the circle, then $\overline{SQ}$ intersects the circle.

Proof Let $\mathscr{C}$ be a circle with center C and radius r. Suppose $CS < r$ and $CQ > r$. If C is on $\overleftrightarrow{SQ}$, then the theorem is trivial by the Segment-Construction Theorem. Suppose $\triangle CSQ$. (See Figure 20.5.) If $\overrightarrow{SQ}$ intersects $\mathscr{C}$, then necessarily $\overline{SQ}$ intersects $\mathscr{C}$ (Theorem 18.17). We shall now show that $\overrightarrow{SQ}$ does intersect $\mathscr{C}$. Let P be the foot of the perpendicular from C to $\overleftrightarrow{SQ}$. Since $CP \leqq CS$ (Theorem 18.22), then $CP < r$ and P is in int $(\mathscr{C})$. Since int $(\mathscr{C})$ is a convex set, $\overrightarrow{SQ}$ intersects $\mathscr{C}$ iff

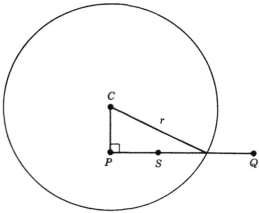

FIGURE 20.5

$\overrightarrow{PQ}$ intersects $\mathscr{C}$. We know $\overrightarrow{PQ}$ intersects $\mathscr{C}$ by the previous theorem. Therefore $\overrightarrow{SQ}$ intersects $\mathscr{C}$ and $\overline{SQ}$ intersects $\mathscr{C}$. ∎

Theorem 20.11 *Line-Circle Theorem* If a line intersects the interior of a circle, then the line intersects the circle exactly twice.

Proof Suppose point S is on line l and in the interior of circle $\mathscr{C}$ with center C and radius r. If C is on l, the theorem is trivial. Suppose C is off l. Let P be the foot of the perpendicular from C to l. (See Figure 20.6.) Then $CP \leqq CS < r$ and P is in int $(\mathscr{C})$. Let Q_1 and Q_2 be points on l such that $Q_1 - P - Q_2$ and $Q_1 P = PQ_2 = r$. So Q_1 and Q_2 are in the exterior of $\mathscr{C}$. Thus both $\overrightarrow{PQ_1}$ and $\overrightarrow{PQ_2}$ intersect $\mathscr{C}$ (Theorem 20.9 or

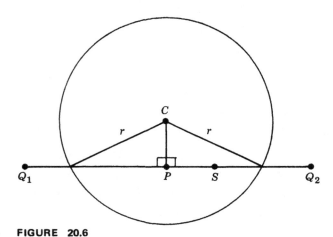

FIGURE 20.6

20.10). Hence l intersects $\mathscr{C}$ at least twice. Since a line intersects a circle in at most two points, line l intersects the circle exactly twice. ∎

Theorem 20.12 A point in the exterior of a circle is on exactly two tangents of the circle. Further, if $\overleftrightarrow{PQ}$ and $\overleftrightarrow{PR}$ are tangents to a circle with Q and R on the circle, then $\overline{PQ} \simeq \overline{PR}$.

Proof Let point P be in the exterior of the circle $\mathscr{C}$ with center C and radius CQ'. Without loss of generality we may suppose $C-Q'-P$. (See Figure 20.7.) Since Q' is in the interior of the circle $\mathscr{C}_2$ with center C and radius CP, the perpendicular to $\overleftrightarrow{CP}$ at Q' must intersect $\mathscr{C}_2$, say at point P', by the Line-Circle Theorem. Let m be the perpendicular bisector of $\overline{PP'}$. Then m contains C since $\overline{PP'}$ is a chord of $\mathscr{C}_2$. So $\rho_m C = C$ and $\rho_m P' = P$. Let $\rho_m Q' = Q$. Then, since $\overleftrightarrow{P'Q'}$ is a tangent of $\mathscr{C}$ (Theorem 20.4), $\overleftrightarrow{PQ}$ must be a tangent of $\mathscr{C}$ with Q on $\mathscr{C}$ (Theorem 20.6). If R is the image of Q under the reflection in $\overleftrightarrow{CP}$, then $\overleftrightarrow{PR}$ is also a tangent of $\mathscr{C}$ and $PR = PQ$. That P lies on at most two tangents is left as Exercise 20.2. ∎

Theorem 20.13 If $\triangle ABC$ has a right angle at C, $\triangle A'B'C'$ has a right angle at C', $AB = A'B'$, and $AC > A'C'$, then $BC < B'C'$.

Proof Let D be such that $C-D-A$ and $CD = C'A'$; let E be the point on $\overrightarrow{CB}$ such that $CE = C'B'$. Then $\triangle DCE \simeq \triangle A'C'B'$ by SAS. Thus $DE = A'B' = AB$ and $EC = B'C'$. Now $B \neq E$, as otherwise $\triangle ABC \simeq$

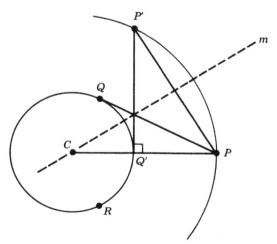

FIGURE 20.7

$\triangle DEC$ by Hypotenuse-Leg Theorem and $AC = DC$. Assume $B - E - C$. Then (Theorem 18.17) we have $DE < BD < AB$, a contradiction. Hence $E - B - C$ and $BC < EC = B'C'$. ∎

Theorem 20.13 is used to prove that of two chords of a circle the one closer to the center is the longer.

Theorem 20.14 If chords $\overline{PR}$ and $\overline{QS}$ of a circle with center C and radius r are perpendicular to a diameter of the circle at X and Y, respectively, such that $C - X - Y$, then $r > PX > QY$.

Proof $\triangle CQY$ has a right angle at Y, and $\triangle CPX$ has a right angle at X. Since $CQ = CP$ and $CY > CX$, we have $QY < PX$ by the preceding theorem. ∎

If $0 < s < a$, then there exist right triangles with hypotenuse of length a and a leg of length s. This result, which is used in the next proof, is a restatement of Theorem 20.9. From this, one can prove directly the existence of isosceles triangles having sides of length r, r, and c when $2r > c > 0$ (Exercise 20.4). When $c = r$ we have a proof of the existence of equilateral triangles with sides of length r for any positive real r, which is Euclid's Proposition I.1. The existence of particular isosceles triangles is a special case of our next theorem, which is Euclid's Proposition I.22.

Theorem 20.15 *Triangle Theorem* Positive real numbers a, b, c are the lengths of the sides of some triangle iff each of the numbers is less than the sum of the other two.

Proof From the Triangle Inequality we already know that the length of any side of a triangle is less than the sum of the lengths of the other two sides. We now prove the converse. Without loss of generality we may suppose $a \leqq b \leqq c$. Let A and B be any two points such that $AB = c$. Let f be a coordinate system for $\overleftrightarrow{AB}$ such that $f(A) = 0$ and $f(B) = c$. Let $\mathscr{C}_A$ be the circle with center A and radius a; let $\mathscr{C}_B$ be the circle with center B and radius b. Let $f(G) = c - b$ and $f(H) = a$. Then G on $\mathscr{C}_B$, H on $\mathscr{C}_A$, and $0 \leqq c - b < a \leqq c$. (See Figure 20.8.)

If $0 \leqq x < a$ and $f(X) = x$, then define $g(x) = XP_x$ where P_x is on $\mathscr{C}_A$ and $\overline{XP_x} \perp \overline{AB}$. Such a point P_x exists by the Line-Circle Theorem since X is an interior point of $\mathscr{C}_A$. Further, we know g is a strictly decreasing function for $0 \leqq x < a$ by the previous theorem. Define $g(a) = 0$. We also know that if $0 < s < a$ then there exists point X such that $0 < f(X) = x < a$ and $g(x) = s$ (Theorem 20.9). Thus g is a strictly decreasing function defined for x such that $0 \leqq x \leqq a$, and g takes on all values v such that $a \geqq v \geqq 0$. In particular, g is a continu-

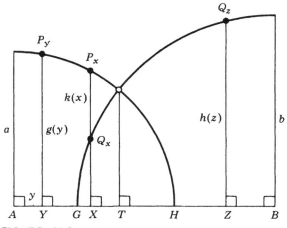

FIGURE 20.8

ous function defined for x when $c - b \leqq x \leqq a$ and such that $g(c - b) > 0$ and $g(a) = 0$.

Paralleling the construction of g, if $c - b < x \leqq c$ and $f(X) = x$, define $h(x) = XQ_x$ where Q_x is on $\mathscr{C}_B$ and $\overline{XQ_x} \perp \overline{AB}$. Define $h(c - b) = 0$. Following through with the same reasoning, we see that h is a continuous (increasing) function defined for x when $c - b \leqq x \leqq a$ and such that $h(c - b) = 0$ and $h(a) > 0$.

Now define $k(x) = g(x) - h(x)$ for $c - b \leqq x \leqq a$. Then k is continuous, $k(c - b) > 0$, and $k(a) < 0$. By the intermediate-value theorem there is a real number t such that $c - b < t < a$ and $k(t) = 0$. Then $g(t) = h(t)$. Letting $f(T) = t$, we see that the perpendicular to $\overline{AB}$ at T must contain a point C on both $\mathscr{C}_A$ and $\mathscr{C}_B$. So $AC = a$, $BC = b$, and $AB = c$. Therefore, $\triangle ABC$ has the desired properties. ∎

Theorem 20.16 *Two-Circle Theorem*   If $\mathscr{C}_A$ is a circle with center A and radius a, $\mathscr{C}_B$ is a circle with center B and radius b, $AB = c$, and each of a, b, c is less than the sum of the other two, then $\mathscr{C}_A$ and $\mathscr{C}_B$ intersect in exactly two points, one on each side of $\overleftrightarrow{AB}$.

Proof   By the Triangle Theorem there exists $\triangle PQR$ such that $a = PR$, $b = RQ$, and $c = PQ$. There exist two points C_1 and C_2, one on each side of $\overleftrightarrow{AB}$, such that $\angle BAC_1$ and $\angle BAC_2$ are congruent to $\angle QPR$ and such that $\overline{AC_1}$ and $\overline{AC_2}$ are congruent to $\overline{PR}$. Thus C_1 and C_2 are on $\mathscr{C}_A$. Further, $\triangle BAC_1 \simeq \triangle QPR$ and $\triangle BAC_2 \simeq \triangle QPR$ by SAS. So $BC_1 = b$ and $BC_2 = b$. Thus C_1 and C_2 are on $\mathscr{C}_B$. Since two circles intersect in at most two points, C_1 and C_2 are the unique points of intersection of $\mathscr{C}_A$ and $\mathscr{C}_B$. ∎

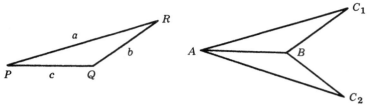

FIGURE 20.9

20.3 EXERCISES

20.1 Theorem 20.6.

20.2 Finish the proof of Theorem 20.12.

● **20.3** Write out an alternative proof of Theorem 20.9 using the following outline: Define $g(x)$ as in the proof of the theorem. Let t be the greatest lower bound of all x such that $g(x) > r$. Assume $g(t) > r$. Let $g(t) = r + h$ and $s + h = t$. Then $g(s) + h > g(t)$, giving contradiction $g(s) > r$. Assume $g(t) < r$. Let $g(t) + h = r$ and $s = t + h$. Then $g(s) < r$. If $0 < x < s$, then $g(x) < r$. So s is a lower bound, a contradiction. Hence $g(t) = r$.

20.4 Prove the existence of an isosceles triangle having sides of length r, r, and c when $2r > c > 0$ directly from Theorem 20.9.

● **20.5** True or False?

    (a) There do not exist two concentric congruent circles.

    (b) The set of all circles is fixed under any isometry.

    (c) The exterior of a circle is a convex set.

    (d) If a point off a circle is on a tangent of the circle, then the point is in the exterior of the circle.

    (e) If two points A and B are on the same side of $\overleftrightarrow{CD}$, then C and D are on the same side of $\overleftrightarrow{AB}$.

    (f) Line l is a line of symmetry for line m iff $l = m$ or $l \perp m$.

    (g) If line l is a line of symmetry for each of two circles, then the circles are concentric.

    (h) Three distinct points lie on some circle.

(i) Three points on a circle determine the center and the radius of the circle.

(j) If line l intersects a circle, then the foot of the perpendicular from the center of the circle to l is in the interior of the circle.

20.6 Read "Mathematics and Creativity" by A. Adler in the February 19, 1972 issue of *The New Yorker* magazine (pp. 39–45). Don't miss the cartoon on page 38!

20.7 Give a different proof of Theorem 20.12 using Theorem 20.4 and Theorem 20.9.

20.8 A circle with radius r has a chord of length x iff $0 < x \leq 2r$.

• **20.9** Let $\mathscr{C}_A$ be the circle with center A and radius a, $\mathscr{C}_B$ be the circle with center B and radius b, and $AB = c$. Find necessary and sufficient conditions on a, b, c for the two circles (i) to be disjoint, (ii) to intersect in exactly one point, (iii) to intersect in exactly two points, and (iv) to intersect in three points.

20.10 Two circles intersect in exactly two points iff one (and hence each) of the circles contains points in the interior and points in the exterior of the other circle.

20.11 Two circles intersect in exactly one point iff the two circles have a common tangent.

• **20.12** Let x, y, u, v be numbers. If $x < y$ implies $u > v$, $x > y$ implies $u < v$, and $x = y$ implies $u = v$, then $x < y$ iff $u > v$, $x = y$ iff $u = v$, and $x > y$ iff $u < v$.

20.13 $A - C - B$ iff point C is on $\overleftrightarrow{AB}$ and there exists a line l containing C such that A and B are on opposite sides of l.

20.14 The union of a circle and its interior is a convex set.

20.15 Which of the statements in Theorems 20.2, 20.4, and 20.5 fail to hold in the Taxicab Plane $(M1, t, m)$? Which statements in the theorems of this chapter hold for the Rational Cartesian Plane $(M2, d, m)$?

20.16 If the vertices of two congruent triangles are all on one circle, then any isometry that maps one of the triangles to the other is a product of at most two reflections.

*__20.17__ Which statements in the theorems of this chapter hold for the Taxicab Plane $(M1, t, m)$?

*__20.18__ Answer Exercise 20.9 for the Rational Cartesian Plane.

GRAFFITI

Our Geometry is an abstract Geometry. The reasoning could be followed by a disembodied spirit who has no idea of a physical point; just as a man blind from birth could understand the Electromagnetic Theory of Light.

Forder

Among them [the Greeks] geometry was held in highest honor: nothing was more glorious than mathematics. But we have limited the usefulness of this art to measuring and calculating.

Cicero

What distinguishes the straight line and circle more than anything else, and properly separates them for the purpose of elementary geometry? Their self-similarity. Every inch of a straight line coincides with every other inch, and of a circle with every other of the same circle. Where, then, did Euclid fail? In not introducing the third curve, which has the same property — the screw. *The right line, the circle, the screw — the representations of translation, rotation, and the two combined — ought to have been the instruments of geometry. With a screw we should never have heard of the impossibility of trisecting an angle, squaring the circle, etc.*

De Morgan

. . . when Gauss was nineteen, she [Gauss' mother] asked his mathematical friend Wolfgang Bolyai whether Gauss would ever amount to anything. When Bolyai exclaimed "The greatest mathematician in Europe!" she burst into tears.

E. T. Bell

It is evident to everyone that the equilateral is the most beautiful of triangles and most akin to the circle.

Proclus

CHAPTER 21

Absolute Geometry and Saccheri Quadrilaterals

21.1 EUCLID'S ABSOLUTE GEOMETRY

Euclid's proof of his first proposition in Book I of the *Elements* is as follows.

Theorem 21.1 *Euclid's Proposition I.1* Given $\overline{AB}$, there exists an equilateral triangle with side $\overline{AB}$.

Proof *(Euclid)* Let $AB = r$. Let C be a point of intersection of the circle with center A and radius r and the circle with center B and radius r. (Such a point C exists by the Two-Circle Theorem.) Since $AC = BC = r$, $\triangle ABC$ is an equilateral triangle. ■

The parenthetic note that such a point C does exist has been added to Euclid's proof. Of course, the existence of C is essential! Without the Two-Circle Theorem or some axiom other than those we suppose were given by Euclid, the proof falls apart. That this flaw is not easily repaired can be seen by considering the rather difficult arguments, such as the proof of Theorem 20.9, that are needed to prove the Two-Circle Theorem.

Although Euclid's second proposition follows directly from the Segment-Construction Theorem, we give Euclid's proof for later reference.

Theorem 21.2 *Euclid's Proposition I.2* Given $\overline{BC}$ and point A, there exists a point L such that $\overline{AL} \simeq \overline{BC}$.

Proof *(Euclid)* If $A = B$, the result is trivial. Suppose $A \neq B$ and point D is such that $\triangle ABD$ is equilateral (Theorem 21.1). So $DA = DB$. (See Figure 21.1.) Let the circle with center B and radius BC intersect $\overleftrightarrow{DB}$ at point G such that $D - B - G$. Then $BG = BC$ and $DG = DB + BG$. Let the circle with center D and radius DG intersect $\overrightarrow{DA}$ at point L. So $DL = DG$. Since $DL = DG > DB = DA$, we have $D - A - L$ and $DL = DA + AL$. Since $DL = DG$, we have $DA + AL = DB + BG$. Finally, since $DA = DB$ and $BG = BC$, we have $AL = BC$. Hence $\overline{AL} \simeq \overline{BC}$, as desired. ■

DEFINITION 21.3 Let $\overleftrightarrow{AB}$, $\overleftrightarrow{CD}$, and $\overleftrightarrow{FG}$ be three distinct lines such that $A - F - B$ and $C - G - D$. Then $\overleftrightarrow{AB}$ and $\overleftrightarrow{CD}$ are *cut* by *transversal* $\overleftrightarrow{FG}$ such that $\angle AFG$ is an *interior angle* (as are $\angle BFG$, $\angle CGF$, and $\angle DGF$) and, if A and D are on opposite sides of $\overleftrightarrow{FG}$, then $\angle AFG$ and $\angle DGF$ are a pair of *alternate interior angles* (as are $\angle BFG$ and $\angle CGF$). Further, if x and y are a pair of alternate interior angles and y and z are a pair of vertical angles, then x and z are a pair of *corresponding angles*.

Theorem 21.4 If two lines are cut by a transversal, then the following are equivalent:

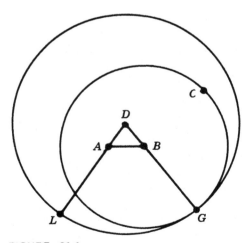

FIGURE 21.1

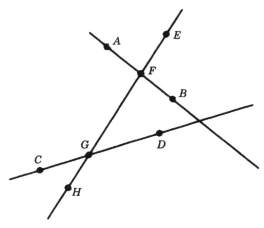

FIGURE 21.2

(a)   The angles in a pair of alternate interior angles are congruent.

(b)   The angles in a pair of corresponding angles are congruent.

(c)   Each of the four pairs of corresponding angles is a pair of congruent angles.

(d)   Each of the two pairs of alternate interior angles is a pair of congruent angles.

(e)   The interior angles intersecting the same side of the transversal are supplementary.

Proof   Since vertical angles are congruent and a linear pair of angles are supplementary, the result follows from the previous definition.   ■

Theorem 21.5   If two lines are cut by a transversal such that a pair of alternate interior angles are congruent, then the two lines have a common perpendicular.

Proof   If a pair of alternate interior angles are right angles, the result is trivial. Also, if one pair of alternate interior angles are obtuse, then another pair is acute. Therefore, suppose A and D are on opposite sides of $\overleftrightarrow{FG}$ and $\angle AFG$ and $\angle DGF$ are congruent and acute. We must show $\overleftrightarrow{AF}$ and $\overleftrightarrow{DG}$ have a common perpendicular. Let M be the midpoint of $\overline{FG}$. Let P be the foot of the perpendicular from M to $\overleftrightarrow{AF}$. Since $\angle AFM$ is acute, P and A are on the same side of $\overleftrightarrow{FG}$. Let Q be the foot of the perpendicular from M to $\overleftrightarrow{DG}$. Since $\angle DGM$ is acute, Q and D are on the same side of $\overleftrightarrow{FG}$. Then, since $\angle MFP \simeq \angle MGQ$, we have $\triangle MFP \simeq \triangle MGQ$ by SAA. So $\angle FMP \simeq \angle GMQ$. Therefore, since $F-M-G$ and

P and Q are on opposite sides of $\overleftrightarrow{FG}$, $\angle FMP$ and $\angle GMQ$ must be vertical angles. Thus P, M, Q are collinear and $\overleftrightarrow{PQ}$ is perpendicular to both $\overleftrightarrow{AF}$ and $\overleftrightarrow{DG}$. ■

Theorem 21.5 is stronger than the two corollaries below since we have not proved that two parallel lines have a common perpendicular.

Corollary 21.6 *Euclid's Proposition I.27* If two lines are cut by a transversal such that a pair of alternate interior angles are congruent, then the two lines are parallel.

Proof By the hypothesis, the two lines have a common perpendicular (Theorem 21.5). Hence, the two lines are parallel (Theorem 18.2). ■

Corollary 21.7 *Euclid's Proposition I.28* If two lines are cut by a transversal such that a pair of corresponding angles are congruent or such that the interior angles intersecting one side of the transversal are supplementary, then the two lines are parallel.

Proof Follows from Theorem 21.4 and Corollary 21.6. ■

We can now say that we have proved all of Euclid's first twenty-eight propositions. A geometry that satisfies our five axioms is called *the absolute plane*. The *"the"* is meant to imply that no other axioms are allowed and not that the axiom system is categorical! It can truly be said that Euclid was the first to write a treatise on absolute geometry. His first twenty-eight propositions comprise such a treatise.

Because of Theorem 21.4, Euclid's Proposition I.27 and Proposition I.28 can be summarized in one sentence: *If A and D are on the same side of $\overleftrightarrow{BC}$ and $m\angle ABC + m\angle BCD = \pi$, then $\overleftrightarrow{AB} \parallel \overleftrightarrow{CD}$.* Also, since Euclid's Proposition I.29 states all the converses of I.27 and I.28, we can state I.29 succinctly as follows: *If A and D are on the same*

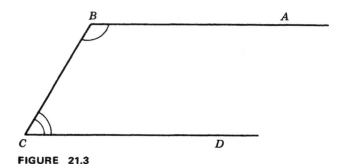

FIGURE 21.3

side of $\overleftrightarrow{BC}$ *and* $\overleftrightarrow{AB} \parallel \overleftrightarrow{CD}$, *then* $m\angle ABC + m\angle BCD = \pi$. In order to prove this proposition Euclid uses *for the first time* his parallel postulate.

   Euclid's Parallel Postulate: If A and D are on the same side of $\overleftrightarrow{BC}$ *and* $m\angle ABC + m\angle BCD < \pi$, *then* $\overrightarrow{BA}$ *and* $\overrightarrow{CD}$ *intersect.* This postulate, which certainly has been one of the most controversial statements ever made, is exactly what is needed to prove Proposition I.29. The argument runs as follows. Suppose two parallel lines are cut by a transversal. Let x and y be the measures of the interior angles intersecting one side of the transversal, and let u and v be the measures of the interior angles intersecting the other side of the transversal. Since the two lines are parallel, from Euclid's Parallel Postulate we have $x + y \geqq \pi$ and $u + v \geqq \pi$. However, since $x + u + y + v = 2\pi$, it follows that $x + y = \pi$ and $u + v = \pi$ as desired.

   Probably the most familiar parallel axiom is Euclid's Proposition I.31: *If point P is off line l, then there exists a unique line through P that is parallel to l.*

   This proposition has come to be known as *Playfair's Parallel Postulate* since the work of John Playfair (1748–1819) was influential in having Euclid's Parallel Postulate replaced by this proposition in most geometry textbooks. The proposition follows from the theorems of absolute geometry and Euclid's Proposition I.29. Let F be the foot of the perpendicular from P to l. Then, by Proposition I.29, for any line $\overleftrightarrow{PQ}$ parallel to l we must have $m\angle QPF = \pi/2$. Therefore (Theorems 18.1 and 18.3) there is a unique line through P that is parallel to l, as desired.

   To show that Playfair's Parallel Postulate is equivalent to Euclid's Parallel Postulate, we now prove Euclid's Parallel Postulate using only the theorems of absolute geometry and Playfair's Parallel Postulate. Suppose A and D are on the same side of $\overleftrightarrow{BC}$ and $m\angle ABC + m\angle BCD < \pi$. Let E be on the same side of $\overleftrightarrow{BC}$ as A and such that $m\angle ABC + m\angle BCE = \pi$. Then $\overleftrightarrow{BA} \parallel \overleftrightarrow{CE}$ by Euclid's Proposition I.28, which is a theorem of absolute geometry. By Playfair's Parallel Postulate $\overleftrightarrow{CE}$ is the only line through C that is parallel to $\overleftrightarrow{BA}$. Since $m\angle BCD < m\angle BCE$, then $\overleftrightarrow{CD} \neq \overleftrightarrow{CE}$ and $\overleftrightarrow{CD}$ must intersect $\overleftrightarrow{BA}$. Since $\overleftrightarrow{CE} \parallel \overleftrightarrow{BA}$, A and D are on the same side of $\overleftrightarrow{CE}$, and A and D are on the same side of $\overleftrightarrow{BC}$, then it follows that $\overrightarrow{CD}$ and $\overrightarrow{BA}$ must intersect, as desired.

   By the arguments above, each of Euclid's Parallel Postulate, Euclid's Proposition I.29, and Playfair's Parallel Postulate is equivalent to the other. We shall avoid the useless discussions about which of these is the most *self-evident*. Playfair's Parallel Postulate does have the advantage of being a simple incidence property. Other such statements that can be added to the list of equivalents are Euclid's Proposition I.30, which states that lines parallel to a given line are parallel, and its contrapositive, which states that a third line intersecting one of two parallel lines intersects the other (Exercise 21.1).

The introduction of the parallel postulate into Euclid's *Elements* was a stroke of pure genius, probably unparalleled in the history of human thought. Yet, from the beginning, the commentators considered the postulate a flaw. It must be remembered that until a little over a hundred years ago a postulate was supposed to be a *self-evident truth*. If one can put emotions and previous indoctrination aside, it is clear that Euclid's Parallel Postulate is not *self-evident*. However, since the postulate was deemed to be necessarily true, there was the problem of showing that the postulate is a theorem. Perhaps more *genius* man-hours have been spent on the problems of the parallel postulate than any other human intellectual endeavor. The mathematical study of the whole problem of the parallel postulate in absolute geometry is aptly called the *theory of parallels*. It may have taken two thousand years for Euclid to be vindicated, but vindicated he was! We now know that any argument given to prove Euclid's Parallel Postulate in absolute geometry is necessarily circular, whether the argument is given by a mathematician, a philosopher, or a theologian.

The problems of the parallel postulate became tied up with the quandary *What is truth?* The ramifications of the theory of parallels have had as important implications on man's view of his relation to his universe and his gods as have either the Copernican theory of heliocentricity or the Darwinian theory of evolution. Because the theory of parallels cannot be intelligently discussed in any depth without assuming a technical vocabulary and an understanding of absolute geometry, the subject is seldom mentioned by general historians. Even though the theory of parallels is inaccessible to most educated people, their modes of thought have been influenced by its development.

The oldest known *proof* of Euclid's Parallel Postulate is that which Proclus attributes to Ptolemy. Ptolemy proves the equivalent Euclid's Proposition I.29 as follows: Suppose $\overleftrightarrow{AB}$ and $\overleftrightarrow{CD}$ are two parallel lines cut by transversal $\overleftrightarrow{FG}$ such that $A-F-B$ and $C-G-D$. (See Figure 21.4.) Now $\overleftrightarrow{AF}$ and $\overleftrightarrow{CG}$ are no more parallel than $\overleftrightarrow{FB}$ and $\overleftrightarrow{GD}$. Therefore, assuming $m\angle AFG + m\angle CGF > \pi$, then $m\angle BFG + m\angle DGF > \pi$; and, likewise, assuming $m\angle AFG + m\angle CGF < \pi$, then $m\angle BFG + m\angle DGF < \pi$. However, each assumption leads to a contradiction since the sum of the measures of the four interior angles is 2π. Hence the sum of the measures of the interior angles intersect-

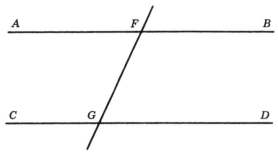

FIGURE 21.4

ing one side of the transversal $\overleftrightarrow{FG}$ is π. Proclus correctly points out Ptolemy's error in assuming that whatever is true about the interior angles intersecting one side of a transversal cutting two parallel lines is necessarily true of the interior angles intersecting the other side of the transversal.

After pointing out the error in Ptolemy's *proof* of Euclid's Parallel Postulate, Proclus gives his own *proof* by first showing that a third line intersecting one of two parallel lines intersects the other. We have previously observed that this proposition is equivalent to Euclid's Parallel Postulate, since it is the contrapositive of the equivalent Euclid's Proposition I.30. For Proclus' *proof*, let $\overleftrightarrow{AB}$ and $\overleftrightarrow{CD}$ be two parallel lines such that $\overleftrightarrow{FG}$ intersects $\overleftrightarrow{AB}$ at F with G on the same side of $\overleftrightarrow{AB}$ as $\overleftrightarrow{CD}$. (See Figure 21.5.) Let r be the distance between $\overleftrightarrow{AB}$ and $\overleftrightarrow{CD}$. Let H be a point on $\overrightarrow{FG}$ such that, if H' is the foot of the perpendicular from H to $\overleftrightarrow{AB}$, then $HH' > r$. Hence H and F are on opposite sides of $\overleftrightarrow{CD}$. Thus $\overleftrightarrow{FG}$ intersects $\overleftrightarrow{CD}$ as desired.

Let's examine Proclus' argument in some detail. Does there actually exist an H on $\overrightarrow{FG}$ such that $HH' > r$? To verify the existence of H, Proclus cites an axiom from Aristotle. Aristotle's Axiom: *Given $r > 0$ and $\angle BFG$, then there is a point H on $\overrightarrow{FG}$ such that the distance*

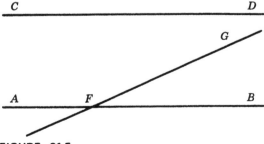

FIGURE 21.5

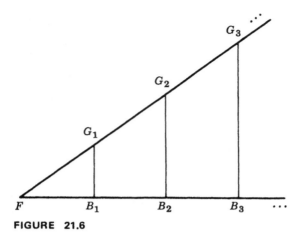

FIGURE 21.6

from H *to* $\overleftrightarrow{BF}$ *is greater than r*. This statement certainly seems reasonable. However, as Proclus says, quoting Geminus, we must not pay attention to plausible imaginings in deciding what propositions are to be accepted. To support his claim that Euclid's Parallel Postulate ought to be struck from the postulates altogether since it is a theorem, Proclus should either prove Aristotle's Axiom or admit that he is introducing a new postulate.

Aristotle's Axiom holds if $\angle BFG$ is a right angle and is true for obtuse angles if true for acute angles. Looking at Figure 21.6, since $\angle B_i FG_i$ is acute, it follows (see Theorem 21.8) that $G_1B_1 < G_2B_2 < G_3B_3 < \cdots$. Does this imply $G_nB_n > r$ for some n? After noting that although implausible and paradoxical there do exist curves that approach each other *indefinitely* but never meet, Proclus then asks whether this may not be possible for lines as well. The question is excellent; an answer is lacking! Since the hyperbola with equation $xy = 1$ is asymptotic to the x-axis in the Cartesian plane, we have $G_1B_1 < G_2B_2 < G_3B_3 < \cdots$ in Figure 21.7 but $G_nB_n < 1$ for all n. Of course a hyperbola is not a line in the Cartesian plane, but the situation is analogous. However, we shall later see that Aristotle's Axiom is a theorem of absolute geometry.

Assuming Aristotle's Axiom is a theorem of absolute geometry, where does Proclus' *proof* fall apart? The trouble is in letting r be *the distance between parallel lines* $\overleftrightarrow{AB}$ *and* $\overleftrightarrow{CD}$. Is r defined? Proclus was tacitly assuming a property of parallel lines that is equivalent to Euclid's Parallel Postulate. Proclus was probably not the first to fall into this pit since he informs us that Posidonius, in the first century B.C., had *defined* two lines to be parallel if all the points of one line are equidistant from the other line. We can only wonder how

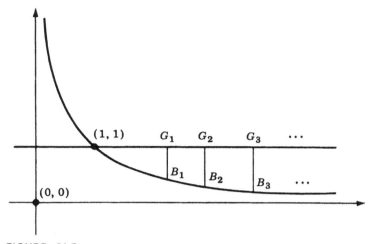

FIGURE 21.7

Posidonius tried to show that two distinct lines are *parallel* if and only if the lines do not intersect. Nor was Proclus the last to be trapped by this pitfall in the theory of parallels. Two thousand years after Posidonius we find in an American journal an article supposedly proving Playfair's Parallel Postulate by using the *lemma* that parallel lines are equidistant. Ironically, this article is next to another on non-Euclidean geometry. The argument for the lemma is as follows. Let $\overleftrightarrow{PQ}$ be perpendicular to $\overleftrightarrow{MN}$ at M; let $\overleftrightarrow{RS}$ be perpendicular to $\overleftrightarrow{MN}$ at N. (See Figure 21.8.) Let B any point on $\overleftrightarrow{PQ}$. Let A be the foot of the perpendicular from B to $\overleftrightarrow{RS}$. Let D be such that $A - N - D$ and $AN = ND$. Let C be such that C is on the same side of $\overleftrightarrow{RS}$ as B, $\overline{CD} \perp \overline{AD}$, and $CD = BA$. Now the reflection in $\overleftrightarrow{MN}$ fixes $\overleftrightarrow{PQ}$ and $\overleftrightarrow{RS}$ (Theorem 19.6) and sends A to D. Since a reflection preserves distance and angle measure, the reflection must take B to C. Hence C is on $\overleftrightarrow{PQ}$ and $CD = BA$. The conclusion is that, since B is any point on $\overleftrightarrow{PQ}$, then all the

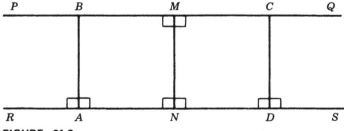

FIGURE 21.8

points of $\overleftrightarrow{PQ}$ are equidistant from $\overleftrightarrow{RS}$. Before reading about the flaws given in the next paragraph, try to find the errors yourself.

The first error is the most subtle. Supposedly we are showing that $\overleftrightarrow{PQ} \parallel \overleftrightarrow{RS}$ implies all the points of $\overleftrightarrow{PQ}$ are equidistant from $\overleftrightarrow{RS}$. Where does $\overleftrightarrow{MN}$ come from? We are tacitly assuming that two parallel lines have a common perpendicular. It can be shown (Corollary 24.14) that this assumption itself is equivalent to Euclid's Parallel Postulate! (We do know that two lines having a common perpendicular are parallel (Theorem 18.2), but we do not know that two parallel lines have a common perpendicular.) The case $B = M$ is not considered in the argument, but this is minor. The most obvious error is assuming that $CD = BA$ implies $\overleftrightarrow{PQ}$ is equidistant from $\overleftrightarrow{RS}$. What is actually proved is that for each point B on $\overleftrightarrow{PQ}$ and on one side of $\overleftrightarrow{MN}$ there exists at least *one* point C on $\overleftrightarrow{PQ}$ and on the opposite side of $\overleftrightarrow{MN}$ such that B and C are equidistant from $\overleftrightarrow{RS}$.

21.2 GIORDANO'S THEOREM

One after another the geometers of Africa, Asia, and Europe followed Posidonius and Proclus into the pitfall of trying to identify the locus of all points equidistant from a line and on one side of the line with a line. The first significant result in this aspect of the theory of parallels does not come until 1680 in an attempt by Giordano Vitale (1633–1711) to prove Euclid's Parallel Postulate. Giordano, as Omar Khayyam (circa 1050–1123) before him, anticipated the first few theorems of Gerolamo Saccheri (1667–1733).

Theorem 21.8 If $\square ABCD$ has right angles at A and D, then $\square ABCD$ is a convex quadrilateral. Further, $\overline{AB} < \overline{CD}$ iff $\angle B > \angle C$, $\overline{AB} \simeq \overline{CD}$ iff $\angle B \simeq \angle C$, and $\overline{AB} > \overline{CD}$ iff $\angle B < \angle C$.

Proof Suppose $\square ABCD$ has right angles at A and D. $\overline{AD}$ and $\overline{BC}$ cannot intersect by definition of a quadrilateral. $\overline{BC}$ cannot intersect $\overleftrightarrow{AD}$ at a point E off $\overline{AD}$ as otherwise one of $\triangle AEB$ or $\triangle DEC$ has two angles of measure at least $\pi/2$, a contradiction (Theorem 18.20). So B, C, and $\overline{BC}$ are on the same side of $\overleftrightarrow{AD}$. Then, $\overline{AD}$ cannot intersect $\overleftrightarrow{BC}$ at a point F as otherwise one of $\triangle AFB$ or $\triangle DFC$ has two angles of measure at least $\pi/2$. So $\overline{AD}$ is on a halfplane of $\overleftrightarrow{BC}$. Since $\overleftrightarrow{AB} \parallel \overleftrightarrow{CD}$ (Theorem 18.2), each side of $\square ABCD$ is on a halfplane of the opposite side. Thus $\square ABCD$ is convex.

Suppose $AB = CD$. Then $\triangle ADC \simeq \triangle DAB$ by SAS and $AC = DB$. So $\triangle ABC \simeq \triangle DCB$ by SSS and $\angle ABC \simeq \angle DCB$. Hence, $AB = CD$ implies $\angle B \simeq \angle C$. Now suppose $AB < CD$. Let E be such that $C - E - D$ and $DE = AB$. Then E is in int $(\angle ABC)$. So $m \angle ABC > m \angle ABE$. Since $m \angle ABE = m \angle DEB$ by the result above and since $m \angle DEB > m \angle DCB$ (Theorem 17.9), we have $m \angle ABC > m \angle DCB$. Hence, $AB < CD$ implies $\angle B > \angle C$. By symmetry, $AB > CD$ implies $\angle C < \angle B$. The theorem follows by the three implications and their contrapositives. ∎

DEFINITION 21.9 If $\square ABCD$ has right angles at A and D with $AB = CD$, then the quadrilateral is denoted by $\boxed{S} ABCD$ and called a *Saccheri quadrilateral.* $\boxed{S} ABCD$ has **legs** $\overline{AB}$ and $\overline{CD}$, **lower base** $\overline{AD}$, **upper base** $\overline{BC}$, and **upper base angles** $\angle B$ and $\angle C$. A quadrilateral having four right angles is a **rectangle**. If $\overleftrightarrow{AB} \parallel \overleftrightarrow{CD}$, then $\overline{AB}$ and $\overline{CD}$ are said to be **parallel**. Line l is **equidistant** from line m if every two points of l are equidistant from m.

Although $\boxed{S} ABCD = \boxed{S} DCBA$ by the symmetry of the definition, it would be a mistake to infer that $\boxed{S} ABCD$ implies $\boxed{S} BCDA$. The order of the letters as they appear in the notation is important. Further, we must avoid having the words *lower* and *upper* lead us astray. For $\boxed{S} RSPQ$ in Figure 21.9, $\overline{QR}$ is the *lower* base and $\angle P$ and $\angle S$ are the *upper* base angles.

We shall usually abbreviate such a phrase as "if $\boxed{S} ABCD$ exists" to "if $\boxed{S} ABCD$." Likewise, we shall frequently use the concise phrase "so $\boxed{S} ABCD$" in place of the longer phrase "so A, B, C, D are points such that $\boxed{S} ABCD$ exists." Thus, the symbol "$\boxed{S} ABCD$" has the double duty of serving as the name of a particular set of points and as an abbreviation for a statement that the points A, B, C, D are in a special relation to each other. This same sort of convention will be applied to the notation for Lambert quadrilaterals, which are introduced in the next chapter.

Theorem 21.10 A Saccheri quadrilateral is a convex quadrilateral.

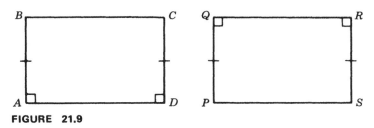

FIGURE 21.9

The upper base angles of a Saccheri quadrilateral are congruent. The diagonals of a Saccheri quadrilateral are congruent. The opposite sides of a rectangle are congruent.

Proof The first, second, and last statements are special cases of Theorem 21.8. If $\boxed{S}\,ABCD$, then $\triangle BAD \simeq \triangle CDA$ by SAS and $\overline{BD} \simeq \overline{CA}$. ∎

We have *defined* rectangles and even proved a property rectangles must have if they exist, but we do not know that a rectangle exists. We can define a *unicorn* and describe properties such an animal must have, but this does not imply that unicorns exist.

Theorem 21.11 The line through the midpoints of the bases of a Saccheri quadrilateral is perpendicular to each base.

Proof Suppose $\boxed{S}\,ABCD$ with M and N the midpoints of $\overline{BC}$ and $\overline{AD}$, respectively. Then $\triangle BAN \simeq \triangle CDN$ by SAS, and $\triangle ABM \simeq \triangle DCM$ by SAS. So $BN = CN$ and $AM = DM$. Hence N is on the perpendicular bisector of $\overline{BC}$, and M is on the perpendicular bisector of $\overline{AD}$ (Theorem 18.5). Thus $\overset{\leftrightarrow}{MN}$ is the perpendicular bisector of each base. ∎

Corollary 21.12 The perpendicular bisector of one base of a Saccheri quadrilateral is the perpendicular bisector of the other base. The bases of a Saccheri quadrilateral are parallel. If two points on $\overset{\leftrightarrow}{PQ}$ are on the same side of $\overset{\leftrightarrow}{RS}$ and are equidistant from $\overset{\leftrightarrow}{RS}$, then $\overset{\leftrightarrow}{PQ} \parallel \overset{\leftrightarrow}{RS}$. The line through the midpoints of the legs of a Saccheri quadrilateral is perpendicular to the line through the midpoints of the bases.

Proof Only the last statement is not a direct consequence of the theorem. If $\boxed{S}\,ABCD$ with F and G the midpoints of $\overline{AB}$ and $\overline{CD}$, respectively, then $\boxed{S}\,AFGD$. So the last statement in the corollary follows from the first. ∎

Theorem 21.13 *Giordano's Theorem* If three points of line l are equidistant from line m, then l is equidistant from m.

Proof If $l = m$, the result is trivial. Suppose l and m are distinct lines such that A, B, C are three points on l that are equidistant from m. Since two of these points must be on the same side of m, then $l \parallel m$ (Corollary 21.12). Thus A, B, C are on the same side of m. Suppose $A - B - C$. Let A', B', C' be the feet of the perpendiculars to m from A, B, C, respectively. Then $\boxed{S}\,A'ABB'$, $\boxed{S}\,B'BCC'$, and $\boxed{S}\,A'ACC'$.

So $\angle B'BA$, $\angle A'AB$, $\angle C'CA$, and $\angle B'BC$ are all congruent (Theorem 21.10), and, since $\angle B'BA$ and $\angle B'BC$ are a linear pair, the four angles are right angles. Hence $\boxed{S}$ $A'ACC'$ is a rectangle.

Suppose P is any point such that $A - P - C$. Let P' be the foot of the perpendicular from P to m. We wish to show $PP' = AA'$. Assume $\angle P'PA$ is acute. So $PP' > AA' = CC'$ (Theorem 21.8) and $\angle P'PC$ is obtuse. Since $\angle P'PC$ is obtuse, then $CC' > PP'$ (Theorem 21.8). Therefore $PP' > PP'$, a contradiction. Likewise, the assumption that $\angle P'PA$ is obtuse leads to the contradiction $PP' < PP'$. Hence $\angle P'PA$ is a right angle. Thus $\square A'APP'$ is a rectangle and $PP' = AA'$, as desired. We also note that we have shown the lemma that if three points are equidistant from a line and a fourth point is between two of the three points, then the four points are equidistant from the line.

Finally, suppose Q is on l but off $\overline{AC}$. Let Q' be the foot of the perpendicular from Q to m. (See Figure 21.10.) We wish to show $QQ' = AA'$. Let R be the image of Q under the reflection in $\overleftrightarrow{AA'}$; let S be the image of Q under the reflection in $\overleftrightarrow{CC'}$. Since these two reflections are isometries that fix both l and m (Theorem 19.6), we have $R - A - Q$ and Q, R, S are three points equidistant from m. By the lemma, we must have $AA' = QQ'$, as desired. Therefore l is equidistant from m. ∎

Theorem 21.14 If line l is equidistant from line m, then m is equidistant from l.

Proof Exercise 21.2. ∎

Given line l it is easy to find a different line m such that *two* points of m are equidistant from l. All we have to do is *construct* a Saccheri quadrilateral with lower base on l and let m be the line containing the upper base. Giordano has shown that if we are to find a line m different from line l such that *every two* points of m are equidistant from l, then it is sufficient to find *three* collinear points on the same side of l that are equidistant from l.

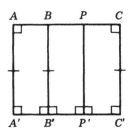 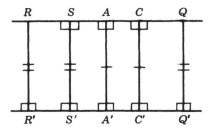

FIGURE 21.10

21.3 EXERCISES

21.1 The following are equivalent in absolute geometry: (i) Playfair's Parallel Postulate. (ii) Lines parallel to a given line are parallel. (iii) If a third line intersects one of two parallel lines, then the third line intersects the other.

21.2 Theorem 21.14.

● **21.3** If $\boxed{\text{S}}\,ABCD$ with M and N the midpoints of $\overline{BC}$ and $\overline{AD}$, respectively, then $\overline{MN}$ is shorter than, congruent to, or longer than $\overline{AB}$ as, respectively, $\angle B$ is acute, right, or obtuse.

● **21.4** The following is not a theorem of absolute geometry: If A, B, C, D are four points such that $\angle BAD$ and $\angle CDA$ are right angles, then $AB < CD$ iff $\angle ABC > \angle DCB$.

● **21.5** True or False?

(a) If A and D are points on opposite sides of $\overleftrightarrow{BC}$ and $\angle ABC \simeq \angle BCD$, then $\overleftrightarrow{AB} \parallel \overleftrightarrow{CD}$.

(b) If A and D are points on the same side of $\overleftrightarrow{BC}$ and $m\angle ABC + m\angle BCD = \pi$, then $\overleftrightarrow{AB} \parallel \overleftrightarrow{CD}$.

(c) If A and D are points on the same side of $\overleftrightarrow{BC}$ and $m\angle ABC + m\angle BCD < \pi$, then $\overrightarrow{BA}$ intersects $\overrightarrow{CD}$.

(d) If A and D are points on the same side of $\overleftrightarrow{BC}$ and $m\angle ABC + m\angle BCD > \pi$, then $\overleftrightarrow{AB}$ intersects $\overleftrightarrow{CD}$.

(e) If A and D are points on opposite sides of $\overleftrightarrow{BC}$ and $m\angle ABC + m\angle BCD > \pi$, then $\overleftrightarrow{AB}$ intersects $\overleftrightarrow{CD}$.

(f) If two lines are parallel, then the lines have a common perpendicular.

(g) If two distinct points A and B are equidistant from line l, then $\overleftrightarrow{AB} \parallel l$.

(h) If lines l and m are parallel, then there are two points on l that are equidistant from m.

(i) If lines l and m intersect, then there are two points on l that are equidistant from m.

(j) Euclid's Proposition I.28 is equivalent to Euclid's Parallel Postulate.

• **21.6** Find the flaw in Aganis' *proof* of Euclid's Parallel Postulate: Suppose A and D are on the same side of $\overleftrightarrow{BC}$, $\angle ABC$ is acute, and $\angle BCD$ is right. Let E be foot of perpendicular from A to $\overleftrightarrow{BC}$, and suppose $B - E - C$. Let F be such that $B - F - E$ and $BC = 2^n BF$ for some integer n. Let the perpendicular to $\overline{BC}$ at F intersect $\overrightarrow{BA}$ at G. Let H be such that $B - G - H$ and $BH = 2^n BG$. Then $\overrightarrow{BA}$ intersects $\overrightarrow{CD}$ at H.

21.7 $\square ABCD$ is a rectangle iff $\boxed{S}\,ABCD$ and $\boxed{S}\,BCDA$.

21.8 $\square ABCD$ is a rectangle iff $\boxed{S}\,ABCD$ and $\boxed{S}\,CDAB$.

21.9 If two lines $\overleftrightarrow{AB}$ and $\overleftrightarrow{CD}$ have a common perpendicular $\overleftrightarrow{BC}$ and M is the midpoint of $\overline{BC}$, then every pair of corresponding angles of a transversal through M is a pair of congruent angles.

21.10 If l and m are distinct common perpendiculars to distinct lines a and b, then l is equidistant from m and a is equidistant from b.

21.11 Prove Corollary 21.7 directly from Theorem 17.9.

21.12 Find several statements equivalent to "AAA is false."

21.13 Read Proclus' commentary on Euclid's Definition 23, Postulate 5, and Propositions 28 through 32 in *Proclus: A Commentary on the First Book of Euclid's Elements* by G. R. Morrow (Princeton, 1970).

21.14 To see how an *angle* can be *equal* to a right angle but not be a right angle read Proclus' commentary on Euclid's Postulate 4 (see Exercise 21.13).

*21.15 Compare Aristotle's I.5 of *De caelo* with Archimedes' *The Sand-Reckoner*.

*21.16 "Two parallel lines have a common perpendicular" is equivalent to Euclid's Parallel Postulate.

*21.17 Euclid's Proposition I.32 is equivalent to Euclid's Parallel Postulate.

GRAFFITI

Genius is a willingness to test the strangest alternatives.

A new scientific truth triumphs, not because it convinces its opponents and makes them see the light, but because the opponents eventually die, and a new generation that is familiar with it grows up.

Max Planck

Lagrange, in the belief that he had settled the problem in the theory of parallels, was giving a lecture on the subject. In the middle of the talk he broke off with the comment, "Il faut que j'y songe encore!"

Mathematics, considered as a science, owes its origin to the idealistic *needs of the Greek philosophers, and not as fable has it, to the* practical *demands of Egyptian economics. . . .*

Hankel

Not until rigour was recaptured in the 19th century did people understand the essence of Greek mathematics.

Freudenthal

CHAPTER 22

Saccheri's Three Hypotheses

22.1 OMAR KHAYYAM'S THEOREM

The role played by Gerolamo Saccheri's book *Euclid Vindicated of all Flaw* (1733) in the development of the history of the theory of parallels will be discussed in Section 24.3 after we have learned something of its contents. In this chapter the earlier propositions from this famous book on absolute geometry are examined. Saccheri's Proposition I is our Theorem 21.10, which states that the upper base angles of a Saccheri quadrilateral are congruent. This leads to three hypotheses of which one, two, or all three might be true.

DEFINITION 22.1 *Hypothesis of the Acute Angle:* There exists a Saccheri quadrilateral with acute upper base angles. *Hypothesis of the Right Angle:* There exists a Saccheri quadrilateral with right upper base angles. *Hypothesis of the Obtuse Angle:* There exists a Saccheri quadrilateral with obtuse upper base angles.

Saccheri's Proposition II is our Theorem 21.11, stating that the line through the midpoints of the bases of a Saccheri quadrilateral is perpendicular to each base. Besides using what we have called *Saccheri quadrilaterals,* Saccheri also used quadrilaterals with three right angles to study the theory of parallels. Since Lambert later used such quadrilaterals, these are usually called *Lambert quadrilaterals.*

DEFINITION 22.2 If $\square ABCD$ has right angles at $A, B,$ and $D,$ then the quadrilateral is denoted by $\boxed{\text{L}}\,ABCD$ and called a ***Lambert quadrilateral.*** A rectangle with adjacent sides congruent is a ***square.*** $\square ABCD \simeq \square EFGH$ iff $\square ABCD$ and $\square EFGH$ are such that $\angle A \simeq \angle E,$ $\angle B \simeq \angle F, \angle C \simeq \angle G, \angle D \simeq \angle H, \overline{AB} \simeq \overline{EF}, \overline{BC} \simeq \overline{FG}, \overline{CD} \simeq \overline{GH},$ and $\overline{DA} \simeq \overline{HE}.$

By the symmetry of the definition, $\boxed{\text{L}}\,ABCD = \boxed{\text{L}}\,ADCB.$ However, the order of the letters in "$\boxed{\text{L}}\,ABCD$" is important as we do not know that $\boxed{\text{L}}\,ABCD$ implies $\boxed{\text{L}}\,DCBA.$ Of course the order of the letters in "$\square ABCD \simeq \square EFGH$" is also important. We always have $\square ABCD = \square BCDA$ but not necessarily $\square ABCD \simeq \square BCDA.$ It is assumed that Definition 22.2 determines the obvious meaning of "$\boxed{\text{L}}\,ABCD \simeq \boxed{\text{L}}\,EFGH$" and "$\boxed{\text{S}}\,ABCD \simeq \boxed{\text{S}}\,EFGH.$"

Theorem 22.3 If $\boxed{\text{S}}\,ABCD,$ M is the midpoint of $\overline{BC},$ and N is the midpoint of $\overline{AD},$ then $\boxed{\text{L}}\,NMBA$ and $\boxed{\text{L}}\,NMCD.$ If $\boxed{\text{L}}\,NMBA,$ then there exist unique points C and D such that $\boxed{\text{S}}\,ABCD$ with M the midpoint of $\overline{BC}$ and N the midpoint of $\overline{AD}.$

Proof For the first statement, since $\overleftrightarrow{NM}$ is perpendicular to both bases of $\boxed{\text{S}}\,ABCD$ (Theorem 21.11), then $\boxed{\text{L}}\,NMBA$ and $\boxed{\text{L}}\,NMCD$ by definition. Suppose $\boxed{\text{L}}\,NMBA.$ By the Midpoint Theorem there are unique points C and D such that M is the midpoint of $\overline{BC}$ and N is the midpoint of $\overline{AD}.$ Since C and D are, respectively, the images of B and A under the reflection in $\overleftrightarrow{NM},$ we have $\boxed{\text{S}}\,ABCD.$ ∎

We shall need to know that if the lower bases and legs of two Saccheri quadrilaterals are respectively congruent then the Saccheri quadrilaterals are congruent. This is covered by (1) in the next theorem which contains all the absolute congruence theorems for Saccheri quadrilaterals and Lambert quadrilaterals. Parts (2), (3), and (4) may be omitted at this time.

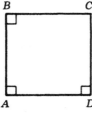

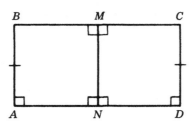

FIGURE 22.1

Theorem 22.4 If $\boxed{\text{L}}\,NMBA$ and $\boxed{\text{L}}\,N'M'B'A'$ such that either (1) $AB=A'B'$ and $AN=A'N'$, (2) $MB=M'B'$ and $MN=M'N'$, (3) $NA=N'A'$ and $NM=N'M'$, or (4) $BA=B'A'$ and $BM=B'M'$, then $\boxed{\text{L}}\,NMBA \simeq \boxed{\text{L}}\,N'M'B'A'$. If $\boxed{\text{S}}\,ABCD$ and $\boxed{\text{S}}\,A'B'C'D'$ with M, M', N, N' the midpoints of $\overline{BC}, \overline{B'C'}, \overline{AD}, \overline{A'D'}$, respectively, are such that either (1) $AB=A'B'$ and $AD=A'D'$, (2) $BC=B'C'$ and $MN=M'N'$, (3) $AD=A'D'$ and $MN=M'N'$, or (4) $AB=A'B'$ and $BC=B'C'$, then $\boxed{\text{S}}\,ABCD \simeq \boxed{\text{S}}\,A'B'C'D'$.

Proof Suppose $\boxed{\text{L}}\,NMBA$ and $\boxed{\text{L}}\,N'M'B'A'$. If (1), then $\triangle BAN \simeq \triangle B'A'N'$ by SAS and so $\triangle BNM \simeq \triangle B'N'M'$ by SAA. If (2), then $\triangle BMN \simeq \triangle B'M'N'$ by SAS and so $\triangle BNA \simeq \triangle B'N'A'$ by SAA. If (3), then $\triangle ANM \simeq \triangle A'N'M'$ by SAS and so $\triangle AMB \simeq \triangle A'M'B'$ by ASA. In each of these three cases the desired result follows easily. Now suppose (4). Let R be on $\overrightarrow{NA}$ such that $NR=N'A'$. If $R=A$, then we are done by (1). Assume $R \neq A$ and let S be on the perpendicular to $\overleftrightarrow{NA}$ at R such that S and B are on the same side of $\overleftrightarrow{NA}$ and $RS=A'B'$. Let T be the foot of the perpendicular from S to $\overleftrightarrow{MN}$. Then $\boxed{\text{L}}\,NTSR \simeq \boxed{\text{L}}\,N'M'B'A'$ by (1). So $SR=B'A'=BA$ and $ST=B'M'=BM$. Now $T \neq M$ as otherwise $S=B$ and $R=A$. Hence $\boxed{\text{S}}\,ABSR$ and $\boxed{\text{S}}\,MBST$. Let V be the midpoint of $\overline{AR}$ and W the midpoint of $\overline{MT}$. Since the perpendicular bisector of the upper base of a Saccheri quadrilateral is the perpendicular bisector of the lower base, we have the perpendicular bisector of $\overline{BS}$ is perpendicular to $\overline{NV}$ at V and also perpendicular to $\overline{NW}$ at W. Thus $\triangle NVW$ has three right angles, a contradiction. Therefore $R=A$ and $\boxed{\text{L}}\,NMBA \simeq \boxed{\text{L}}\,N'M'B'A'$. These results for the Lambert quadrilaterals imply the corresponding results for the Saccheri quadrilaterals by the previous theorem. ∎

Theorem 22.5 If $\boxed{\text{L}}\,ABCD$, then $\overline{BC}$ is longer than $\overline{AD}$ iff $\angle C$ is acute, $\overline{BC}$ is congruent to $\overline{AD}$ iff $\angle C$ is right, and $\overline{BC}$ is shorter than $\overline{AD}$ iff $\angle C$ is obtuse.

Proof Special case of Theorem 21.8. ∎

Theorem 22.6 *Omar Khayyam's Theorem* If $\boxed{\text{S}}\,ABCD$, then $\overline{BC} > \overline{AD}$ iff $\angle B$ is acute, $\overline{BC} \simeq \overline{AD}$ iff $\angle B$ is right, and $\overline{BC} < \overline{AD}$ iff $\angle B$ is obtuse.

Proof Follows directly from Theorems 22.3 and 22.5. ∎

Saccheri's Proposition III and its converse Proposition IV are contained in Omar Khayyam's Theorem. (More on Omar Khayyam in the next chapter.) Then Saccheri has our Theorems 22.5 and 21.8

as corollaries. As does Saccheri, we next turn to the problem of whether two or all three of the hypotheses regarding the upper base angles of a Saccheri quadrilateral can hold. For example, it certainly seems possible to have $\boxed{S}\,ABCD$ with $\angle B$ acute and $\boxed{S}\,EFGH$ with $\angle F$ obtuse. Theorem 22.10 will answer this problem.

Theorem 22.7 If $\boxed{S}\,ABCD$, $B\text{-}P\text{-}C$, $A\text{-}Q\text{-}D$, and $\overline{PQ} \perp \overline{AD}$, then $\overline{PQ} < \overline{AB}$ iff $\angle B$ is acute, $\overline{PQ} \simeq \overline{AB}$ iff $\angle B$ is right, and $\overline{PQ} > \overline{AB}$ iff $\angle B$ is obtuse.

Proof Since $\boxed{S}\,ABCD$, then $AB=CD$ and $\angle B \simeq \angle C$. Suppose $PQ < AB$. So $PQ < CD$. Then $m\angle ABC < m\angle QPB$ and $m\angle DCB < m\angle QPC$ (Theorem 21.8). Since $\angle QPB$ and $\angle QPC$ are supplementary and $m\angle ABC = m\angle DCB$, it follows that $2m\angle ABC < \pi$. As a first result we have $PQ < AB$ implies $\angle B$ is acute. If $PQ = AB$, then $m\angle ABC = m\angle QPB$ and $m\angle DCB = m\angle QPC$ (Theorem 21.8). In this case $2m\angle ABC = \pi$. As a second result we have $PQ = AB$ implies $\angle B$ is right. Suppose $PQ > AB$. Then $m\angle ABC > m\angle QPB$ and $m\angle DCB > m\angle QPC$ (Theorem 21.8). Thus $2m\angle ABC > \pi$. As a third result we have $PQ > AB$ implies $\angle B$ is obtuse. The theorem follows from the three results and their contrapositives. ∎

Theorem 22.8 If $\boxed{S}\,ABCD$, $B\text{-}C\text{-}R$, $A\text{-}D\text{-}S$, and $\overline{RS} \perp \overline{AD}$, then $\overline{RS} > \overline{AB}$ iff $\angle B$ is acute, $\overline{RS} \simeq \overline{AB}$ iff $\angle B$ is right, and $\overline{RS} < \overline{AB}$ iff $\angle B$ is obtuse.

Proof Since $\boxed{S}\,ABCD$, then $AB=CD$ and $\angle B \simeq \angle C$. Suppose $RS > CD$. Let J be such that $S\text{-}J\text{-}R$ and $SJ = DC$. So $\boxed{S}\,DCJS$ and $\boxed{S}\,ABJS$. (See Figure 22.2.) Since B is in int $(\angle SJC)$ and the upper base angles of a Saccheri quadrilateral are congruent, we have $m\angle DCJ = m\angle SJC > m\angle SJB = m\angle ABJ$. Then, since $m\angle JCR > m\angle JBC$ (Theo-

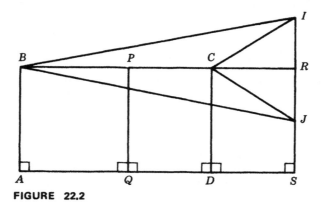

FIGURE 22.2

rem 17.9), we have $m\angle DCR = m\angle DCJ + m\angle JCR > m\angle ABJ + m\angle JBC = m\angle ABC = m\angle DCB$. Therefore, $\angle DCB$ is acute as $\angle DCR$ and $\angle DCB$ are a linear pair. Thus $\angle ABC$ is acute.

If $RS = CD$, then $\boxed{S}\,ABRS$ and it follows from the previous theorem that $\angle ABC$ is a right angle. Now suppose $RS < CD$. Let I be such that $S-R-I$ and $SI = DC$. So $\boxed{S}\,DCIS$ and $\boxed{S}\,ABIS$. Hence $m\angle DCI = m\angle SIC < m\angle SIB = m\angle ABI$. Then, since $m\angle ICR > m\angle IBC$, we have $m\angle DCR = m\angle DCI - m\angle ICR < m\angle ABI - m\angle IBC = m\angle ABC = m\angle DCB$. Since $\angle DCR$ and $\angle DCB$ are a linear pair, it follows that $\angle DCB$ is obtuse. Thus $\angle ABC$ is obtuse.

We have $RS > AB$ only if $\angle B$ is acute, $RS = CD$ only if $\angle B$ is right, and $RS < AB$ only if $\angle B$ is obtuse. These three results with their contrapositives complete the proof. ∎

The next theorem brings some of the previous results together. What is particularly interesting is that in Figure 22.3 $\angle QPM$ and $\angle SRM$ are either both acute, both right, or both obtuse.

Theorem 22.9 Suppose $\boxed{L}\,NMPQ$, $\boxed{L}\,NMRS$, $M-P-R$, and $N-Q-S$. Then, the following are equivalent: (1) $\angle SRM$ is acute, (2) $PQ < RS$, (3) $\angle QPM$ is acute, and (4) $MN < PQ$. Also, the following are equivalent: (1) $\angle SRM$ is right, (2) $PQ = RS$, (3) $\angle QPM$ is right, and (4) $MN = PQ$. Further, the following are equivalent: (1) $\angle SRM$ is obtuse, (2) PQ > RS, (3) $\angle QPM$ is obtuse, and (4) MN > PQ.

Proof Let the images of P, Q, R, S under the reflection in $\overleftrightarrow{MN}$ be P', Q', R', S', respectively. Since $\boxed{S}\,S'R'RS$ and $R'-P-R$ with $\overline{PQ} \perp \overline{S'S}$, (1) and (2) are equivalent in each case by Theorem 22.7. Since $\boxed{S}\,Q'P'PQ$ and $P'-P-R$ with $\overline{RS} \perp \overline{Q'Q}$, (2) and (3) are equivalent in each case by Theorem 22.8. Since $\boxed{L}\,NMPQ$, (3) and (4) are equivalent in each case by Theorem 22.5. ∎

Theorem 22.10 *Saccheri's Propositions V, VI, and VII* If there exists one Saccheri quadrilateral with acute upper base angles, then

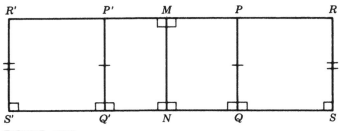

FIGURE 22.3

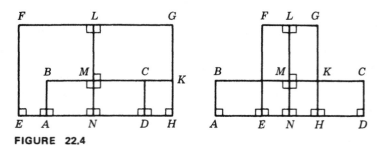

FIGURE 22.4

every Saccheri quadrilateral has acute upper base angles. If one Saccheri quadrilateral is a rectangle, then every Saccheri quadrilateral is a rectangle. If there exists one Saccheri quadrilateral with obtuse upper base angles, then every Saccheri quadrilateral has obtuse upper base angles.

Proof Suppose $\boxed{\text{S}}\, ABCD$ with M the midpoint of $\overline{BC}$ and N the midpoint of $\overline{AD}$; suppose $\boxed{\text{S}}\, A'B'C'D'$ with M' the midpoint of $\overline{B'C'}$ and N' the midpoint of $\overline{A'D'}$. Without loss of generality suppose $M'N' \cong MN$. Let E be on $\overrightarrow{NA}$ such that $2NE = A'D'$. Let F be on the same side of $\overleftrightarrow{AD}$ as B and such that $\overline{FE} \perp \overline{AD}$ and $EF = A'B'$. Let L be the foot of the perpendicular from F to $\overleftrightarrow{NM}$. Then $\boxed{\text{L}}\, NLFE$. Let G and H be the unique points such that $\boxed{\text{S}}\, EFGH$ with $\overleftrightarrow{LN}$ the perpendicular bisector of each base (Theorem 22.3). (See Figure 22.4.) Since $\boxed{\text{L}}\, NLGH \simeq \boxed{\text{L}}\, N'M'C'D'$ and $\boxed{\text{S}}\, EFGH \simeq \boxed{\text{S}}\, A'B'C'D'$ (Theorem 22.3), the theorem follows if we can show upper base angle $\angle DCM$ of $\boxed{\text{S}}\, ABCD$ and upper base angle $\angle HGL$ of $\boxed{\text{S}}\, EFGH$ are either both acute, both right, or both obtuse.

Since $LN = M'N' \cong MN$, either G is on $\overleftrightarrow{BC}$ or G and H are on opposite sides of $\overleftrightarrow{BC}$. So we may let $\overrightarrow{MC}$ intersect $\overline{GH}$ at point K. Since $\boxed{\text{L}}\, NMCD$ and $\boxed{\text{L}}\, NMKH$, then $\angle DCM$ and $\angle HKM$ are either both acute, both right, or both obtuse by the previous theorem. Also, since $\boxed{\text{L}}\, NHKM$ and $\boxed{\text{L}}\, NHGL$, $\angle HKM$ and $\angle HGL$ are either both acute, both right, or both obtuse by the previous theorem. Therefore, $\angle DCM$ and $\angle HGL$ are either both acute, both right, or both obtuse. ∎

22.2 SACCHERI'S THEOREM

We have just seen that Saccheri's three hypotheses are mutually exclusive (Theorem 22.10). Hence exactly one of the hypotheses must hold for any particular model of Σ. We know the Hypothesis of the Right Angle is one possibility. However, this does not deny the possibility of absolute planes for which one of the other hypotheses holds. In this section we shall see that one of these hypotheses is in fact impossible.

Theorem 22.11 *Saccheri's Proposition VIII* Given $\boxed{S}\,ABCD$, then $\angle ABD < \angle BDC$ iff $\angle B$ is acute, $\angle ABD \simeq \angle BDC$ iff B is right, and $\angle ABD > \angle BDC$ iff $\angle B$ is obtuse.

Proof Follows directly from Euclid's Propositions I.24 and I.25 (Theorems 18.18 and 18.19) and Saccheri's Propositions III and IV (Theorem 22.6). ■

Theorem 22.12 *Saccheri's Proposition IX* Let $\triangle ABC$ have a right angle at C. Under the Hypothesis of the Acute Angle, $m\angle A + m\angle B < \pi/2$. Under the Hypothesis of the Right Angle, $m\angle A + m\angle B = \pi/2$. Under the Hypothesis of the Obtuse Angle, $m\angle A + m\angle B > \pi/2$.

Proof There is a unique point E such that $\boxed{S}\,CAEB$. Since $m\angle ABE + m\angle CBA = \pi/2$, the result follows from the previous theorem. ■

Corollary 22.13 Under the Hypothesis of the Acute Angle, the sum of the measures of the angles of a triangle is less than π. Under the Hypothesis of the Right Angle, the sum of the measures of the angles of a triangle is π. Under the Hypothesis of the Obtuse Angle, the sum of the measures of the angles of a triangle is greater than π.

Proof Exercise 22.1. ■

Corollary 22.14 *Saccheri's Proposition XV* If the sum of the measures of the angles of one triangle is, respectively, less than, equal to, or greater than π, then the sum of the measures of the angles of any triangle is, respectively, less than, equal to, or greater than π.

Proof Let s be the sum of the measures of the angles of $\triangle ABC$. Let t be the sum of the measures of the angles of any other triangle. If $s < \pi$, then neither the Hypothesis of the Right Angle nor the Hypothesis of the Obtuse Angle can hold (Corollary 22.13). So $t < \pi$ (Corollary 22.13). By analogous reasoning $s = \pi$ implies $t = \pi$ and $s > \pi$ implies $t > \pi$. ■

DEFINITION 22.15 If $\triangle ABC$, then $\delta\triangle ABC$ is the *defect* of $\triangle ABC$ where $m\angle A + m\angle B + m\angle C + \delta\triangle ABC = \pi$. If $\square ABCD$ is convex, then $\delta\square ABCD$ is the *defect* of $\square ABCD$ where $m\angle A + m\angle B + m\angle C + m\angle D + \delta\square ABCD = 2\pi$.

Since $\delta\triangle ABC = \pi - (m\angle A + m\angle B + m\angle C)$, every statement

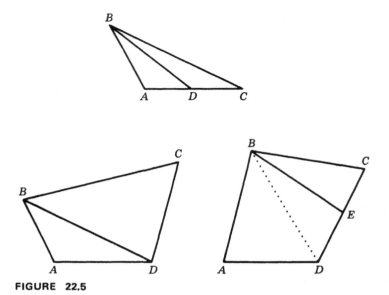

FIGURE 22.5

about the sum of the measures of the angles of some triangle can be translated to a statement about the defect of the triangle and conversely. As with the words *irrational* and *imaginary*, we should not allow the nontechnical usage of the word *defect* to prejudice our thinking. To say a triangle has nonzero *defect* does not mean there is something *wrong* with the triangle.

Theorem 22.16 If $\triangle ABC$ and $A-D-C$, then $\delta\triangle ABC = \delta\triangle ABD + \delta\triangle BDC$. If $\square ABCD$ is convex, then $\delta\square ABCD = \delta\triangle ABD + \delta\triangle BDC$. If $\square ABCD$ is convex and $C-E-D$, then $\delta\square ABCD = \delta\triangle BCE + \delta\square ABED$.

Proof Suppose $\triangle ABC$ and $A-D-C$. Then D is in int $(\angle ABC)$ and $m\angle ABD + m\angle DBC = m\angle ABC$. Also $\angle ADB$ and $\angle BDC$ are a linear pair. So adding equations $m\angle A + m\angle ABD + m\angle ADB + \delta\triangle ABD = \pi$ and $m\angle C + m\angle DBC + m\angle BDC + \delta\triangle BDC = \pi$, we have the first statement of the theorem (Definition 22.15).

Suppose $\square ABCD$ is convex. Then D is in int $(\angle ABC)$ and B is in int $(\angle ADC)$, (Theorem 13.23). So $m\angle ABD + m\angle DBC = m\angle ABC$ and $m\angle ADB + m\angle BDC = m\angle ADC$. The second statement of the theorem now follows from the definitions. The third statement follows from the first two. ■

The Hypothesis of the Acute Angle is equivalent to the existence of any triangle with defect greater than zero. Likewise, the Hypothesis of the Right Angle is equivalent to the existence of any triangle with defect zero, and the Hypothesis of the Obtuse Angle is equivalent to the existence of any triangle with defect less than zero. The *Theorem of Thales*, stating that an angle inscribed in a semicircle is a right angle, is certainly one of the oldest *theorems* in mathematics. We shall now see that the Theorem of Thales is equivalent to the Hypothesis of the Right Angle.

Theorem 22.17 *Saccheri's Proposition XVIII* Let C be a point off $\overline{AB}$ but on the circle with diameter $\overline{AB}$. Then $\angle ACB$ is an acute angle iff the Hypothesis of the Acute Angle holds, $\angle ACB$ is a right angle iff the Hypothesis of the Right Angle holds, and $\angle ACB$ is an obtuse angle iff the Hypothesis of the Obtuse Angle holds.

Proof Let M be the midpoint of $\overline{AB}$. By the Pons Asinorum $m\angle MAC = m\angle MCA$ and $m\angle MBC = m\angle MCB$. So $\delta\triangle MAC = \pi - (m\angle AMC + 2m\angle ACM)$ and $\delta\triangle MBC = \pi - (m\angle BMC + 2m\angle BCM)$. Since $\delta\triangle ABC = \delta\triangle MAC + \delta\triangle MBC$ and $\angle AMC$ and $\angle BMC$ are a linear pair, it follows that $\delta\triangle ABC = \pi - 2m\angle ACB$. Hence $m\angle ACB = \pi/2 - \frac{1}{2}\delta\triangle ABC$. ■

Theorem 22.18 The lower base of a Saccheri quadrilateral is not longer than the upper base.

Proof Suppose $\boxed{S}\,A_0B_0B_1A_1$ and $l = \overleftrightarrow{A_0A_1}$. Let n be any positive integer. Let $A_1, A_2, \ldots, A_n$ be the n distinct points on l such that A_i is the midpoint of A_{i-1} and A_{i+1}. So $A_iA_{i+1} = A_0A_1$. Let B_i be the point on the same side of l as B_0 such that B_iA_i is perpendicular to l at A_i and $B_iA_i = B_0A_0$. (See Figure 22.7. We do *not* know all the B_i are collinear.) Since $\triangle B_iA_iA_{i+1} \cong \triangle B_0A_0A_1$ by SAS, we have $B_iA_{i+1} = B_0A_1$ and $m\angle B_iA_{i+1}A_i = m\angle B_0A_1A_0$. So $m\angle B_iA_{i+1}B_{i+1} = m\angle B_0A_1B_1$, for $i = 1, 2, \ldots, n-1$, and $\triangle B_iA_{i+1}B_{i+1} \cong \triangle B_0A_1B_1$ by SAS. Thus $B_iB_{i+1} = B_0B_1$. Now, from these equalities and the Polygonal Inequality, we have

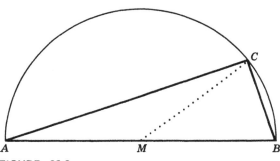

FIGURE 22.6

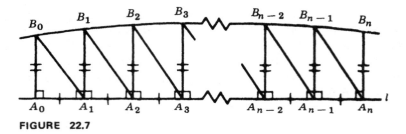

FIGURE 22.7

$$nA_0A_1 = A_0A_n \leqq A_0B_0 + (B_0B_1 + B_1B_2 + \cdots + B_{n-1}B_n) + A_nB_n$$
$$= 2A_0B_0 + nB_0B_1.$$

So $n(A_0A_1 - B_0B_1) \leqq 2A_0B_0$ for arbitrarily large positive integer n. This is a contradiction if $A_0A_1 - B_0B_1 > 0$ (Archimedes' axiom). Hence $A_0A_1 - B_0B_1 \leqq 0$ and $A_0A_1 \leqq B_0B_1$. ■

Suppose $\boxed{\text{S}}ABCD$. We have just shown $BC \geqq AD$. From Omar Khayyam's Theorem we know $\angle B$ is obtuse iff $BC < AD$. Hence, as a corollary of the previous theorem, we have *the Hypothesis of the Obtuse Angle leads to a contradiction.* It would be a pity not to state this important result in Saccheri's own colorful words:

Corollary 22.19 *Saccheri's Proposition XIV* The Hypothesis of the Obtuse Angle is absolutely false because it destroys itself.

For any model of absolute geometry either the Hypothesis of the Acute Angle must hold or else the Hypothesis of the Right Angle must hold. Of the many corollaries that follow, one in particular should be called *Saccheri's Theorem*.

Corollary 22.20 *Saccheri's Theorem* If $\triangle ABC$, then $m\angle A + m\angle B + m\angle C \leqq \pi$.

Corollary 22.21 $0 \leqq \delta\triangle ABC < \pi$. $0 \leqq \delta\square ABCD < 2\pi$. If $\boxed{\text{S}}ABCD$, then $m\angle B = m\angle C \leqq \pi/2$ and $BC \geqq AD$. If $\boxed{\text{L}}ABCD$, then $m\angle C \leqq \pi/2$, $BC \geqq AD$, and $CD \geqq AB$.

Corollary 22.22 *Absolute Exterior Angle Theorem* The measure of an exterior angle of a triangle is greater than or equal to the sum of the measures of the remote interior angles.

Proof Exercise 22.2. ■

Theorem 22.23 *Saccheri's Propositions XIX and XX* Let $\triangle ABC$

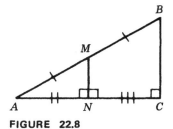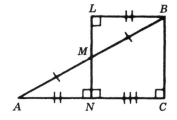

FIGURE 22.8

have a right angle at C. Let M be the midpoint of $\overline{AB}$ and N the foot of the perpendicular from M to $\overline{AC}$. Then the Hypothesis of the Acute Angle implies $AC < 2AN$ but $BC > 2MN$, and the Hypothesis of the Right Angle implies $AC = 2AN$ and $BC = 2MN$.

Proof Since A and B are on opposite sides of $\overleftrightarrow{MN}$ and $\overleftrightarrow{MN} \parallel \overleftrightarrow{BC}$, we have $A-N-C$ and $AC = AN + NC$. Let L be such that $L-M-N$ and $LM = MN$. Then $\triangle BML \cong \triangle AMN$ by SAS. So $BL = AN$ and $\angle BLM$ is a right angle. Hence $\boxed{\text{L}}NLBC$. Under the Hypothesis of the Acute Angle we have $NC < BL = AN$ and $BC > LN$. Then $AC < 2AN$ and $BC > 2MN$. However, under the Hypothesis of the Right Angle, we have $NC = BL = AN$ and $BC = LN$. Then $AC = 2AN$ and $BC = 2MN$. ∎

Referring to Figure 22.8 in Euclidean geometry, we have $\sin m\angle A = BC/AB = MN/AM$, $\cos m\angle A = AC/AB = AN/AM$, and $\tan m\angle A = BC/AC = MN/AN$. However, under the Hypothesis of the Acute Angle, we find $BC/AB > MN/AM$, $AC/AB < AN/AM$, and $BC/AC > MN/AN$.

Theorem 22.24 *Saccheri's Proposition XXI; Aristotle's Axiom* Given $r > 0$ and $\angle BFG$, there exists H on $\overrightarrow{FG}$ such that the distance from H to $\overleftrightarrow{BF}$ is greater than r.

Proof Let D be the foot of the perpendicular from G to $\overleftrightarrow{BF}$. Let n be a positive integer such that $2^n GD > r$ (Archimedes' axiom). Let H_1 be such that $F-G-H_1$ and $FG = GH_1$. Let $H_2, H_3, \ldots, H_n$ be such that $F-H_{i-1}-H_i$ and $FH_{i-1} = H_{i-1}H_i$. Let E_i be the foot of the perpendicular

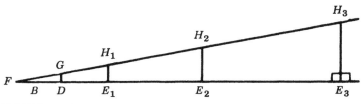

FIGURE 22.9

from H_i to $\overleftrightarrow{BF}$. (See Figure 22.9.) Then, by the previous theorem (or trivially if $D = F$), we have $H_n E_n \cong 2H_{n-1}E_{n-1}, \ldots, H_3 E_3 \cong 2H_2 E_2$, $H_2 E_2 \cong 2H_1 E_1$, and $H_1 E_1 \cong 2GD$. Hence $H_n E_n \cong 2^n GD > r$. With $H = H_n$ the distance from H to $\overleftrightarrow{BF}$ is greater than r. ■

22.3 EXERCISES

22.1 Corollary 22.13.

22.2 Corollary 22.22.

22.3 Define $\mu \triangle ABC$ and $\mu \square ABCD$ to be the sum of the measures of the angles of $\triangle ABC$ and convex $\square ABCD$, respectively. Restate Theorem 22.16 using this notation.

● **22.4** Saccheri's Proposition XVII: Playfair's Parallel Postulate does not hold under the Hypothesis of the Acute Angle.

● **22.5** State and prove some theorems about equiangular quadrilaterals. (A quadrilateral that is both equiangular and equilateral is called a *Gersonides quadrilateral*.)

● **22.6** True or False?

(a) An equilateral quadrilateral is a Gersonides quadrilateral. (See Exercise 22.5)

(b) If $\boxed{S}ABCD$ and $\boxed{L}NMCD$, then M, B, C are collinear.

(c) If $\boxed{S}ABCD$ and $\boxed{L}NMCD$, then N, A, D are collinear.

(d) The defect of a triangle is positive.

(e) An angle inscribed in a semicircle is a right angle.

(f) If $\boxed{L}PQRS$, then $PQ \cong RS$.

(g) If $\boxed{S}ABCD$ and $\boxed{L}PQRS$, then $\boxed{S}DCBA$ and $\boxed{L}PSRQ$.

(h) If $\boxed{L}ABCD$ and $\boxed{S}PQRS$, then $\boxed{L}DCBA$ or $\boxed{S}PSRQ$.

(i) The Hypothesis of the Acute Angle is absolutely false because it destroys itself.

(j) Playfair's Parallel Postulate implies the Hypothesis of the Right Angle.

● **22.7** Saccheri's Proposition X: If $A - B - M$ and $\overline{DB} \perp \overline{AB}$, then $DM > AD$ iff $BM > BA$.

- **22.8**   Saccheri's Proposition XI: If P and Q are points on the same side of $\overleftrightarrow{AB}$, $\angle PAB$ acute, and $\angle ABQ$ right, then $\overrightarrow{AP}$ intersects $\overrightarrow{BQ}$ under the Hypothesis of the Right Angle.

22.9   Saccheri's Proposition XII: If P and Q are points on the same side of $\overleftrightarrow{AB}$, $\angle PAB$ acute, and $\angle ABQ$ right, then $\overrightarrow{AP}$ intersects $\overrightarrow{BQ}$ under the Hypothesis of the Obtuse Angle.

22.10   Saccheri's Proposition XIII: Under the Hypothesis of the Right Angle or the Hypothesis of the Obtuse Angle, Euclid's Parallel Postulate holds.

22.11   Saccheri's Proposition XVI: Let $\square ABCD$ be convex. Then $\delta\square ABCD > 0$ iff the Hypothesis of the Acute Angle, $\delta\square ABCD = 0$ iff the Hypothesis of the Right Angle, and $\delta\square ABCD < 0$ iff the Hypothesis of the Obtuse Angle.

- **22.12**   If $\angle PQR$ is a right angle, then Q is either on the circle with diameter $\overline{PR}$ or in the interior of the circle.

22.13   Show that the hypothesis $N-Q-S$ can be omitted from Theorem 22.9.

- **22.14**   Draw the figures, other than those in Figure 22.4, that could accompany Theorem 22.10.

22.15   All the possible congruence theorems for Saccheri quadrilaterals and Lambert quadrilaterals in absolute geometry are contained in Theorem 22.4.

- **22.16**   The line through the midpoints of two sides of a triangle is parallel to the third side.

22.17   Theorem 22.17 could have followed Definition 21.9. Why didn't it? Would this rearrangement have saved any work?

*__22.18__   Saccheri's Proposition XXII: If $\square ABCD$, $\overline{AB} \perp \overline{AD}$, $\overline{AD} \perp \overline{CD}$, $\angle ABC$ acute, and $\angle BCD$ acute, then $\overleftrightarrow{BC}$ and $\overleftrightarrow{AD}$ have a common perpendicular intersecting $\overline{BC}$.

GRAFFITI

The nature of absolute truth cannot be but one and the same same at Maros-Vasarhely as at Kamschatka and on the Moon, or, in a word, anywhere in the world; and what one reasonable being discovers, that can also quite possibly be discovered by another.

Bolyai

> *How shall a man know there be strait lines which shall never meet though both ways infinitely produced?*
>
> **Hobbs (1655)**

From Edward FitzGerald's translation The Rubaiyat of Omar Khayyam:

> *A Book of Verses underneath the Bough,*
> *A Jug of Wine, a Loaf of Bread — and Thou*
> *   Beside me singing in the Wilderness —*
> *Ah, Wilderness were Paradise enow!*

> *For "Is" and "Is-not" though with Rule and Line,*
> *And, "Up-and-Down" without, I could define,*
> *   I yet in all I only cared to know,*
> *Was never deep in anything but — Wine.*

> *The treatise itself [Bolyai's* The Science of Absolute Space*], therefore, contains only twenty-four pages — the most extraordinary two dozen pages in the whole history of thought!*
>
> **Halsted**

> *On 20 October 1983 the International Bureau of Weights and Measures defined the speed of light to be 229,792,458 meters per second. Thus the meter, a unit of distance, is now defined in terms of the second, a unit of time. The change was made in large part because time-measuring methods are far more precise than those applied to distances. The speed of light is now a constant, and any change in the experimental determination of the duration of the second automatically entails a change in the length of the meter. At least for very precise measurements, scientist now measure distance with clocks instead of rulers.*

CHAPTER 23

Euclid's Parallel Postulate

23.1 EQUIVALENT STATEMENTS

We know several statements equivalent to Euclid's Parallel Postulate (Chapter 21), and we know several statements equivalent to Saccheri's Hypothesis of the Right Angle (Chapter 22). We shall now confirm your expectation that the two propositions are themselves equivalent.

Theorem 23.1 Playfair's Parallel Postulate implies Saccheri's Hypothesis of the Right Angle.

Proof Suppose $\boxed{\text{S}}\, ABCD$. Let $\overleftrightarrow{BE} \perp \overleftrightarrow{AB}$. So $\overleftrightarrow{BC} \parallel \overleftrightarrow{AD}$ and $\overleftrightarrow{BE} \parallel \overleftrightarrow{AD}$. By Playfair's Parallel Postulate, we have $\overleftrightarrow{BE} = \overleftrightarrow{BC}$ and $\angle B$ is right. ∎

Saccheri showed that Euclid's Parallel Postulate holds under either the Hypothesis of the Right Angle or the Hypothesis of the Obtuse Angle by following the idea of Aganis in Exercise 21.6 and using Theorem 22.4 (see Exercises 22.7 through 22.10). In this way, he proved the converse of the preceding theorem and demonstrated the impossibility of the Hypothesis of the Obtuse Angle. Taking a different approach we shall prove the converse of Theorem 23.1 from the next theorem.

Theorem 23.2 If $r > 0$ and $\overline{PQ} \perp \overline{QR}$, then there exists a point S on $\overrightarrow{QR}$ such that $m\angle PSQ < r$.

Proof Suppose $m\angle ABC = r < \pi/2$. By Aristotle's Axiom we may suppose $\overline{AC} \perp \overline{BC}$ and $AC > PQ$. Let D be such that Q-P-D and and $QD = AC$. Let S be on $\overrightarrow{QR}$ such that $QS = CB$. So $m\angle QSD = r$ since $\triangle DQS \simeq \triangle ACB$ by SAS. (See Figure 23.1.) Since P is in int $(\angle DSQ)$, we have $m\angle PSQ < r$. ∎

Theorem 23.3 Saccheri's Hypothesis of the Right Angle implies Playfair's Parallel Postulate.

Proof Let point P be off $\overleftrightarrow{QR}$. We may suppose $\overline{PQ} \perp \overline{QR}$. It is sufficient to show $\overleftrightarrow{PT}$ intersects $\overleftrightarrow{QR}$ unless $\overleftrightarrow{PT} \perp \overleftrightarrow{PQ}$. We may suppose T and R are on the same side of $\overleftrightarrow{PQ}$ and $\angle QPT$ is acute. Let $r = \pi/2 - m\angle QPT$. By the previous theorem there exists a point S on $\overrightarrow{QR}$ such that $m\angle QSP < r$. (See Figure 23.2.) Since $m\angle QPS + m\angle PSQ = \pi/2$ under the Hypothesis of the Right Angle, then $m\angle QPS > m\angle QPT$. Hence T is in int $(\angle QPS)$ and $\overrightarrow{PT}$ intersects $\overline{QS}$ by Crossbar. Thus the perpendicular to $\overline{PQ}$ at P is the only line through P that is parallel to $\overleftrightarrow{QR}$. ∎

Corollary 23.4 Euclid's Parallel Postulate is equivalent to Saccheri's Hypothesis of the Right Angle.

Since Saccheri's great book was published in 1733, we might assume there are exactly two thousand years between Euclid and Saccheri. During all this time there was little advancement in the theory of parallels.

The Hejira, Mohammed's flight from Mecca to Medina in 622, marks the beginning of the Mohammedan era. In 641 Alexandria fell to the Arabs who went on to conquer the lands from southern Spain to India. After a century of upheaval, the Muslim world was stable enough to become interested in the civilizations it had overrun. Baghdad became a new Alexandria. This center of learning had its

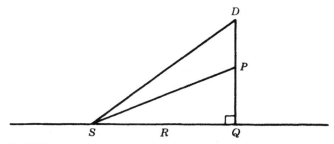

FIGURE 23.1

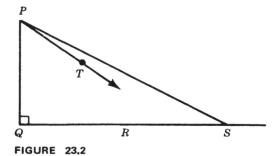

FIGURE 23.2

House of Wisdom, which could be compared to Alexandria's famed Museum. Here al-Khowarizmi wrote the *Al-jabr* about 825. Centuries later Europe would learn about *algebra.*

At the time of the height of the Muslim empire. Gerbert, who was born in France about 940, was teaching calculation with Hindu-Arabic numerals in Europe – a couple hundred years before such numerals would be widely accepted there. As the early Mohammedans had not needed the Alexandrian Library because they had the *Koran,* so most European Christians did not need any Arabic knowledge because they had the *Gospel.* However, Gerbert, who was a distinguished statesman and the most accomplished European scholar of his age, had taken some of the first steps in stimulating the transmission of learning from the East to the West. Among the books written by Gerbert was his *Geometria,* where we find "Two straight lines distinct from each other by the same space continually, and never meeting each other when indefinitely produced, are called parallel, that is, equidistant." From 999 until his death in 1003, Gerbert was Pope Sylvester II.

In mathematics the Arabs excelled in algebra, arithmetic, and trigonometry. The theory of parallels was studied but not significantly advanced. Ibn-al-Haitham (circa 965 – 1039), known as Alhazen in the West, gave a *proof* of Euclid's Parallel Postulate by first showing that all four angles of what we have called a *Lambert quadrilateral* are right angles. To do this he assumed the collinearity of the locus of a point that moves so as to remain equidistant from a given line. Alhazen had followed many of his predecessors into the *equidistance-trap.*

To the Arab world Omar Khayyam (circa 1050 – 1123) is known for his accomplishments in astronomy and mathematics. Omar's significant contributions to mathematics were in algebra. In geometry he criticized Alhazen's work on the grounds that Aristotle had forbidden the use of motion in geometry. Omar's own efforts in the theory of parallels may be found in "On the Truth of Parallels and Discussion of the Famous Doubt," which is Part I of his *Discussion of the Difficulties in Euclid*. Here we find eight propositions. The first two prove that the upper base angles of a Saccheri quadrilateral are congruent and that the perpendicular bisector of the lower base of a Saccheri quadrilateral is the perpendicular bisector of the upper base. Over six hundred years later Saccheri would have the same first two propositions. Had Saccheri's work not been so extensive we might be calling a *Saccheri quadrilateral* an *Omar Khayyam quadrilateral*. Omar's third proposition states that a Saccheri quadrilateral is a rectangle. In the rather circuitous argument for the third proposition he does establish Saccheri's Proposition III. For this reason, we called Theorem 22.6 *Omar Khayyam's Theorem*. Omar then denies the Hypothesis of the Acute Angle and the Hypothesis of the Obtuse Angle on the grounds that distance between parallel lines "does neither expand nor contract." To this he adds the enticing comment "This indeed is what a philosopher believes." His third proposition follows. The remaining propositions lead to the last, which is Euclid's Parallel Postulate. As does Proclus, Omar makes the inexplicable statement "If we are satisfied with a proposition or its converse but not both, then that proposition is not proved."

Omar Khayyam lived during the time of the first Crusade, when Arab science and mathematics were already entering a state of decline. After his several accomplishments in scientific areas, Omar took up philosophy. To most people in the West he is not known for his work on quadrilaterals but rather for his work on quatrains. *Rubai* means *quatrain,* and Omar Khayyam is the celebrated author of the *Rubaiyat*.

Nasir Eddin (1201–1274) was the astronomer to Hulagu Khan, brother of Kublai Khan and grandson of Genghis Khan. In Nasir Eddin's *Elements of Euclid* between Euclid's Proposition I.28 and I.29, we find a *proof* of Euclid's Parallel Postulate. He assumed our Theorem 21.8 and incorrectly assumed that, if $\square ABCD$ has right angles at A and D, then the angle at B is acute iff the angle at C is obtuse and, further, $AB > CD$ when the angle at B is acute. (See Figure 23.3.) Since the assumptions deny both the Hypothesis of the Acute Angle and the Hypothesis of the Obtuse Angle, Nasir Eddin could go on to prove Euclid's Parallel Postulate. He would not be the last to assume a line cannot approach and then diverge from another line without crossing.

By the twelfth century the Arabic excellence in learning had run its course. Fortunately, at the same time, the Renaissance in Europe was getting under way. Adelard of Bath (circa 1075–1160) translated Euclid's *Elements* from Arabic into Latin about 1142. Levi ben Gerson (1288–1344), a rabbi from Avignon and known as Gersonides, was probably the first person in the West to discuss the parallel postulate.

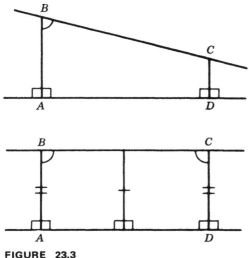

FIGURE 23.3

He considered quadrilaterals that are both equilateral and equiangular, known today as *Gersonides quadrilaterals* (see Exercise 22.5). Euclid's *Elements* was first translated into English in 1570 by Sir Henry Billingsley, later Lord Mayor of London. Descartes and Saccheri learned their geometry from *Euclid's Elements* as edited by Christoph Clavius (1537–1612). By an argument similar to that of Nasir Eddin, Clavius demonstrates Euclid's Parallel Postulate by first arguing that a curve equidistant from a line is a line. In 1680, Giordano Vitale (1633–1711) failed to give an accurate proof of Euclid's Parallel Postulate but did make the significant contribution we have called Giordano's Theorem (Theorem 21.13).

John Wallis (1616–1703) translated Nasir Eddin's commentary on the theory of parallels into Latin. Being aware of the equidistance-trap, Wallis proves Euclid's Parallel Postulate by assuming a new axiom: *To every figure there exists a similar figure of arbitrary magnitude.* In a scholium following his Proposition XXI Saccheri pointed out that Wallis could have proved Euclid's Parallel Postulate by assuming only the existence of *two* similar but noncongruent triangles.

DEFINITION 23.5 If $\triangle ABC$ and $\triangle DEF$ are such that $\angle A \simeq \angle D$, $\angle B \simeq \angle E$, and $\angle C \simeq \angle F$, then $\triangle ABC \sim \triangle DEF$. We say $\triangle ABC$ and $\triangle DEF$ are **similar** if $\triangle ABC \sim \triangle DEF$, $\triangle BCA \sim \triangle DEF$, $\triangle CAB \sim \triangle DEF$, $\triangle CBA \sim \triangle DEF$, $\triangle ACB \sim \triangle DEF$, or $\triangle BAC \sim \triangle DEF$.

Theorem 23.6 Euclid's Parallel Postulate follows from the existence of two similar but noncongruent triangles.

Proof Suppose $\triangle ABC \sim \triangle DEF$ but not $\triangle ABC \simeq \triangle DEF$. Then $AB \neq DE$ by ASA. We may suppose $AB > DE$. Let G be such that $A-G-B$ and $AG = DE$. Let H be on $\overrightarrow{AC}$ such that $AH = DF$. Then $\triangle GAH \simeq \triangle EDF$ by SAS. (See Figure 23.4.) So $\angle AGH \simeq \angle ABC$ and $\angle AHG \simeq \angle ACB$. Then $\overleftrightarrow{GH} \| \overleftrightarrow{BC}$ by Euclid's Proposition I.28 (Corollary 21.7). Hence $A-H-C$ and $\square BCHG$ is convex. Since $m\angle BGH = \pi - m\angle GBC$ and $m\angle CHG = \pi - m\angle HCB$, then $\delta\square BCHG = 0$. So $\delta\triangle BCG = 0$ (Theorem 22.16). Hence the Hypothesis of the Right Angle holds, and Euclid's Parallel Postulate must hold. ■

The major contribution of Gottfried Wilhelm Leibniz (1646–1716) to the theory of parallels was to add confusion by using the idea of *direction*. The problem is one of definition. Supposedly two lines would have the *same direction* if they can be cut by a transversal so that corresponding angles are congruent. The trap comes from assuming that if two lines have the same direction then the corresponding angles are congruent for any transversal to the two lines. This means assuming Euclid's Proposition I:29. Suppose $\boxed{S}\,ABCD$. Since $\overleftrightarrow{AD}$ is perpendicular to both $\overleftrightarrow{AB}$ and $\overleftrightarrow{CD}$, then $\overleftrightarrow{AB}$ and $\overleftrightarrow{CD}$ have the same direction. However, assuming $\overleftrightarrow{BD}$ cuts $\overleftrightarrow{AB}$ and $\overleftrightarrow{CD}$ so that $\angle ABD$ and $\angle BDC$ are congruent is equivalent to assuming the Hypothesis of the Right Angle (Theorem 22.11). Although *same direction,* as defined above, has its intuitive meaning in Euclidean geometry, it turns out that under the Hypothesis of the Acute Angle the intuitive meaning is given only by defining two lines to have the *same direction* if they are parallel but there does *not* exist a transversal to the two lines such that corresponding angles are congruent.

Adrian Marie Legendre (1752–1833) was a mathematician of great prominence and perhaps the most indefatigable pursuer of a proof of Euclid's Parallel Postulate. His various attempts appear in the twelve editions of his influential *Éléments de géométrie* from 1794 to 1823. A final monograph *Réflexions sur différentes manières de démontrer la théorie des parallèles ou le théorème sur la somme des trois angles du triangle* appeared in 1833, exactly one hundred years

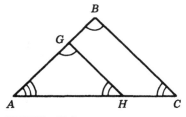
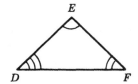

FIGURE 23.4

after the publication of Saccheri's book. Although Saccheri's work had created some excitement when it first appeared, most mathematicians of Legendre's time were unaware of its contents! Thus we find Legendre rediscovering theorems already proved by Saccheri. Our proof of Theorem 22.18, which had as a corollary the elimination of the Hypothesis of the Obtuse Angle, imitates one of Legendre's proofs that $\delta\triangle ABC \geqq 0$. This is why *Saccheri's Theorem* (Corollary 22.20) came to be known as *Legendre's First Theorem*. Later, Legendre gave a different proof of Saccheri's Theorem (Exercise 23.10). That the existence of one triangle such that the sum of the measures of its angles is π implies the sum of the measures of the angles of *any* triangle is π came to be known as *Legendre's Second Theorem*. Of course this result is contained in Saccheri's Proposition XV (Corollary 22.14). In proving our Theorem 23.3, Legendre gave a different proof (Exercise 23.11) for our Theorem 23.2. If Legendre could have proved $\delta\triangle ABC \leqq 0$ for any $\triangle ABC$, he would have had a proof of Euclid's Parallel Postulate. We shall be interested in one of Legendre's flawed proofs that occurs in the third through eighth editions. (Another attempt from the twelfth edition is indicated in Exercise 23.12.) *You should study the following argument carefully and discover the flaw for yourself.*

   Legendre's *proof* of Euclid's Parallel Postulate: Assume $\triangle ABC$ is such that $\delta\triangle ABC = t > 0$. Let n be an integer such that $2^n t > \pi$. If we like we may suppose $m\angle A \leqq \pi/3$ by taking $\angle A$ to be a smallest angle of $\triangle ABC$. In any case, let D be the point such that A and D are on opposite sides of $\overleftrightarrow{BC}$, $\angle DBC \simeq \angle ACB$, and $BD = AC$. (See Figure 23.5.) So $\triangle ACB \simeq \triangle DBC$, $\overleftrightarrow{BD} \parallel \overleftrightarrow{AC}$, and $\overleftrightarrow{AB} \parallel \overleftrightarrow{CD}$. Let l be a line through D, off A, but intersecting $\overrightarrow{AB}$ at B_1 and $\overrightarrow{AC}$ at C_1. Assume $A - B_1 - B$. Then A and B_1 are on the same side of $\overleftrightarrow{BD}$. Since $\overleftrightarrow{AC} \parallel \overleftrightarrow{BD}$, then A, C, C_1 are on the same side of $\overleftrightarrow{BD}$. Thus B_1 and C_1 are on the same side of $\overleftrightarrow{BD}$, contradicting $B_1 - D - C_1$. Therefore, since $B_1 \neq B$, we must have $A - B - B_1$. Similarly, $A - C - C_1$. Since $\delta\triangle ABC = t$ and

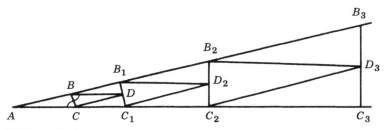

FIGURE 23.5

$\delta\triangle DCB = t$, we must have $\delta\triangle AB_1C_1 > 2\delta\triangle ABC$. Repeating the argument, we have $\delta\triangle AB_2C_2 > 2\delta\triangle AB_1C_1$. So $\delta\triangle AB_2C_2 > 2^2t$. After n steps we have $\delta\triangle AB_nC_n > 2^nt$. Hence $\delta\triangle AB_nC_n > \pi$, a contradiction. Therefore, $\delta\triangle ABC \leq 0$. Since we know $\delta\triangle ABC \geq 0$, we must have $\delta\triangle ABC = 0$ and Euclid's Parallel Postulate follows. (Give the argument a careful second reading if you did not catch the flaw.)

Here is a *proof* of Playfair's Parallel Postulate: Let point P be off line l and F the foot of the perpendicular from P to l. We wish to show $\overrightarrow{PA}$ intersects l whenever $\angle APF$ is acute. Let $m = \overleftrightarrow{PF}$ and $\rho_m\overrightarrow{PA} = \overrightarrow{PB}$. So l intersects $\overrightarrow{PA}$ iff l intersects $\overrightarrow{PB}$. Let n be any line through F, off P, but intersecting $\overrightarrow{PA}$ at C and $\overrightarrow{PB}$ at D. Then $C-F-D$ and l intersects $\overline{PC}$ or $\overline{PD}$ by PASCH. Hence $\overrightarrow{PA}$ intersects l. We have proved Playfair's Parallel Postulate under the same tacit assumption made by Legendre in the preceding paragraph.

Another argument of Legendre runs as follows. Given $\triangle ABC$, it follows from ASA that $m\angle A$ is a function of $m\angle B$, BC, and $m\angle C$. Unless $m\angle A$ depends on $m\angle B$ and $m\angle C$ alone and is independent of BC, we would have BC is a function of $m\angle A$, $m\angle B$, and $m\angle C$. Since BC is simply a number, this is impossible *provided* we assume that no unit of length can be associated with some unit of angle measure. In that case $m\angle A$ depends only on $m\angle B$ and $m\angle C$. Letting C' be the midpoint of $\overline{BC}$ and A' on $\overline{AB}$ such that $\angle BC'A' \simeq \angle BCA$, it then follows that $\angle BA'C' \simeq \angle BAC$. Thus we have Wallis' assumption of two similar triangles that are not congruent, and Euclid's Parallel Postulate follows.

The literature of the theory of parallels is not without its amusing aspects. While Legendre was still trying to eliminate Euclid's Parallel Postulate in 1833, Thomas Perronet Thompson was busy eliminating *all* the axioms in his *Geometry without Axioms*.

The following theorem lists many propositions that are equivalent to Euclid's Parallel Postulate. It cannot be overemphasized that no one of these propositions will be proved without assuming another one of the propositions.

Theorem 23.7 The following propositions are equivalent:

Proposition A. Euclid's Parallel Postulate: If A and D are points on the same side of $\overleftrightarrow{BC}$ such that $m\angle ABC + m\angle BCD < \pi$, then $\overrightarrow{BA}$ intersects $\overrightarrow{CD}$.

Proposition B. Euclid's Proposition I.29: If A and D are points on the same side of $\overleftrightarrow{BC}$ and $\overleftrightarrow{BA} \parallel \overleftrightarrow{CD}$, then $m\angle ABC + m\angle BCD = \pi$.

Proposition C. Euclid's Proposition I.30: $l \parallel m$ and $m \parallel n$ implies $l \parallel n$ for lines l, m, n. (Lines parallel to a given line are parallel.)

Proposition D. A third line intersecting one of two parallel lines intersects the other. (Contrapositive of Euclid's Proposition I.30.)

Proposition E. Euclid's Proposition I.31; Playfair's Parallel Postulate: If point P is off line l, then there exists a unique line through P that is parallel to l.

Proposition F. A line perpendicular to one of two parallel lines is perpendicular to the other.

Proposition G. $l \parallel m$, $r \perp l$, and $s \perp m$ implies $r \parallel s$ for lines l, m, r, s.

Proposition H. The perpendicular bisectors of the sides of a triangle are concurrent.

Proposition I. There exists a circle passing through any three noncollinear points.

Proposition J. There exists a point equidistant from any three noncollinear points.

Proposition K. A line intersecting and perpendicular to one ray of an acute angle intersects the other ray.

Proposition L. Through any point in the interior of an angle there exists a line intersecting both rays of the angle not at the vertex.

Proposition M. Euclid's Proposition I.32: The sum of the measures of the angles of any triangle is π. The measure of an exterior angle of a triangle is equal to the sum of the measures of the remote interior angles.

Proposition N. Theorem of Thales: If point C is off $\overline{AB}$ but on the circle with diameter $\overline{AB}$, then $\angle ACB$ is right.

Proposition O. If $\angle ACB$ is right, then C is on the circle with diameter $\overline{AB}$.

Proposition P. The perpendicular bisectors of the legs of a right triangle intersect.

Proposition Q. $l \perp r$, $r \perp s$, and $s \perp m$ implies l intersects m for lines l, m, r, s.

Proposition R. There exists an acute angle such that every line intersecting and perpendicular to one ray of the angle intersects the other ray.

Proposition S. There exists an acute angle such that every point in the interior of the angle is on a line intersecting both rays of the angle not at the vertex.

Proposition T. There exists one triangle such that the sum of the measures of the angles of the triangle is π.

Proposition U. There exists one triangle with defect zero.

Proposition V. Saccheri's Hypothesis of the Right Angle: There exists a rectangle.

Proposition W. There exist two lines l and m such that l is equidistant from m.

Proposition X. If three angles of a quadrilateral are right, then so is the fourth.

Proposition Y. There exists some line l and there exists some point P off l such that there is a unique line through P that is parallel to l.

Proposition Z. There exists a pair of similar, noncongruent triangles.

Proof From Section 21.1 we already know that Propositions A, B, C, D, and E are equivalent. We continue by showing each proposition implies the next. Since Proposition F follows easily from Proposition D and E and since Proposition D is equivalent to Proposition E, it follows that Proposition E implies Proposition F. Proposition F implies Proposition G by Euclid's Proposition I.28 (Corollary 21.7). Since we have not shown Proposition G implies any of the previous propositions, in proving Proposition G implies Proposition H we can assume *only* Proposition G and any of our previous theorems.

Proposition G implies Proposition H: Suppose l and m are the perpendicular bisectors of sides $\overline{AC}$ and $\overline{BC}$, respectively, of $\triangle ABC$. Let $r = \overleftrightarrow{AC}$ and $s = \overleftrightarrow{BC}$. By Proposition G, $l \parallel m$ implies the contradiction $r \parallel s$. Hence l and m intersect at point P which is equidistant from A and C and is equidistant from B and C (Theorem 18.5). So P is also on the perpendicular bisector of $\overline{AB}$.

Proposition H implies Proposition I trivially; Proposition I implies Proposition J trivially.

Proposition J implies Proposition K: Suppose $\angle ABC$ is acute and $\overleftrightarrow{DC} \perp \overleftrightarrow{BC}$. Let R and S be such that R is the midpoint of $\overleftrightarrow{BC}$ and C is the midpoint of $\overline{RS}$. Let T be the image of R under the reflection in $\overleftrightarrow{AB}$. Then, by Proposition J, there exists a point P equidistant from

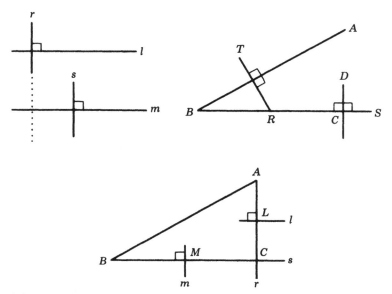

FIGURE 23.6

R, S, and T. Since P must be on both $\overleftrightarrow{AB}$ and $\overleftrightarrow{CD}$ (Theorem 18.5), it follows that $\overleftrightarrow{CD}$ intersects $\overrightarrow{BA}$.

Proposition K implies Proposition L: Given $\angle ABE$, let $\overrightarrow{BC}$ be the angle bisector. So $\angle ABC$ and $\angle EBC$ are acute. By Proposition K, the perpendicular from any point in int $(\angle ABE)$ to $\overleftrightarrow{BC}$ intersects both $\overrightarrow{BA}$ and $\overrightarrow{BE}$.

Proposition L implies Proposition M: Did you find the flaw in Legendre's *proof* that $\delta\triangle ABC = 0$? Legendre tacitly assumed Proposition L. Under this assumption Legendre's proof is valid. Our Proposition M contains two statements because Euclid put them together in his Proposition I.32. The two statements are themselves equivalent as if $\angle ACD$ is an exterior angle of $\triangle ABC$, then $m\angle ACD = m\angle A + m\angle B + \delta\triangle ABC$.

Proposition M implies Proposition N since each is equivalent to the Hypothesis of the Right Angle (Theorem 22.17). Since the Hypothesis of the Right Angle implies Euclid's Parallel Postulate, we know at this time that propositions A through N are equivalent. However, we have *not* proved any one of these propositions! We do know that if any one holds, then all of them hold. It also follows that if any one is false then all of them are false!

Proposition N implies Proposition O: Suppose $\angle ACB$ is right. Let D be the midpoint of $\overline{AB}$. Let $\overrightarrow{DC}$ intersect the circle with diameter

$\overline{AB}$ at point E. Then $\angle AEB$ is right by Proposition N. Since $\angle ACB$ is also right, both $D-E-C$ and $D-C-E$ are impossible (Theorem 18.16). Therefore, we must have $C=E$.

Proposition O implies Proposition P: If $\triangle ABC$ has a right angle at C, then the hypotenuse $\overline{AB}$ is the diameter of a circle through C by Proposition O. Thus the perpendicular bisector of the legs of a right triangle must intersect at the midpoint of the hypotenuse.

Proposition P implies Proposition Q: If $l=s$ or $m=r$, then the result is trivial. So suppose L, C, M distinct points where $\{L\}=l \cap r$, $\{C\}=r \cap s$, and $\{M\}=m \cap s$. Let A and B be such that L is the midpoint of $\overline{AC}$ and M is the midpoint of $\overline{BC}$. Then l and m are the perpendicular bisectors of the legs of right triangle $\triangle ABC$. Therefore, l intersects m by Proposition P.

Proposition Q implies Proposition R: We shall prove the contrapositive, the negation of Proposition R implies the negation of Proposition Q. The negation of Proposition Q states that there exist four lines l, m, r, s such that $l \perp r, r \perp s, s \perp m$, and $l \parallel m$. The negation of Proposition R states that *for every* acute angle *there exists* some line intersecting and perpendicular to one ray of the angle but parallel to the other ray. (Note the negation of Proposition K merely says that *there exists* some acute angle such that *there exists* some line intersecting and perpendicular to one ray of the angle but parallel to the other ray.) By the negation of Proposition R, we may suppose $\angle ACN$ has measure $\pi/4$ and l is a line perpendicular to $\overrightarrow{CA}$ at A but parallel to $\overleftrightarrow{CN}$. Let $r=\overleftrightarrow{AC}$, so $l \perp r$. Let m and s be the images of l and r, respectively, under the reflection in $\overleftrightarrow{NC}$. So $s \perp m$. Also, since $m \angle ANC = \pi/4$, we have $r \perp s$. Further, since l is on one side of $\overleftrightarrow{NC}$ and m must be on the other side of $\overleftrightarrow{NC}$, we have $l \parallel m$. Thus, $l \perp r, r \perp s, s \perp m$, and $l \parallel m$, as desired.

Proposition R implies Proposition S trivially. Proposition S implies Proposition T by using Legendre's proof again. We know from Chapter 22 that Propositions T and U are each equivalent to Proposition V. That each of Propositions W, X, and Y is equivalent to Proposition V is left for Exercise 23.1. Since Proposition V implies Proposition A (Corollary 23.4), it follows that the first twenty-five propositions are equivalent. Since Proposition Z implies Proposition A (Theorem 23.6), there only remains to show Proposition Z follows from any of the other propositions.

Suppose $\triangle ABC$ has a right angle at C. Let F be the foot of the perpendicular from C to $\overline{AB}$. By Proposition M, $m \angle ACF = \pi/2 - m \angle CAF = m \angle ABC$. So $\triangle ACF$ and $\triangle ABC$ are similar right triangles but not congruent since $AC \neq AB$. ∎

One of the first things a child learns to say is *NO*. However, stating the negation of a proposition often gives students trouble. In Exercise 23.2 you are asked to state the negation of each of the twenty-six propositions in Theorem 23.7. These negations are important because they will be *theorems* once we deny Euclid's Parallel Postulate.

By the beginning of the nineteenth century the theory of parallels was not in much better shape than in antiquity. The advancement can be summarized by the statement "$\delta \triangle ABC \cong 0$." In 1832 John Bolyai published what the mathematical historian G. B. Halsted called "the most extraordinary two dozen pages in the whole history of thought!" It would take a couple generations for the world to learn about the invention of non-Euclidean geometry by Bolyai and Loba-chevsky. In the meantime there were many so-called proofs of Euclid's Parallel Postulate. Legendre's 1833 *Reflections* contains some half dozen itself. The Graffiti section in this chapter contains some of the more interesting attempts at proving Euclid's Parallel Postulate. Some of these require the definition of a *biangle*. We follow Legendre for the next definition.

DEFINITION 23.8   If A and D are points on the same side of $\overleftrightarrow{BC}$ such that $\overrightarrow{BA} \parallel \overleftrightarrow{CD}$, then $\llcorner ABCD = \overrightarrow{BA} \cup \overline{BC} \cup \overrightarrow{CD}$. $\llcorner ABCD$ is a *biangle* with *vertices* B and C, *angles* $\angle ABC$ and $\angle BCD$, *sides* $\overrightarrow{BA}$ and $\overrightarrow{CD}$, and *base* $\overline{BC}$. The *interior* of a biangle is the intersection of the interiors of its two angles; int $(\llcorner ABCD)$ = int $(\angle ABC) \cap$ int $(\angle BCD)$. If P is a point in the interior of a biangle with vertex B, then $\overrightarrow{BP}$ is an *interior ray* of the biangle. If $\llcorner ABCD$ and $B - C - E$, then $\angle DCE$ is an *exterior angle* of the biangle with *remote interior angle* $\angle ABC$. If $\llcorner ABCD$ and $\llcorner PQRS$ are such that $\angle B \simeq \angle Q$, $\overline{BC} \simeq \overline{QR}$, and $\angle C \simeq \angle R$, then $\llcorner ABCD \simeq \llcorner PQRS$.

Suppose point B is off $\overleftrightarrow{CD}$. Under the Hypothesis of the Acute Angle, there exist many rays $\overrightarrow{BA}$ such that $\llcorner ABCD$. Under the Hypothesis of the Right Angle, there is exactly one ray $\overrightarrow{BA}$ such that $\llcorner ABCD$.

23.2 INDEPENDENCE

In this section we shall show Euclid's Parallel Postulate can never be deduced from the axioms for absolute geometry. This is accomplished by producing a model that satisfies all the axioms for absolute geometry and for which Saccheri's Hypothesis of the Acute Angle holds. Such a model is called a *hyperbolic geometry*. You will hardly have to think to read this section. Unlike the preceding section, there are no subtleties here. Everything is placed before you except the verification of four *simple* substitutions, which require only ninth grade algebra but are very, very long and tedious.

In Section 9.2 we added a distance h to the Cayley–Klein Incidence Plane M13 such that (M13, h) satisfies our first two axioms and $A-B-C$ in (M13, h) only if $A-B-C$ in the Cartesian plane. It follows that (M13, h) must also satisfy PSP, our third axiom. The halfplanes of the line with equation $Ax+By+C=0$ are given by $Ax+By+C>0$ and $Ax+By+C<0$. Let P and Q be distinct points in M13 where $P=(x_1,y_1)$ and $Q=(x_2,y_2)$. Then $\overleftrightarrow{PQ}$ has an equation $Ax+By+C=0$ where $A=y_1-y_2$, $B=x_2-x_1$, and $C=x_1y_2-x_2y_1$. The Cartesian line with this equation intersects the Cartesian unit circle at two Cartesian points S and T which are

$$\left(\frac{-AC\pm B(A^2+B^2-C^2)^{1/2}}{A^2+B^2},\frac{-BC\mp A(A^2+B^2-C^2)^{1/2}}{A^2+B^2}\right).$$

In general, $Ax+By+C=0$ is an equation of a line in M13 iff $A^2+B^2>C^2$, and (x,y) is a point in M13 iff $x^2+y^2<1$. By *simple* substitution the formula for $h(P,Q)$ from Section 9.2 now becomes

$$h(P,Q)=\tfrac{1}{2}\ln\frac{1-x_1x_2-y_1y_2+[(x_2-x_1)^2+(y_2-y_1)^2-(x_1y_2-x_2y_1)^2]^{1/2}}{1-x_1x_2-y_1y_2-[(x_2-x_1)^2+(y_2-y_1)^2-(x_1y_2-x_2y_1)^2]^{1/2}}$$

when $P=(x_1,y_1)$ and $Q=(x_2,y_2)$.

For each line l in M13 with equation $Ax+By+C=0$, define mapping σ_l on the points of M13 by $\sigma_l(x,y)=(x',y')$ where

$$x'=\frac{(A^2+B^2-C^2)x-2A(Ax+By+C)}{(A^2+B^2-C^2)+2C(Ax+By+C)},$$

$$y'=\frac{(A^2+B^2-C^2)y-2B(Ax+By+C)}{(A^2+B^2-C^2)+2C(Ax+By+C)}.$$

Since $(A^2+B^2-C^2)+2C(Ax+By+C)=(A+xC)^2+(B+yC)^2+C^2(1-x^2-y^2)$, then (x',y') is a Cartesian point. That σ_l is a mapping into the points of M13 follows from

$$(x')^2+(y')^2=1-\frac{(A^2+B^2-C^2)^2(1-x^2-y^2)}{[(A+xC)^2+(B+yC)^2+C^2(1-x^2-y^2)]^2}<1.$$

By *simple* substitution, $\sigma_l(x', y') = (x, y)$. So σ_l is an involution and must be a bijection on the points of M13.

From the identity $h((x_1', y_1'), (x_2', y_2')) = h((x_1, y_1), (x_2, y_2))$, established by *simple* substitution, it follows that σ_l is a collineation of (M13, h) that preserves distance. That σ_l fixes l pointwise and interchanges the halfplanes of l follows from the identities

$$x' = x - \frac{2(A + xC)(Ax + By + C)}{(A^2 + B^2 - C^2) + 2C(Ax + By + C)},$$

$$y' = y - \frac{2(B + yC)(Ax + By + C)}{(A^2 + B^2 - C^2) + 2C(Ax + By + C)},$$

$$Ax' + By' + C = -(Ax + By + C)\frac{A^2 + B^2 - C^2}{(A + xC)^2 + (B + yC)^2 + C^2(1 - x^2 - y^2)}.$$

An angle measure function n for (M13, h) is introduced next. Because betweenness in (M13, h) depends on betweenness in the Cartesian plane, $\angle PVQ$ in (M13, h) determines the angle $\angle PVQ$ in the Cartesian plane in the obvious way. There should be no confusion in using "$\angle PVQ$" for the two different angles. If $V = (0, 0)$, let $n\angle PVQ = m\angle PVQ$; so angles in (M13, h) that have vertex $(0, 0)$ inherit their measure from the Cartesian plane. Because Protractor Postulate holds in the Cartesian plane where m is the usual Euclidean angle measure function given by

$$\cos m\angle POQ = \frac{x_1 x_2 + y_1 y_2}{(x_1^2 + y_1^2)^{1/2}(x_2^2 + y_2^2)^{1/2}}$$

when $O = (0, 0)$, $P = (x_1, y_1)$, and $Q = (x_2, y_2)$, the Protractor Postulate as stated in Section 14.1 holds for (M13, h, n) when V is $(0, 0)$ and "m" is replaced by "n". Now suppose $V = (x_0, y_0) \neq (0, 0)$ with V in M13. Then there is a unique line t in M13 with equation $Ax + By + C = 0$ such that $A = x_0$, $B = y_0$, and $C = -1 + (1 - x_0^2 - y_0^2)^{1/2}$. The mapping σ_t interchanges (x_0, y_0) and $(0, 0)$. Hence, defining $n\angle PVQ = m\angle P'V'Q'$ where $V' = \sigma_t V$, $P' = \sigma_t P$, and $Q' = \sigma_t Q$, it follows that n also satisfies the Protractor Postulate when $V \neq (0, 0)$. So (M13, h, n) satisfies our first four axioms. Model (M13, h, n) is called the *Cayley–Klein Model*.

Finally, consider the formula

$$\cos n\angle PVQ =$$
$$\frac{(x_1 - x_0)(x_2 - x_0) + (y_1 - y_0)(y_2 - y_0) - (x_1 y_0 - y_1 x_0)(x_2 y_0 - y_2 x_0)}{[(x_1 - x_0)^2 + (y_1 - y_0)^2 - (x_1 y_0 - y_1 x_0)^2]^{1/2}[(x_2 - x_0)^2 + (y_2 - y_0)^2 - (x_2 y_0 - y_2 x_0)^2]^{1/2}}$$

when $V = (x_0, y_0)$, $P = (x_1, y_1)$, and $Q = (x_2, y_2)$. The formula is easily seen to be correct when $(x_0, y_0) = (0, 0)$. To show that the formula is also correct when $(x_0, y_0) \neq (0, 0)$ and, at the same time, to show that for any line l in M13 the mapping σ_l preserves angle measure n, call the right-hand side of the equation $g(x_0, y_0, x_1, y_1, x_2, y_2)$ and verify by *simple* substitution that this is equal to $g(x_0', y_0', x_1', y_1', x_2', y_2')$ when $(x_0', y_0') = \sigma_l(x_0, y_0)$, $(x_1', y_1') = \sigma_l(x_1, y_1)$, and $(x_2', y_2') = \sigma_l(x_2, y_2)$. The Cayley–Klein Model is completely determined by M13, the formula for $h(P, Q)$ above, and the formula for $\cos n\angle PVQ$ above.

For any line l in the Cayley-Klein Model, we have shown σ_l is a collineation that preserves distance and angle measure, fixes l pointwise, and interchanges the halfplane of l. Thus (M13, h, n) satisfies the Mirror Axiom of Section 16.2 and so satisfies SAS by Theorem 16.3. Therefore the Cayley–Klein Model satisfies all the axioms for absolute geometry. Further, it is obvious that Euclid's Parallel Postulate does not hold for the Cayley–Klein Model, verifying the result whose proof consumed two thousand years of effort: EUCLID'S PARALLEL POSTULATE IS INDEPENDENT OF THE AXIOMS FOR ABSOLUTE GEOMETRY.

The Cayley–Klein Model is a model of hyperbolic geometry, since (M13, h, n) satisfies the axioms for absolute geometry and Saccheri's Hypothesis of the Acute Angle. Hence the existence of the Cayley–Klein Model also confirms the fact that THE AXIOMS FOR HYPERBOLIC GEOMETRY ARE AS CONSISTENT AS THE AXIOMS FOR EUCLIDEAN GEOMETRY.

The angle measure function for the Cayley–Klein Model is related to an incidence property of the Cartesian plane as follows. If l is a line in the Cayley–Klein Model with equation $Ax + By + C = 0$ and $C \neq 0$, then $(-A/C, -B/C)$ is not a point of the model but is a Cartesian point. It turns out that the lines in the Cayley–Klein Model that are perpendicular to l are exactly those that are subsets of the Cartesian lines through the Cartesian point $(-A/C, -B/C)$. This and the situation for $C = 0$ are illustrated in Figure 23.7.

Other models frequently used to show the independence of Euclid's Parallel Postulate and the consistency of hyperbolic geometry can be obtained from the Cayley–Klein Model. For abbreviation we suppose below that t is $(1 - x^2 - y^2)^{1/2}$. The mapping from the Cayley–Klein Incidence Plane M13 onto the Poincaré Incidence Plane M11 that sends (x, y) to $(x/(1 + t), y/(1 + t))$ is an isomorphism (a collineation) from M13 onto M11. If distance function h_1 and angle measure function n_1 are defined for M11 such that the collineation is an isomorphism from (M13, h, n) onto (M11, h_1, n_1), then (M11, h_1, n_1) is called the *Poincaré Model*. The collineation above can be obtained as the composite of the following four mappings. (1) Imbed M13 into

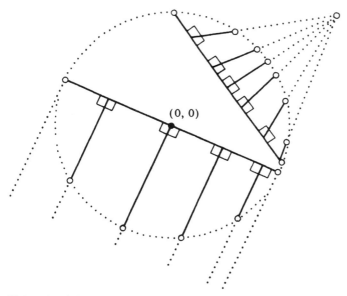

FIGURE 23.7

Cartesian three-space such that (x, y) goes to $(x, y, 0)$. (2) By a projection parallel to the z-axis send $(x, y, 0)$ onto the northern hemisphere of the unit sphere. So $(x, y, 0)$ goes to (x, y, t). The lines of M13 have now been mapped to the semicircles on the northern hemisphere that are orthogonal to the equator. (3) Then from the south pole $(0, 0, -1)$ project (x, y, t) to the x-y-plane. In Figure 23.8, P_2 goes to P_3. Just from similar triangles we see that (x, y, t) goes to $(x', y', 0)$ where $x' = x/(1 + t)$ and $y' = y/(1 + t)$. (4) Finally, we get back to subsets of the Cartesian plane by sending $(x', y', 0)$ to (x', y').

The *Poincaré Halfplane*, another model of hyperbolic geometry, is obtained from the Cayley–Klein Model in analogous fashion. This is the model (M12, h_2, n_2) defined such that the collineation from M13 onto the Poincaré Halfplane Incidence Plane M12 given by sending (x, y) to (x', y') where $x' = x/(1 - y)$ and $y' = t/(1 - y)$ is an isomorphism from (M13, h, n) onto (M12, h_2, n_2). This collineation can also be obtained as the composite of four mappings. The first two are the same as (1) and (2) above. For the third, this time project from $(0, 1, 0)$ to the x-z-plane, thus sending (x, y, t) to $(x/(1 - y), 0, t/(1 - y))$. Finally, to get back to subsets of the Cartesian plane, send $(x/(1 - y), 0, t/(1 - y))$ to $(x/(1 - y), t/(1 - y))$.

In general, lines do not have linear equations in either of the Poincaré models of hyperbolic geometry. However, each of these two models does have the property of being *conformal*, which means angle

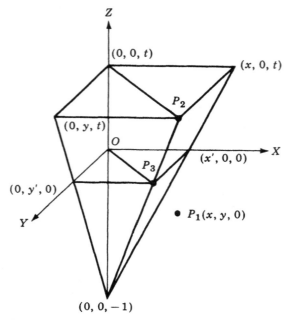

FIGURE 23.8

measure is Euclidean in the following sense. If two intersecting lines of one of the models are considered as two intersecting curves in the Cartesian plane, then the hyperbolic measure of the angles between the intersecting lines of the model is the same as the Euclidean measure of the angles between the curves in the Cartesian plane. The Cayley–Klein Model arises in a natural way from the study of real projective geometry and the Poincaré models from the study of inversive geometry (the geometry of complex numbers).

23.3 EXERCISES

● **23.1** Propositions V, W, X, and Y of Theorem 23.7 are equivalent.

23.2 State the negation of each of the propositions of Theorem 23.7.

● **23.3** Where is a parallel axiom hidden in Birkhoff's axioms in Section 14.1?

● **23.4** True or False?

(a) If $\angle ABC$ and line l is perpendicular to $\overrightarrow{BC}$, then l intersects $\overrightarrow{BA}$ under the Hypothesis of the Right Angle.

(b)   There exists an acute angle such that there exists a line perpendicular and intersecting one ray of the angle and intersecting the other ray.

(c)   For every angle there exists a point P in its interior such that every line through P intersects one ray.

(d)   A line through a point in the interior of an angle intersects at least one ray of the angle.

(e)   For every angle there .exists a point P in its interior such that there exist two lines through P that each intersect both rays.

(f)   For every angle and for any point P in its interior there exists a line through P that intersects both rays.

(g)   For any angle and any point P in its interior there exist two lines through P that each intersect both rays.

(h)   If $\boxed{S}\,ABCD$, then $\sqcup ABCD$, $\sqcup BADC$, and $\sqcup DABC$.

(i)   If $\sqcup ABCD$ and $\sqcup PQRS$ such that $\angle B \simeq \angle Q$ and $\overline{BC} \simeq \overline{QR}$, then $\sqcup ABCD$ is congruent to $\sqcup PQRS$.

(j)   If $\boxed{S}\,ABCD$ and $A-D-E$, then int $(\angle CDE)$ is a proper subset of int $(\angle BAD)$ and is congruent to int $(\angle BAD)$.

23.5   Read "Euclid, Omar Khayyam, and Saccheri" by D. E. Smith in the journal *Scripta Mathematica*, Vol. 3 (1935) pp. 5 – 10.

23.6   The following are equivalent to Euclid's Parallel Postulate:

(a)   The diagonals of a Saccheri quadrilateral bisect each other.

(b)   Any three lines have a common transversal.

(c)   There do not exist three lines such that each two are on the same side of the third.

(d)   If $\triangle ABC$ with M the midpoint of $\overline{AB}$ and N the midpoint of $\overline{AC}$, then $MN = \frac{1}{2} BC$.

23.7   The diagonals of a Saccheri quadrilateral intersect on the line joining the midpoints of the bases.

● **23.8**   Does the point of intersection of the diagonals of a Saccheri quadrilateral bisect the segment joining the midpoints of the bases?

23.9   Nasir Eddin's assumptions deny the Hypothesis of the Acute Angle and the Hypothesis of the Obtuse Angle.

23.10 Complete Legendre's second proof that $\delta\triangle ABC \geq 0$ given the following outline. Assume $\delta\triangle ABC < 0$. Suppose $\angle A$ a smallest angle. Let M_1 be midpoint of $\overline{BC}$ and of $\overline{AC_1}$. Then $\delta\triangle ABC_1 = \delta\triangle ABC$ and $\triangle ABC_1$ has a smallest angle of measure at most $\frac{1}{2}m\angle CAB$. Repeating the construction, eventually obtain a triangle which has the same defect as $\triangle ABC$ and a smallest angle of measure less than $-\delta\triangle ABC$, a contradiction.

23.11 Complete Legendre's proof of Theorem 23.2 given the following outline. Let R_1 be on $\overrightarrow{QR}$ such that $PQ = QR_1$. Let R_i be midpoint of Q and R_{i+1}. By Pons Asinorum and Absolute Exterior Angle Theorem, $2m\angle QR_{i+1}P \leq m\angle QR_iP$. So $2^nm\angle QR_nP \leq \pi$ when $2^nr > \pi$.

• **23.12** Find the flaw in another of Legendre's arguments that $\delta\triangle ABC \leq 0$ as outlined below. Suppose $AB \geq AC \geq BC$ with D_1 midpoint of $\overline{BC}$. Let C_1 on $\overrightarrow{AD_1}$ and B_1 on $\overrightarrow{AB}$ such that $AC_1 = AB$ and $AB_1 = 2AD_1$. Then $\delta\triangle AB_1C_1 = \delta\triangle ABC$, $AB_1 \geq AC_1 \geq B_1C_1$, and $2m\angle B_1AC_1 < m\angle BAC$. Repeating, $\delta\triangle AB_nC_n = \delta\triangle ABC$, $AB_n \geq AC_n \geq B_nC_n$, and $2^nm\angle B_nAC_n < m\angle BAC$. Since $m\angle C_nAB_n$ and $m\angle AB_nC_n$ approach 0, C_n approaches $\overleftrightarrow{AB}$. So $m\angle AC_nB_n$ approaches π. So $\delta\triangle AB_nC_n$ approaches 0. Thus $\delta\triangle ABC = 0$.

23.13 Legendre's argument that all the points equidistant from a line and on one side of the line are collinear may be found in his *Reflections*. Almost all of the *Reflections* is reproduced and translated into Interlingue in *Le Axiome de Paralleles* edited by C. E. Sjöstedt, (Interlingue-Fundation, 1968).

23.14 Read Omar Khayyam's proof of Euclid's Parallel Postulate in his "Discussion of Difficulties in Euclid" as translated by A. R. Amir-Moez in *Scripta Mathematica* Vol. 24 (1959) pp. 275–303.

23.15 $C(x^2 + y^2) + 2Ax + 2By + C = 0$ is an equation of a line in M11 iff $A^2 + B^2 > C^2$, and every line has such an equation.

23.16 $(C + B)(x^2 + y^2) + 2Ax + (C - B) = 0$ is an equation of a line in M12 iff $A^2 + B^2 > C^2$, and every line has such an equation.

23.17 $Ax + By + C = 0$ is an equation of a line in M13 iff $A^2 + B^2 > C^2$, and every line has such an equation.

23.18 The mapping that sends (x, y) to $(x/(1 + (1 - x^2 - y^2)^{1/2}), y/(1 + (1 - x^2 - y^2)^{1/2}))$ is a collineation from M13 onto M11.

23.19 The mapping that sends (x, y) to $(x/(1 - y), (1 - x^2 - y^2)^{1/2}/(1 - y))$ is a collineation from M13 onto M12.

23.20 That two parallel lines have two common perpendiculars is equivalent to Euclid's Parallel Postulate.

*23.21   That each two parallel lines have a common perpendicular is equivalent to Euclid's Parallel Postulate.

*23.22   Given $\triangle ABC$ there exists $\triangle DEF$ such that each vertex of $\triangle ABC$ lies on exactly one side of $\triangle DEF$.

*23.23   That any four points lie in the interior of some triangle is equivalent to Euclid's Parallel Postulate.

GRAFFITI

Some of the attempts at proving Euclid's Parallel Postulate are amusing. Perhaps you will find the following selection interesting.

In the "Notes" to the edition of his Elements of Geometry *written in 1813, Playfair gives a proof of Euclid's Parallel Postulate. Playfair's actual wording of his parallel postulate is "Two straight lines which intersect one another, cannot be both parallel to the same straight line." Playfair, like C. L. Dodgson (Lewis Carroll) and many others, wished to improve Legendre's proofs because the proofs were deemed too difficult for students just beginning to study geometry and not because the proofs were wrong! The idea of Playfair's proof was anticipated by B. F. Thibaut in 1809. The argument runs as follows.*

Suppose $\triangle ABC$ with B–A–D, A–C–E, and C–B–F. See Figure 23.9. At A rotate $\overleftrightarrow{AB}$ to $\overleftrightarrow{AC}$ through $\angle DAC$; then at C rotate $\overleftrightarrow{AC}$ to $\overleftrightarrow{BC}$ through $\angle ECB$; and, finally, at B rotate $\overleftrightarrow{BC}$ to $\overleftrightarrow{AB}$ through $\angle FBA$. Since $\overleftrightarrow{AB}$ has returned to itself (but, we might add, translated along itself), $\overleftrightarrow{AB}$ has been rotated through four right angles. So adding up the measures of the angles we have the formula $(\pi - m\angle A) + (\pi - m\angle C) + (\pi - m\angle B) = 4(\pi/2)$. Hence $\delta\triangle ABC = 0$ and Euclid's Parallel Postulate follows.

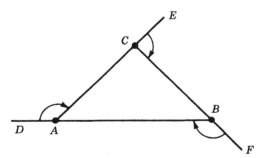

FIGURE 23.9

Of course one problem is that we do not yet have a definition of a rotation. Suppose we can give a reasonable definition. Even so there is the tacit assumption that we can ignore the translation of $\overleftrightarrow{AB}$ and add the rotations at A, C, and B as if all three rotations were about A. To give a flawless proof we would have to find a justification for these assumptions. Some intuition of the difficulty is easily obtained by testing the proof with a triangle on a sphere, where the sum of the measures of the angles is greater than π.

Louis Bertrand gave the following argument in 1778 where A, B, C, A_i, B_i are as in Figure 23.10 with $B_iB_{i+1} = BB_i$. The area within $\angle ABC$ is a finite fraction of the area of the entire plane, in fact, $(m\angle ABC)/(2\pi)$. However, the area within $\llcorner ABB_iA_i$ is an infinitesimal fraction of the whole plane since for no positive integer n does the union of the interiors of n biangles congruent to $\llcorner ABB_iA_i$ cover the plane. It follows that int $(\angle ABC)$ cannot be contained in int $(\llcorner ABB_iA_i)$. Thus $\overrightarrow{BC}$ must intersect $\overrightarrow{B_iA_i}$, and Euclid's Parallel Postulate follows.

Legendre picks up Bertrand's argument in his Reflections *using the language that int $(\angle ABC)$ cannot be contained in int $(\llcorner ABB_iA_i)$ because the first is infinite of the second order and the second is only infinite of the first order. Another of Legendre's arguments based on infinite areas refers back to Figure 23.9 again. Since the area of the entire plane equals the area within four right angles and since the finite area of $\triangle ABC$ can be disregarded when considering infinities of the second order, it follows that the sum of the measures of the three indicated exterior angles must equal four times $\pi/2$, giving the same formula obtained by Playfair.*

In 1850 Victor Bunaikovskij criticized the proofs of Bertrand and Legendre. He gave the following direct stab at Euclid's Parallel

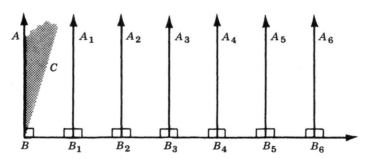

FIGURE 23.10

Postulate. Suppose ⌐ABCD *with* $\overline{AB} \perp \overline{BC}$ *and* B–C–E. *Assume* ∠BCD *is acute. Then, since* ∠DCE > ∠ABC, *the infinite sector* int (∠DCE) *is of greater area than the infinite sector* int (∠ABC). *However, the infinite sector* int (∠DCE) *is of less area than the infinite sector* int (∠ABC) *because the first is wholly contained within the second. The contradiction gives the desired result.*

All these arguments based on infinite areas *are nonsense, however interesting. Further, even arguments based on* finite areas *should be scrupulously examined to see what assumptions are made concerning the meaning of the word* area.

Legendre's attempt at proving that a Lambert quadrilateral is a rectangle is most interesting. The following is based on an argument in the Reflections. *Suppose* ⌐L⌐ABCD. *Let other points be as in Figure 23.11 where* △AQP ≃ △DQC *and* △PCB ≃ △RCE. *Since* △AQP ≃ △DQC, *then* int (⌐MPCL) *is piecewise congruent to* int (⌐MADL). *Roughly speaking,* int (⌐MPCL) *can be cut up by chopping off* int (△AQP) *and the pieces put back together again to form* int (⌐MADL) *by superimposing* int (△AQP) *on* int (△DQC). *The idea of piecewise congruence, which can be made precise, gets around having to talk nonsense about infinite areas. Since* △PCB ≃ △RCE, *we also have that* int (⌐MPRN) *is piecewise congruent to* int (⌐MBEN). *Further,* int (⌐MPCL) *and* int (⌐LCRN) *are piecewise congruent since they are congruent. So* int (⌐MBEN) *is piecewise congruent to* int (⌐MPRN), *which is piecewise congruent to two disjoint copies of* int (⌐MPCL). *Thus* int (⌐MBEN) *is piecewise congruent to two disjoint copies of* int (⌐MADL). *Hence* int (⌐MBEN) *is piecewise congruent to the interior of a biangle with two right angles and a base of length* 2AD. *Therefore (?),* BE = 2AD. *Since* BE = 2BC, *we have* AD = BC. *So* ⌐S⌐ADCB *with* ∠D *and* ∠C *right angles. Hence* ⌐L⌐ABCD *is a rectangle.*

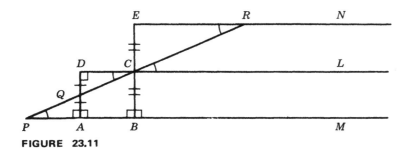

FIGURE 23.11

CHAPTER 24

Biangles

24.1 CLOSED BIANGLES

A *biangle* is defined in Definition 23.8. For Euclidean geometry every interior ray of a biangle intersects both sides of the biangle. However, this is not the case under the Hypothesis of the Acute Angle. For example, if $\llcorner PQRS$ has right angles at Q and R, the negation of Playfair's Parallel Postulate implies that some interior ray $\overrightarrow{QT}$ is parallel to $\overleftrightarrow{RS}$ and so does not intersect $\overrightarrow{RS}$. If every interior ray $\overrightarrow{BE}$ of $\llcorner ABCD$ does intersect $\overrightarrow{CD}$, we shall say the biangle is *closed from B*. We shall prove that a biangle closed from one vertex is necessarily closed from the other vertex. Although trivially true for Euclidean geometry, this result is not trivial for absolute geometry.

DEFINITION 24.1 If every interior ray $\overrightarrow{BE}$ of $\llcorner ABCD$ intersects $\overrightarrow{CD}$, then the biangle is *closed from vertex B*. If a biangle is closed from both vertices, then the biangle is *closed*. If $\llcorner ABCD$ with $\angle B \simeq \angle C$, then the biangle is *isosceles*. $\overrightarrow{AB}$ is *equivalent* to $\overrightarrow{PQ}$ if either $\overrightarrow{AB}$ contains $\overrightarrow{PQ}$ or $\overrightarrow{PQ}$ contains $\overrightarrow{AB}$, in which case we write $\overrightarrow{AB} \sim \overrightarrow{PQ}$. If $\llcorner ABCD$, $\overrightarrow{BA} \sim \overrightarrow{QP}$, and $\overrightarrow{CD} \sim \overrightarrow{RS}$, then $\llcorner ABCD$ is *equivalent* to $\llcorner PQRS$, in which case we write $\llcorner ABCD \sim \llcorner PQRS$.

Theorem 24.2 If $\llcorner ABCD$ is isosceles and equivalent to $\llcorner AQRD$

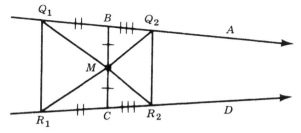

FIGURE 24.1

where Q and R are on the same side of $\overleftrightarrow{BC}$ with $BQ = CR$, then $\llcorner AQRD$ is isosceles.

Proof Let M be the midpoint of $\overline{BC}$. (See Figure 24.1.) $\triangle MBQ \simeq \triangle MCR$ by SAS. So $\angle BQM \simeq \angle CRM$ and $\overline{MQ} \simeq \overline{MR}$. Then $\angle MQR \simeq \angle MRQ$ by the Pons Asinorum. It follows that $\angle BQR \simeq \angle CRQ$ and $\llcorner AQRD$ is isosceles. ∎

Theorem 24.3 If $\llcorner ABCD$ is closed from B, then the biangle is equivalent to an isosceles biangle with vertex B.

Proof Since $\llcorner ABCD$ is closed from B, the angle bisector of $\angle B$ intersects $\overrightarrow{CD}$ at some point E. Then the angle bisector of $\angle C$ intersects $\overline{BE}$ at some point P by Crossbar. Let Q, R, S be the feet of the perpendiculars from P to $\overleftrightarrow{AB}$, $\overleftrightarrow{CD}$, $\overleftrightarrow{BC}$, respectively. Q, R, and S are on $\llcorner ABCD$ since the foot of the perpendicular from a point on one side of an acute angle to the line containing the other side is on the other side of the angle. By SAA, $\triangle PBQ \simeq \triangle PBS$ and $\triangle PCS \simeq \triangle PCR$. So $PQ = PS = PR$. Without loss of generality, suppose A and D are such that $B - Q - A$ and $C - R - D$. If P is on $\overline{QR}$, then $\llcorner AQRD$ is isosceles. Otherwise $\angle PQR \simeq \angle PRQ$ by the Pons Asinorum, which implies $\angle BQR \simeq \angle CRQ$. So, in either case, $\llcorner AQRD$ is isosceles. By the previous theorem there is a point C' on $\overleftrightarrow{CD}$ such that $\llcorner ABCD$ is equivalent to isosceles $\llcorner ABC'D$. ∎

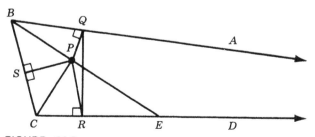

FIGURE 24.2

In proving that a biangle closed from one vertex is closed, we shall use the following lemma.

Theorem 24.4 Given $\angle GCD$ and given point P off $\overleftrightarrow{AB}$, there exists a point E on $\overrightarrow{AB}$ such that $m\angle PEA < m\angle GCD$.

Proof By the Absolute Exterior Angle Theorem, this theorem is only a restatement of Theorem 23.2. ∎

Theorem 24.5 A biangle is closed iff the biangle is closed from one vertex.

Proof We need to show only that a biangle closed from one vertex is also closed from the other vertex. To do this, we shall prove the contrapositive: a biangle not closed from one vertex is not closed from the other vertex. Suppose $\sqcup ABCD$ is not closed from C. Then $\sqcup ABCD$ has an interior ray $\overrightarrow{CE}$ that does not intersect $\overrightarrow{BA}$. (See Figure 24.3.) By the lemma, we may suppose $m\angle BEC < m\angle ECD$ without loss of generality. Hence, if $B-E-F$, then $\overrightarrow{EF}$ and $\overrightarrow{CD}$ cannot intersect by the Absolute Exterior Angle Theorem. So $\overrightarrow{BE}$ and $\overrightarrow{CD}$ do not intersect. Hence, $\sqcup ABCD$ is not closed from B. ∎

Theorem 24.6 If $\overrightarrow{QD} \sim \overrightarrow{CD}$ and $\sqcup ABCD$ is closed from B, then $\sqcup ABQD$ is closed from B.

Proof If Q is on $\overrightarrow{CD}$, then the result is trivial. If $Q-C-D$, then the result follows from Crossbar. ∎

Theorem 24.7 A biangle equivalent to a closed biangle is closed.

Proof It is sufficient to show that if $\overrightarrow{PA} \sim \overrightarrow{BA}$, $\overrightarrow{QD} \sim \overrightarrow{CD}$, and $\sqcup ABCD$ is closed, then $\sqcup APQD$ is closed. Since $\sqcup ABCD$ is closed from B, then $\sqcup ABQD$ is closed from B by the previous theorem. So $\sqcup ABQD$

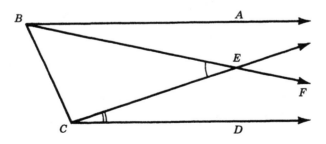

FIGURE 24.3

is closed from Q. Hence, $\llcorner APQD$ is closed from Q by the previous theorem. Therefore, $\llcorner APQD$ is closed. ∎

We finish the section with a congruence theorem for closed biangles. The proof is quite like the proof of ASA. The idea is to *copy* one figure onto the other. In the proof below, the isometry σ merely plays the role of an express copying machine.

Theorem 24.8 *Angle-Base Theorem* If $\llcorner ABCD$ and $\llcorner PQRS$ are closed, $\angle ABC \simeq \angle PQR$, and $\overline{BC} \simeq \overline{QR}$, then $\llcorner ABCD \simeq \llcorner PQRS$.

Proof We need to show $\angle BCD \simeq \angle QRS$. Without loss of generality suppose $\overline{AB} \simeq \overline{PQ}$. Since $\triangle ABC \simeq \triangle PQR$ by SAS, then there exists an isometry σ such that $\sigma P = A$, $\sigma Q = B$, and $\sigma R = C$. Let $\sigma S = E$. Since σ preserves incidence, we must have $\llcorner ABCE$ is closed. Since $\llcorner ABCD$ and $\llcorner ABCE$ are both closed from C, it follows from the definition of a biangle closed at a vertex that $\overrightarrow{CD} = \overrightarrow{CE}$. Thus $\angle BCD \simeq \angle QRS$. ∎

24.2 CRITICAL ANGLES AND ABSOLUTE LENGTHS

We first show that each segment can be associated with an angle of some particular measure. Under the Hypothesis of the Right Angle this association is less than exciting as the angle associated with any segment is always a right angle. However, under the Hypothesis of the Acute Angle, a segment is associated with an acute angle of some particular measure, and this association is critical.

Theorem 24.9 If $\overline{BC} \perp \overline{CD}$, then there exists unique $\overrightarrow{BA}$ such that $\llcorner ABCD$ is closed.

Proof Let S be the set of all positive real numbers $m\angle CBP$ where P

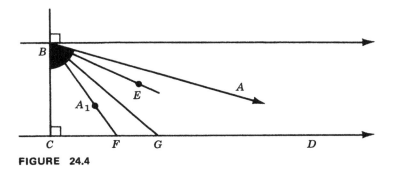

FIGURE 24.4

is a point in int $(\overrightarrow{CD})$. Certainly S is not the empty set as $m\angle CBD$ is in S. Further, every element of S is less than $\pi/2$ since the perpendicular to $\overline{BC}$ at B is parallel to $\overleftrightarrow{CD}$. Since S is a bounded, nonempty set of real numbers, then S has a least upper bound m_0. So $0 < m_0 \leqq \pi/2$. Let A be on the same side of $\overleftrightarrow{BC}$ as D such that $m\angle ABC = m_0$. Assume $\overrightarrow{BA}$ intersects $\overrightarrow{CD}$ at some point F. Let C-F-G. Then $m\angle CBG$ is in S and $m\angle CBG > m_0$, contradicting $m_0 = $ lub S. Hence $\overrightarrow{BA}$ does not intersect $\overrightarrow{CD}$, and we have $\llcorner\!\lrcorner ABCD$. Now assume $\overrightarrow{BE}$ is an interior ray of $\llcorner\!\lrcorner ABCD$ that does not intersect $\overrightarrow{CD}$. Since int $(\overrightarrow{CD})$ is contained in int $(\angle CBE)$, every element of S is less than $m\angle CBE$. This contradicts $m_0 = $ lub S since $m\angle CBE < m_0$. Hence $\llcorner\!\lrcorner ABCD$ is closed at B. $\overrightarrow{BA}$ is necessarily unique by the definition of closure from B. ■

Corollary 24.10 If point P is off $\overleftrightarrow{AB}$, then there exists unique $\overrightarrow{PQ}$ such that $\llcorner\!\lrcorner ABPQ$ is closed.

Proof Exercise 24.1. ■

In the proof of Theorem 24.9 the number m_0 depends on $\overline{BC}$. As an application of the Angle-Base Theorem, then m_0 depends on at most BC. (Note $BC \neq \overline{BC}$.) It is quite possible that m_0 is even independent of BC, as in the Euclidean case where m_0 is always $\pi/2$ regardless of the length of $\overline{BC}$. In any case, Theorems 24.8 and 24.9 assure the next definition is well-defined.

DEFINITION 24.11 Π, the function from the positive reals into the positive reals defined by $\Pi(BC) = m\angle ABC$ if $\llcorner\!\lrcorner ABCD$ is closed with $\angle C$ right, is called the **critical function**. $\angle PQR$ is a **critical angle** for $\overline{ST}$ and $m\angle PQR$ is the **critical value** of ST if $m\angle PQR = \Pi(ST)$.

The critical function is aptly named, as in Figure 24.5 the lines

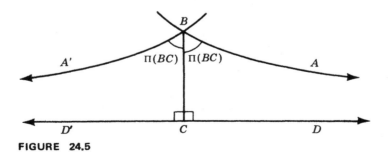

FIGURE 24.5

$\overleftrightarrow{BA}$ and $\overleftrightarrow{BA'}$ are the *bounds* for the parallels to $\overleftrightarrow{CD}$ that pass through B. Of course these lines coincide under the Hypothesis of the Right Angle.

Theorem 24.12 If $\Pi(x_0) = \pi/2$ for some positive real x_0, then $\Pi(x) = \pi/2$ for every positive real x and Euclid's Parallel Postulate holds. If $\Pi(x_0) < \pi/2$ for some positive real x_0, then $\Pi(x) < \pi/2$ for every positive real x and the Hypothesis of the Acute Angle holds.

Proof Exercise 24.2. ∎

The next theorem and its corollary are a slight digression to settle a matter introduced back in Section 21.1.

Theorem 24.13 If $\Pi(BC) = m\angle ABC < \pi/2 = m\angle BCD$, then $\overleftrightarrow{AB}$ and $\overleftrightarrow{CD}$ do not have a common perpendicular.

Proof We may suppose A and D are points on the same side of $\overleftrightarrow{BC}$. Then $\sqcup ABCD$ is closed by definition of the critical function. Assume $\overrightarrow{QR}$ is perpendicular to $\overleftrightarrow{AB}$ at Q and perpendicular to $\overleftrightarrow{CD}$ at R. Let $\overrightarrow{QP} \sim \overrightarrow{BA}$ and $\overrightarrow{RS} \sim \overrightarrow{CD}$. So $\sqcup PQRS$ is equivalent to $\sqcup ABCD$. Since $\sqcup ABCD$ is closed, then $\sqcup PQRS$ is closed (Theorem 24.7). Hence $\Pi(QR) = \pi/2$, contradicting $\Pi(BC) < \pi/2$. So $\overleftrightarrow{AB}$ and $\overleftrightarrow{CD}$ have no common perpendicular. ∎

Corollary 24.14 That every two parallel lines have a common perpendicular is equivalent to Euclid's Parallel Postulate.

Proof That Euclid's Parallel Postulate implies every two parallel lines have a common perpendicular is trivial. The converse is not trivial! However, from the theorem above, the Hypothesis of the Acute Angle implies there exist two parallel lines that have no common perpendicular. The contrapositive of this is the missing implication. ∎

Theorem 24.15 Under the Hypothesis of the Right Angle, the critical function is constant. Under the Hypothesis of the Acute Angle, the critical function is strictly decreasing.

Proof Since we already know the first statement (Theorem 24.12), suppose the Hypothesis of the Acute Angle holds. We must show $d_0 < d_1$ implies $\Pi(d_1) < \Pi(d_0)$ where d_0 and d_1 are any two positive real numbers. Suppose $B-E-C$ with $BE = d_0$ and $BC = d_1$. Let $\overline{CD} \perp \overline{BC}$, $\overline{EF} \perp \overline{BC}$, and $m\angle ABC = \Pi(BC)$ with A, D, and F on the same side of $\overleftrightarrow{BC}$. Let G be such that $\llcorner FECG$ is closed. Since $m\angle GCE = \Pi(EC) < \pi/2$, $\overrightarrow{CG}$ is an interior ray of $\llcorner ABCD$ intersecting $\overrightarrow{BA}$ at some point H. Because $\overleftrightarrow{EF} \| \overleftrightarrow{CG}$, $\overrightarrow{EF}$ must intersect $\overline{BH}$ by PASCH. Hence $\overrightarrow{BA}$ intersects $\overrightarrow{EF}$ and $m\angle ABC < \Pi(BE)$. So $\Pi(BC) < \Pi(BE)$, and $\Pi(d_1) < \Pi(d_0)$. ∎

Suppose the Hypothesis of the Acute Angle holds. Then Theorem 24.15 implies segments of different lengths are associated by the critical function with acute angles of different angle measure, since a strictly decreasing function is necessarily one-to-one. Thus some acute angles determine segments of unique length! We shall now show that every acute angle determines a segment under the Hypothesis of the Acute Angle.

Theorem 24.16 Under the Hypothesis of the Acute Angle, if $0 < m_0 < \pi/2$, then there exists unique positive number d_0 such that $\Pi(d_0) = m_0$.

Proof The uniqueness of such a number d_0 follows from the preceding theorem. We must show d_0 exists. Suppose $m\angle ABE = m_0$. Since the Hypothesis of the Acute Angle holds, we may suppose without loss of generality that the perpendicular to $\overleftrightarrow{BE}$ at E does not intersect $\overrightarrow{BA}$ (Theorem 23.7, Proposition R). (See Figure 24.7.)

Let S be the set of all positive real numbers BF where F is the foot of the perpendicular to $\overleftrightarrow{BE}$ from a point P in int $(\overrightarrow{BA})$. Certainly S is not empty as the perpendicular to $\overleftrightarrow{BE}$ from every point on $\overrightarrow{BA}$ intersects $\overrightarrow{BE}$. If line l is perpendicular to $\overleftrightarrow{BE}$ at point L such that $B-E-L$, then l and $\overrightarrow{BA}$ do not intersect as they are on opposite sides of the line perpendicular to $\overleftrightarrow{BE}$ at E. So every number in S is less than BE. Since S is a bounded, nonempty set of real numbers, then S has a

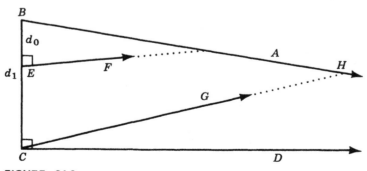

FIGURE 24.6

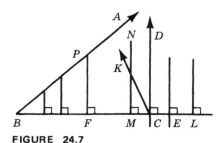 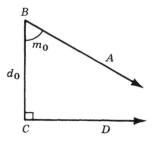

FIGURE 24.7

least upper bound d_0. So $d_0 > 0$. Let C be on $\overrightarrow{BE}$ such that $BC = d_0$. Let $\overline{CD} \perp \overline{BC}$ with A and D on the same side of $\overleftrightarrow{BC}$. Assume $\overleftrightarrow{CD}$ intersects $\overrightarrow{BA}$ at some point H. Let $B-H-I$, and let J be the foot of the perpendicular from I to $\overleftrightarrow{BE}$. Since $\overleftrightarrow{CH} \| \overleftrightarrow{IJ}$ and $B-H-I$, we have $B-C-J$. So BJ is in S and $BC < BJ$, contradicting $d_0 = \operatorname{lub} S$. Therefore, $\sqcup ABCD$ with $\overline{BC} \perp \overline{CD}$.

We now show $\sqcup ABCD$ is closed from C. Let $\overrightarrow{CK}$ be an interior ray of $\sqcup ABCD$. Assume $\overrightarrow{CK}$ does not intersect $\overrightarrow{BA}$. Let M be the foot of the perpendicular from K to $\overleftrightarrow{BE}$ and $M-K-N$. So $B-M-C$, and $\overline{MK}$ does not intersect $\overrightarrow{BA}$ as M and K are on the same side of $\overleftrightarrow{BA}$. Also, int $(\overrightarrow{KN})$ and $\overrightarrow{BA}$ do not intersect as they are on opposite sides of $\overleftrightarrow{CK}$. Thus $\overrightarrow{MN}$ and $\overrightarrow{BA}$ do not intersect. It follows that every number in S is less than BM with $BM < BC$, contradicting $BC = \operatorname{lub} S$. Thus $\sqcup ABCD$ is closed from C. Since we are fortunate enough to know that a biangle closed from one vertex is closed from the other vertex, we now have $\sqcup ABCD$ is closed with $\angle C$ right. Hence $m \angle ABC = \Pi(BC) = \Pi(d_0)$. ∎

Corollary 24.17 Under the Hypothesis of the Acute Angle, Π is a strictly decreasing, continuous function such that $\lim_{x \to 0^+} \Pi(x) = \pi/2$ and $\lim_{x \to \infty} \Pi(x) = 0$.

DEFINITION 24.18 If $\angle PQR$ is a critical angle for $\overline{ST}$, then ST is the *absolute length* for $\angle PQR$.

In 1305 Edward I standardized units of length in England by decreeing "three grains of barley dry and round make an inch; twelve inches make a foot; three feet make a yard." It was understood the barley corn was to come from the middle of the ear and was to be laid end to end. The king's official *yard* then became the end-to-end measure of an iron bar, which differed from today's yard by at most 0.1%. By 1878 the Imperial Standard Yard was accepted. This was the distance at 62°F between two fine lines on gold plugs in a bronze bar at Westminster, England. Should this standard be destroyed, the *yard* was regained as the length of a pendulum beating seconds at sea level in the latitude of London. It was learned that this bronze bar was shrinking about one part in a million every twenty-three years. In 1866 the United States Congress defined a *yard* to be 3600/3937 meters. Today a *yard* is defined to be exactly 0.9144 meters; so an *inch* is now exactly 2.54 centimeters.

A *meter* was originally intended to be one ten-millionth of the distance from the earth's equator to a pole measured along a meridian. From 1889 to 1960 a *meter* was the distance between two lines on a platinum–iridium bar preserved at atmospheric pressure and 0°C at the International Bureau of Weights and Measures near Paris. All lengths were compared with this International Prototype Meter. The National Bureau of Standards in Washington, D.C. maintained a copy. As technology improved there was a demand for a more precise definition. Presently* a *meter* is defined as exactly 1,650,763.73 wavelengths of the orange-red radiation line of krypton 86 under certain conditions. Now, no national or international bureau keeps a standard right angle tucked away in its archives. Shouldn't there be an International Prototype Right Angle against which all other angles could be compared? Why not?

Suppose the Hypothesis of the Acute Angle holds. Then there exists a d_0 such that $\Pi(d_0) = \pi/4$. Hence d_0 or some fraction of d_0 could be defined as the *standard length* against which all other lengths could be compared. We can multiply all distances in our theory by some positive constant t without changing the content of the theory. In other words, if $(\mathscr{P}, \mathscr{L}, d, m)$ is a model of Σ, then so is $(\mathscr{P}, \mathscr{L}, td, m)$ when t is a positive constant. We would say that we had just changed the *scale* for distance. For reasons that are not at all apparent now, it turns out to be mathematically convenient to pick a scale such that $\Pi(1)$ is $2 \arctan e^{-1}$. In applying this to the physical world, there is little difficulty in determining A, B, C such that $\angle ABC$ has measure quite close to $2 \arctan e^{-1}$. However, how *long* is a segment of *length* 1? That is, how many *meters long* is it? Since a meter has nothing to do with the axioms of our geometry, the question is a valid one. Although it may be really neat to have a geometry that provides for a standard angle determining a standard length, all physical experiments indicate that $\Pi(x)$ could noticeably differ from $\pi/2$ only for very large astronomical distances x. So a physical segment of length 1 would be very, very *long* indeed. The Hypothesis of the Acute Angle cannot be verified by an experiment devised to determine the measure of a critical angle because this would involve a physical proof that two lines never meet. Any verification that one hypothesis or the other

---

* See page 268 for the subsequent definition.

applies to physical space must depend on experiments with finite objects, such as triangles.

Each acute angle of an isosceles right triangle with legs of length d must have angle measure less than $\Pi(d)$. See Figure 24.8. By pasting two such triangles together we can obtain a triangle such that the sum of the measures of its angles is less than $4\Pi(d)$. Suppose Π is not constant. Then, by taking d *big enough*, we have a triangle with the sum of the measures of its angles as close to 0 as we like. In other words, we can obtain a triangle whose defect is as close to π as we like.

Theorem 24.19 Under the Hypothesis of the Acute Angle, if $0 < t < \pi$, then there exists $\triangle ABC$ such that $\delta\triangle ABC > t$.

Proof Since $\pi > t$, let d be such that $\Pi(d) = \frac{1}{4}(\pi - t)$. Let $\overline{AD} \perp \overline{BC}$ with D the midpoint of $\overline{BC}$ and $AD = BD = DC = d$. Since $\overrightarrow{AC}$ intersects $\overrightarrow{DC}$, it follows that each of the congruent angles $\angle DAC$, $\angle DAB$, $\angle DCA$, and $\angle DBA$ has measure less than $\Pi(d)$. Hence $\delta\triangle ABC > \pi - 4\Pi(d) = t$. ∎

In order to draw a figure of a triangle with defect close to π so that the angle measures *look* correct, we must draw something like the right-hand part of Figure 24.8. This is necessary because the sides are just too long to get on paper. The largest physical triangles that can be accurately measured are astronomical. Let E stand for the Earth, S for the Sun, and V for the brilliant blue star Vega. $\angle SEV$ can be measured from the Earth when $\angle ESV$ is right. Using this measurement and the fact that $\delta\triangle SEV$ is less than $\pi/2 - m\angle SEV$, one

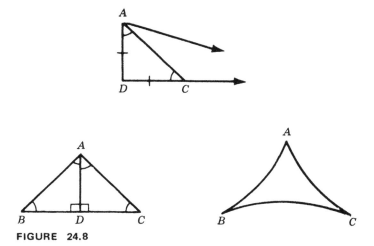

FIGURE 24.8

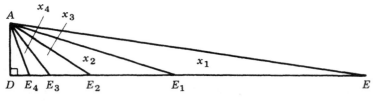

FIGURE 24.9

obtains $\delta \triangle SEV < 0.0000004$. Hence, assuming the Hypothesis of the Acute Angle, the defect of what most of us would consider a large physical triangle would be very small. It follows (Exercise 24.10) from the proof of the next theorem that if all the sides of a triangle are *short enough* then the defect of the triangle is close to 0. $\overline{EV}$ above or any physical segment may be *short enough* so that any difference between 0 and the measured defect of a physical triangle can be attributed to error in measurement. Since measurement of physical angles can never be exact, this brings home the impossibility of proving the applicability of the Hypothesis of the Right Angle to physical space.

Theorem 24.20 If $0 < t < \pi$, then there exists $\triangle ABC$ such that $\delta \triangle ABC < t$.

Proof Suppose $\triangle ADE$ with $\angle D$ right. Let E_1 be the midpoint of $\overline{DE}$, and let E_{i+1} be the midpoint of $\overline{DE_i}$. (See Figure 24.9.) Let $\delta \triangle AEE_1 = x_1$ and $\delta \triangle AE_iE_{i+1} = x_{i+1}$. So $x_1 + x_2 + \cdots + x_n < \delta \triangle ADE$ (Theorem 22.16). Assume $t \leqq x_n$ for each positive integer n. Then $nt < \delta \triangle ADE$ for each positive integer n, contradicting Archimedes' axiom. Hence, for some n, $t > x_n$. So $\delta \triangle AE_{n-1}E_n < t$. With $B = E_{n-1}$ and $C = E_n$, we have $\delta \triangle ABC < t$. ∎

Our theory of the hyperbolic plane continues at the beginning of Chapter 26.

24.3 THE INVENTION OF NON-EUCLIDEAN GEOMETRY

About 300 B.C. Euclid was the first to abolish the *petitio principii* involving parallels by formulating the famous Postulate 5 in his *Elements* (see Section 11.1). Although the postulate created controversy for the next two millennia (see Section 23.1), there was little progress in the theory of parallels until **1733**. It was then that *Euclides ab Omni Naevo Vindicatus* by Gerolamo Saccheri (1667–1733) appeared. In the second of the two books in this work, Saccheri proves a tacit assumption made by Euclid in his proof of Proposition V.18 on the theory of proportion. In Book I, Saccheri apparently proves Euclid's fifth postulate. Ironically, had Saccheri actually succeeded in proving that Euclid's Parallel Postulate is a theorem of absolute geometry, then his efforts would not have vindicated Euclid of all fault. On the contrary, Euclid is vindicated by the existence of hyperbolic geometry, which demonstrates the independence of Euclid's Parallel Postulate!

Where did Saccheri go wrong? In the third corollary to his Proposition XXVI, Saccheri uses a point at infinity (i.e., a limit point not in the plane) as if it were a point in the plane. In fact, he even talks about a point "beyond" the limit point. However, he then adds that the corollary is unnecessary! Is the superfluous corollary with its blatant error simply a *mistake,* or is it a *signal* to pay careful attention to the sequel? The answer to this question is the key to Saccheri's attitude toward non-Euclidean geometry. By Proposition XXXII, Saccheri has proved many of the theorems of elementary hyperbolic geometry. In particular, he has proved that if $\text{II}(BC) = m\angle ABC$ as in Figure 24.10 then $m\angle BAD$ approaches $\pi/2$ and AD approaches 0 as CD approaches infinity. Thus having "disproved the hostile hypothesis of acute angle by a manifest falsity, since it must lead to the recognition of two straight lines which at one and the same point have in the same plane a common perpendicular," Saccheri states his Proposition XXXIII: The hypothesis of acute angle is abolutely false, being repugnant to the nature of the straight line. Then Part I of Book I ends with the following scholium, which we quote from G. B. Halsted's translation "Euclid Freed of Every Fleck" in his *Girolamo Saccheri's Euclides Vindicatus* (Open Court, 1920). "And here I might safely stop. But I do not wish to leave any stone unturned, that I may show the hostile hypothesis of acute angle, torn out by the very roots, contradictory to itself. However, this will be the single aim of the subsequent theorems of this Book."

In Part II of Book I, Saccheri investigates the properties of an *equidistant curve,* which is the locus of all points on one side of $\overset{\leftrightarrow}{AB}$ and

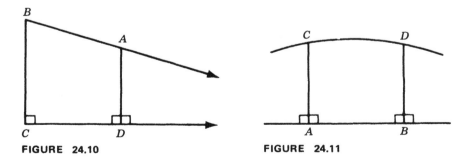

FIGURE 24.10 **FIGURE 24.11**

of distance r from $\overleftrightarrow{AB}$. See Figure 24.11 and suppose s is the length of the arc from C to D on the equidistant curve. Based on an incorrect use of infinitesimals, Saccheri's Proposition XXXVII states that $s = AB$ under the Hypothesis of the Acute Angle. Interestingly enough, the "proof" contradicts Corollary III of Proposition III, which essentially states that Omar Khayyam's Theorem holds even when considering infinitesimals and which is followed by a separate paragraph stating "this indeed ought opportunely to be noted in remaining subsequent propositions." Following the "proof" of Proposition XXXVII Saccheri adds, "But perchance [the proof] will seem to some one by no means evident" and then goes on to give two more proofs in two scholia. The "proof" in the first scholium implies the circumferences of any two circles are equal, and the second scholium gives a false physical argument involving time and motion. All three arguments are fallacious. Nowhere else does Saccheri do anything like this.

Saccheri's Proposition XXXVIII states "The hypothesis of acute angle is absolutely false because it destroys itself." The proof involves *correctly* showing that $s > AB$, which contradicts the previous proposition. The next, and last, proposition is Euclid's Parallel Postulate. In recapitulation Saccheri points out that the refutation of the Hypothesis of the Obtuse Angle is "clear as midday light" but the demise of the Hypothesis of the Acute Angle depends on the proof of Proposition XXXVII and its two scholia.

Saccheri's book, which has the full title *Euclid Vindicated of all Flaw or A Geometric Endeavor in which are Established the Fundamental Principles of Universal Geometry,* received the imprimatur of the Inquisition on July 13, 1773 and of the Provincial of the Jesuits on August 16, 1773. Saccheri died October 25, 1773. Had Saccheri practiced the motto of George Washington (1732 – 1799) – the result justifies the deed – and implied what he certainly would not have been allowed to print? Is Halsted correct in saying that Saccheri's work "may be looked upon as something like the stucco for the king's inspection with which the immortal architect [Sostratos] in Egypt covered the stone bearing his own name"? In any case, by being the first to investigate the logical conclusions that follow from denying Euclid's Parallel Postulate – the method that would eventually settle the whole problem – Saccheri *invented* non-Euclidean geometry. To what extent he was aware of his invention is still debated.

Saccheri's work attracted considerable attention at the time of its publication, and mention of it is made in German and French histories of mathematics during the eighteenth century. Nevertheless, it was soon forgotten in France and Italy, but, as we shall see, not in Germany. However, most mathematicians were ignorant of this early masterpiece until it resurfaced in 1899.

The number π was first proved to be irrational in 1761 by Johann Heinrich Lambert (1728–1777). Five years later, in **1766,** he wrote the paper *Theorie der Parallellinien.* Since it is almost certain that Lambert was familiar with the work of Saccheri, we cannot credit Lambert with the invention of non-Euclidean geometry. Lambert did not progress very far beyond Saccheri's results and did not publish the paper, which was edited and published posthumously in **1786.**

Saccheri had proved that given $d_0 > 0$ there exists a unique m_0 such that $\Pi(d_0) = m_0$. Lambert proved that, conversely, given $0 < m_0 < \pi/2$ there exists a unique d_0 such that $\Pi(d_0) = m_0$. Lambert explicitly mentions that the Hypothesis of the Acute Angle implies the existence of an absolute unit of length. This also follows directly from the fact that the angles of one equilateral triangle cannot be congruent to the angles of another equilateral triangle with sides of different length, as Saccheri pointed out in his lengthy commentary on the work of Wallis. Lambert also went a step beyond Saccheri's observation that defect is additive (Theorem 22.16) and argued that the area of a polygonal region must be proportional to its defect under the Hypothesis of the Acute Angle. It follows that area must be proportional to *excess* under the Hypothesis of the Obtuse Angle, as it is on a sphere. Lambert then makes the astoundingly prescient remark "From this I must almost conclude that the third hypothesis must occur on an imaginary sphere." (The "third hypothesis" is the Hypothesis of the Acute Angle; a sphere, looked at from an algebraic point of view, with radius i where $i^2 = -1$ is an "imaginary sphere.") Perhaps it was this remark that caused Lambert to study the trigonometric functions of $i\theta$ where θ is real. This led to the first comprehensive presentation of the hyperbolic functions.

Between 1776, the year the Declaration of Independence was adopted, and 1803, the year President Jefferson made the Louisiana Purchase, the three principal characters in the next part of our story were born. They are Gauss (Carl Friedrich Gauss, 1777–1855), Lobachevsky (Nicolai Ivanovitch Lobachevsky, 1792–1856), and Bolyai (Bolyai Janos, 1802–1860). Gauss was recognized during his own lifetime as one of the greatest mathematicians to have ever lived. Most of the events that precede the publications of Nicholas Lobachevsky and John Bolyai are told in the correspondence of Gauss.

Wolfgang Bolyai (Bolyai Farkas, 1775–1856), the father of (John) Bolyai and whom Gauss called "the rarest spirit I ever knew," proved that Euclid's Parallel Postulate is equivalent to the assumption that three noncollinear points lie on a circle. On December 16, 1799, Gauss wrote to Wolfgang Bolyai that he was sorry they had not discussed the theory of parallels during their student days together (1796–1798). Concerning the parallel postulate, Gauss added, "It is true I have found much which most would accept as proof but which in my eyes proves as much as *Nothing;* for example, if it can be shown that there is a triangle whose area is greater than that of any given surface, then I can rigorously establish the whole of geometry." The example given by Gauss is obvious from Lambert's conclusions, since the defect of a triangle is less than π. We know Gauss was familiar with the results of Saccheri and Lambert. Marginalia in his books show Gauss paid particular attention to passages on the theory of parallels that quoted the works of Saccheri and Lambert. Further, Gauss checked Lambert's work out of the library in 1795 and again in 1797.

In **1813** Gauss wrote, "In the theory of parallels we have advanced no farther than Euclid." Without reference to Saccheri or Lambert, Gauss wrote in **1816** to his student, the astronomer C. L. Gerling, "It is easy to prove that, if Euclid's geometry is not the true one, then there are absolutely no similar figures. . . . It would even be desirable that Euclid's geometry should be false, because we would then have *a priori* a universal unit of length. . . ." In the same year, Gauss gave a critical review of the pseudo proofs contained in two papers on the parallel axiom. For this he was "subjected to vulgar attack."

Friedrick Ludwig Wachter (1792–1817), another student of Gauss, called the geometry obtained by denying Euclid's Parallel Postulate *anti-Euclidean geometry*. In a letter written to Gauss in December of **1816**, Wachter proves the very remarkable result that the surface to which a sphere through a given point tends as its radius approaches infinity is not a plane in anti-Euclidean geometry but that the geometry of this surface is identical with a Euclidean plane. (This surface is called a *horosphere* and is the three-dimensional analogue of the *horocircle*, which we shall study later.) Wachter had great insight into the theory of parallels and must have greatly influenced Gauss. On April 3, 1817, Wachter took his customary evening walk but never returned. The riddle of the sudden disappearance of this young man who might very well have become the "inventor of non-Euclidean geometry" has never been solved. In 1817, Lobachevsky was giving his students "proofs" of the parallel postulate in his lectures. On April 28 of the same year, Gauss wrote to the astronomer H. W. M. Oblers, "I am becoming more and more convinced that the necessity of our geometry cannot be proved, at least not by *human* intellect nor for the human intellect."

By **1816** Ferdinand Karl Schweikart (1780–1859) had advanced further than the stage Gauss reached in 1817. Schweikart, a professor of law, summarized the work he had done earlier in a one-page *Memorandum* in December **1818** and asked his colleague Gerling at Marburg to send it to Gauss for comment. Schweikart's *Memorandum* begins, "There exists a two-fold geometry, — a geometry in the strict sense — the Euclidean geometry; and an astral geometry." Although the remainder follows from the work of Saccheri, with which he also was familiar, the *Memorandum* is an important document in that it is probably the first pronouncement of the actual existence of a non-Euclidean geometry. (Schweikart's *astral geometry* is hyperbolic geometry and gets its name from the fact that a segment having a critical angle of measure $\pi/4$ would have to be astronomically long.) Gauss replied to Gerling, "Professor Schweikart's Memorandum has given me the greatest pleasure, and I ask that you convey to him my hearty congratulations. To me it is as if almost all my innermost thoughts have been put on paper." Schweikart did not publish any of his results on astral geometry.

In **1820** Bolyai informed his father that he was interested in the theory of parallels. Wolfgang Bolyai advised John Bolyai, "Don't waste an hour on that problem. Instead of reward, it will poison your whole life. The world's greatest geometers have pondered the problem for hundreds of years and not proved the parallel postulate without a new axiom. I believe that I myself have investigated all the possible ideas. . . . [Gauss] affirmed that he had meditated fruitlessly about it." However, Bolyai did not follow his father's advice. The twenty-one year old Hungarian artillary officer wrote to his father on November 3, **1823,** "I have resolved to publish a work on the theory of parallels as soon as I have arranged the material and my circumstances allow it. I have not completed the work, but the path I have followed makes it almost certain that the goal will be attained, if that is at all possible; the goal is not yet reached, but I have made such wonderful discoveries that I have almost been overwhelmed by them, and it would be the cause of constant regret if they were lost. When you see them, my dear father, you too will understand. At present I can say nothing except this: *I have created a new universe from nothing.* All that I have sent to you till now is but a house of cards in comparison with a tower. I'm fully persuaded that this will bring me honor, as if I had already completed the discovery." In reply, W. Bolyai now advised his son to finish the work and publish as soon as possible ". . . first because ideas pass easily from one to another, who can then publish them, and, secondly, there is some truth in the fact that many things have an epoch in which they are discovered at the same time in several places, just as the violets appear on every side in spring." At the time, Wolfgang was working on his *Tentamen Juventutem Studiosam in Elementa Matheseos* (Essays on the Elements of Mathematics for Studious Youths) and invited John to include his results in the *Tentamen.*

By 1823, Bolyai had invented non-Euclidean geometry and determined the formula for the critical function, which is the key to all of hyperbolic geometry. At this time, Lobachevsky, who would be the first to publish, had not even started on the path that would lead him to success. Although Gauss may have *found* non-Euclidean geometry in the work of others more than actually *inventing* it himself, Gauss also held the key to the problem and was acutely aware that the very foundations of nineteenth century mathematics, science, and philosophy were at stake. The fullest information on Gauss' views is contained in a letter to Taurinus on November 8, **1824.** After pointing out the error in Taurinus' proof of the parallel postulate, Gauss says, "the assumption that the sum of the three angles of a triangle is less than 180° leads to a peculiar geometry completely different from ours, a completely self-consistent geometry that I have developed for myself perfectly satisfactorily, so that I can solve any problem in it with the assumption of a value for a constant that cannot be ascertained *a priori.* The greater this constant is assumed to be the closer Euclidean geometry is approached. . . . All my efforts to find a contradiction, an inconsistency in this Non-Euclidean geometry have been fruitless . . . in any case, please regard this as a private communication, of which no public use or use leading to publicity is to be made in any way. If at some time I have more leisure than now, I may publish my investigations."

Franz Adolf Taurinus (1794 – 1874) was encouraged by his uncle, Schweikart, to study the theory of parallels. Taurinus was always convinced of the absolute truth of Euclid's Parallel Postulate. His *Geometriae Prima Elementa,* published in **1826,** discards the Hypothesis of the Acute Angle on the grounds that an absolute unit of length is impossible. Although Taurinus did not progress beyond Wallis in his thinking, his work completely develops Lambert's idea of studying the trigonometry of an imaginary sphere. Because logarithms appear so often in his calculations, Taurinus called this analytic geometry the *logarithmic-spherical geometry.* In retrospect, the logarithmic-spherical geometry provides a proof of the relative consistency of hyperbolic geometry. However, the *Elementa* attracted no attention, and Taurinus burned the remaining copies in disgust.

Also in **1826,** on February 12, Lobachevsky presented a paper which is unaccountably lost and which had the title *Exposition succincte des principes de la géométrie avec une démonstration rigoureuse du théorème des parallèles.* In spite of the ominous "rigorous proof of the theorem on parallels," it is likely that the lecture did present the beginnings of hyperbolic geometry.

In **1829,** Gauss wrote to the astronomer F. Bessel that he would not publish his extensive investigations on the theory of parallels ". . . since I fear the cry of the Boeotians were I to *completely* express my views." Although Bolyai had sent out abstracts of his work in

1825, his manuscript was not delivered to his father until 1829. Wolfgang Bolyai did not understand why John's formulas contained an indeterminate constant, but it was agreed that the new theory of space would be an appendix to the first volume of his *Tentamen*. The *Kasan Messenger* for the year **1829** contains Lobachevsky's *On the foundations of geometry,* the first *published* work that presents a non-Euclidean geometry. This monumental paper is in Russian and had no impact, even though it contains the complete development of hyperbolic geometry, which Lobachevsky unfortunately named *imaginary geometry.*

After many delays, the first volume of W. Bolyai's *Tentamen* was finally published in **1832**. It contains the forever famous *Appendix* by John Bolyai with the title "The science of absolute space with a demonstration of the independence of the truth or falsity of Euclid's parallel postulate (which cannot be decided a priori) and, in addition, the quadrature of the circle in case of its falsity." Hyperbolic geometry is simply called S in the *Appendix.*

An advance copy of Bolyai's *Appendix* was sent to Gauss in June **1831** but did not reach its destination. A month earlier, Gauss had written to the astronomer H. C. Schumacher, "In the past few weeks I have begun to write down some of my own meditations [on the theory of parallels]. . . . I wished that they should not perish with me." A second copy of the *Appendix* was sent in January of 1832. On February 14, Gauss wrote to Gerling, "Let me add further that I have this day received from Hungary a little work on the Non-Euclidean geometry, in which I find all *my own ideas and results* developed with greater elegance, although in a form so concise as to offer great difficulty to anyone not familiar with the subject. . . . I regard this young geometer Bolyai as a genius of the first order." On March 6, 1832, Gauss wrote to W. Bolyai, ". . . Now a word about your son's work. If I begin by saying *that I cannot praise it* you will be surprised for a moment, but I cannot do otherwise. To praise it would be to praise myself, for the content of the work, the approach your son has taken, and the results to which he is led coincide almost exactly with my own meditations which I partly carried out thirty to thirty-five years ago. In fact, I'm extremely surprised by it. . . ."

The only influence Gauss has on the inventions of Lobachevsky and Bolyai was that each author knew that Gauss had failed earlier at the problem of proving the parallel postulate. Bolyai and Lobachevsky each made his momentous invention independent of anyone else. Although Bolyai did not publish further, Lobachevsky, in an effort to make his invention more widely known, published the French paper *Géométrie imaginaire* in Crelle's journal in 1837 and a little German book *Geometrische Untersuchungen* in 1840. A year before his death in 1856, the now blind Lobachevsky dictated and published in both Russian and French his complete exposition *Pangéométrie*. Lobachevsky's new name *pangeometry* was certainly more attractive than his earlier, self-deprecating name *imaginary geometry*. In 1842, Gauss saw to it that Lobachevsky became a member of the Royal Society at Göttingen. Apparently Lobachevsky seems never to have heard of Bolyai, although Bolyai learned in 1848 that he had to share the honor of his invention with Lobachevsky. However, there was very little honor to share on his part. Just a line or two from Gauss in any scientific journal would have made Bolyai famous. The ungenerous Gauss, the so-called "Prince of Mathematicians," never gave Bolyai any public mention. Bolyai died long before his work received the recognition it deserved.

Certainly, Euclid invented absolute geometry. Certainly, Saccheri invented hyperbolic geometry. However, since each of Bolyai and Lobachevsky independently invented hyperbolic geometry *and* each published a work claiming the consistency of hyperbolic geometry, the credit for the invention of the Saccheri-Lambert-Wachter-Schweikart-Gauss-Taurinus-Bolyai-Lobachevsky geometry is usually divided between John Bolyai and Nicholas Lobachevsky.

For a century the invention of non-Euclidean geometry was a recurring event! Only the highlights of the long struggle the mathematical community had in discovering the existence of and, in particular, the significance of the invention will be told next. Compared to the story of its *invention,* the story of the *discovery* of non-Euclidean geometry is far more complicated. Bolyai and Lobachevsky had declared, "Here it is!" The question "But what is it?" now had to be tackled.

For thirty-five years the works of Bolyai and Lobachevsky were essentially ignored. Thought was dominated by the Kantian theory that space exists intuitively in the human mind and the axioms of Euclidean geometry are *a priori* judgments imposed on the mind without which no consistent reasoning about space is possible. The great change started about **1866.** The expository papers of Jules Hoüel (1823–1886) that followed his translations of the works of Lobachevsky (in 1866) and Bolyai (in 1867) were paramount in focusing attention on the new geometry. Expository articles and translations in other languages quickly followed. Most influential were the paper *On the facts that underlie the foundation of geometry* (in German, dated May 22, 1866) and subsequent expository articles by Herman von Helmholtz (1821–1894). The results of Helmholtz were essentially anticipated by Bernhard Riemann (1826–1866). Riemann's lecture on differential geometry, *On the hypotheses that underlie the foundation of* *geometry* (in German), which is rather vague because of its intended general audience, was delivered on January 10, 1854, but first published in 1866. Both Helmholtz and Riemann begin by considering space as a coordinate geometry. However, Riemann's paper does not mention non-Euclidean geometry and was overshadowed by the work of Helmholtz for many years.

The whole perspective was changed in **1868** by *Essay on the interpretation of non-euclidean geometry* (in Italian), a paper by Eugenio Beltrami (1835 – 1900). To understand Beltrami's result we shall need some background from differential geometry. (See Exercise 24.16.) The *geodesics* on a surface in Euclidean space are the curves of shortest length that connect two given points. The *intrinsic geometry* of the surface is the geometric interpretation of the surface as a *plane* where the geodesics are used to define *lines* and *distance. Angle measure* is the Euclidean angle measure on the surface. We can see that these intrinsic geometries provide many models of planes that are not Euclidean. For example, in the intrinsic geometry of a sphere, the lines are the great circles on the sphere and the Hypothesis of the Obtuse Angle holds. For another example, in the intrinsic geometry of a circular cylinder, the Hypothesis of the Right Angle holds but the geometry is quite different from the Euclidean plane. Wachter had shown that the Euclidean plane is the intrinsic geometry of a horosphere in hyperbolic three-space. That the hyperbolic plane can never be the intrinsic geometry of a surface in Euclidean three-space would not be discovered until the twentieth century. A *pseudosphere* is the Euclidean surface obtained by revolving a tractrix about its asymptote. (A *tractrix* with x-axis as asymptote has the differential equation $(dy/dx)^2 = y^2/(a^2 - y^2)$.) Certain regions of a pseudosphere are isomorphic to regions of the hyperbolic plane. See Figure 24.12. This is the result given by Beltrami in his *Essay.* This partial representation of the hyperbolic plane on a familiar Euclidean surface is what convinced many that the hyperbolic plane is consistent. For this reason, it is often stated that the (relative) consistency of the hyperbolic plane

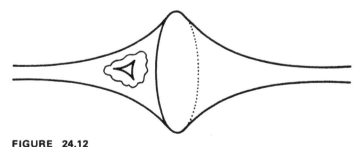

FIGURE 24.12

was proved in 1868 by Beltrami. Since a *proof* is a convincing argument, that today *we* would be convinced by the algebraic models that go back to Taurinus, Bolyai, and Lobachevsky is irrelevant.

Surprisingly, Beltrami's *Essay* contains a proof that the hyperbolic plane is isomorphic to the Cayley–Klein Model. But, Beltrami draws no conclusion from this fact! It was the construction of the new system as a "geometry" that carried the day. The Cayley-Klein Model also appeared in an 1859 paper by Arthur Cayley (1821–1895). However, it was Felix Klein (1849–1925) who in his *On the so-called noneuclidean geometry* of **1871** and subsequent papers pointed out that the Cayley–Klein Model is actually a representation of the hyperbolic plane. Although the hyperbolic plane is consistent, it was thought that hyperbolic three-space might (must) be impossible. Bolyai, who had greater insight into hyperbolic geometry than Lobachevsky, considered this and even thought he had found an inconsistency until he found his error. Klein's three-dimensional analogue of the Cayley–Klein Model showed that hyperbolic three-space is also (relatively) consistent.

The names *hyperbolic geometry, parabolic geometry,* and *elliptic geometry* are due to Klein and are taken from the context of projective geometry. Parabolic geometry is Euclidean geometry. The *elliptic plane,* which is also called the *Riemann plane,* is really due to Klein. Riemann had pointed out that we must distinguish between the *unboundedness* of lines and the *infinite extent* of lines. For example, a line has infinite extent but is bounded in the intrinsic geometry of a sphere. That the geometry of a sphere is a sort of non-Euclidean geometry would not even have surprised Euclid. (Circles and spheres were a Greek specialty.) Klein's ingenious contribution was essentially to deduce our M15, the Riemann Incidence Plane, from our M14, the Sphere Incidence Plane. By identifying antipodal Euclidean points on a sphere as one *point,* Klein obtained a model such that two points determine a line. (To see how this relates to projective geometry, consider how we defined the real projective plane in Section 4.2.) To obtain the *elliptic plane,* we add distance and angle measure as follows. If $P = \{P_1, P_2\}$ and $Q = \{Q_1, Q_2\}$ are two points of M15, define the distance PQ from P to Q to be the minimum of the measures of the Euclidean angles $\angle P_1OQ_1$ and $\angle P_1OQ_2$ where O is the center of the sphere. See Figure 24.13. Angle measure between lines in the elliptic plane is inherited from Euclidean angle measure on the sphere. So $\pi/2$ is an upper bound for the distance between any two points in the elliptic plane. Projective geometry also provides higher dimensional analogues of the elliptic plane. The Hypothesis of the Obtuse Angle holds in elliptic spaces.

As a result of the efforts of Hoüel, Helmholtz, Beltrami, Klein,

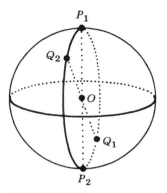 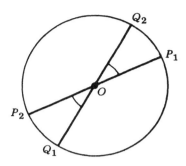

FIGURE 24.13

and others, knowledge of non-Euclidean geometry became widespread. Confusion was still the order of the day, however. It was not easy to think that the words *space* and *geometry* could have a plural. The models of Klein did not diminish the confusion. On the contrary, they further obscured the logic of geometry. Mathematicians, including Klein, did not then understand the logical function of a model. It was not clear how hyperbolic geometry and elliptic geometry could be shown to be "true" because they could be demonstrated by Euclidean geometry. Of course, only a few asked, "Is Euclidean geometry true?" To add more confusion to the scene, the idea of *n*-dimensional space for $n > 3$ was now abroad. Both four-dimensional space and non-Euclidean geometry became the domain of crackpots and mystics.

Soon there was a plethora of *not-Euclidean geometries*. (The name *non-Euclidean geometry* started out to mean only hyperbolic geometry. Its meaning was then extended by many to include elliptic geometry. Since the elliptic plane is not the result of denying the parallel postulate alone among Euclid's postulates, some use the name *non-Euclidean geometry* to describe any geometry in which Playfair's Parallel Postulate fails; others use this name to describe any geometry other than Euclidean geometry. In general, you must rely on the context.) The non-Archimedean geometries of Veronese provide examples of *non-Legendrian geometry* where the Hypothesis of the Obtuse Angle holds although through a point off a given line there are infinitely many lines parallel to the given line and examples of *semi-Euclidean geometry* where the Hypothesis of the Right Angle holds although through a point off a given line there are infinitely many lines parallel to the given line. Even more startling were the finite geometries of Fano and Moore. The realization that an axiom should be a *stated assumption* rather than some supposedly self-evident truth was slow in developing. There was a great deal written about the nature of the axioms of geometry without stating just what "the axioms" were. Even Pasch's great achievement of **1882** was clouded by his expository papers on the nature of axioms. Gradually it became evident that *geometry* is not only the mathematical theory of physical space. Although the Italian phalanx of Peano, Pieri, Padoa, and Fano placed geometry on an absolutely axiomatic basis, it was Hilbert's *Grundlagen der Geometrie* that was most influential in spreading the axiomatic viewpoint. In one sense, geometry, not only non-Euclidean geometry, was discovered at Paris in **1900**. (See Section 12.2.)

Poincaré asserted that the laws of physics would be changed, if necessary, in order to preserve Euclidean geometry as the model of physical space. The assertion has proven to be false. Non-Euclidean geometry became a necessary tool for the scientist, beginning with Einstein's work of **1916**. Einstein later wrote, "The axioms are voluntary creations of the human mind. . . . To this interpretation of geometry I attach great importance because if I had not been acquainted with it, I would never have been able to develop the theory of relativity."

We must terminate our story here, with suggestions for further reading. Bibliophiles will be interested in *Bibliography of Non-Euclidean Geometry* by D. M. Y. Sommerville (Chelsea, 1970). *The* history in English of non-Euclidean geometry is *Non-Euclidean Geometry* by Roberto Bonola. The Dover paperback of Bonola's history is an exceptionally good buy since it also contains English translations of *The Science of Absolute Space* by Bolyai and *The Theory of Parallels* by Lobachevsky. The coffeetable book for the theory of parallels is not inexpensive (about U.S. $40) but also deserves special mention. Edited by C. E. Sjöstedt and published in 1968 by Interlingue-Fundation (Sweden), this book can be enjoyed by anyone who can understand its full title, *Le Axiome de Paralleles de Euclides a Hilbert, Un Probleme Cardinal in le Evolution del Geometrie, Excerptes in facsimile ex le principal ovres original e traduction in le lingue international auxiliari Interlingue.*

24.4 EXERCISES

24.1 Corollary 24.10.

● **24.2** Theorem 24.12.

24.3 If $\overleftrightarrow{AB} \parallel \overleftrightarrow{CD}$, $\overleftrightarrow{CD} \parallel \overleftrightarrow{EF}$, and $A - C - E$, then $\overleftrightarrow{AB} \parallel \overleftrightarrow{EF}$.

24.4 If $\sqcup ABCD$ is closed, then $\sqcup ABCD$ is equivalent to an isosceles closed biangle with vertex B.

● **24.5** True or False?

(a) $\overleftrightarrow{AB} \parallel \overleftrightarrow{CD}$ and $\overleftrightarrow{CD} \parallel \overleftrightarrow{EF}$ implies $\overleftrightarrow{AB} \parallel \overleftrightarrow{EF}$.

(b) If $\sqcup ABCD$ is closed, then $\angle B \approx \angle C$.

(c) That there exist two parallel lines without a common perpendicular is equivalent to the Hypothesis of the Acute Angle.

(d) That a biangle closed from one vertex must be closed from the other vertex is equivalent to the Hypothesis of the Right Angle.

(e) That every biangle is closed is equivalent to Euclid's Parallel Postulate.

(f) Under the Hypothesis of the Right Angle, any two biangles are equivalent.

(g) If $\sqcup ABCD$ is closed, then every ray in int $(\angle ABC)$ intersects $\overrightarrow{CD}$.

(h) If $\sqcup ABCD$, then $\sqcup CDAB$ and $\sqcup ABDC$.

(i) If $\sqcup ABCD$ is closed, then $\sqcup BADC$ is closed.

(j) If $0 < t < \pi$, then there exists d such that $\delta \triangle ABC > t$ when $AB, BC,$ and AC are each greater than d.

24.6 Read "The main trends in the foundations of geometry in the 19th century" by H. Freudenthal in *Logic, Methodology and Philosophy of Science* edited by E. Nagel *et al.* (Stanford, 1962).

24.7 Read "The Copernicus of Geometry," Chapter 16 of E. T. Bell's classic *Men of Mathematics.*

24.8 Equivalence of rays is an equivalence relation on the set of all rays; equivalence of biangles is an equivalence relation on the set of all biangles.

● **24.9** If s and t are positive numbers, then there exists a right triangle with one leg of length greater than s and with defect less than t.

24.10 If $0 < t < \pi$, then there exists positive number d such that $\delta \triangle PQR < t$ for any $\triangle PQR$ with all sides of length less than d.

24.11 State and prove some theorems regarding the sum of the measures of the angles of a closed biangle.

24.12 If the angle bisector of one angle of $\sqcup ABCD$ intersects both sides of the biangle, then $\sqcup ABCD$ is equivalent to an isosceles biangle with vertex C.

24.13 That a biangle is equivalent to an isosceles biangle with right angles is equivalent to Euclid's Parallel Postulate.

24.14 Read "Traveling on Surfaces" and "Space Curvature," Chapters II and XIV of *Famous Problems of Mathematics* by H. Tietze (Graylock, 1965).

24.15 Read the articles by Helmholtz and Clifford in Volume I of *The World of Mathematics*, edited by J. R. Newman (Simon and Schuster, 1956).

24.16 A closed biangle may have an obtuse angle.

24.17 Does Euclid's Proposition I.1 hold in the Riemann plane?

*24.18 The "squaring of the circle" is not impossible in the Riemann plane.

GRAFFITI

$\mathscr{E}$: *The German word for* plane *is* Ebene.

The space constant may be different in different places. It may also vary with time.

Riemann

But neither thirty years, nor thirty centuries, affect the clearness, or the charm, of Geometrical truths. Such a theorem as "the square of the hypotenuse of a right-angled triangle is equal to the sum of the squares of the sides" is as dazzlingly beautiful now as it was in the day when Pythagoras first discovered it, and celebrated its advent, it is said, by sacrificing a hecatomb of oxen — a method of doing honor to Science that has always seemed to me slightly exaggerated and uncalled-for. One can imagine oneself, even in these degenerate days, marking the epoch of some brilliant scientific discovery by inviting a convivial friend or two, to join one in a beefsteak and a bottle of wine. But a hecatomb of oxen! It would produce a quite inconvenient supply of beef.

C. L. Dodgson

229,792,458 m/sec
IT'S THE LAW.

CHAPTER 25

Excursions

25.1 PROSPECTUS

Our objective from the beginning has been to develop that geometry that is very like the Euclidean plane but avoiding any parallel postulate for as long as is reasonably possible. After the last two chapters you may think we have already stretched the bounds of reasonableness. It is becoming a nuisance to begin most of our theorems with something equivalent to either "Under the Hypothesis of the Right Angle" or else "Under the Hypothesis of the Acute Angle." It is time to decide to accept one or the other of the two hypotheses. In the next section of this chapter, we look at what would happen if we were to choose the Hypothesis of the Right Angle. You may not be surprised to learn that this path leads to the familiar Euclidean plane. Hopefully you are curious about the other path — where you can be sure there are vistas totally different from what you are used to. This new path is explored in Part Two, which begins with the next chapter. For those who would like a preview of Part Two, a summary of its nine chapters follows.

In *Chapter* 26 the Hyperbolic Parallel Postulate is taken as our last axiom. The resulting axiom system is called *the Bolyai–Lobachevsky plane.* (The *"the"* is meant to imply only that no other axioms are allowed and not that the axiom system is necessarily categorical, in much the same way we talk about *the* absolute plane. The names "the Bolyai–Lobachevsky plane" and "the hyperbolic plane" are interchangeable.) The results in the chapter are mostly a consolidation of results already obtained in Chapters 22, 23, and 24. In particular, the material on biangles is used to distinguish two types of parallelness for lines. Essentially, of all the lines through a given point P that are parallel to a given line l off P, the two lines that *bound* all the others will be said to be *horoparallel* to l while those lines that are *over* one of these two boundaries will be *hyperparallel* to l. (The prefixes *horo* and *hyper* will be used often.) We need two names to distinguish these two very different types of parallelness.

Chapter 27 deals with things called *brushes* and *cycles*. The brushes in the Euclidean plane are the pencils (set of all lines through a given point) and the parallel pencils (set of all lines parallel to a given line). Since there are two types of parallelness for lines, we can expect three types of brushes in the Bolyai–Lobachevsky plane, namely the *pencils*, the *horopencils*, and the *hyperpencils*. Associated with each type of brush is a family of curves. These curves are the cycles. Specifically, a cycle is the set of all points obtained by reflecting a point in the lines of a given brush. In particular, the *circles* are associated with the pencils, *horocircles* are associated with the horopencils, and *hypercircles* are associated with the hyperpencils. We might mention that a hypercircle is also called an *equidistance curve* since it is the set of all points on one side of and equidistant from a given line.

Chapter 28 and *Chapter* 29 are, in some sense, the two most important chapters in this book. To fully understand any mathematical system the automorphisms of that system must be studied. The isometries of the Bolyai–Lobachevsky plane are studied in these two short chapters, and a classification of all the isometries is determined. Further, with only a few minor changes, the results are applicable to the Euclidean plane as well. So while learning about the isometries of the Bolyai–Lobachevsky plane, it is also possible to formalize a rigorous approach to the isometries of the Euclidean plane.

Chapter 30 is a pleasant digression into some applications of the material on isometries. The chapter deals with two aspects of symmetry. From a result obtained by Leonardo da Vinci in studying the possible floor plans for a building, the finite groups of symmetries are related to groups of symmetries for polygons. Secondly, although there is infinite variety in the possible subject matter for the frieze of a building, it is established that any frieze design that repeats a basic pattern must fall into exactly one of seven different classes. The results of the chapter are applicable to the Euclidean plane as well as to the Bolyai–Lobachevsky plane.

Chapter 31 is mostly concerned with showing that a bounded arc on a horocircle has a finite *length*. This is done rigorously without assuming knowledge of advanced calculus. (You may wish to read only the statements of the definitions and of the theorems, skipping the proofs, in order to get on to the more exciting aspects of hyperbolic geometry.) Then, the existence of a constant k, called the *distance scale*, is established. For any positive number k there exists a model of the Bolyai–Lobachevsky plane having k as its distance scale. Hence, a numerical value of the distance scale k cannot be determined in the theory.

In *Chapter* 32 the formula $\Pi(x) = 2 \arctan e^{-x/k}$ is established, where Π is the critical function and k is the distance scale. Also, the trigonometric formulas for the Bolyai–Lobachevsky plane are determined. To mention the hyperbolic analogue of the Pythagorean theorem, suppose the distance scale is 1 and a right triangle has hypotenuse of length c and legs of length a and b. Then these lengths are related by the equation $\cosh c = \cosh a \cosh b$. (Recall that $\cosh x$ is $(e^x + e^{-x})/2$. For other trig formulas see Corollary 32.13 where the distance scale is assumed to be 1 and the notation refers to Figure 32.12.) Further, the Euclidean plane is shown to be a limiting case of the Bolyai–Lobachevsky plane.

Chapter 33 contains two mutually independent sections, one on analytic geometry and another on area. Once the analytic geometry for the Bolyai–Lobachevsky plane is developed, the question of categoricalness can be considered. It turns out that the axiom system we are calling the Bolyai–Lobachevsky plane is not categorical. Two models are shown to be isomorphic iff they have the same distance scale. However, any two models are *similar* in that the distance scale of a model can be arbitrarily changed merely by multiplying all distances by the same positive constant. (To obtain a categorical axiom system it is only necessary to add an additional axiom such as "The distance scale is 1.") In the second section of the chapter it is shown that area *must* be defined to be proportional to defect. We also prove Bolyai's theorem that two polygonal regions are equivalent by triangulation iff the regions have the same area. The theory behind dissection puzzles is a consequence of our results on area and is also applicable to the Euclidean plane.

Chapter 34 contains three mutually independent sections. In the first section the hyperbolic analogues of the classical theorems of Menelaus, Ceva, Desargues, and Pappus are proved. In the second section the techniques of the calculus are applied to problems in the Bolyai–Lobachevsky plane. Finally, in the third section, we look at the ruler and compass game of Euclidean constructions as it is played in the Bolyai–Lobachevsky plane. Several construction problems are posed and some are solved. For example, the construction of a triangle given only angles congruent to its angles is considered, as is the old problem of "squaring the circle."

If it is assumed that area is defined to be proportional to defect, then Chapter 34 depends on Chapter 33 only to the extent that the end of Section 34.1 uses the similarity of models of the Bolyai – Loba-chevsky plane. Therefore, the sections after Chapter 32 are essentially independent of each other – excluding the Exercises, of course.

25.2 EUCLIDEAN GEOMETRY

For this section, in addition to the axioms for the absolute plane, we assume the Hypothesis of the Right Angle. Unless the real number system is itself inconsistent, this augmented axiom system cannot be inconsistent due to the existence of the Cartesian plane. Any model of this axiom system is called *a Euclidean plane*. It is certainly conceivable that there are nonisomorphic models of a Euclidean plane as just defined. However, this is not the case. After proving the Pythagorean theorem, we shall introduce coordinates in order to show that any two models of a Euclidean plane are isomorphic. We will then know that we have a categorical axiom system for *the* Euclidean plane.

We begin with a special case of a theorem stating that the projections of congruent collinear segments are congruent. See Figure 25.1. Suppose $\angle LVL''$ is acute and $V-L-M-N$ with $LM = MN$. Suppose L'', M'', N'' are, respectively, the feet of the perpendiculars from L, M, N to $\overleftrightarrow{VL''}$. Let N' be the foot of the perpendicular from L to $\overleftrightarrow{NN''}$. Let M' be the intersection of $\overleftrightarrow{LN'}$ and $\overleftrightarrow{MM''}$. Under the Hypothesis of the Right Angle, $\angle L''LN'$ and $\angle M''M'N'$ are right angles. Hence, by Theorem 22.23, we have $LM' = M'N'$. Also, since opposite sides of a rectangle are congruent, we have $L''M'' = M''N''$.

The result above is used to show $VA/VB = VC/VD$ where V, A, B, C, D are points as in Figure 25.2. This will follow if it can

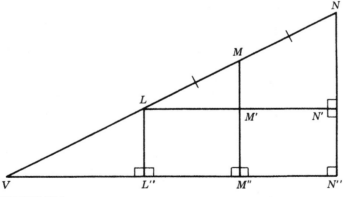

FIGURE 25.1

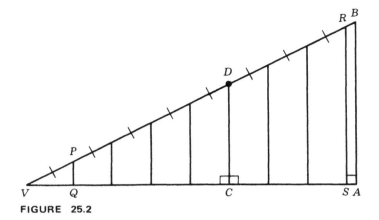

FIGURE 25.2

be shown that the numbers VB/VD and VA/VC are equal or, equivalently, that their difference is 0. Since 0 is the only nonnegative number less than every positive real number, it is sufficient to prove the absolute value of the difference of these numbers is less than every positive real number. Suppose ε is any positive number. Let n be an integer such that $n\varepsilon > 2$. Now let P be the point on $\overrightarrow{VD}$ such that $VD = nVP$. Then there exist a point R on $\overline{VB}$ and an integer m such that $VB = mVP + RB$ where $0 < RB \leq VP$. Let Q and S be the feet of the perpendiculars from P and R, respectively, to $\overleftrightarrow{VA}$. It follows from the previous paragraph that $VC = nVQ$ and $VA = mVQ + SA$ where $0 < SA \leq VQ$. By substitution, we have the result

$$\left|\frac{VB}{VD} - \frac{VA}{VC}\right| \leq \frac{1}{n}\left(\frac{RB}{VP} + \frac{SA}{VQ}\right) \leq \frac{2}{n} < \varepsilon.$$

So $VA/VB = VC/VD$, as desired. In general, since the acute angles of a right triangle are complementary under the Hypothesis of the Right Angle, it then follows that the lengths of the sides of a right triangle are proportional to the lengths of the corresponding sides of any similar right triangle.

Now suppose $\triangle ABC$ has a right angle at C. Let D be the foot of the perpendicular from C to $\overleftrightarrow{AB}$. Then $A-D-B$. From the last result above, we have $AD/AC = AC/AB$ and $BD/BC = BC/BA$. Hence,

$$(AC)^2 + (BC)^2 = (AB)(AD + BD) = (AB)^2,$$

proving the Pythagorean theorem for right triangles in a Euclidean plane.

Point coordinates are introduced next. Let l and m be two fixed lines that are perpendicular. Let g and h be fixed coordinate systems for l and m, respectively. For any point P, let P' and P'' be the feet of the perpendiculars from P to l and m, respectively. If $g(P') = x$ and $h(P'') = y$, then P is said to have *coordinates* (x, y). So every point has a unique pair of coordinates. Conversely (Exercise 25.2), every ordered pair of real numbers is the coordinates of a unique point. For fixed real number a, under the Hypothesis of the Right Angle, the set of all points with coordinates (a, y) is a line. We say such a line has equation $x = a$. Every line parallel to m has such an equation. In particular, m has equation $x = 0$. Further (Exercise 25.3), every line that intersects m only at a point with coordinates $(0, b)$ has an equation $y = sx + b$, and conversely. Therefore, every line has an equation of the form $Ax + By + C = 0$, and every such equation is an equation of some line provided $A^2 + B^2 \neq 0$. A line with equation $x = a$ is perpendicular to a line with equation $y = b$ for every a and b. Let points P and Q have coordinates (x_1, y_1) and (x_2, y_2), respectively. We claim

$$PQ = [(x_2 - x_1)^2 + (y_2 - y_1)]^{1/2}.$$

To prove this, let R be the point with coordinates (x_1, y_2). If P, Q, R are collinear, then the result is trivial. Otherwise, $\triangle PQR$ has a right angle at R and the result follows from the Pythagorean theorem.

We are now prepared to show that any two models of a Euclidean plane are isomorphic. It is sufficient to show that any model of a Euclidean plane is isomorphic to the Cartesian plane. We suppose we have any model of a Euclidean plane and that this model is coordinatized as above. Let f be the mapping from the set of points of the model into the set of the points of the Cartesian plane that sends the point with coordinates (x, y) to the point (x, y). (In the model, (x, y) is the *name* of a point; in the Cartesian plane, (x, y) *is* a point.) The mapping f is obviously a bijection that preserves lines. Further, f preserves distance. (The distance formula in the preceding paragraph is a *theorem* applicable to our arbitrary model of a Euclidean plane; the distance formula for the Cartesian plane is a *definition*.) Since f preserves distance, then f preserves betweenness. So f preserves segments, rays, angles, and triangles. Since f preserves congruence between segments, then f preserves congruence between triangles by SSS. Hence, f preserves congruence between angles. Since the lines l and m used in coordinatizing the arbitrary model are perpendicular, then f preserves right angles. So f must preserve angles of measure $\pi/2^n$, where angle measurement in the Cartesian plane is assumed to be radian measurement. Therefore (Exercise 25.4), it follows that f must preserve angle measure. Hence f is an isomorphism from the arbitrary model onto the Cartesian plane. We have achieved our purpose.

IF TO OUR FIVE AXIOMS FOR THE ABSOLUTE PLANE WE ADD ANY AXIOM EQUIVALENT TO EUCLID'S PARALLEL POSTULATE, THEN WE HAVE A CATEGORICAL AXIOM SYSTEM FOR THE EUCLIDEAN PLANE.

25.3 HIGHER DIMENSIONS

In this section, we give a set of axioms for absolute three-space and a set of axioms for absolute four-space. This is surprisingly easy to do. Starting with our axioms for the absolute plane, we need only to restate PSP and to augment the Incidence Axiom. We begin by defining an *incidence three-space* to be an ordered triple $(\mathscr{P}, \mathscr{L}, \mathscr{E})$ satisfying the Three-Space Incidence Axiom as defined next.

Three-Space Incidence Axiom: $\mathscr{P}$, $\mathscr{L}$, and $\mathscr{E}$ are sets, the elements of which are respectively called *points, lines,* and *planes.* A line is a set of points, containing at least two points. A plane is a set of points, containing at least three points not all in one line. Two points determine a line. Three points not in one line determine a plane. If two points are in a plane, then the line containing these points is a subset of that plane. If two planes have a point in common, then their intersection is a line. There exist four points that are not all in one plane and not all in one line.

The usual language denoting incidence is assumed for any incidence three-space. However, by convention, lines l and m are defined to be *parallel* if either $l = m$ or else l and m are coplanar lines that do not intersect. Nonintersecting lines that are not coplanar are *skew* to each other. Planes E and F are *parallel* if either $E = F$ or else E and F do not intersect. Line l and plane E are *parallel* if either l is on E or else l and E do not intersect.

Among others, the following theorems can be proved for any incidence three-space: (1) Neither $\mathscr{P}$, $\mathscr{L}$, nor $\mathscr{E}$ is empty. (2) A plane is not a line. (3) Two lines intersect in at most one point. (4) If a line intersects a plane not containing the line, then the line and the plane intersect at exactly one point. (5) If line l is on plane E, then there exists a point in E that is off l. (6) If point P is in plane E, then there exists a line on E that is off P. (7) A line and a point off the line determine a plane. (8) Two parallel lines determine a plane. (9) Two intersecting lines determine a plane. (10) Given a plane, there exists a point off that plane; given a point, there exists a plane off that point. (11) If point P is on line l, then there exists a plane that intersects l only at P. (12) There exist two planes through a given line. (13) If plane E intersects both of two parallel planes F and G, then the intersections of E with F and with G are parallel lines. (14) If l and m are two parallel lines and point P is off the plane containing l and m, then there exists a unique line through P that is parallel to both l and m. (15) If each of two intersecting lines is parallel to a third line, then the three lines are coplanar.

Of the theorems just mentioned, the last three are probably the most interesting. The last of these deserves special attention. Assuming there are such things as parallel lines, we can deduce from (15) that if there is anything exciting in regard to the theory of parallels then all the excitement concerns what takes place in a plane! It is for this reason that we have not bothered to carry a three-space structure throughout the development of our geometry in the other chapters.

Our Ruler Postulate together with our definitions of *between* and *segment* make sense when applied to an incidence three-space. However, our PSP then becomes an absurd statement. One consequence (Theorem 12.15) of our PSP was to limit our geometry to what would conventionally be called a plane. Except for this aspect, we maintain the original intent of PSP in a statement called the Spatial PSP as defined next.

Spatial PSP: If line l is on plane E, then there exist convex sets H_1 and H_2 such that $E \setminus l$ is the union of H_1 and H_2 and such that if P and Q are two points with P in H_1 and Q in H_2 then $\overline{PQ}$ intersects l.

We are now in a position to state a set of *axioms for absolute three-space* with undefined terms $\mathscr{P}$, $\mathscr{L}$, $\mathscr{E}$, d, m. Assuming the necessary definitions from our theory of the absolute plane, the axioms are: (I) Three-Space Incidence Axiom, (II) Ruler Postulate, (III) Spatial PSP, (IV) Protractor Postulate, and (V) SAS.

That's all there is to it! It might be emphasized that II, IV, and V above refer to the exact statements for our axioms for the absolute plane. For example, "SAS" above means the statement of our Axiom 5 exactly as it is given in Section 17.2. Therefore "SAS" in the context of absolute three-space does not require that the triangles involved be in the same plane. Of course this is intentional and makes absolute three-space homogeneous.

All of our definitions and theorems for the absolute plane hold for absolute three-space *provided* the phrase "In a plane" is tacitly assumed to begin each definition and theorem. In some cases we can remove this tacit assumption. Most notably, we define "$\triangle ABC \simeq \triangle DEF$" exactly as before (Definition 17.1) with no reference to triangles being in the same plane. Then it seems we must give new proofs for all the congruence theorems (Section 17.3) for triangles when the triangles are in different planes. However, the old proofs are still valid for absolute three-space, and the old congruence theorems still hold in the new context. In general, the tacit assumption is necessary. For example, our Theorem 18.2 states "If two lines are perpendicular to the same line, then the two lines are parallel." This statement is hardly true in absolute three-space unless it is assumed the three lines are in a plane.

A three-dimension analogue of PSP need not be included among the axioms for absolute three-space. Imitating the proof of Theorem 12.7, we can prove the *Three-Space Separation Theorem:* For any plane E there exist convex sets H_1 and H_2 such that $\mathscr{P} \setminus E$ is the union of H_1 and H_2 and such that if P and Q are two points with P in H_1 and Q in H_2 then $\overline{PQ}$ intersects E. Such sets H_1 and H_2 are called *sides* or *halfflats* of the plane E.

Since the real Cartesian three-space exists, the set of axioms for absolute three-space is *consistent* if the real number system is consistent. Using Euclidean spheres, we can extend the models of Section 23.2 to provide a model of absolute three-space in which Euclid's Parallel Postulate fails. Such examples of *hyperbolic three-space* demonstrate that the set of axioms for absolute three-space is *not categorical.*

We say line l is *perpendicular* to plane E if l intersects E at a point P such that every line on E that passes through P is perpendicular to l. Among others, the following theorems can be proved: (1) If two points of a line are equidistant from two given points, then every point of the line is equidistant from the two points. (2) If a line is perpendicular to each of two intersecting lines at their point of intersection, then the line is perpendicular to the plane containing the two lines. (3) All the lines perpendicular to a given line at a given point are in one plane. (4) The locus of all points equidistant from two points A and B is a plane perpendicular to $\overleftrightarrow{AB}$ at the midpoint of A and B. (5) Given a point and a line, there is a unique plane through the point that is perpendicular to the line. (6) Given a point and a plane, there is a unique line through the point that is perpendicular to the plane. (7) Two lines perpendicular to the same plane are parallel, and two planes perpendicular to the same line are parallel. (8) If m and n are coplanar lines each perpendicular to a line on plane E and m is perpendicular to E, then n is perpendicular to E.

A *dihedral angle* with *edge l* is the union of line l and two half-planes of l that are not coplanar. A *plane angle* of a dihedral angle is the intersection of the dihedral angle with a plane perpendicular to its edge. It is well known that any two plane angles of a given dihedral angle are congruent in Euclidean three-space. Since the usual proof of this fact involves the Euclidean theory of parallels, we shall indicate how the theorem can be proved for absolute three-space. Suppose $\angle APC$ and $\angle BQD$ are two plane angles of a given dihedral angle. We may suppose the points are taken as labeled in Figure 25.3 where $\angle LMN$ is the plane angle containing the midpoint of P and Q. From $\boxed{S}\,PABQ \simeq \boxed{S}\,PCDQ$ we conclude AL, LB, CN, ND are all equal, $\overleftrightarrow{AB} \perp \overleftrightarrow{ML}$, and $\overleftrightarrow{CD} \perp \overleftrightarrow{MN}$. By (8) above, each of $\overleftrightarrow{AB}$ and $\overleftrightarrow{CD}$ is perpendicular to the plane containing $\angle LMN$ and, hence, to $\overleftrightarrow{LN}$. Then $\boxed{S}\,LACN \simeq \boxed{S}\,LBDN$. So $AC = BD$. Hence $\triangle APC \simeq \triangle BQD$ by SSS, and the desired result follows.

We now define two planes to be *perpendicular* if their union contains a dihedral angle having a right plane angle. Et cetera, et cetera, et cetera.

The existence of Euclidean n-space is accepted today by most students of linear algebra without a moment's hesitation. What was fantastic less than a century ago is now held to be commonplace. By employing hyperspheres from Euclidean $(n+1)$-space, we can extend models of hyperbolic n-space to models of hyperbolic $(n+1)$-space. From this we deduce the existence of absolute n-space. Absolute four-space by itself should make an excellent subject for a seminar or for an independent study course. (See Exercise 25.10.) Teachers of high school mathematics would surely benefit from a detailed study of the isometries of Euclidean four-space. The introduction to absolute four-space below goes no further than to parallel the introduction to absolute three-space above. We begin with an incidence axiom.

Four-Space Incidence Axiom: $\mathscr{P}$, $\mathscr{L}$, $\mathscr{E}$, and $\mathscr{F}$ are sets, the elements of which are respectively called *points, lines, planes,* and *flats.* A line is a set of points, containing at least two points. A plane is a set of points, containing at least three points not in one line. A flat is a set of points, containing at least four points that are neither all in one

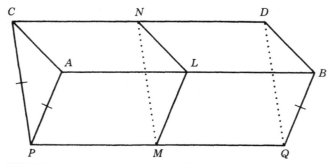

FIGURE 25.3

plane nor all in one line. Two points determine a line. Three points not in one line determine a plane. Four points not in one plane determine a flat. If two points are in one plane, then the line containing the two points is a subset of that plane. If three points are in one flat but not in one line, then the plane containing the three points is a subset of that flat. If two planes in the same flat have a point in common, then their intersection is a line. If a plane and a flat have a point in common, then their intersection contains a line. There exist five points that are neither all in one line, all in one plane, nor all in one flat.

Assuming the necessary definitions from the absolute plane, a set of *axioms for absolute four-space* is: (I) Four-Space Incidence Axiom, (II) Ruler Postulate, (III) Spatial PSP, (IV) Protractor Postulate, and (V) SAS.

The usual language denoting incidence is assumed for absolute four-space. By convention, line l and plane E are *parallel* if either l is on E or else l and E are coflat and nonintersecting. A line and a plane that are neither coflat nor intersecting are *skew*. Planes E_1 and E_2 are *parallel* if either $E_1 = E_2$ or else E_1 and E_2 are coflat planes that do not intersect. Two planes that are neither coflat nor intersecting are *skew*. Line l and flat f are *parallel* if either l is on f or else l and f do not intersect. Plane E and flat f are *parallel* if either E is on f or else E and f do not intersect. Flats f_1 and f_2 are *parallel* if either $f_1 = f_2$ or else f_1 and f_2 do not intersect.

Among others, the following theorems can be proved for absolute four-space: (1) If two points of a line are on a flat, then the line is on the flat. (2) A plane and a point off that plane determine a flat. (3) Three concurrent noncoplanar lines determine a flat. (4) Two planes with a common line determine a flat. (5) Two skew lines determine a flat. (6) Two planes not in a flat have at most one point in common. (7) There exist two planes that intersect at exactly one point. (8) If two flats have a point in common, then their intersection is a plane. There is also a Four-Space Separation Theorem, the statement of which is left for Exercise 25.7.

We say line l is *perpendicular* to flat f if l intersects f at a point P such that every line on f that passes through P is *perpendicular* to l. (As before, two planes are *perpendicular* if their union contains a dihedral angle having a right plane angle.) Two planes having a point P in common are *absolutely perpendicular* at P if every line through P on one plane is perpendicular to every line through P on the other plane. A plane E and a flat f intersecting at point P are *perpendicular* at P if there exist a plane in f that is absolutely perpendicular to E at P.

Among others, the following theorems can be proved: (1) If line l

intersects flat f at point P and l is perpendicular at P to each of three noncoplanar lines on f, then l is perpendicular to f. (2) Two lines perpendicular to a flat are parallel. (3) Given point P and flat f, there exists a unique line through P that is perpendicular to f. (4) Given point P and line l, there exists a unique flat through P that is perpendicular to l. (5) The intersection of two absolutely perpendicular planes is a point. (6) If two planes are absolutely perpendicular, then the two planes are not perpendicular. (7) Given a point P and a plane E, there exists a unique plane through P that is absolutely perpendicular to E. (8) If line l is on plane E, then all the planes absolutely perpendicular to E at points on l are in one flat. (9) If planes E_1 and E_2 are absolutely perpendicular at point P and E is a plane through P that is perpendicular to E_1, then E is perpendicular to E_2. (10) A third plane having a line in common with each of two absolutely perpendicular planes is perpendicular to each of the two planes. (11) If a plane is perpendicular to a flat at one point of their intersection, then the plane is perpendicular to the flat at every point on the line of their intersection. (12) If two flats are perpendicular to a plane E at point P, then the two flats intersect in a plane that is absolutely perpendicular to E at P. (13) If line l is not perpendicular to flat f, then there exists a unique plane through l that is perpendicular to f. (14) If line l is not in a plane absolutely perpendicular to plane E, then there exists a unique flat through l that is perpendicular to E. (15) Two lines not in the same plane have a common perpendicular line.

A *diflat angle* with *face E* is the union of a plane E with two half-flats of E that are not coflat; these two halfflats of E are the *sides* of the diflat angle. If P is any point on the face E of a diflat angle and points A and B are on different sides of the diflat angle such that $\overleftrightarrow{AP}$ and $\overleftrightarrow{BP}$ are perpendicular to E, then $\angle APB$ is a *plane angle* of the diflat angle. So the intersection of a diflat angle with the plane absolutely perpendicular to the face of the diflat angle at any point on the face is a plane angle of the diflat angle. It can be shown that any two plane angles of a given diflat angle are congruent. We then define two flats to be *perpendicular* if their union contains a diflat angle having a right plane angle. Et cetera, et cetera, et cetera.

Et cetera.

25.4 EXERCISES

● **25.1** What would be the difficulty in replacing the proof of the Pythagorean theorem in Section 25.2 by one of the simpler area proofs known to high school students?

25.2 For the coordinatization in Section 25.2, every ordered pair of real numbers is the coordinates of a unique point.

25.3 For the coordinatization in Section 25.2, every line that intersects m only at one point has an equation $y = sx + b$, and, conversely, a line with an equation $y = sx + b$ intersects m at exactly one point.

• **25.4** The mapping f in Section 25.2 preserves angle measure.

• **25.5** What happens if we omit the last six words in the statement of the Three-Space Incidence Axiom?

25.6 Prove directly from the Three-Space Incidence Axiom that if each of two intersecting lines is parallel to a third line, then the three lines are coplanar.

25.7 State a Four-Space Separation Theorem.

25.8 Suppose the vertices of $\triangle ABC$ together with the vertices of $\triangle A'B'C'$ are six noncoplanar points of absolute three-space. Suppose $\overleftrightarrow{AB}$ and $\overrightarrow{A'B'}$ are either parallel or else intersect at point C''. Suppose $\overleftrightarrow{AC}$ and $\overleftrightarrow{A'C'}$ are either parallel or else intersect at point B''. Suppose $\overleftrightarrow{BC}$ and $\overleftrightarrow{B'C'}$ are either parallel or else intersect at point A''. Then $\overleftrightarrow{AA'}$, $\overrightarrow{BB'}$, and $\overleftrightarrow{CC'}$ are either concurrent or else mutually parallel. Further, if A'', B'', C'' all exist, then they are collinear.

25.9 Two lines are *Clifford parallel* if the lines are in different planes and each is equidistant from the other. Show that the existence of two lines that are Clifford parallel implies the Hypothesis of the Obtuse Angle. (Such lines exist in elliptic geometry.)

25.10 Read "The Geometry of Four Dimensions," which is Chapter 3 of *Foundations of Geometry* by C. R. Wylie (McGraw-Hill, 1964). For more on absolute four-space read the book *Geometry of Four Dimensions* by H. P. Manning (Dover, 1956).

25.11 *Flatland* (A romance of many dimensions, By the author A Square) was written by E. A. Abbot, a Shakespearean scholar having mathematics as a hobby. Read this short classic, which was published in 1884 and is still in print (Dover). Read *Sphereland* (A Fantasy About Curved Spaces and an Expanding Universe) by D. Burger (Crowell, 1965).

25.12 "If two planes in the same flat have a point in common, then their intersection is a line" can be omitted from the Four-Space Incidence Axiom in the axioms for absolute four-space.

GRAFFITI

Suppose, for example, a world enclosed in a large sphere and subject to the following laws: The temperature is not uniform; it is greatest at the center, and gradually decreases as we move towards the circumference of the sphere, where it is absolute zero. The law of this temperature is as follows: If R *be the radius of the sphere, and* r *the distance of the point considered from the center, the absolute temperature will be proportional to* $R^2 - r^2$. *Further, I shall suppose that in this world all bodies have the same coefficient of dilatation, so that the linear dilatation of any body is proportional to its absolute temperature. Finally, I shall assume that a body transported from one point to another of different temperature is instantaneously in thermal equilibrium with its new environment. There is nothing in these hypotheses either contradictory or unimaginable. A moving object will become smaller and smaller as it approaches the circumference of the sphere. Let us observe, in the first place, that although from the point of view of our ordinary geometry this world is finite, to its inhabitants it will appear infinite. As they approach the surface of the sphere they become colder, and at the same time smaller and smaller. The steps they take are therefore also smaller and smaller, so that they can never reach the boundary of the sphere. If to us geometry is only the study of the laws according to which invariable solids move, to these imaginary beings it will be the study of the laws of motion of solids* deformed by the differences of temperature *alluded to.*

No doubt, in our world, natural solids also experience variations of form and volume due to differences of temperature. But in laying the foundations of geometry we neglect these variations; for besides being but small they are irregular, and consequently appear to us to be accidental. In our hypothetical world this will no longer be the case, the variations will obey very simple and regular laws. On the other hand, the different solid parts of which the bodies of these inhabitants are composed will undergo the same variations of form and volume.

Let me make another hypothesis: suppose that light passes through media of different refractive indices, such that the index of refraction is inversely proportional to $R^2 - r^2$. *Under these conditions it is clear that the rays of light will no longer be rectilinear but circular. To justify what has been said, we have to prove that certain changes in the position of external objects may be corrected by correlative movements of the beings which inhabit this imaginary world; and in such a way as to restore the primitive aggregate of the impressions experienced by these sentient beings. Suppose, for example, that an object is displaced and deformed, not like an invariable solid, but like a solid subjected to unequal dilatations in exact conformity with the law of temperature assumed above. To use an abbreviation, we shall call such a movement a non-Euclidean displacement.*

If a sentient being be in the neighborhood of such a displacement of the object, his impressions will be modified; but by moving in a suitable manner, he may reconstruct them. For this purpose, all that is required is that the aggregate of the sentient being and the object, considered as forming a single body, shall experience one of those special displacements which I have just called non-Euclidean. This is possible if we suppose that the limbs of these beings dilate according to the same laws as the other bodies of the world they inhabit.

Although from the point of view of our ordinary geometry there is a deformation of the bodies in this displacement, and although their different parts are no longer in the same relative position, nevertheless we shall see that the impressions of the sentient being remain the same as before; in fact, though the mutual distances of the different parts have varied, yet the parts which at first were in contact are still in contact. It follows that tactile impressions will be unchanged. On the other hand, from the hypothesis as to refraction and the curvature of the rays of light, visual impressions will be unchanged. These imaginary beings will therefore be led to classify the phenomena they observe, and to distinguish among them the "changes of position," which may be corrected by a voluntary corrective movement, just as we do.

If they construct a geometry, it will not be like ours, which is the study of the movements of our invariable solids; it will be the study of the changes of position which they will have thus distinguished, and will be "non-Euclidean displacements," and this will be non-Euclidean geometry. *So that beings like ourselves, educated in such a world, will not have the same geometry as ours.*

Poincaré

Part Two

NON-EUCLIDEAN GEOMETRY

We now center our attention on the hyperbolic plane of Bolyai and Loba-chevsky. However, many of our results, such as the classification of the isometries, have an immediate application to the Euclidean plane as well. A summary of each chapter in Part Two appears in Section 25.1.

CHAPTER 26

Parallels and the Ultimate Axiom

26.1 AXIOM 6: HPP

Let's do it!

> **Axiom 6** HPP If point P is off line l, then there exist two lines through P that are parallel to l.

Our axiom system, now called the *Bolyai–Lobachevsky plane*, is as consistent as the Euclidean plane or the real numbers (Section 23.2).

Axiom 6, the Hyperbolic Parallel Postulate, could be weakened to require only the existence of nonincident point P_0 and line l_0 such that there exist two lines through P_0 that are parallel to l_0. This follows from Proposition Y of Theorem 23.7. On the other hand, Axiom 6 could be replaced by our next theorem.

Theorem 26.1 If point P is off line l, then there exist an infinite number of lines through P that are parallel to l.

Proof Let F be the foot of the perpendicular from P to l. By HPP there exist two lines $\overleftrightarrow{PA}$ and $\overleftrightarrow{PB}$ parallel to l where A and B are on the same side of $\overleftrightarrow{PF}$. The result follows from the fact that $\overleftrightarrow{PE}$ is parallel to l whenever $m\angle FPE$ is between $m\angle FPA$ and $m\angle FPB$. ∎

Theorem 26.2

(a) There does not exist a rectangle.
(b) AAA.
(c) If $\triangle ABC$, then $m\angle A + m\angle B + m\angle C < \pi$.
(d) If $\triangle ABC$ and $A-C-D$, then $m\angle BCD > m\angle A + m\angle B$.
(e) $\delta\triangle ABC > 0$; $\delta\square ABCD > 0$.
(f) If point C is off $\overline{AB}$ but on the circle with diameter $\overline{AB}$, then $\angle ACB$ is acute.
(g) If $\boxed{S}\,ABCD$, then $m\angle B = m\angle C < \pi/2$ and $BC > AD$.
(h) If $\boxed{L}\,ABCD$, then $m\angle C < \pi/2$, $CB > AD$, and $CD > AB$.
(i) No line is equidistant from a second line.
(j) The critical function is strictly decreasing.

Proof From our study of the equivalents of Euclid's Parallel Postulate, we know each of the ten statements is actually equivalent to HPP. ■

Any theorem of the Euclidean plane that is false under the assumption of Axiom 6 must be equivalent to Euclid's Parallel Postulate. Also, any theorem that we now prove and that is false for the Euclidean plane must be equivalent to HPP. For either parallel postulate, the list of statements equivalent to that postulate is endless.

That AAA, the negation of Proposition Z in Theorem 23.7, is a theorem has an amusing consequence. In a hyperbolic world one could not build exact small scale models, say of a building. The angles would have to be distorted in any model or otherwise the model would be congruent to the building. Of course, due to the limitations of measurement, it is impossible to build an *exact* model in any kind of a physical world.

The next two theorems are mostly for *shock* value. Certainly one must depend more on his head than his emotions to feel secure in the Bolyai–Lobachevsky plane.

Theorem 26.3 There exist four lines l, m, r, and s such that $l \perp r$, $r \perp s$, $s \perp m$, and $l \parallel m$. The inside of any angle is a proper subset of the interior of the angle. The interior of any angle contains two perpendicular lines.

Proof The first statement follows from Proposition Q of Theorem 23.7. (To find such lines, see Figure 26.1 where $\Pi(PQ) = \Pi(QR) = \pi/4$.) Suppose $\angle AVB$. Let $\overrightarrow{VD}$ be the angle bisector of $\angle AVB$ where $\Pi(VD) = \frac{1}{2}m\angle AVB$ (Theorem 24.16). Let p be the perpendicular to

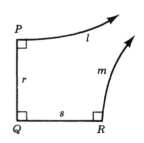

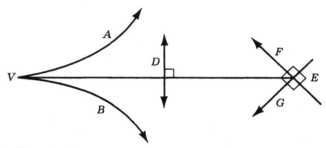

FIGURE 26.1

$\overleftrightarrow{VD}$ at D. By the definition of the critical function, since $\Pi(VD) =$ $m\angle DVA = m\angle DVB$, then p is parallel to $\overleftrightarrow{VA}$ and to $\overleftrightarrow{VB}$. So D is in the interior of $\angle AVB$ but not in the inside of $\angle AVB$. Let E be the unique point such that $V-D-E$ and $\Pi(DE) = \pi/4$. Let F and G be points on opposite sides of $\overleftrightarrow{VD}$ such that $m\angle DEF = m\angle DEG = \pi/4$. Then $\overleftrightarrow{EF}$ and $\overleftrightarrow{EG}$ are both parallel to p and $\overleftrightarrow{EF} \perp \overleftrightarrow{EG}$. Since V and E are on opposite sides of p, then $\overleftrightarrow{EF}$ and $\overleftrightarrow{EG}$ are two perpendicular lines in int $(\angle AVB)$. ∎

Theorem 26.4 For any positive integer n, there exist $n+1$ distinct

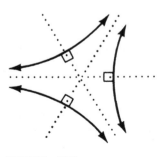 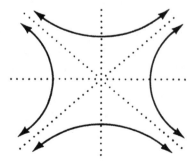

FIGURE 26.2

lines such that any n of the lines are on the same halfplane of the remaining line.

Proof   Exercise 26.1.   ■

Theorem 26.5   An exterior angle of a closed biangle is larger than its remote interior angle.

Proof   Suppose $\llcorner ABCD$ and $B-C-E$. Assume $\angle ABC \simeq \angle DCE$. Then $\overleftrightarrow{AB}$ and $\overleftrightarrow{CD}$ have a common perpendicular (Theorem 21.5). So $\llcorner ABCD$ is equivalent to a closed biangle with two right angles, contradicting $\Pi(x) < \pi/2$ for all positive numbers x. Now assume $\angle ABC > \angle DCE$. By Euclid's Proposition I.28 there is a point F such that $\llcorner FBCD$ with $\angle FBC \simeq \angle DCE$. (See Figure 26.3.) Since $\angle ABC > \angle FBC$, then F is in int $(\angle ABC)$. Thus $\overrightarrow{BF}$ is an interior ray of $\llcorner ABCD$ that does not intersect $\overleftrightarrow{CD}$, contradicting $\llcorner ABCD$ is closed from B. Therefore, $\angle ABC < \angle DCE$.   ■

Corollary 26.6   The sum of the measures of the angles of a closed biangle is less than π. The angles of an isosceles closed biangle are acute.

Theorem 26.7   *Angle-Angle Theorem*   If $\llcorner ABCD$ and $\llcorner PQRS$ are closed, $\angle B \simeq \angle Q$, and $\angle C \simeq \angle R$, then $\llcorner ABCD \simeq \llcorner PQRS$.

Proof   The result follows by symmetry if we can show the assumption $BC > QR$ leads to a contradiction. Since $\angle ABC$ is the image of $\angle PQR$ under an isometry, we may suppose $P = A$, $Q = B$, and $B-R-C$. (See Figure 26.4.) $\llcorner SRCD$ by Euclid's Proposition I.28. If $\overrightarrow{CE}$ intersects $\overrightarrow{BA}$, then $\overrightarrow{CE}$ intersects $\overrightarrow{RS}$ since C and $\overrightarrow{BA}$ are on opposite sides of $\overleftrightarrow{RS}$. Thus, if E in int $(\llcorner SRCD)$, then $\overrightarrow{CE}$ intersects $\overrightarrow{RS}$ because

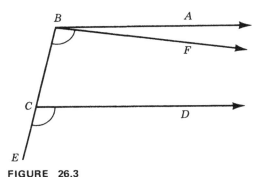

FIGURE   26.3

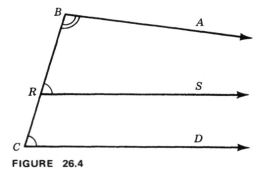

FIGURE 26.4

$\llcorner\lrcorner ABCD$ is closed from C. Hence $\llcorner\lrcorner SRCD$ is closed from C. We now have $\llcorner\lrcorner SRCD$ is closed with exterior angle $\angle BRS$ congruent to its remote interior angle, contradicting the preceding theorem. ∎

Theorem 26.8 Suppose $\boxed{S}ABCD$ and $\boxed{S}A'B'C'D'$. If (1) $AD = A'D'$ and $BC = B'C'$, (2) $AB = A'B'$ and $m\angle B = m\angle B'$, (3) $AD = A'D'$ and $m\angle B = m\angle B'$, or (4) $BC = B'C'$ and $m\angle B = m\angle B'$, then $\boxed{S}ABCD \simeq \boxed{S}A'B'C'D'$.

Proof (1) Assume $\boxed{S}AQRD \simeq \boxed{S}A'B'C'D'$ with $A - Q - B$ or $A - B - Q$. Let M and T be the midpoints of $\overline{BC}$ and $\overline{QR}$, respectively. Then $\boxed{S}TQBM$. If $A - Q - B$, then $\angle TQA$ and $\angle TQB$ are a linear pair of acute angles; if $A - B - Q$, then $\angle MBA$ and $\angle MBQ$ are a linear pair of acute angles. In either case we have a contradiction. (2) Assume $\boxed{S}ABRS \simeq \boxed{S}A'B'C'D'$ with $B - R - C$ or $B - C - R$. Then $\delta\square SRCD = 0$, a contradiction. (3) Assume $\boxed{S}AQRD \simeq \boxed{S}A'B'C'D'$ with $A - B - Q$ or $A - Q - B$. Then $\delta\square BQRC = 0$, a contradiction. (4) Assume $\boxed{S}PBCS \simeq \boxed{S}A'B'C'D'$ with $B - P - A$ or $B - A - P$. Then $\delta\square APSD = 0$, a contradiction. ∎

26.2 PARALLEL LINES

Parallelism is not an equivalence relation on the set of lines. By Proposition C of Theorem 23.7 there exist lines l, m, n such that $l \parallel m$, $m \parallel n$, but $l \nparallel n$. Suppose point P is off line m. If l and n are two lines through P that are parallel to line m, then l and n are certainly not parallel. We know there are two lines that are the bounds for the lines through P that are parallel to m. These bounds are of critical importance. (*Horos* means "boundary" in Greek.)

DEFINITION 26.9 If $\overrightarrow{AB}$ and $\overrightarrow{CD}$ are equivalent or the sides of a closed biangle, then we say the rays are **horoparallel** or **critically parallel** and write $\overrightarrow{AB} \mid \overrightarrow{CD}$. If a ray on line l is horoparallel to a ray on line m, then we say the lines are **horoparallel** or **critically parallel** and write $l \mid m$.

So $\overrightarrow{AB} \mid \overrightarrow{CD}$ iff either $\overrightarrow{AB} \sim \overrightarrow{CD}$ or $\llcorner BACD$ is closed. $\overrightarrow{AB} \mid \overrightarrow{CD}$ implies $\overleftrightarrow{AB} \parallel \overleftrightarrow{CD}$, but not conversely. Critical parallelism for lines is of secondary importance to critical parallelism for rays. A comparison of the next theorem with the following statement should suggest the importance of critical parallelism for rays. *Playfair's Parallel Postulate:* If point P is off $\overleftrightarrow{AB}$, then there exists unique $\overleftrightarrow{PQ}$ such that $\overleftrightarrow{PQ} \parallel \overleftrightarrow{AB}$.

Theorem 26.10 If point P is off $\overleftrightarrow{AB}$, then there exists unique $\overrightarrow{PQ}$ such that $\overrightarrow{PQ} \mid \overleftrightarrow{AB}$.

Proof If P is on $\overleftrightarrow{AB}$ the result is trivial. For P off $\overleftrightarrow{AB}$ the result is a restatement of Corollary 24.10. ■

Corollary 26.11 If point P is off line l, then there exist exactly two lines through P that are horoparallel to l.

A glance back at Figure 26.2 should show that the following theorem does have some content.

Theorem 26.12 If $\overleftrightarrow{AB}$, $\overleftrightarrow{CD}$, $\overleftrightarrow{EF}$ are three lines such that $\overrightarrow{AB} \mid \overrightarrow{CD}$ and $\overrightarrow{AB} \mid \overrightarrow{EF}$, then there exists a fourth line which intersects the three lines.

Proof Since $\overleftrightarrow{CE}$ intersects both $\overleftrightarrow{CD}$ and $\overleftrightarrow{EF}$, we may suppose C and E are on the same side of $\overleftrightarrow{AB}$ as otherwise $\overleftrightarrow{CE}$ also intersects $\overleftrightarrow{AB}$ by PSP. Then there exists $\overrightarrow{AH}$ with H in int $(\angle BAC)$ and in int $(\angle BAE)$ by the Protractor Postulate. Since $\llcorner BACD$ and $\llcorner BAEF$ are both closed, $\overrightarrow{AH}$ must intersect $\overleftrightarrow{CD}$ and $\overleftrightarrow{EF}$. Hence $\overleftrightarrow{AH}$ is a desired fourth line. ■

That $\overrightarrow{AB} \mid \overrightarrow{AB}$ and that $\overrightarrow{AB} \mid \overrightarrow{CD}$ implies $\overrightarrow{CD} \mid \overrightarrow{AB}$ are trivial observations from Definition 26.9. We shall use these facts in proving the following nontrivial theorem, which states the remaining requirement for critical parallelism to be an equivalence relation on the set of all rays.

Theorem 26.13 If $\overrightarrow{AB} \mid \overrightarrow{CD}$ and $\overrightarrow{CD} \mid \overrightarrow{EF}$, then $\overrightarrow{AB} \mid \overrightarrow{EF}$.

Proof We may suppose no two of $\overrightarrow{AB}$, $\overrightarrow{CD}$, $\overrightarrow{EF}$ are equivalent as otherwise the result follows immediately from the fact that a biangle equivalent to a closed biangle is closed. So $\overleftrightarrow{AB}$, $\overleftrightarrow{CD}$, $\overleftrightarrow{EF}$ are three lines, $\llcorner BACD$ is closed, $\llcorner DCEF$ is closed, and we want to show $\llcorner BAEF$ exists and is closed. Since the three lines have a common transversal by the preceding theorem and since a biangle equivalent to a closed biangle is closed, we may suppose A, C, E are collinear without loss of generality. There are three cases.

First suppose $A - C - E$. (See Figure 26.5.) Then $\overleftrightarrow{AB}$ and $\overleftrightarrow{EF}$ are on opposite sides of $\overleftrightarrow{CD}$. So $\llcorner BAEF$. Let $\overrightarrow{AG}$ be any interior ray of $\angle BAC$. Since $\llcorner BACD$ is closed from A, then $\overrightarrow{AG}$ intersects $\overrightarrow{CD}$ at some point H. Let I and J be such that $A - H - I$ and $\overrightarrow{HJ} \sim \overrightarrow{CD}$. So $\llcorner JHEF \sim \llcorner DCEF$ and $\llcorner JHEF$ is closed from H. Hence $\overrightarrow{HI}$ intersects $\overrightarrow{EF}$. So $\overrightarrow{AG}$ intersects $\overrightarrow{EF}$. So $\llcorner BAEF$ is closed from A. Therefore, $\llcorner BAEF$ is closed and $\overrightarrow{AB} \mid \overrightarrow{EF}$.

Now suppose $A - E - C$. Let $\overrightarrow{AK}$ be the unique ray (Theorem 26.10) such that $\overrightarrow{AK} \mid \overrightarrow{EF}$. Then $\overrightarrow{AK} \mid \overrightarrow{EF}$, $\overrightarrow{EF} \mid \overrightarrow{CD}$, and $A - E - C$. This is exactly the situation of the first case (only with different letters). Hence $\overrightarrow{AK} \mid \overrightarrow{CD}$. Then, since $\overrightarrow{AB} \mid \overrightarrow{CD}$, we must have $\overrightarrow{AK} = \overrightarrow{AB}$. Therefore $\overrightarrow{AB} \mid \overrightarrow{EF}$.

Finally, suppose $E - A - C$. Let $\overrightarrow{EL}$ be the unique ray such that $\overrightarrow{EL} \mid \overrightarrow{AB}$. Then $\overrightarrow{EL} \mid \overrightarrow{AB}$, $\overrightarrow{AB} \mid \overrightarrow{CD}$, and $E - A - C$. Again, this is exactly

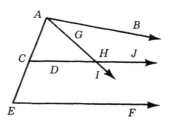

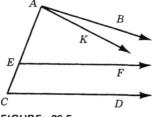

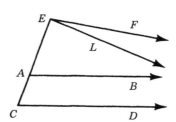

FIGURE 26.5

the situation of the first case. Hence $\overrightarrow{EL} \mid \overrightarrow{CD}$. Then, since $\overrightarrow{EF} \mid \overrightarrow{CD}$, we must have $\overrightarrow{EL} = \overrightarrow{EF}$. Therefore $\overrightarrow{EF} \mid \overrightarrow{AB}$ and $\overrightarrow{AB} \mid \overrightarrow{EF}$. ∎

Corollary 26.14 Critical parallelism is an equivalence relation on the set of all rays.

Two distinct horoparallel lines cannot have a common perpendicular because a critical angle is always acute. Of course, two parallel lines can have at most one common perpendicular because rectangles do not exist.

DEFINITION 26.15 If lines l and m have a common perpendicular then we say the lines are *hyperparallel.*

Our next theorem shows that two parallel lines must be either horoparallel or hyperparallel. (*Hyper* means "over" in Greek.) One could write "$l \mid\mid\mid m$" as a blackboard shorthand for "l hyperparallel to m." An historical argument against this is that Bolyai used this notation to denote critical parallelism. (We have been more conservative than Bolyai by two bars.) The notation was selected by Bolyai to express the fact (Theorem 26.19) that horoparallels are asymptotic. We now show that there are only the two kinds of parallelism in our geometry. The theorem and its corollary are equivalent to Saccheri's Proposition XXIII.

Theorem 26.16 If two lines are parallel but not horoparallel, then the two lines are hyperparallel.

Proof Suppose lines m and n are parallel but not horoparallel. So $m \neq n$. Let $n = \overleftrightarrow{PQ}$ and let A be a point on m. (See Figure 26.6.) Let B and C be such that $\overrightarrow{AB} \mid \overrightarrow{QP}$ and $\overrightarrow{AC} \mid \overrightarrow{PQ}$ (Theorem 26.10). Since m is not horoparallel to n, then B and C are two points on the same side of

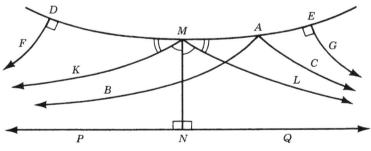

FIGURE 26.6

m as n. If $\overleftrightarrow{AB} \perp m$, let $D=A$; otherwise, let D be on m such that $m\angle BAD = \Pi(AD)$ (Theorem 24.16). If $\overleftrightarrow{AC} \perp m$, let $E = A$; otherwise, let E be on m such that $m\angle CAE = \Pi(AE)$. $D \neq E$ because $\overrightarrow{AB} \neq \overrightarrow{AC}$. Let F and G be points on the same side of m as n such that $\overrightarrow{DF} \perp m$ and $\overrightarrow{EG} \perp m$. Then $\overrightarrow{DF} \mid \overrightarrow{AB}$ and $\overrightarrow{EG} \mid \overrightarrow{AC}$ (Definition 24.11). Since $\overrightarrow{DF} \mid \overrightarrow{AB}$ and $\overrightarrow{AB} \mid \overrightarrow{QP}$, then $\overrightarrow{QP} \mid \overrightarrow{DF}$ (Corollary 26.14). Likewise $\overrightarrow{PQ} \mid \overrightarrow{EG}$.

Let M be the midpoint of $\overline{DE}$. Let N be the foot of the perpendicular from M to n. Let K and L be such that $\overrightarrow{MK} \mid \overrightarrow{QP}$ and $\overrightarrow{ML} \mid \overrightarrow{PQ}$. Now $\angle NMK \simeq \angle NML$ because $\angle NMK$ and $\angle NML$ are critical angles for $\overline{MN}$. Since $\overrightarrow{MK} \mid \overrightarrow{QP}$ and $\overrightarrow{QP} \mid \overrightarrow{DF}$, then $\overrightarrow{MK} \mid \overrightarrow{DF}$. Since $\angle MDF$ is right, then $\angle KMD$ is a critical angle for $\overline{DM}$. Likewise, $\overrightarrow{ML} \mid \overrightarrow{EG}$ and $\angle LME$ is a critical angle for $\overline{ME}$. From $\overline{DM} \simeq \overline{ME}$, it follows that $\angle KMD \simeq \angle LME$. Hence $\angle NMD$ and $\angle NME$ are a linear pair of congruent angles. So $\angle NMD$ is a right angle. Therefore $\overleftrightarrow{MN}$ is perpendicular to both m and n. ∎

Corollary 26.17 If l and m are distinct lines, then exactly one of the following holds: (i) l and m intersect, (ii) l and m are hyperparallel, (iii) l and m are horoparallel.

If two lines m and n intersect at point C, then the distance from a point P on m to line n increases without limit as CP increases (Aristotle's Axiom). Now suppose m and n are two hyperparallel lines with common perpendicular $\overleftrightarrow{MN}$ where M is on m and N is on n. Then MN is the shortest distance from a point on m to line n, and the distance from a point P on m to line n increases as MP increases (Theorem 22.9). We shall now show this distance increases without limit. Thus two hyperparallel lines are often described as converging toward their common perpendicular and then diverging.

Theorem 26.18 If $t > 0$ and $\boxed{\text{L}}\, NMAB$, then there exists a point P on $\overrightarrow{MA}$ whose distance to $\overleftrightarrow{NB}$ is greater than t.

Proof Let $\overrightarrow{NE} \mid \overrightarrow{MA}$. So $\angle BNE$ is acute. (See Figure 26.7.) By Aristotle's Axiom, let R be a point on $\overrightarrow{NE}$ whose distance to $\overleftrightarrow{NB}$ is greater than t. Let Q be the foot of the perpendicular from R to $\overleftrightarrow{NB}$. Let $N-R-S$ and $Q-R-T$. Since $\overleftrightarrow{RQ} \parallel \overleftrightarrow{MN}$, then T is in int $(\angle MRS)$. Since $\sqcup SNMA$ is closed, then $\sqcup SRMA$ is closed and $\overrightarrow{RT}$ intersects $\overrightarrow{MA}$ at some point P. So $P-R-Q$ with $RQ > t$. Therefore $PQ > t$. ∎

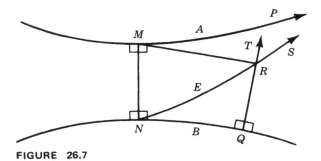

FIGURE 26.7

Finally, suppose $\overleftrightarrow{AB}$ and $\overleftrightarrow{CD}$ are two horoparallel lines with $\overrightarrow{BA} \mid \overrightarrow{CD}$, $\overline{BC} \perp \overline{CD}$, and $\overline{AD} \perp \overline{CD}$. (See Figure 26.8.) Since critical angles are acute, then $\angle CBA$ is acute and $\angle DAB$ is obtuse. So $BC > AD$ (Theorem 21.8). It follows that the distance from a point S on $\overrightarrow{AB}$ to line $\overleftrightarrow{CD}$ increases as AS increases. That this distance increases without limit is proved in much the same way as Theorem 26.18. In the other *direction*, if P is on $\overrightarrow{BA}$, then the distance from P to $\overleftrightarrow{CD}$ decreases as BP increases. It seems quite possible that this distance is always greater than some constant. However this is not the case. Two horoparallel lines are asymptotic (Saccheri's Proposition XXV)!

Theorem 26.19 Let t be any positive real number. If m and n are two horoparallel lines, then there exists a unique point P on m whose distance to line n is t.

Proof Suppose $\llcorner\lrcorner ABCD$ is closed with $\overline{BC} \perp \overline{CD}$. (See Figure 26.8.) Let R be such that $\overrightarrow{CR} \mid \overrightarrow{AB}$. So $\angle BCR$ is acute (Corollary 26.6). By Aristotle's Axiom we may suppose the distance from R to $\overleftrightarrow{CD}$ is greater than t. Let T be the foot of the perpendicular from R to $\overleftrightarrow{CD}$. So $RT > t$. By the same reasoning as in the proof of the preceding theorem, we know $\overrightarrow{TR}$ must intersect $\overrightarrow{AB}$ at some point S and $ST > t$.

Let E be such that $S-E-T$ with $TE = t$. (See Figure 26.9.) Let F and G be such that $\overrightarrow{EF} \mid \overrightarrow{DC}$ and $\overrightarrow{EG} \mid \overrightarrow{CD}$. Since $\overrightarrow{EG} \mid \overrightarrow{CD}$ and $\overrightarrow{CD} \mid \overrightarrow{SA}$,

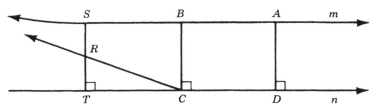

FIGURE 26.8

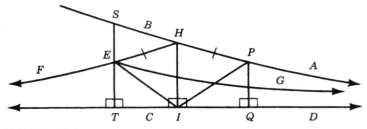

FIGURE 26.9

then $\sqcup GESA$ is closed from E. So $\overrightarrow{FE}$ intersects $\overrightarrow{SA}$ at some point H. Let P be the point such that $S-H-P$ and $\overline{PH} \simeq \overline{EH}$. Let I be the foot of the perpendicular from H to $\overset{\leftrightarrow}{CD}$. $\angle IHE \simeq \angle IHP$ because $\angle IHE$ and $\angle IHP$ are both critical angles for $\overline{HI}$. Hence $\triangle PHI \simeq \triangle EHI$ by SAS. With Q the foot of the perpendicular from P to $\overset{\leftrightarrow}{CD}$, it follows that $\triangle PIQ \simeq \triangle EIT$ by SAA. So $\overline{PQ} \simeq \overline{ET}$. Therefore $PQ = t$. Thus P is a point whose distance to $\overset{\leftrightarrow}{CD}$ is t.

P is unique as otherwise $\overset{\leftrightarrow}{AB}$ contains the upper base of a Saccheri quadrilateral whose lower base is on $\overset{\leftrightarrow}{CD}$. In that case $\overset{\leftrightarrow}{AB}$ and $\overset{\leftrightarrow}{CD}$ would be both hyperparallel and horoparallel, a contradiction (Corollary 26.17). ∎

Although every line in the Bolyai–Lobachevsky plane is isomorphic to a Euclidean line, the plane has properties quite different from those of the Euclidean plane.

26.3 EXERCISES

- **26.1** Theorem 26.4.

- **26.2** There exist three lines with no common transversal.

- **26.3** Is it true that lines m and n are hyperparallel iff the lines contain equivalent rays or the lines contain the sides of a biangle that is not closed?

26.4 If $\angle ABC$ and $\angle PQR$ are any angles, then int $(\angle ABC)$ contains an angle congruent to $\angle PQR$.

- **26.5** True or False?

   (a) $\overrightarrow{AB} \mid \overrightarrow{CD}$ and $\overrightarrow{CD} \mid \overrightarrow{EF}$ implies $\overrightarrow{AB} \mid \overrightarrow{EF}$.

   (b) $\overset{\leftrightarrow}{AB} \parallel \overset{\leftrightarrow}{CD}$ and $\overset{\leftrightarrow}{CD} \parallel \overset{\leftrightarrow}{EF}$ implies $\overset{\leftrightarrow}{AB} \parallel \overset{\leftrightarrow}{EF}$.

(c) $\overleftrightarrow{AB} \mid \overleftrightarrow{CD}$ and $\overleftrightarrow{CD} \mid \overleftrightarrow{EF}$ implies $\overleftrightarrow{AB} \parallel \overleftrightarrow{EF}$.

(d) $\overleftrightarrow{AB} \mid \overleftrightarrow{CD}$ and $\overleftrightarrow{CD} \mid \overleftrightarrow{EF}$ implies $\overleftrightarrow{AB} \mid \overleftrightarrow{EF}$.

(e) If $\llcorner ABCD$ is closed, then $\Pi(BC) = m\angle ABC$.

(f) If $\llcorner ABCD$ is closed, then $\angle B$ or $\angle C$ is acute.

(g) If $\llcorner ABCD$ is closed, then $\angle B$ and $\angle C$ are acute.

(h) The perpendicular bisectors of the sides of a triangle are concurrent.

(i) The angle bisectors of the angles of a triangle are concurrent.

(j) Any three points not on a line are on some circle.

26.6 If $\overleftrightarrow{AB}$ and $\overleftrightarrow{CD}$ are two horoparallel lines, then exactly one of the following holds: $\overrightarrow{AB} \mid \overrightarrow{CD}$, $\overrightarrow{AB} \mid \overrightarrow{DC}$, $\overrightarrow{BA} \mid \overrightarrow{CD}$, $\overrightarrow{BA} \mid \overrightarrow{DC}$.

26.7 Every biangle is equivalent to an isosceles biangle.

26.8 If a biangle is not closed, then its interior contains a line.

26.9 If a line intersects the interior of a closed biangle, then the line intersects the biangle.

26.10 If $\llcorner ABCD$ is not closed, then int $(\llcorner ABCD)$ contains two perpendicular lines each of which is hyperparallel to both $\overleftrightarrow{AB}$ and $\overleftrightarrow{CD}$.

26.11 The Angle-Base Theorem and the Angle-Angle Theorem both fail for biangles in general.

• **26.12** Define $\overrightarrow{AB}$ to be *hyperparallel* to $\overrightarrow{CD}$ if the rays are equivalent or the sides of a biangle equivalent to an isosceles biangle with right angles. This hyperparallelism is not an equivalence relation on the set of all rays.

26.13 Two Saccheri quadrilaterals with the same defect need not be congruent.

26.14 If $0 < s < \pi/2$ and $\llcorner ABCD$ is closed, then the biangle is equivalent to a biangle with one angle right and the other of measure s.

26.15 Draw the figure for Theorem 26.16 that covers the case when $\angle BAH$ is obtuse where H is on m and H and P are on opposite sides of $\overleftrightarrow{AB}$. Show that this case is possible.

26.16 A line in the interior of an angle does not intersect the inside of the angle.

26.17 Give a reasonable definition for *direction*.

• **26.18** If P, Q, R, S are, respectively, the midpoints of sides $\overline{AB}$, $\overline{BC}$, $\overline{CD}$, $\overline{DA}$ of $\square ABCD$, then $\square PQRS$. We call $\square PQRS$ a *Varignon quadrilateral*, since Varignon showed that in the Euclidean plane such a quadrilateral is a parallelogram. State as many results about Varignon quadrilaterals as you can.

*26.19 Can SAS and HPP be replaced by AAA in the axioms for the Bolyai–Lobachevsky plane?

GRAFFITI

Either for the data from Figure 26.10 or for the data from Figure 26.11, with Definitions 31.1 and 32.18 it follows that

$$\Pi(l) + \Pi(c+m) = \Pi(b),$$

$$\Pi(l+b) + \Pi(m-a) = \Pi(0),$$

$$\Pi(l) + \Pi(b) = \Pi(c-m).$$

Further,

$$\Pi(m) + \Pi(c+l) = \Pi(a), \qquad \Pi(c) + \Pi(l+a^*) = \Pi(b),$$

$$\Pi(m+a) + \Pi(l-b) = \Pi(0), \qquad \Pi(b+c) + \Pi(a^* - m^*) = \Pi(0),$$

$$\Pi(m) + \Pi(a) = \Pi(c-l), \qquad \Pi(c) + \Pi(b) = \Pi(l-a^*).$$

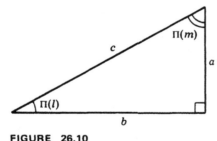

FIGURE 26.10   **FIGURE 26.11**

Also,

$$\Pi(c) + \Pi(m+b^*) = \Pi(a), \qquad \Pi(0) = \Pi(c+a) + \Pi(b^* - l^*),$$

$$\Pi(c) + \Pi(a) = \Pi(m-b^*), \qquad \Pi(0) = \Pi(c-a) + \Pi(b^* + l^*),$$

$$\Pi(l) + \Pi(a^* + b^*) = \Pi(m^*), \qquad \Pi(0) = \Pi(a^* - c^*) + \Pi(m+l^*),$$

$$\Pi(l) + \Pi(m^*) = \Pi(a^* - b^*), \qquad \Pi(0) = \Pi(a^* + c^*) + \Pi(m-l^*).$$

CHAPTER 27

Brushes and Cycles

27.1 BRUSHES

In Euclidean geometry the set of all lines passing through a point is called a *pencil*. (The term comes from the former use of the word for an artist's paintbrush.) The set of all lines parallel to a given line in the Euclidean plane is called a *parallel pencil*. To be useful this second idea must be refined for the Bolyai–Lobachevsky plane where we have two distinct types of parallelism.

DEFINITION 27.1 The set of all lines through a point C is a **pencil** with **center** C. The set of all lines perpendicular to a line c is a **hyperpencil** with **center** c. The set b of all lines containing a ray horoparallel to a given ray is a **horopencil** with **center** b. A **brush** is any one of a pencil, a hyperpencil, or a horopencil.

The center of a pencil is obviously unique. The center of a hyperpencil is also unique since there do not exist rectangles. Since pencils and hyperpencils have natural centers, it would seem unfair for a horopencil not to have a unique center too. Because horoparallelism is an equivalence relation for rays, no particular ray or line could be designated as the unique center of a horopencil. Thus we have defined the center of a horopencil to be itself. Although this may seem a little odd at first, the convention will make the statement of some theorems easier.

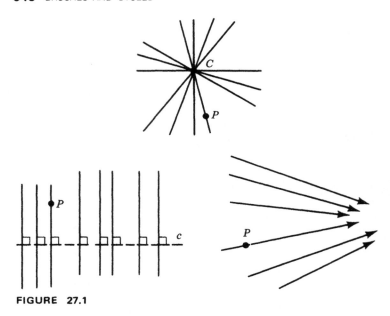

FIGURE 27.1

Every brush has a unique center. The center of a pencil is a point. The center of a hyperpencil is a set of points, namely a line. The center of a horopencil is a set of lines, namely the horopencil itself. We shall frequently use the phrase "point P is off the center of brush b." If b is a pencil with center C, then this means $P \neq C$. If b is a hyperpencil with center c, then the phrase means P is off c. If b is a horopencil, then the phrase places no restriction on P since the center of a horopencil does not contain any points but is a set of lines.

Theorem 27.2 If point P is off the center of brush b, then there is a unique line through P that is in b. Any two distinct lines are in a unique brush.

Proof If b is a pencil with center C, then $\overleftrightarrow{PC}$ is the unique line through P in b by the Incidence Axiom. If b is a hyperpencil with center c, then the perpendicular from P to c is the unique line through P in b (Theorem 18.1). If b is the horopencil of all lines containing a ray horoparallel to $\overrightarrow{AB}$ and $\overrightarrow{PQ} \mid \overrightarrow{AB}$, then $\overleftrightarrow{PQ}$ is the unique line through P in b (Theorem 26.10). The second statement in the theorem is essentially a restatement of Corollary 26.17. ∎

Theorem 27.3 Any two horopencils have a unique line in common.

Proof Exercise 27.4. ∎

Other theorems concerning the intersection of two brushes are relegated to the exercises. We shall turn to some theorems that will be more useful later.

Theorem 27.4 If $\llcorner BACD$ is closed and isosceles, then the perpendicular bisector of the base $\overline{AC}$ is in the same horopencil as $\overleftrightarrow{AB}$ and $\overleftrightarrow{CD}$.

Proof Let M be the midpoint of $\overline{AC}$. Let $\overrightarrow{MN}$ be the unique ray such that $\overrightarrow{MN} \mid \overrightarrow{AB}$. (See Figure 27.2.) Hence $\overrightarrow{MN} \mid \overrightarrow{CD}$ as $\overrightarrow{AB} \mid \overrightarrow{CD}$. Thus $\overleftrightarrow{AB}$, $\overleftrightarrow{CD}$, and $\overleftrightarrow{MN}$ are in the same horopencil. Also, $\llcorner BAMN$ and $\llcorner DCMN$ are closed. Since $\overline{AM} \simeq \overline{MC}$ and $\angle A \simeq \angle C$, we have $\llcorner BAMN \simeq \llcorner DCMN$ by the Angle-Base Theorem (Theorem 24.8). So $\angle AMN$ and $\angle CMN$ are a linear pair of congruent angles. Thus $\overleftrightarrow{MN}$ is the perpendicular bisector of $\overline{AC}$. ∎

Theorem 27.5 If three distinct lines have a common transversal and each of the three lines is horoparallel to the other two, then the three lines are in one horopencil.

Proof Suppose $A-C-E$, $\overrightarrow{AB} \mid \overrightarrow{CD}$, and $\overleftrightarrow{EF}$ is horoparallel to both $\overleftrightarrow{AB}$ and $\overleftrightarrow{CD}$. We wish to show that $\overleftrightarrow{AB}$, $\overleftrightarrow{CD}$, and $\overleftrightarrow{EF}$ are in a horopencil. Let $B'-A-B$ and $D'-C-D$. (See Figure 27.3.) If either $\overrightarrow{EF} \mid \overrightarrow{AB}$ or $\overrightarrow{EF} \mid \overrightarrow{CD}$, then each of $\overleftrightarrow{AB}$, $\overleftrightarrow{CD}$, and $\overleftrightarrow{EF}$ is horoparallel to the other two and the three lines are in one horopencil by definition. Assuming otherwise, we must have $\overrightarrow{EF} \mid \overrightarrow{AB'}$ and $\overrightarrow{EF} \mid \overrightarrow{CD'}$. So $\overrightarrow{AB'} \mid \overrightarrow{CD'}$. Hence $\llcorner BACD$ and $\llcorner B'ACD'$ are both closed from A. Therefore, $\overleftrightarrow{AB}$ is the only line through A parallel to $\overleftrightarrow{CD}$, contradicting HPP. ∎

We know the perpendicular bisectors of the sides of a triangle are not necessarily concurrent (Proposition H of Theorem 23.7). How-

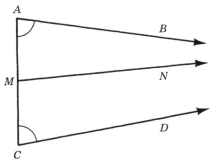

FIGURE 27.2

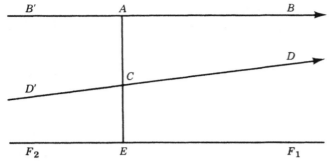

FIGURE 27.3

ever, if the perpendicular bisectors are not in a pencil, then they must be in either a hyperpencil or a horopencil.

Theorem 27.6 The three perpendicular bisectors of the sides of a triangle are in a unique brush.

Proof Let a, b, c be the perpendicular bisectors of the three sides of $\triangle ABC$ opposite A, B, C, respectively. Now a, b, c are distinct as otherwise there is a triangle with two right angles. Let b be the unique brush containing a and b. We must show c is in b. If b is a pencil with center Q, then Q is equidistant from A, B, C. Then Q must be on c. Thus c is in b if b is a pencil.

Suppose b is a hyperpencil with center l. Let D and E be the midpoints of $\overline{BC}$ and $\overline{AC}$, respectively. A, B, C, D, E are off l since a triangle has at most one right angle. Let A', B', C', D', E' be the feet of the perpendiculars to l from A, B, C, D, E, respectively. (See Figure 27.4.) Again, since a triangle has at most one right angle, A', B', C', D', E'

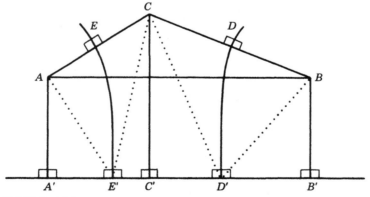

FIGURE 27.4

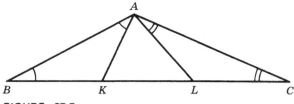

FIGURE 27.5

are distinct. $\overleftrightarrow{AA'}, \overleftrightarrow{BB'}, \overleftrightarrow{CC'}, \overleftrightarrow{DD'}, \overleftrightarrow{EE'}$ are in b. So $A-E-C$ and $C-D-B$ implies $A'-E'-C'$ and $C'-D'-B'$. Thus A is in int $(\angle A'E'E)$, B is in int $(\angle B'D'D)$, and C is in both int $(\angle C'E'E)$ and int $(\angle C'D'D)$. By SAS, $\triangle AEE' \simeq \triangle CEE'$ and $\triangle CDD' \simeq \triangle BDD'$. Then, by SAA, $\triangle AE'A' \simeq \triangle CE'C'$ and $\triangle CD'C' \simeq \triangle BD'B'$. So $AA'=CC'=BB'$. Hence $\boxed{S}\,A'ABB'$, and c, the perpendicular bisector of $\overline{AB}$, must be the perpendicular bisector of $\overline{A'B'}$. Since $l=\overleftrightarrow{A'B'}$, we have $c \perp l$. Therefore, c is in b if b is a hyperpencil.

Finally, suppose b is a horopencil. By the previous cases, c must be horoparallel to both a and b. To show c is in b, it is sufficient to show a, b, c have a common transversal (Theorem 27.5). Suppose $\overline{BC}$ is a longest side of $\triangle ABC$. Then $m\angle A \gtrsim m\angle B$ and $m\angle A \gtrsim m\angle C$. So there exist points K and L on $\overline{BC}$ such that $\angle BAK \simeq \angle B$ and $\angle CAL \simeq \angle C$. (See Figure 27.5.) It follows that K is on the perpendicular bisector of $\overline{AB}$ and that L is on the perpendicular bisector of $\overline{AC}$. Hence $\overline{BC}$ intersects a, b, and c. Therefore, c is in b if b is a horopencil. ∎

27.2 CYCLES

Each brush can be used to define a particular relation on the set of all points.

DEFINITION 27.7 If P and Q are points, then P *is equivalent to Q with respect to brush* b if there is a line l in b such that $\rho_l P = Q$. That point P is equivalent to point Q with respect to a given brush is written $P \sim Q$.

The following restatement of the definition is designated as a theorem for emphasis.

Theorem 27.8 If P and Q are two distinct points, then $P \sim Q$ with respect to brush b iff the perpendicular bisector of $\overline{PQ}$ is in b. $P \sim P$ with respect to brush b for any point P.

Proof Follows from the definition of a reflection (Definition 19.4). ■

Theorem 27.9 If P, Q, R are three distinct points on line c such that $P \sim Q$ and $Q \sim R$ with respect to brush b, then b is the hyperpencil with center c and $P \sim R$ with respect to b.

Proof Let l, m, n be the perpendicular bisectors of $\overline{PQ}$, $\overline{QR}$, $\overline{PR}$, respectively. Since P, Q, R are distinct and on c, then l, m, n are distinct lines perpendicular to c. Since $P \sim Q$ with respect to brush b, then l is in b. Since $Q \sim R$ with respect to brush b, then m is in b. Hence b is the hyperpencil with center c (Theorem 27.2). Further, $P \sim R$ with respect to b because n is in b. ■

Theorem 27.10 Equivalence of points with respect to a given brush is an equivalence relation on the set of all points.

Proof Equivalence of points with respect to a given brush b is reflexive and symmetric (Theorem 27.8). We need to show that if P, Q, R are distinct points such that $P \sim Q$ and $Q \sim R$ with respect to b, then $P \sim R$ with respect to b. If P, Q, R are collinear, then this transitivity is given by the preceding theorem. If $\triangle PQR$, then the perpendicular bisectors of $\overline{PQ}$ and $\overline{QR}$ are in b (Theorem 27.8). Hence the perpendicular bisector of $\overline{PR}$ is in b because the perpendicular bisectors of the sides of a triangle are in one brush (Theorem 27.6). Therefore, $P \sim R$ with respect to b. ■

Theorem 27.11 If l and m are in the horopencil b and point P is on l, then there exists a unique point Q on m such that $P \sim Q$ with respect to b.

Proof If $l = m$, then $P \sim Q$ iff $P = Q$ since the perpendicular bisector of a segment on a line of b cannot be in b. (Two lines of a horopencil are parallel and cannot be perpendicular.) Suppose $l \neq m$. By the hypothesis we may suppose $l = \overleftrightarrow{PA}$, $m = \overleftrightarrow{QD}$, and $\llcorner\!\lrcorner APQD$ is closed. We may further suppose $\llcorner\!\lrcorner APQD$ is isosceles since a closed biangle with vertex P is equivalent to an isosceles closed biangle with vertex P (Theorem 24.3). Hence the perpendicular bisector of $\overline{PQ}$ is in b (Theorem 27.4). So $P \sim Q$ with respect to b. Assume Q_2 is a second point on m corresponding to P. Then $\triangle PQQ_2$ with the perpendicular bisector of $\overline{PQ_2}$ in b. Since the perpendicular bisectors of the sides of a triangle are in one brush (Theorem 27.6), we have the contradiction that the perpendicular bisector of $\overline{QQ_2}$ is in horopencil b and perpendicular to a line of b. Therefore, Q is the unique point on m such that $P \sim Q$ with respect to horopencil b. ■

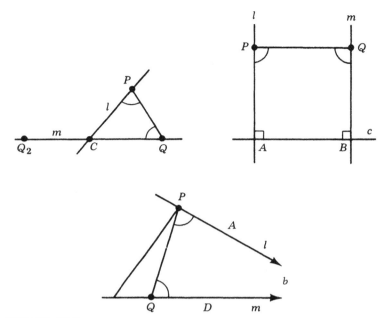

FIGURE 27.6

Theorem 27.12 If $\overleftrightarrow{AP}$ and $\overleftrightarrow{BQ}$ are in the hyperpencil b with center $\overleftrightarrow{AB}$, then $P \sim Q$ with respect to b iff $\boxed{S}\, APQB$. If l and m are in the hyperpencil b and point P is on l, then there exists a unique point Q on m such that $P \sim Q$ with respect to b.

Proof Suppose $\overleftrightarrow{AP}$ and $\overleftrightarrow{BQ}$ are in the hyperpencil b with center $\overleftrightarrow{AB}$ and $P \sim Q$ with respect to b. Then the perpendicular bisector of $\overline{PQ}$ is perpendicular to $\overleftrightarrow{AB}$ (Theorem 27.8), say at point N. Let M be the midpoint of $\overline{PQ}$. Then $A-N-B$ follows from $P-M-Q$. Further, $\boxed{L}\, NMPA$ and $\boxed{L}\, NMQB$ implies $\boxed{S}\, APQB$. Conversely, if $\boxed{S}\, APQB$, then $P \sim Q$ with respect to b since the perpendicular bisector of the lower base of a Saccheri quadrilateral is the perpendicular bisector of the upper base. The second statement in the theorem follows from the first when l and m are distinct. When $l = m$, $P \sim Q$ iff $P = Q$ because the perpendicular bisector of a segment on a line cannot be hyperparallel to that line. ∎

Theorem 27.13 $P \sim Q$ with respect to the pencil with center C iff $CP = CQ$.

Proof If $P = C$, then $P \sim Q$ iff $Q = C$. If $P \neq C$, then $P \sim Q$ iff C is on the perpendicular bisector of $\overline{PQ}$ (Definition 27.7). Hence $P \sim Q$ iff $CP = CQ$ (Theorem 18.5). ∎

Let b be the pencil with center C. Then point C is equivalent only to itself with respect to b. Thus $\{C\}$ is an equivalence class (see Section 1.3) for the relation of equivalence with respect to b. This set, which contains just one point, could be considered a *degenerate circle*. The equivalence classes determined by b that contain a point off the center of b are the circles with center C.

Theorem 27.14 Let b be a pencil with center C. The equivalence classes for the relation of equivalence with respect to b are the concentric circles with center C and $\{C\}$.

Proof The theorem follows from the preceding theorem and the definition of a circle. ■

The first two statements in the next definition are analogous to the following: If P is a point off the center C of *pencil b* then the set of all points Q such that $P \sim Q$ with respect to b is a *circle* with *center C*.

DEFINITION 27.15 If P is a point off the center c of hyperpencil b then the set of all points Q such that $P \sim Q$ with respect to b is a **hypercircle** with **center** c. If P is a point and b is a horopencil, then the set of all points Q such that $P \sim Q$ with respect to b is a **horocircle** with **center** b. A **cycle** is any one of a circle, a hypercircle, or a horocircle. Cycles having the same center are **concentric**. A cycle and a brush having the same center are also said to be **concentric**. If A and B are distinct points on a cycle, then $\overline{AB}$ is a **chord** of the cycle.

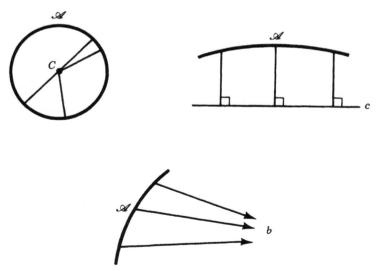

FIGURE 27.7

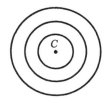 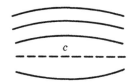 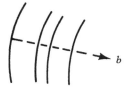

Concentric circles   Concentric hypercircles   Concentric horocircles

FIGURE 27.8

From Definition 27.15 we immediately have two theorems analogous to Theorem 27.14.

Theorem 27.16 Let b be a hyperpencil with center c. The equivalence classes for the relation of equivalence with respect to b are the concentric hypercircles with center c and the line c.

Theorem 27.17 Let b be a horopencil. The equivalence classes for the relation of equivalence with respect to b are the concentric horocircles with center b.

Suppose C is a point, c is a line, and b is a horopencil. Every point P with $P \neq C$ is on a unique circle with center C. Every point P off c is on a unique hypercircle with center c. Every point P is on a unique horocircle with center b. All these statements are contained in the following theorem.

Theorem 27.18 Every point off the center of a brush is on a unique cycle concentric with the brush.

Proof Equivalence classes are either identical or disjoint. (See Section 1.3.) ∎

Theorem 27.19 Three distinct noncollinear points are on a unique cycle. Two cycles intersect in at most two points. No three points of a cycle are collinear. A cycle is exactly one of a circle, a hypercircle, or a horocircle.

Proof Suppose P, Q, R are distinct points. We can have $P \sim Q$ and $Q \sim R$ with respect to one and only one brush since two lines determine a unique brush (Theorem 27.2) and the perpendicular bisectors of $\overline{PQ}$ and $\overline{QR}$ are in the brush (Theorem 27.8). Hence, if P, Q, R are not collinear, then the three points lie on a unique cycle (Theorem 27.18). If P, Q, R are on line c, then the brush is the unique hyperpencil with center c (Theorem 27.9), but c is not a cycle by definition. ∎

In Euclidean geometry a set of points is said to be *concyclic* if the points lie on one *circle*. Of course, any three noncollinear points are concyclic in the Euclidean plane. For the hyperbolic plane, we know that three noncollinear points do not necessarily lie on a *circle*. However, such points do lie on a *cycle*. So, in our geometry, any three noncollinear points are *concyclic* provided the word means that the points lie on a *cycle*. Because of the possible confusion, we shall avoid the word *concyclic* altogether in the theory.

Theorem 27.20 The locus of all points equidistant from line c and on one side of c is a hypercircle with center c.

Proof Restatement of Theorem 27.12. ∎

Because of Theorem 27.20, a hypercircle is often called an *equidistant curve*. A hypercircle is also called an *ultracircle* or a *hypercycle*. A horocircle is also called a *limiting curve*, a *critical circle*, or a *horocycle*.

27.3 EXERCISES

● **27.1** There exists a pentagon with five right angles.

● **27.2** Describe the brushes in the Cayley–Klein Model.

27.3 If l and m are lines and σ is an isometry, then σl and σm intersect iff l and m intersect, are hyperparallel iff l and m are hyperparallel, and are horoparallel iff l and m are horoparallel.

● **27.4** Any two horopencils have a unique line in common.

● **27.5** True or False?

(a) Two lines cannot have two common perpendiculars.

(b) Each of the set of all pencils, the set of all hyperpencils, the set of all horopencils, the set of all circles, the set of all hypercircles, and the set of all horocircles is fixed under any isometry.

(c) If $\sqcup ABCD$ is closed and isosceles, then $\Pi(\frac{1}{2}BC) = m\angle ABC$.

(d) A hyperpencil is not a pencil.

(e) A pencil is a brush, but a brush may not be a pencil.

(f) A line intersects a cycle in at most two points, and two cycles intersect in at most two points.

(g)   If l is a line and b is a brush, then $\rho_l b = b$ iff l is in b.

(h)   The perpendicular bisector of a chord of a cycle concentric with brush b is in b.

(i)   Two concentric cycles are disjoint.

(j)   If each of three lines is horoparallel to the other two, then the three lines are in a horopencil iff the three lines have a common transversal.

27.6   Let l and m be two hyperparallel lines. All the transversals to l and m that form congruent corresponding angles with l and m lie in a pencil.

27.7   Each of the three cases in the proof of Theorem 27.6 is possible.

27.8   If l and m are in the pencil b with center C and point P is on l with $P \neq C$, then there exist exactly two points on m that are equivalent to P with respect to b.

27.9   Let l and m be distinct lines in brush b. Let P and Q be points on l and m, respectively. Then, $P \sim Q$ with respect to b if $P = Q$ or $\overleftrightarrow{PQ}$ is a transversal to l and m such that the interior angles intersecting one side of $\overleftrightarrow{PQ}$ are congruent.

● **27.10**   Give the exact value of n for each of the following statements, but write "∞" for "an infinite number of."

(a)   If two lines intersect, there are n lines horoparallel to both.

(b)   If two lines are hyperparallel, there are n lines horoparallel to both.

(c)   If two lines are horoparallel, there are n lines horoparallel to both.

(d)   There are n circles through two distinct points.

(e)   There are n hypercircles through two distinct points.

(f)   There are n horocircles through two distinct points.

(g)   There are n lines common to two pencils.

(h)   There are n lines common to two hyperpencils whose centers are hyperparallel.

(i)   There are n lines common to two hyperpencils whose centers are not hyperparallel.

(j)   There are n lines common to two horopencils.

●**27.11** What is the intersection of two brushes?

27.12 Describe the brushes in the Poincaré Model.

27.13 If the vertices of $\square ABCD$ are on a cycle, then $m\angle A + m\angle C = m\angle B + m\angle D$.

27.14 Through any point P on a circle there is exactly one line that intersects the circle only at P. Through any point P on a hypercircle there are an infinite number of lines that intersect the hypercircle only at P.

27.15 Through any point P on a horocircle there are exactly two lines that intersect the horocycle only at P.

27.16 Suppose point P is on a cycle concentric with brush b. Let l be the line in b through P. Let t be the perpendicular to l at P. Then every point of the cycle except P is on the same side of t. Further, t is the unique line with this property.

27.17 Why would an etymologist prefer either *ultracircle* or *hypercycle* to *hypercircle*?

***27.18** Are the three *medians* of a triangle concurrent? Are the three *altitudes* of a triangle concurrent?

***27.19** Let $\overline{AB}$ be a longest side of $\triangle ABC$. Then A, B, C are on a circle, a horocircle, or a hypercircle iff $m\angle C$ is respectively less than, equal to, or greater than $\Pi(AC/2) + \Pi(BC/2)$.

***27.20** Find the cycles in the Poincaré Model.

GRAFFITI

The most frequently told story about János Bolyai concerns the succession of duels he fought with thirteen of his brother officers. As a consequence of some friction, these thirteen officers simultaneously challenged János, who accepted with the proviso that between duels he should be permitted to play a short piece on his violin. The concession granted, he vanquished in turn all thirteen of his opponents. What is seldom told is what happened very shortly after the batch of duels. János was promoted to a captaincy on the condition that he immediately retire with the pension assigned his new rank. The government felt bound to consult its interests, for it could hardly suffer the possibility of such an event recurring.

Howard Eves

*Hypercircles and horocircles have many properties in common
with circles. Whether one verifies the result (Exercise 27.20) or not,
it is interesting to see the cycles in the Poincaré Model.
Surprisingly, the circles are the Euclidean circles in the model, and
conversely. (The Euclidean center and the hyperbolic center coincide
only when both are at the origin.) Every cycle is a subset of a
Euclidean circle or of a Euclidean line. Concentric cycles are
illustrated in Figure 27.9.*

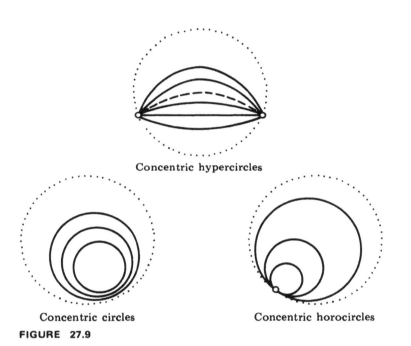

Concentric hypercircles

Concentric circles

Concentric horocircles

FIGURE 27.9

CHAPTER 28

Rotations, Translations, and Horolations

28.1 PRODUCTS OF TWO REFLECTIONS

In this chapter we consider products of reflections in lines from one brush. In particular, such isometries fix those cycles that are concentric with the brush. Check Definition 20.1 if you don't recall what a *line of symmetry* is.

Theorem 28.1 If cycle $\mathcal{A}$ and brush b are concentric, then l is a line of symmetry for $\mathcal{A}$ iff l is in b.

Proof If l is a line of symmetry for $\mathcal{A}$, then l is the perpendicular bisector of a chord of $\mathcal{A}$ and hence (Theorem 27.8) is in b. Conversely, if l is in b, then l is a line of symmetry for $\mathcal{A}$ by the definition of $\mathcal{A}$. ∎

Figure 28.1 should explain the statement of the next theorem. If b is a pencil with center C, then every line in b intersects a circle with center C at *two* points. (Compare this with Theorems 27.11 and 27.12.) This explains the necessity of some additional hypothesis in the theorem when b is a pencil.

Theorem 28.2 Let $\mathcal{A}$ and $\mathcal{B}$ be two cycles concentric with brush b. Let m and m' be two lines in b such that m intersects $\mathcal{A}$ and $\mathcal{B}$ at A and B, respectively, and m' intersects $\mathcal{A}$ and $\mathcal{B}$ at A' and B', respec-

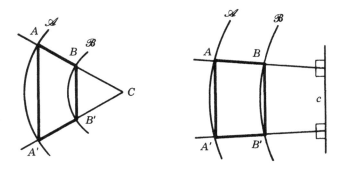

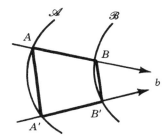

FIGURE 28.1

tively. If b is a pencil with center C, suppose $A-B-C$ and $A'-B'-C$. Then $AB=A'B'$ and the perpendicular bisector of $\overline{AA'}$ is the perpendicular bisector of $\overline{BB'}$.

Proof Let l be the perpendicular bisector of $\overline{AA'}$. So $\rho_l A=A'$, and l is in b (Theorem 27.8). Then, since l and m are in b with $m=\overset{\leftrightarrow}{AB}$, we have $\rho_l m$ is the line in b through A'. Thus $\rho_l m=m'$ and $\rho_l B$ is on m'. Further, since $\rho_l \mathscr{B}=\mathscr{B}$ by the preceding theorem, we have $\rho_l B$ is on $\mathscr{B}$. Hence $\rho_l B=B'$. Therefore, l is the perpendicular bisector of $\overline{BB'}$ and $AB=A'B'$. ∎

Complete familiarity with Theorem 19.10 will be assumed without further mention throughout our study of isometries. Let's recall the three important statements from this theorem: *Every isometry is a product of at most three reflections. If an isometry fixes a point, then the isometry is either a reflection or a product of two reflections. If an isometry fixes two points on line l then the isometry is either the reflection in l or the identity.*

If a is a line, then $\rho_a \rho_a$ is the identity. We now consider $\rho_b \rho_a$ where a and b are distinct lines. Remember $\rho_b \rho_a$ is ρ_a *followed by* ρ_b.

There are three cases depending on which type of brush is determined by a and b. The next definitions are crucial. Everything depends on knowing the meaning of the words we use.

DEFINITION 28.3 Let a and b be two distinct lines in brush b. If b is a pencil with center C, then $\rho_b\rho_a$ is a **rotation** with **center** C. If b is a hyperpencil with center c, then $\rho_b\rho_a$ is a **translation** with **center** c. If b is a horopencil with center b, then $\rho_b\rho_a$ is a **horolation** with **center** b.

If a and b are two intersecting lines, then $\rho_b\rho_a$ is a rotation. If a and b are two hyperparallel lines, then $\rho_b\rho_a$ is a translation. If a and b are two horoparallel lines, then $\rho_b\rho_a$ is a horolation. A translation may also be called a *hyperlation*. A horolation may also be called a *criticallation* or a *parallel displacement*. (Pronouncing the "h" in "horolation" is optional.) By our choice of definitions, the identity isometry is neither a rotation, a translation, nor a horolation.

At present all we know about rotations, translations, and horolations is their definitions! The terms *rotation* and *translation* are familiar. Do not fall into the intellectual trap of *assuming* these isometries have properties that are associated with the words in Euclidean geometry. Of course, they do or we would use different names! However we must *prove this*. Pedagogically it might be preferable to use

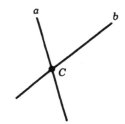

$\rho_b\rho_a$ is a rotation.

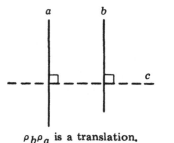

$\rho_b\rho_a$ is a translation.

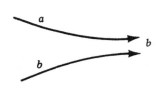

$\rho_b\rho_a$ is a horolation.

FIGURE 28.2

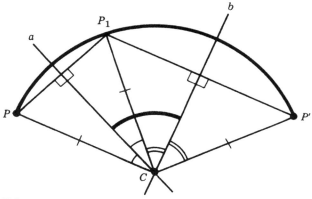

FIGURE 28.3

hyperlation instead of *translation*, to call *rotations* by some other name such as *pencillations*, and *then* to *discover* the properties of these mappings.

Let's see that our definitions of *rotation* and *translation* are reasonable in relation to the Euclidean plane. Suppose a and b are two lines intersecting at point C in the Euclidean plane. Let P be any point, $P_1 = \rho_a P$, and $P' = \rho_b(\rho_a P) = \rho_b \rho_a P$. From Figure 28.3 it might be seen that $\rho_b \rho_a$ *moves* point P about C through *twice the directed angle* from a to b. (Figure 28.3 is something of a hoax in that P is nicely situated unlike Figure 28.4.) In any case, $\rho_b \rho_a$ does what a *rotation* should do. How do you *define* a *rotation* for the Euclidean plane in the first place? *Now* you have a very nice definition, namely a *rotation* is the product of the two reflections in two intersecting lines.

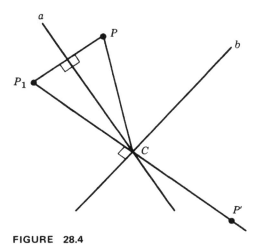

FIGURE 28.4

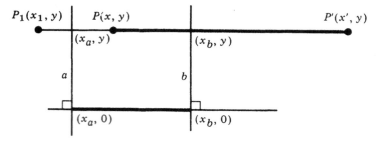

FIGURE 28.5

Suppose a and b are two lines perpendicular to line c in the Euclidean plane. We may suppose c is the x-axis without loss of generality. Let a intersect c at $(x_a, 0)$; let b intersect c at $(x_b, 0)$. Let $P = (x, y)$, $P_1 = \rho_a P = (x_1, y_1)$, and $P' = \rho_b \rho_a P = (x', y')$. So $y = y_1 = y'$. See Figure 28.5. Since (x_a, y) must be the midpoint of P and P_1 we have $x_a = \frac{1}{2}(x_1 + x)$. So $x_1 = 2x_a - x$. Since (x_b, y) must be the midpoint of P_1 and P', we have $x_b = \frac{1}{2}(x_1 + x')$. So $x' = 2x_b - x_1 = x + 2(x_b - x_a)$. Thus $\rho_b \rho_a$ sends (x, y) to $(x + h, y)$ where $h = 2(x_b - x_a)$. So $\rho_b \rho_a$ is the *translation* that *moves* any point through *twice the directed distance* from a to b. In Euclidean geometry a *translation* is the product of the two reflections in two parallel lines. This last statement is either a definition or a theorem of Euclidean geometry depending on how one chooses the definitions.

As the theory of isometries develops, you should keep an eye out to see which theorems are valid for the absolute plane and, hence, for the Euclidean plane. This way you will be learning about isometries for two geometries at the same time!

All of the discussion since Definition 28.3 has not advanced our *theory* one bit. We have only shown that our choice of terminology is not unreasonable. All we know about rotations, translations, and horolations is their definitions. The next theorem follows as an immediate consequence of the definitions.

Theorem 28.4 If b is a pencil with center C, then a rotation with center C or any product of reflections in lines of b is an isometry that fixes C, b, and every circle with center C. If b is a hyperpencil with center c, then a translation with center c or any product of reflections in lines of b is an isometry that fixes c, b, and every hypercircle with center c. If b is a horopencil, then a horolation with center b or any product of reflections in lines of b is an isometry that fixes b and every horocircle with center b.

28.2 REFLECTIONS IN LINES OF A BRUSH

If you think about it, you will realize that we do not know that rotations, translations, and horolations are necessarily distinct. An important method in studying mappings is to distinguish them by their fixed points and fixed lines. We shall use this approach in studying the isometries.

The first of the following algebraic techniques is used in the next theorem. $\rho_a P = \rho_b Q$ follows from $\rho_b \rho_a P = Q$ since $\rho_b Q = \rho_b(Q) = \rho_b(\rho_b\rho_a P) = \rho_b\rho_b\rho_a P = \rho_a P$. Note $\rho_b\rho_b = \iota$ and $\rho_b^{-1} = \rho_b$ since ρ_b is an involution. Further, if $\rho_b\rho_a = \sigma$, then $\rho_a = \rho_b\sigma$ follows from multiplying both sides of the equations $\rho_b\rho_a = \sigma$ by ρ_b on the *left* and $\rho_b = \sigma\rho_a$ follows from multiplying both sides of the equation $\rho_b\rho_a = \sigma$ by ρ_a on the *right*.

Theorem 28.5 A rotation fixes exactly one point, its center. Neither a translation nor a horolation fixes a point.

Proof Suppose a and b are distinct lines and $\rho_b\rho_a$ fixes point P. Then $\rho_b\rho_a P = P$. So $\rho_b P = \rho_b(\rho_b\rho_a P) = \rho_a P$. Let $P' = \rho_a P = \rho_b P$. We must have $P' = P$, as otherwise a and b are two distinct lines each of which is the perpendicular bisector of $\overline{PP'}$. Thus P is on both a and b. The theorem now follows from the definitions. ■

If we replace the second sentence of Theorem 28.5 by "A translation does not fix a point," then the theorem and its proof are valid for the Euclidean plane. The *center* of translation $\rho_b\rho_a$ in the Euclidean plane is the parallel pencil of all lines perpendicular to a and, hence, also to b. The Euclidean analogue for the next theorem states that the lines fixed by a translation are exactly those in its center.

Theorem 28.6 A translation fixes exactly one line, its center. A horolation does not fix a line.

Proof Let b be the unique brush containing distinct parallel lines a and b. Let $\sigma = \rho_b\rho_a$ and assume σ fixes line l. If b is a hyperpencil with center c, we also assume $l \neq c$ as we already know (Theorem 28.4) that σ fixes c. To prove the theorem we shall obtain a contradiction. Let P be a point on l, and let $\mathscr{A}$ be the unique cycle concentric with b that passes through P (Theorem 27.18). Let $P' = \sigma P$. Since σ fixes no point (Theorem 28.5) and since σ fixes both l and $\mathscr{A}$, we have $P' \neq P$ and P' is on both l and $\mathscr{A}$. For the same reasons, it follows that $\sigma P' = P$ as a line intersects a cycle in at most two points. Hence σ interchanges

P and P'. Therefore, σ fixes the midpoint of P and P' (Theorem 19.11), a contradiction (Theorem 28.5). ■

Theorem 28.7   A horolation fixes exactly one brush, its center.

Proof   Any isometry fixing a brush must fix the unique center of that brush. Let σ be a horolation. Since σ fixes no points or lines, then σ cannot fix a pencil or a hyperpencil. Further, σ cannot fix two horopencils since σ would then have to fix the unique line common to the two horopencils (Theorem 27.3). Hence, the only brush fixed by σ is the horopencil that is the center of σ. ■

Theorem 28.8   The product of the two reflections in two distinct lines is exactly one of the following: a rotation, a translation, a horolation. The center of the product of the two reflections in two distinct lines is unique.

Proof   Follows directly from the previous three theorems. ■

Theorem 28.9   An isometry fixes exactly one point iff the isometry is a rotation.

Proof   Suppose isometry σ fixes exactly one point. Since σ fixes at least one point, σ is a reflection or a product of two reflections. Since σ fixes at most one point, σ is neither the identity nor a reflection. Hence σ must be the product of two distinct reflections. Since σ can be neither a translation nor a horolation (Theorem 28.5), σ must be a rotation. Conversely, a rotation fixes exactly one point (Theorem 28.5). ■

Although the next theorem is only a summary of previous results (Theorem 19.10 and Theorem 28.9), the theorem is important because it gives a classification of *all* the isometries that have fixed points.

Theorem 28.10   If an isometry fixes a point, then the isometry is either a rotation, a reflection, or the identity.

In the phrase "a product of n reflections," the "n" refers to the number of terms in the product counting repetitions. For example, $\rho_a \rho_a$ is a product of two reflections, and $\rho_a \rho_b \rho_a$ is a product of three reflections. Of course, a product of m reflections could equal a product of n reflections when $m \neq n$. For example, $\rho_a \rho_a \rho_a = \rho_a$ but $3 \neq 1$.

Theorem 28.11   The product of two reflections is not a reflection.

Proof   Let a and b be lines. If $a = b$, then $\rho_b \rho_a$ fixes every point and is

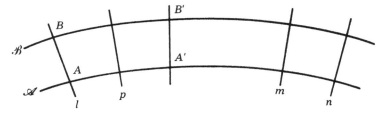

FIGURE 28.6

not a reflection. If $a \neq b$, then $\rho_b \rho_a$ fixes at most one point and is not a reflection. ■

Theorem 28.12 The product of three reflections in lines of a brush is a reflection in a line of that brush.

Proof Suppose l, m, n are lines in brush b. Let A and B be distinct points on l and off the center of b. Let $\mathscr{A}$ and $\mathscr{B}$ be the unique cycles concentric with b that pass through A and B, respectively. (See Figure 28.6.) Let $\sigma = \rho_n \rho_m \rho_l$, $A' = \sigma A$, and $B' = \sigma B$. If $A' \neq A$, let p be the perpendicular bisector of $\overline{AA'}$; if $A' = A$, let $p = l$. Since A' is on $\mathscr{A}$, we know p is in b in either case. Also, $\rho_p \sigma$ fixes A. Since $\rho_p \sigma$ fixes both A and b, then $\rho_p \sigma$ fixes l. If b is not a pencil, in which case l and $\mathscr{B}$ intersect only at B, then $\rho_p \sigma$ fixes B since $\rho_p \sigma$ fixes both l and $\mathscr{B}$. If b is a pencil with center C, then $\rho_p \sigma$ fixes the two points A and C. In any case, $\rho_p \sigma$ fixes two points on l. Hence, either $\rho_p \sigma = \rho_l$ or $\rho_p \sigma = \iota$. However, $\rho_p \sigma = \rho_l$ implies $\rho_p = \rho_n \rho_m$, contradicting the preceding theorem. Therefore, $\sigma = \rho_p$ with p in b. ■

For a proof of Theorem 28.12 in the Euclidean plane, the cycles in the proof above degenerate to lines when the brush is a parallel pencil.

Corollary 28.13 If l, m, n are lines in brush b then there exist unique lines p and q in b such that $\rho_m \rho_l = \rho_n \rho_p = \rho_q \rho_n$. Conversely, if $\rho_m \rho_l = \rho_n \rho_p$ or, equivalently, $\rho_n \rho_m \rho_l = \rho_p$, then the lines l, m, n, p are all in one brush.

Proof By the theorem there exist lines p and q in b such that $\rho_n \rho_m \rho_l = \rho_p$ and $\rho_m \rho_l \rho_n = \rho_q$. The lines p and q are necessarily unique since $\rho_a = \rho_b$ implies $a = b$. The first part of the corollary follows. Since two distinct brushes cannot have the same center, the second part is essentially a restatement of Theorem 28.8. ■

If σ_1 and σ_2 are isometries, what is the inverse of the isometry $\sigma_2 \sigma_1$? Since $(\sigma_2 \sigma_1)(\sigma_1^{-1} \sigma_2^{-1}) = \iota$ and $(\sigma_1^{-1} \sigma_2^{-1})(\sigma_2 \sigma_1) = \iota$, it follows that $(\sigma_2\sigma_1)^{-1} = \sigma_1^{-1}\sigma_2^{-1}$. Likewise, $(\sigma_3\sigma_2\sigma_1)(\sigma_1^{-1}\sigma_2^{-1}\sigma_3^{-1}) = \iota$ implies $(\sigma_3\sigma_2\sigma_1)^{-1} = \sigma_1^{-1}\sigma_2^{-1}\sigma_3^{-1}$. In fact, for any group, it follows that *the inverse of a product is the product of the inverses in reverse order.* Combining this fact with Theorem 28.12, we can prove the next theorem.

Theorem 28.14 If l, m, n are lines in a brush, then $\rho_l\rho_m\rho_n = \rho_n\rho_m\rho_l$.

Proof There exists a line p such that $\rho_n\rho_m\rho_l = \rho_p$. Then $\rho_n\rho_m\rho_l = \rho_p = \rho_p^{-1} = (\rho_n\rho_m\rho_l)^{-1} = \rho_l^{-1}\rho_m^{-1}\rho_n^{-1} = \rho_l\rho_m\rho_n$. ∎

Theorem 28.15 An isometry fixes a cycle concentric with brush b iff the isometry is a product of reflections in lines of b.

Proof We already know that any product of reflections in lines of a brush fixes the cycles concentric with that brush (Theorem 28.4). Conversely, suppose isometry σ fixes cycle $\mathscr{A}$. Let A, B, C be three points on $\mathscr{A}$. Then $\triangle ABC$. (The (nondegenerate) cycles in the Euclidean plane are the circles.) Let $D = \sigma A$, $E = \sigma B$, and $F = \sigma C$. Then D, E, F are on $\mathscr{A}$, and $\triangle ABC \simeq \triangle DEF$. Let $\sigma = \sigma_3\sigma_2\sigma_1$ where $\sigma_1, \sigma_2, \sigma_3$ are as in the proof of Theorem 19.9, which showed that σ is the unique isometry such that $\sigma A = D$, $\sigma B = E$, and $\sigma C = F$. Following through that proof (which we shall not repeat here), we see that each of $\sigma_1, \sigma_2, \sigma_3$ is either the identity or the reflection in some line of b, since the perpendicular bisector of any chord of $\mathscr{A}$ is in b (Theorem 27.8) and a line of symmetry for $\mathscr{A}$ (Theorem 28.1). Thus σ is a product of reflections in lines of b. ∎

28.3 EXERCISES

● **28.1** For each of the theorems in Section 28.2, state the analogous theorem for the Euclidean plane.

● **28.2** If σ is a bijection on the set of all points and preserves angle measure, then σ is an isometry.

● **28.3** Every rotation is the product of two translations, and every horolation is the product of two translations.

● **28.4** Let P and P' be distinct points. Then there exist an infinite number of rotations taking P to P', there exist an infinite number of translations taking P to P', but there exist exactly two horolations taking P to P'.

● **28.5** True or False?

    (a) If S is a set of points, then l is a line of symmetry for S iff $\rho_l S \subset S$.

(b)   A rotation with center C may be expressed as $\rho_b\rho_a$ where either one of the lines a or b is an arbitrarily chosen line through C and the other is then uniquely determined.

(c)   A translation with center c may be expressed as $\rho_b\rho_a$ where either one of the lines a or b is an arbitrarily chosen line perpendicular to c and the other is then uniquely determined.

(d)   A horolation with center b may be expressed as $\rho_b\rho_a$ where either one of the lines a or b is an arbitrarily chosen line in b and the other is then uniquely determined.

(e)   The perpendicular bisector of a chord of a cycle is a line of symmetry for the cycle.

(f)   A rotation is determined by a point and its image.

(g)   A translation is determined by a point and its image.

(h)   A horolation is determined by a point and its image.

(i)   If a and b are two lines perpendicular to line c, then $\rho_c\rho_b\rho_a$ is a translation.

(j)   A translation fixes a line pointwise but a horolation does not fix a line.

28.6   If a rotation with center C takes point P to Q with $P \neq Q$, then the perpendicular bisector of $\overline{PQ}$ is in the pencil with center C. If a translation with center c takes point P to Q, then the perpendicular bisector of $\overline{PQ}$ is in the hyperpencil with center c. If a horolation with center b takes point P to Q, then the perpendicular bisector of $\overline{PQ}$ is in b.

28.7   Each of a rotation, a translation, and a horolation is determined by two points and their images.

28.8   Every translation is a product of two rotations, and every horolation is a product of two rotations.

28.9   Every rotation is a product of two horolations, and every translation is a product of two horolations.

28.10   Possibly $\mathscr{A} = \mathscr{B}$ in the proof of Theorem 28.12.

28.11   Why does a mirror interchange *right* and *left* but not *above* and *below?*

28.12   The rotations with a given center together with the identity isometry form a group.

28.13 The translations with a given center together with the identity form a group. The horolations with a given center together with the identity form a group.

28.14 Find the equivalence classes for the relation on the set of capital roman letters where letters are equivalent iff they have the same number of lines of symmetry. (ABCDEFGHIJKLMNOPQRSTU VWXYZ)

28.15 Let l, m, and n be the perpendicular bisectors of $\overline{AB}$, $\overline{BC}$, and $\overline{AC}$, respectively. Then $\rho_n \rho_m \rho_l$ is the reflection in a line through A.

28.16 If lines a, b, c are not in a brush, then $\rho_c \rho_b \rho_a$ is not a reflection.

28.17 How is a *point of symmetry* defined?

*28.18 Find the necessary and sufficient conditions for the product of two translations to be a translation or the identity.

*28.19 Converse of Theorem 28.14.

*28.20 What are the involutory isometries?

GRAFFITI

The meridians of a globe pass through the north pole and are perpendicular to the equator. From this simple observation we can deduce some facts about the Riemann plane. We suppose we are now considering elliptic geometry. Since there exists a unique line perpendicular to all the lines of a given pencil, every pencil is a hyperpencil. Conversely, since the perpendiculars to a given line are concurrent, every hyperpencil is a pencil. In the elliptic plane, every brush is a pencil. *Every rotation fixes at least one point and at least one line. The involutory rotation with center* P *is called the halfturn about* P *and is denoted by* η_P. *(Such rotations "of 180 degrees" are studied in Chapter 29.) If all the lines perpendicular to line* p *pass through point* P, *then* $\rho_p = \eta_P$. *Every isometry is the product of two halfturns; every isometry is the product of two reflections.* In the elliptic plane, every nonidentity isometry is a rotation.

The Classification of Isometries

29.1 INVOLUTIONS

Before continuing with our results on isometries that apply to the absolute plane and, hence, to the Euclidean plane as well as the Bolyai – Lobachevsky plane, we have a definition and a theorem that are relevant only to the hyperbolic plane. These complement the absolute theorem that two circles are congruent iff they have the same radius (distance).

DEFINITION 29.1 If P is a point on hypercircle $\mathscr{A}$ with center c and $\overline{PQ}$ is perpendicular to c at Q, then $\overline{PQ}$ is a *radius* of $\mathscr{A}$ and PQ is the *radius* of $\mathscr{A}$.

Theorem 29.2 Two hypercircles are congruent iff they have the same radius (distance). However, any two horocircles are congruent.

Proof We leave the proof of the first statement as Exercise 29.1. So suppose $\mathscr{A}_1$ is the horocircle through point A and with center b_1, the horopencil of all lines containing a ray horoparallel to $\overrightarrow{AB}$. Suppose $\mathscr{A}_2$ is the horocircle through point C and with center b_2, the horopencil of all lines containing a ray horoparallel to $\overrightarrow{CD}$. Let σ be either one of the two isometries such that $\sigma(\overrightarrow{AB}) = \overrightarrow{CD}$. It follows that $\sigma(A) = C$ and $\sigma(b_1) = b_2$. Thus $\sigma(\mathscr{A}_1)$ is the horocircle through C and with center b_2 (Theorem 27.18). Therefore, $\sigma(\mathscr{A}_1) = \mathscr{A}_2$, as desired. ∎

An isometry that is a product of an even number of reflections is said to be *even,* and an isometry that is a product of an odd number of reflections is said to be *odd.* There is an intellectual trap here that is used to advantage by politicians and the advertising industry. If a thing is called by a familiar name, it is fairly natural to assume the thing has the properties we otherwise associate with the name. Although admittedly natural, this is not logical. The definition above of *even* and *odd* for isometries is given in terms of *even* and *odd* as applied to the integers. Although an integer cannot be both even and odd, it would be a logical error to assume without proof that an isometry cannot be both even and odd.

DEFINITION 29.3　An isometry that is a product of an even number of reflections is **even;** an isometry that is a product of an odd number of reflections is **odd.**

The even isometries are also called *proper, positive,* or *direct.* The odd isometries are also called *improper, negative, indirect,* or *opposite.*

The trick of replacing $\rho_m \rho_l$ by $\rho_m \rho_n \rho_n \rho_l$ is used so often in studying isometries that this *trick* should be regarded as a *method,* analogous to the method of multiplying by 1 in some particular form that is used so often in elementary algebra. The next proof depends on using this trick twice. When reading the proof from top to bottom it seems that the lines l, m, n appear out of the sky for no good reason. However, if you look at the proof from bottom to top, you will see that the mysterious lines have been purposefully selected.

Theorem 29.4　A product of four reflections is equal to a product of two reflections.

Proof　We wish to show $\rho_d \rho_c \rho_b \rho_a$ is a product of two reflections. Let P be any point on line a. Let l be a line through P in a brush containing b and c. (Line l is unique if b and c are two lines not both on P. See Figure 29.1.) Since l, b, c are in a brush, there is a line m such that $\rho_c \rho_b \rho_l = \rho_m$ (Theorem 28.12). Let n be a line through P in a brush containing d and m. Now

$$\rho_d \rho_c \rho_b \rho_a = \rho_d (\rho_c \rho_b \rho_l) \rho_l \rho_a$$

$$= \rho_d \rho_m \rho_l \rho_a$$

$$= (\rho_d \rho_m \rho_n)(\rho_n \rho_l \rho_a).$$

Since a, l, n are in the pencil with center P, then $\rho_n \rho_l \rho_a$ is a reflection. Since n, m, d are in one brush, then $\rho_d \rho_m \rho_n$ is also a reflection. Therefore, $\rho_d \rho_c \rho_b \rho_a$ is equal to a product of two reflections. ■

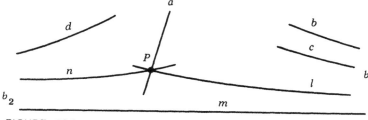

FIGURE 29.1

Theorem 29.5 Neither a reflection nor a product of three reflections is equal to a product of two reflections.

Proof We already know a reflection is not equal to a product of two reflections (Theorem 28.11). Assume $\rho_c\rho_b\rho_a = \rho_m\rho_l$. Then $\rho_m\rho_c\rho_b\rho_a = \rho_l$. Thus, by the preceding theorem, ρ_l is a product of two reflections, a contradiction (Theorem 28.11). ∎

Theorem 29.6 An even isometry is a product of two reflections. An odd isometry is a reflection or a product of three reflections. No isometry is both even and odd.

Proof The statements follow from the preceding two theorems, since a product of n reflections is equal to a product of $n-2$ reflections whenever $n \geqq 4$. ∎

Corollary 29.7 An even isometry is exactly one of the following: the identity, a rotation, a translation, or a horolation.

Proof Restatement of Theorem 28.8 ∎

For the Euclidean plane, an even isometry is exactly one of the identity, a rotation, or a translation.

The next theorem introduces the handy equation $\rho_{\sigma a} = \sigma\rho_a\sigma^{-1}$

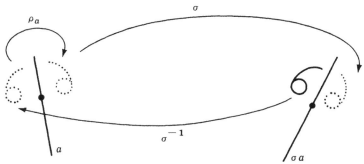

FIGURE 29.2

for line a and isometry σ. Let's see what this equation says. Since σ is an isometry and a is a line, then σa is a line and $\rho_{\sigma a}$ is simply the reflection in line σa. Thus the equation tells us that reflecting in line σa is the same thing as first *undoing* σ, then reflecting in line a, and finally *doing* σ. Perhaps following the "6" in Figure 29.2 will give you a feeling that the equation at least makes sense.

Theorem 29.8 If σ is an isometry and a is a line, then $\rho_{\sigma a} = \sigma \rho_a \sigma^{-1}$.

Proof Isometry $\sigma \rho_a \sigma^{-1}$ is not the identity because $\rho_a \neq \iota$. Suppose P' is any point on σa and $\sigma P = P'$. Then P must be on a. So $\sigma \rho_a \sigma^{-1} P' = \sigma \rho_a P = \sigma P = P'$. Since $\sigma \rho_a \sigma^{-1}$ is not the identity but fixes line σa pointwise, it follows (Theorem 28.10) that $\sigma \rho_a \sigma^{-1}$ is the reflection in σa. ∎

The equation above and Theorem 28.14 are used to prove a theorem that gives further insight into Corollary 28.13. Check back to see what Corollary 28.13 and Theorem 28.14 are. Also, recall that $(\rho_m \rho_l)^{-1} = \rho_l \rho_m$.

Theorem 29.9 If σ is a product of an even number of reflections in lines of brush b such that $\sigma a = c$ and $\sigma b = d$ for lines a and b in b, then $\rho_b \rho_a = \rho_d \rho_c$.

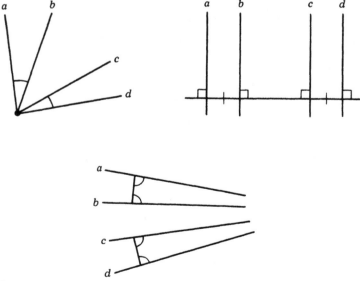

FIGURE 29.3

Proof We may suppose that $\sigma = \rho_m \rho_l$ where l and m are lines in b (Theorem 29.6). By the previous theorem, $\rho_c = \rho_{\sigma a} = \sigma \rho_a \sigma^{-1} = \rho_m \rho_l \rho_a \rho_l \rho_m$ and $\rho_d = \rho_m \rho_l \rho_b \rho_l \rho_m$. Thus, we have (Theorem 28.14)

$$\rho_d \rho_c = (\rho_m \rho_l \rho_b \rho_l \rho_m)(\rho_m \rho_l \rho_a \rho_l \rho_m)$$

$$= (\rho_m \rho_l \rho_b)(\rho_a \rho_l \rho_m)$$

$$= (\rho_b \rho_l \rho_m)(\rho_m \rho_l \rho_a)$$

$$= \rho_b \rho_a. \blacksquare$$

When do two reflections commute? Which even isometries are involutions? These questions are answered by our next theorem. Recall (Theorem 19.6) that $a = \rho_b a$ for line a iff $a = b$ or $a \perp b$.

Theorem 29.10 For lines a and b, $\rho_b \rho_a = \rho_a \rho_b$ iff $a = b$ or $a \perp b$. Further, $\rho_b \rho_a$ is an involution iff lines a and b are perpendicular.

Proof Suppose $a = b$ or $a \perp b$. Then $a = \rho_b a$. Let $\sigma = \rho_b$. So $\rho_a = \rho_{\sigma a} = \sigma \rho_a \sigma^{-1} = \rho_b \rho_a \rho_b$ (Theorem 29.8). Hence $\rho_b \rho_a = \rho_a \rho_b$. Conversely, if $\rho_b \rho_a = \rho_a \rho_b$, then $\rho_a = \rho_b \rho_a \rho_b = \rho_{\sigma a}$ where $\sigma = \rho_b$. So $a = \sigma a = \rho_b a$. Hence $a = b$ or $a \perp b$. Finally, by definition, $\rho_b \rho_a$ is an involution iff $\rho_b \rho_a = (\rho_b \rho_a)^{-1} = \rho_a \rho_b \neq \iota$. Therefore, $\rho_b \rho_a$ is an involution iff $a \perp b$. $\blacksquare$

DEFINITION 29.11 A *halfturn* about point P is an involutory rotation with center P. η_P always denotes a halfturn about P.

Theorem 29.12 The halfturn about point P is unique and is the product of the two reflections in any two perpendicular lines through P. For any point A, the midpoint of A and $\eta_P A$ is P.

Proof Since $\rho_b \rho_a$ is an involution iff lines a and b are perpendicular, the rest of the theorem follows if we prove the last statement. Let η_P be a halfturn about P. So $\eta_P P = P$ by definition. Suppose point A is distinct from P and $l = \overleftrightarrow{AP}$. Then there exists line m through P such that $\eta_P = \rho_m \rho_l$ (Corollary 28.13) and $l \perp m$ (Theorem 29.10). Hence $\eta_P A = \rho_m \rho_l A = \rho_m A$. So m is the perpendicular bisector of the segment with endpoints A and $\eta_P A$. Therefore the midpoint of A and $\eta_P A$ is P. $\blacksquare$

We could have used the last statement in Theorem 29.12 to *define* a halfturn and then proved Definition 29.11 as a theorem along with the rest of Theorem 29.12. In this case it would be natural to call a halfturn about P the *reflection in point P*. We shall not use this language to avoid possible confusion between a *reflection* and a *reflection in a point*.

Since we have certainly not considered all the possible isometries, it is a pleasant surprise to find that we do know all the involutory isometries.

Theorem 29.13 An isometry is an involution iff the isometry is a reflection or a halfturn.

Proof The identity isometry is not an involution by definition. Suppose σ is an involutory isometry that does not fix some point A. Let $\sigma A = B$. Since $\sigma A = B$ and $\sigma B = \sigma^2 A = A$, we have $\sigma M = M$ where M is the midpoint of $\overline{AB}$ (Theorem 19.11). If σ is not a reflection, then σ must be a rotation with center M (Theorem 28.10). In this case, since σ is an involution, $\sigma = \eta_M$. Therefore an involutory isometry is either a reflection or a halfturn. Conversely, reflections and halfturns are involutions by definition. ∎

A halfturn is the product of the two reflections in two perpendicular lines. What is the product of two halfturns? The answer is not at all clear from looking at Figure 29.4. However, a glance at Figure 29.5 should make everything crystal clear.

Theorem 29.14 The product of two distinct halfturns is a translation. Conversely, every translation is the product of two halfturns.

Proof Given halfturns η_A and η_B with $A \neq B$, let $l = \overleftrightarrow{AB}$, a be the perpendicular to l at A, and b be the perpendicular to l at B. So a and b are distinct and hyperparallel. Further, $\eta_B \eta_A = (\rho_b \rho_l)(\rho_l \rho_a) = \rho_b \rho_a$. Therefore, $\eta_B \eta_A$ is a translation. Conversely, if $\rho_b \rho_a$ is a translation, then a and b are distinct hyperparallel lines having a common perpendicular l. Let l intersect a at A and intersect b at B. Then $\rho_b \rho_a = \rho_b \rho_l \rho_l \rho_a = \eta_B \eta_A$. ∎

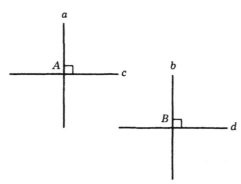

FIGURE 29.4

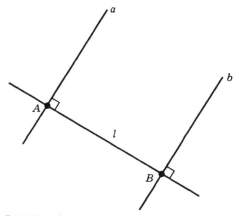

FIGURE 29.5

Our next theorem does not hold in the absolute plane. In the Euclidean plane, the product of three halfturns is always a halfturn, and, if A, B, C are not collinear and $\eta_C \eta_B \eta_A = \eta_D$, then $\square ABCD$ is a parallelogram. It follows that any product of halfturns in the Euclidean plane is the identity, a translation, or a halfturn. We have a different result for the Bolyai–Lobachevsky plane.

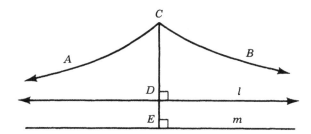

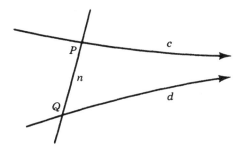

FIGURE 29.6

Theorem 29.15 Every even isometry is a product of halfturns.

Proof We know every translation is a product of two halfturns. The theorem follows if we show every rotation is a product of translations and every horolation is a product of rotations. Suppose $\rho_b\rho_a$ is a rotation with center C where $a = \overset{\leftrightarrow}{CA}$ and $b = \overset{\leftrightarrow}{CB}$. Let $\overset{\rightarrow}{CD}$ be the angle bisector of $\angle ACB$ and $\Pi(CD) = \tfrac{1}{2}m\angle ACB$. (See Figure 29.6.) Let E be such that $C-D-E$. Let l and m be the perpendiculars to $\overset{\leftrightarrow}{CD}$ at D and E, respectively. Then m is parallel to both a and b. Further, m is not horoparallel to either a or b since $\Pi(CE) < \tfrac{1}{2}m\angle ACB$. Thus m is hyperparallel to both a and b. Hence $\rho_b\rho_m$ and $\rho_m\rho_a$ are both translations, and $\rho_b\rho_a = (\rho_b\rho_m)(\rho_m\rho_a)$, as desired. Now suppose $\rho_d\rho_c$ is a horolation. Let P be any point on c, Q any point on d, and $n = \overset{\leftrightarrow}{PQ}$. Then $\rho_d\rho_n$ and $\rho_n\rho_c$ are both rotations, and $\rho_d\rho_c = (\rho_d\rho_n)(\rho_n\rho_c)$. ∎

29.2 THE CLASSIFICATION THEOREM

We know that every isometry in the absolute plane is a product of at most three reflections. There remains to consider $\rho_c\rho_b\rho_a$ where a, b, c are three lines not in one brush. Lines a and b might intersect, might be hyperparallel, or might be horoparallel. Suppose a and b are in the hyperpencil with center l. Now c can intersect a, be hyperparallel to a, or be horoparallel to a. Likewise, c can intersect b, be hyperparallel to b, or be horoparallel to b. All possible combinations have to be considered. Further, we have not even considered the relation between c and l. Even if there are not a "thousand" cases, it seems that there are so many that the task ahead is overwhelming. Undaunted, we begin with a very special case where a and b are in the hyperpencil with center c.

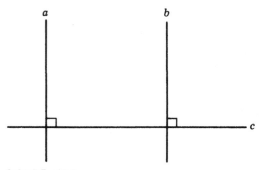

FIGURE 29.7

DEFINITION 29.16 If a and b are distinct lines perpendicular to line c, then $\rho_c \rho_b \rho_a$ is a *glide reflection* with *center c*.

To see where the name comes from, write the glide reflection above as $\rho_c(\rho_b \rho_a)$. The translation $\rho_b \rho_a$ contributes the "glide" while ρ_c contributes the "reflection." Figure 29.8 might suggest a common glide reflection. The next theorem says the "gliding" and the "reflecting" may be done in either order.

Theorem 29.17 $\rho_c(\rho_b \rho_a) = (\rho_b \rho_a)\rho_c$ if c is a common perpendicular to lines a and b.

Proof Since c is perpendicular to both a and b, then ρ_c commutes with both ρ_a and ρ_b (Theorem 29.10). Hence $\rho_c \rho_b \rho_a = \rho_b \rho_c \rho_a = \rho_b \rho_a \rho_c$. ∎

Some authors allow a reflection to be a special case $(a = b)$ of a glide reflection. We have not done so.

Theorem 29.18 The center of a glide reflection is unique. A glide reflection fixes exactly one line, its center. A glide reflection is not a reflection.

Proof Suppose $\sigma = \rho_c \rho_b \rho_a$ where c is perpendicular to the two lines a and b. Each of ρ_a, ρ_b, and ρ_c fixes c. So σ fixes c. If l is a line parallel to c, then $\sigma l \neq l$ because l and σl are on opposite sides of c. If line l intersects c exactly once, say at point P, then $\sigma l \neq l$ as $\sigma P = \rho_b \rho_a P \neq P$ (Theorem 28.5) but $\sigma c = c$. Hence σ fixes exactly the one line c. The rest of the statements in the theorem follow from this. ∎

Theorem 29.19 A glide reflection is a reflection in some line l followed by a halfturn whose center is off l. A glide reflection is a halfturn about some point P followed by a reflection in a line off P. Conversely, if $\sigma = \eta_P \rho_l$ where point P is off line l, then σ and σ^{-1} are two glide reflections and $\sigma^{-1} = \rho_l \eta_P$.

Proof Suppose σ is a glide reflection. Then there exist three lines a, b, c such that $\sigma = \rho_c \rho_b \rho_a$ where $c \perp a$ and $c \perp b$. Let c intersect a and b at A and B, respectively. (See Figure 29.9.) Since $b \perp c$, we have $\sigma = \rho_b \rho_c \rho_a$ (Theorem 29.10). Thus $\sigma = (\rho_b \rho_c)\rho_a = \eta_B \rho_a$ and $\sigma = \rho_b(\rho_c \rho_a) = \rho_b \eta_A$, proving the first two statements of the theorem.

FIGURE 29.8

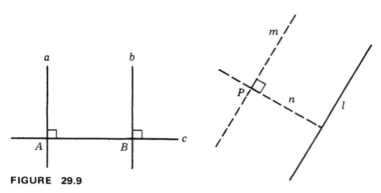

FIGURE 29.9

Now suppose point P is off line l and $\sigma = \eta_P \rho_l$. Let n be the perpendicular from P to l, and let m be the perpendicular to n at P. Then n is the common perpendicular to the two lines l and m. So $\sigma = \eta_P \rho_l = \rho_n \rho_m \rho_l$ and $\sigma^{-1} = \rho_l \eta_P = \rho_l \rho_m \rho_n$. Hence σ and σ^{-1} are glide reflections. Since glide reflection σ is an odd isometry that is not a reflection, then σ is not an involution (Theorem 29.13). Therefore $\sigma \neq \sigma^{-1}$. ■

Let's take a peek at one of the remaining "999" cases mentioned at the beginning of this section. Suppose two lines m and c are on point P but line l is not, as in Figure 29.10. Consider $\rho_c \rho_m \rho_l$. Letting p be the perpendicular from P to l, there is a line q through P such that $\rho_c \rho_m = \rho_q \rho_p$. (See Corollary 28.13.) So $\rho_c \rho_m \rho_l = \rho_q \rho_p \rho_l = \rho_q \eta_F$. We recognize $\rho_q \eta_F$ is a glide reflection. This observation says we can replace "halfturn" by "rotation" in Theorem 29.19. That's interesting, but there is a much more important consequence.

Given $\rho_c \rho_b \rho_a$ with a, b, c not in a brush, we can always replace $\rho_b \rho_a$ by $\rho_m \rho_l$ where m is chosen to intersect c as in Figure 29.11. (Corollary 28.13 again.) Thus we are back to considering $\rho_c \rho_m \rho_l$ as in the last paragraph. In other words, when we put all this together we should have a proof that $\rho_c \rho_b \rho_a$ is a glide reflection! The "999" cases that seemed so overwhelming will just melt away.

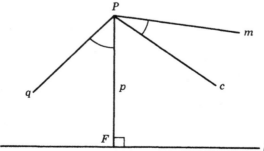

FIGURE 29.10

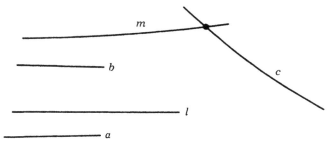

FIGURE 29.11

Theorem 29.20 Any product of the three reflections in three lines not in a brush is a glide reflection.

Proof Suppose $\sigma = \rho_c \rho_b \rho_a$ where a, b, c are not in a brush. Let b be the unique brush containing a and b. Let P be any point on c, and let p be the pencil with center P. Let m be the unique line common to b and p (i.e., the line in b through P). We have $m \neq c$ as c is not in b. (See Figure 29.12.) Since a, b, m are lines in b, there is a line l in b such that $\rho_m \rho_l = \rho_b \rho_a$ with $l \neq m$. Let p be in p and perpendicular to l at point F. We have $F \neq P$ as l is not in p. Since m, c, p are in p, there is a line q in p such that $\rho_c \rho_m = \rho_q \rho_p$ with $p \neq q$. Since q is on P and $q \neq p$, we have F is off q. Now

$$\sigma = \rho_c \rho_b \rho_a = \rho_c \rho_m \rho_l = \rho_q \rho_p \rho_l = \rho_q \eta_F.$$

Since F is off q, then σ is a glide reflection. ∎

Corollary 29.21 An odd isometry is either a reflection or a glide reflection but not both.

Theorem 29.20 and its corollary are a little surprising. Besides the identity isometry, the even isometries for the Euclidean plane are the rotations and the translations. For the Bolyai–Lobachevsky plane we pick up the horolations as even isometries, but we do not pick up

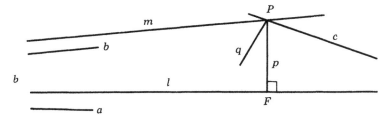

FIGURE 29.12

any new types of odd isometries. The theorem and its corollary hold for both geometries.

Combining Corollary 29.7 and Corollary 29.21, we have a complete classification of all the isometries for the Bolyai–Lobachevsky plane.

Theorem 29.22 An isometry is exactly one of the following: the identity, a rotation, a translation, a horolation, a reflection, or a glide reflection.

29.3 EXERCISES

- **29.1** Theorem 29.2.

- **29.2** A line is hyperparallel to its image under a halfturn.

29.3 If a rotation fixes a line, then the rotation is a halfturn about some point P and fixes exactly those lines that are in the pencil with center P.

- **29.4** $\eta_C \eta_B \eta_A$ is a halfturn iff points A, B, C are collinear.

- **29.5** Let P and P' be distinct points. There exist exactly two involutory isometries taking P to P', but there exist an infinite number of glide reflections taking P to P'.

- **29.6** True or False?

    (a) If isometry $\rho_c \eta_B \rho_a$ is an involution, then the isometry is a halfturn.

    (b) If isometry $\eta_C \rho_b \eta_A$ is an involution, then the isometry is a halfturn.

    (c) $\rho_c \rho_b \rho_a$ is a reflection iff lines a, b, c are in a brush.

    (d) $\rho_c \rho_b \rho_a = \rho_a \rho_b \rho_c$ iff lines a, b, c are in a brush.

    (e) A glide reflection and its inverse have different centers.

    (f) The square of a horolation is a horolation.

    (g) The square of a translation is a translation.

    (h) The square of a rotation is a rotation.

    (i) The square of a glide reflection is a translation.

    (j) Every even isometry other than the identity is the product of two halfturns.

29.7 Fill in the table which has columns headed Fixed Points, Fixed Lines, Lines Fixed Pointwise, Fixed Brushes, Brushes Fixed Linewise, and Fixed Cycles and which has rows headed Halfturn about C, Noninvolutory Rotation with Center C, Translation with Center c, Horolation with Center b, Reflection in l, Glide Reflection with Center c.

● **29.8** If $\boxed{\text{L}}\,ABCD$, $a=\overset{\leftrightarrow}{AB}$, $b=\overset{\leftrightarrow}{BC}$, $c=\overset{\leftrightarrow}{CD}$, and $d=\overset{\leftrightarrow}{DA}$, then $\rho_d\rho_c\rho_b\rho_a$ and $\rho_d\rho_b\rho_c\rho_a$ are translations.

● **29.9** If σ is an isometry and A is a point, then $\eta_{\sigma A}=\sigma\eta_A\sigma^{-1}$.

29.10 Let A, B, C be points and a, b, c be lines. Express each of the following equations in an equivalent form that does not involve isometries.

1 $\rho_b\rho_a=\rho_a\rho_b$.		**2** $\eta_B\eta_A=\eta_A\eta_B$.	
3 $\rho_b\rho_a=\rho_c\rho_b$.		**4** $\eta_B\eta_A=\eta_C\eta_B$.	
5 $\rho_b\eta_A=\eta_B\rho_b$.		**6** $\eta_B\rho_a=\rho_b\eta_B$.	
7 $\rho_a\eta_A=\eta_A\rho_a$.		**8** $\eta_B\rho_a=\rho_a\eta_B$.	

29.11 If σ is a noninvolutory rotation with center C, then there exists a positive number r such that, if P is any point different from C, then $m\angle PCP'=r$ where $P'=\sigma P$. If σ is a translation, then there does not exist a positive number s such that $PP'=s$ for every point P where $P'=\sigma P$.

29.12 If τ is a translation, then there exist A and B such that $\tau=\eta_B\eta_A$. Is either A or B arbitrary?

29.13 Every even isometry is the product of two rotations; every even isometry is the product of two translations; and every even isometry is the product of two horolations.

29.14 Every translation is a product of two glide reflections. Every isometry is a product of glide reflections.

29.15 Isometry σ commutes with ρ_l iff σ fixes line l; isometry σ commutes with η_P iff σ fixes point P.

29.16 Find three noncollinear points A, B, C (1) such that $\eta_C\eta_B\eta_A$ is a rotation, (2) such that $\eta_C\eta_B\eta_A$ is a translation, and (3) such that $\eta_C\eta_B\eta_A$ is a horolation.

29.17 Every even nonidentity isometry is the product of three halfturns.

29.18 Argue that the three perpendicular bisectors of the sides of a triangle are in one brush because the product of the three reflections in these three lines fixes a point.

29.19 If a, b, c, d are lines in some brush other than the pencil with center P and $\rho_b\rho_a P = \rho_d\rho_c P$, then $\rho_b\rho_a = \rho_d\rho_c$.

● **29.20** *Hjelmslev's Theorem:* If l and m are lines and σ is an isometry such that $m = \sigma l$, then there exists a line n such that n contains the midpoint of P and σP for every point P on l.

*29.21 Suppose $\triangle ABC$, τ_1 is the translation with center $\overleftrightarrow{AB}$ that takes A to B, τ_2 is the translation with center $\overleftrightarrow{BC}$ that takes B to C, and τ_3 is the translation with center $\overleftrightarrow{CA}$ that takes C to A. Then $\tau_3\tau_2\tau_1$ is a rotation. Further, the value of r in Exercise 29.11 for this rotation is $\delta\triangle ABC$.

*29.22 Is every glide reflection the product of the three reflections in the three lines containing the sides of some triangle?

*29.23 The product, in any order, of the four reflections in the four lines containing the sides of a Lambert quadrilateral is a translation.

*29.24 What can be said about the products of the four reflections in the four lines containing the sides of a Saccheri quadrilateral?

*29.25 A bijection α on the set of all points which is a collineation having the property that αl is parallel to l for every line l is called a *dilatation*. (The *dilation* sending (x, y) to $(2x, 2y)$ is a dilatation for the Cartesian plane.) What are the dilatations for the Bolyai–Lobachevsky plane?

GRAFFITI

For the Euclidean plane, *every isometry is exactly one of the following: the identity, a rotation, a translation, a reflection, or a glide reflection. The involutory isometries are the reflections and the halfturns. If an isometry fixes a point, then the isometry is either the identity, a reflection, or a rotation. A rotation fixes exactly one point. A rotation fixes a line iff the rotation is a halfturn about some point* P, *in which case only the lines through* P *are fixed. A translation fixes each line in some unique parallel pencil. A glide reflection fixes exactly one line. A product of two distinct halfturns is a translation. Conversely, every translation is a product of two halfturns. A product of three halfturns is a halfturn. A product of two translations is a translation or the identity. A noninvolutory rotation is not a product of halfturns. A product of three reflections* in lines through point P is a reflection in a line through P; a product of three reflections in lines parallel to line l is a reflection in a line parallel to l; and a product of three reflections in lines that are neither concurrent nor parallel is a glide reflection.

Still considering the Euclidean plane, let C be a point and r > 0. A stretch *of ratio* r *about* C is the mapping that fixes C and otherwise sends point P to P′ where P′ is the unique point on $\overrightarrow{CP}$ such that CP′ = rCP. (We allow the identity to be a stretch.) A dilation *about point* C is a stretch about C or a stretch about C followed by the halfturn about C. Then a dilatation (see Exercise 29.25) is either a translation or a dilation. A stretch reflection *is defined to be a nonidentity stretch about some point* C *followed by the reflection in some line through* C; a stretch rotation *is defined to be a nonidentity stretch about some point* C *followed by a rotation about* C. Then every similarity is exactly one of the following: an isometry, a nonidentity dilation other than a halfturn, a stretch rotation, or a stretch reflection.

REFLECTION ИOITƆƎ⅃ⱯƎᴚ

ᴚƎⱯⱢƐƆⱢIOИ ИOITƆƎ⅃ⱧƎᴚ

(Theorem 29.12.)

CHAPTER 30

Symmetry

30.1 LEONARDO'S THEOREM

All the theorems of this chapter hold for the absolute plane. The proofs and results are applicable to the Euclidean plane as well as the Bolyai–Lobachevsky plane.

The *symmetries* of a set S of points are the isometries that fix S. So isometry α is a symmetry of set S of points iff $\alpha S = S$. The identity isometry ι is a symmetry of every set of points. If α and β are symmetries of S, then α^{-1} and $\beta\alpha$ are also symmetries of S. So the symmetries of a set S of points form a subgroup of the group of all isometries. All of the finite groups of isometries are determined in this section. The principal result, Theorem 30.17, was proved by Leonardo da Vinci (1452–1519). Two types of groups will play a central role in our study. These are defined below.

DEFINITION 30.1 Let S be a set of points. If $\eta_p S = S$, then P is a *point of symmetry* for S. If α is an isometry such that $\alpha S = S$, then α is a *symmetry* of S. If group G has exactly n elements, then G is *finite* and n is the *order* of G; if group G does not have a finite number of elements, then G is *infinite*. If every element of group G is a product of the elements $\alpha, \beta, \ldots, \gamma$ in G, then G is *generated* by $\alpha, \beta, \ldots, \gamma$ and we write $G = \langle \alpha, \beta, \ldots, \gamma \rangle$. A group generated by one of its elements is a *cyclic* group. If group G is of order $2n$, has a cyclic subgroup H of order n generated by σ, and has an element ρ not in H such that both ρ and $\sigma\rho$ are involutions, then G is a ***dihedral*** group.

Suppose σ is an element of any group G with identity element ι. If $\sigma^m = \sigma^n$ for integers m and n with $m > n$, then $\sigma^{m-n} = \iota$. If there is a smallest positive integer r such that $\sigma^r = \iota$, then the elements σ, σ^2, . . . , σ^r are distinct and form a cyclic subgroup of order r. The identity element generates the trivial cyclic subgroup of order 1. If there is no positive integer r such that $\sigma^r = \iota$, then the elements σ^m and σ^n are distinct for distinct integers m and n. In this case, $\langle \sigma \rangle$, the subgroup generated by σ, is an infinite cyclic group. So every element of a group G generates a cyclic subgroup of G. Since $\sigma^m \sigma^n = \sigma^n \sigma^m$ for any integers m and n, a cyclic group is necessarily *abelian*, that is, any two elements of a cyclic group commute. Suppose σ and ρ are distinct elements of group G and ρ is an involution. Then $\rho\sigma^{-1} = \rho^{-1}\sigma^{-1} = (\sigma\rho)^{-1}$. Hence, $\sigma\rho$ is an involution iff $\sigma\rho = \rho\sigma^{-1}$ when ρ is an involution and $\rho \neq \sigma$.

Theorem 30.2 If ρ and $\sigma\rho$ are involutions and ρ is not a power of σ, then $\sigma^m\rho$ and $\rho\sigma^m$ are involutions for every integer m.

Proof Since $\sigma\rho = \rho\sigma^{-1}$, then $\sigma^{-1}\sigma\rho\sigma = \sigma^{-1}\rho\sigma^{-1}\sigma$ and $\rho\sigma = \sigma^{-1}\rho$. From $\sigma\rho = \rho\sigma^{-1}$ and $\sigma^{-1}\rho = \rho\sigma$ it follows that $\sigma^r\rho = \rho\sigma^{-r}$ for every integer r. Hence $\sigma^m\rho$ and $\rho\sigma^m$ are involutions or the identity. ∎

Theorem 30.3 Let C be any point and m be an integer greater than 2. Suppose $m\angle ACB$ is $2\pi/m$ and $CA = CB$. Let σ be the rotation with center C that takes A to B. Let ρ be the reflection in $\overleftrightarrow{CA}$. Let $C_1 = \langle \iota \rangle$, $C_2 = \langle \eta_C \rangle$, and $C_m = \langle \sigma \rangle$. Let $D_1 = \langle \rho \rangle$, $D_2 = \langle \eta_C, \rho \rangle$, and $D_m = \langle \sigma, \rho \rangle$. Then for every positive integer n, C_n is a cyclic group of order n and D_n is a dihedral group of order $2n$.

Proof The result for C_n is easy to see. We shall prove the result for D_n. For n equal to 1 or 2, take σ to be ι or η_C, respectively. For $n > 2$, let $n = m$. Now ρ and σ^r are distinct since ρ is odd and σ^r is even. Both ρ and $\sigma\rho$ are involutions since they are reflections (Theorem 28.12). So $\sigma\rho = \rho\sigma^{-1}$. Hence ever element of D_n can be uniquely written $\rho^j\sigma^i$ where i is 1, 2, . . . , or n and j is 1 or 2. Therefore D_n is a dihedral group of order $2n$. ∎

DEFINITION 30.4 For any positive integer n, let C_n and D_n denote the groups in Theorem 30.3.

For $n > 1$, C_n is generated by a rotation and D_n is generated by a rotation and a reflection. D_1 and C_2 are isomorphic since each is generated by an involution.

Theorem 30.5 If $n > 1$, then a dihedral group of order $2n$ is not cyclic.

Proof Let ρ and σ be as in the definition of a dihedral group G of order $2n$. Let α be any element of G. Then $\alpha = \rho^j \sigma^i$. If $j = 1$, then $\langle \alpha \rangle$ has order 2. If $j = 2$, then $\langle \alpha \rangle$ is a subgroup of $\langle \sigma \rangle$ and has order at most n. Hence no element of G generates G. ∎

C_n and D_n are groups of isometries for each positive integer n. It may not be obvious that each of these groups is the group of symmetries of some *polygon*.

DEFINITION 30.6 Let n be an integer greater than 2. Let A_1, A_2, . . . , A_n be n distinct points, $A_{n+1} = A_1$, and $A_{n+2} = A_2$. If the interiors of the n segments $\overline{A_i A_{i+1}}$ are mutually disjoint, then the union of these segments is a *polygon* with *vertices* A_i, *sides* $\overline{A_i A_{i+1}}$, and *angles* $\angle A_i A_{i+1} A_{i+2}$. A polygon with n sides is also called an *n-gon*. If, for each side s of n-gon P, $P \setminus s$ is on one halfplane of the line containing s, then P is *convex* and the intersection of these n halfplanes is the *interior* of P. The interior of n-gon P is int (P). A *regular polygon* is a convex polygon with all its sides congruent and all its angles congruent.

The *Greek cross* in Figure 30.1 illustrates that a polygon with all its sides congruent and all its angles congruent may not be a regular polygon. (Greek crosses exist in both the Euclidean plane and the Bolyai–Lobachevsky plane.)

Let n be an integer greater than 2. Suppose $m\angle VCV_1 = 2\pi/n$ and $CV_1 = CV$. Let σ be the rotation with center C that takes V to V_1. Let ρ be the reflection in $\overleftrightarrow{CV}$. Then σ generates a group C_n. Also, ρ and σ generate a group D_n. Let $V_i = \sigma^i V$. So $V = V_0 = V_n$. The union of the n

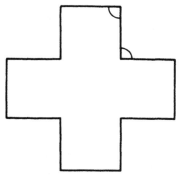

FIGURE 30.1

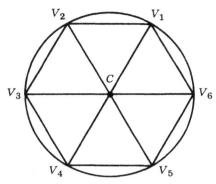

FIGURE 30.2

segments $\overline{V_i V_{i+1}}$ is a regular n-gon P inscribed in the circle with center C and radius CV. So P is fixed by both ρ and σ. Let G be the group of symmetries of P. Then G contains at least the $2n$ isometries in D_n. Under any isometry of P, V_1 can go to any of the n verticies V_i. However, then V_2 can go to only one of the two vertices V_{i-1} or V_{i+1}. In general, an n-gon has at most $2n$ symmetries. So P has the maximum number of symmetries possible for an n-gon. Therefore $G = D_n$.

Let n, V, C, and σ be as above. Suppose $C - V - W$. Let $W_i = \sigma^i W$. The union of all the segments $\overline{V_i W_i}$ and $\overline{W_i V_{i+1}}$ is a $2n$-gon Q, called a *ratchet polygon*. See Figure 30.3. Let H be the group of symmetries of Q. Then H contains C_n, the cyclic group of order n generated by σ. Since any symmetry of Q fixes C, the only possible nonidentity elements of H are rotations with center C and reflections in lines through C. Since W must go to some W_i, we already have all the possible rotations. Assume H contains some reflection ρ fixing C and sending W_2 to W_i. Then $\sigma^{2-i}\rho$ is a reflection fixing C and W_2. This reflection also sends $\overline{W_2 V_3}$ to $\overline{W_2 V_1}$. We have a contradiction since $\overline{W_2 V_3}$ is on Q but $\overline{W_2 V_1}$ is not. Therefore $H = C_n$.

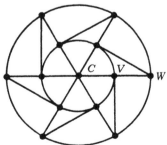

FIGURE 30.3

Theorem 30.7 If n is a positive integer, then there exist polygons with a symmetry group C_n and there exist polygons with a symmetry group D_n.

Proof The case $n > 2$ has been considered above. The case $n \leqq 2$ is left for Exercise 30.1. ∎

Theorem 30.8 If a and b are two parallel lines, then $\langle \rho_b \rho_a \rangle$ is infinite cyclic.

Proof Let $\sigma = \rho_b \rho_a = \rho_c \rho_b = \rho_d \rho_c = \rho_e \rho_d = \cdots$ (Corollary 28.13). The lines $c, d, e, \cdots$ are in the same brush as a and b and are on the same halfplane of a as b. Also, $\sigma^2 = \rho_c \rho_a$, $\sigma^3 = \rho_d \rho_a$, $\sigma^4 = \rho_e \rho_a$, $\cdots$. It follows that σ^r is never the identity for positive integer r. So $\sigma^m \neq \sigma^n$ for distinct integers m and n, as otherwise σ^{n-m} and σ^{m-n} would be the identity. The group generated by σ is cyclic by definition but does not have a finite number of elements. ∎

Corollary 30.9 The only even isometries in a finite group of isometries are rotations and the identity.

Theorem 30.10 The square of a glide reflection with center c is a translation with center c. If τ is a translation with center c, then there exists a unique glide reflection γ with center c such that $\gamma^2 = \tau$.

Proof Let a be a line perpendicular to c. For every translation τ with center c, there exists a unique line b perpendicular to c and different from a such that $\tau = \rho_b \rho_a$. So, for every glide reflection γ with center c, there exists a unique line x perpendicular to c and different from a such that $\gamma = \rho_c \rho_x \rho_a$. Then (Theorems 29.17 and 29.8), $\gamma^2 = \rho_z \rho_a$ where $z = \rho_x a$. Both statements in the theorem now follow. ∎

Corollary 30.11 A glide reflection generates an infinite cyclic group. A finite group of isometries does not contain a glide reflection.

The only odd isometries in a finite group of isometries must be reflections. Further, by Theorem 30.8, these reflections must be in lines of one pencil.

Theorem 30.12 A group of isometries containing rotations with different centers necessarily contains a translation.

Proof Suppose group G of isometries contains rotations σ_1 with center A and σ_2 with center B where $A \neq B$. Let $l = \overleftrightarrow{AB}$. There exist lines m and n such that $\sigma_1 = \rho_l \rho_m$ and $\sigma_2 = \rho_l \rho_n$. Group G must also contain

$\sigma_2^{-1}\sigma_1^{-1}\sigma_2\sigma_1$, which is $(\rho_n\rho_l\rho_m)^2$. Since $m, l,$ and n are not in one brush, then $\rho_n\rho_l\rho_m$ is a glide reflection. So G contains the square of a glide reflection. Therefore, G contains a translation. ■

Summarizing most of the preceding three theorems and their corollaries, we have the next theorem.

Theorem 30.13 A finite group G of isometries can contain only rotations, reflections, and the identity. Also, all the rotations in G must have the same center, and all the reflections in G must be in lines of one pencil. Further, if G contains both rotations with center C and reflections in lines in pencil p, then C is the center of p.

Corollary 30.14 Every finite group of isometries fixes some point, which is unique if the order of the group is greater than 2.

Theorem 30.15 If a finite group G of isometries contains only rotations and the identity, then G is a cyclic group C_n.

Proof Suppose G has n elements. If n is 1 or 2, the result is trivial. Suppose $n > 2$ and all the rotations have center C (Theorem 30.13). Let $P_1, P_2, \ldots, P_n$ be all the distinct images of some point P different from C. Since $n > 2$, not all the P_i are collinear with C. So, without loss of generality, we may suppose $m\angle P_1CP_2$ is the minimum of all possible numbers $m\angle P_iCP_j$. If $\alpha P = P_1$ and $\beta P = P_j$, then $\beta\alpha^{-1}$ is an element of G taking P_1 to P_j. So each P_i is the image of P_1 under some element of G. Let σ be an element of G that takes P_1 to P_2. There is a smallest positive integer m such that σ^m is the identity. So $\sigma, \sigma^2, \ldots, \sigma^m$ are distinct elements in G. We wish to show these are the only elements in G. Let $V_{i+1} = \sigma^iP_1$. Then $V_1 = P_1$, $V_2 = \sigma P_1 = P_2$, and the m points V_i are among the n points P_j. So $m\angle V_iCV_{i+1}$, $m\sigma^{i-1}\angle P_1CP_2$, and $m\angle P_1CP_2$ are equal. By the minimality of $m\angle P_1CP_2$, no P_j can be in int $(\angle V_iCV_{i+1})$. Since the V_i are the vertices of a regular m-gon inscribed in the circle with center C and radius CP_1, it then follows that each P_i must be a V_j. Hence $n = m$ and G is the cyclic group generated by σ. ■

Theorem 30.16 If a finite group G of isometries contains a reflection, then G is a dihedral group D_n.

Proof Suppose G has n even isometries and m odd isometries. The n even isometries by themselves form a subgroup H of G. Since H contains only rotations and the identity, H must be cyclic of order n. Suppose H is generated by σ and G contains the reflection ρ. Since each of the odd isometries multiplied by ρ on the left is an even isometry, then $m \leqq n$; since each of the even isometries multiplied by ρ on the left is an odd isometry, then $n \leqq m$. Hence $n = m$ and G has order $2n$. Since $\sigma\rho$ is a reflection (Theorems 30.13 and 28.12), then ρ and $\sigma\rho$ are involutions. Therefore G is a dihedral group D_n. ∎

Theorem 30.17 *Leonardo's Theorem* The only finite groups of isometries are the cyclic groups C_n and the dihedral groups D_n.

Proof Summary of Theorems 30.13, 30.15 and 30.16. ∎

Corollary 30.18 Every finite group of isometries is the group of symmetries of some polygon.

Hermann Weyl has pointed out (see Exercise 30.10) that Leonardo da Vinci systematically determined the symmetries of a building floorplan in studying how to add chapels and niches without destroying the symmetry of the nucleus. Leonardo's results are essentially what we have called Leonardo's Theorem.

30.2 FRIEZE PATTERNS

From the preceding section only Definition 30.1 and Theorems 30.2, 30.8, and 30.10 are required for this section. We begin with some elementary theorems for the absolute plane. All the theory in this section holds for the absolute plane.

Theorem 30.19 If A, B, C are points on line l, then $\eta_C\eta_B\eta_A$ is a half-turn about some point on l.

Proof Let a, b, c be the lines perpendicular to l at A, B, C, respectively. Then there exists a line d perpendicular to l at a point D such that $\rho_c\rho_b\rho_a = \rho_d$. Then $\eta_C\eta_B\eta_A = \rho_c\rho_b\rho_a\rho_l = \rho_d\rho_l = \eta_D$. ∎

Corollary 30.20 If A, B, C are collinear points, then $\eta_C\eta_B\eta_A = \eta_A\eta_B\eta_C$.

Proof $\eta_C\eta_B\eta_A = \eta_D = \eta_D^{-1} = \eta_A\eta_B\eta_C$ for some point D collinear with A, B, C. ∎

Theorem 30.21 The translations with center c together with the identity form an abelian group.

Proof Suppose τ_1 and τ_2 are translations with center c. Then there exist points A, B, C, D, E on c such that $\tau_1 = \eta_B\eta_A$, $\tau_2 = \eta_D\eta_C$, and $\eta_C\eta_B\eta_A = \eta_E$. Since $\tau_1^{-1} = \eta_A\eta_B$ and $\tau_2\tau_1 = \eta_E\eta_D$, then the translations with center c together with the identity form a group. By the preceding corollary, $\tau_2\tau_1 = \eta_D\eta_C\eta_B\eta_A = \eta_D\eta_A\eta_B\eta_C = \eta_B\eta_A\eta_D\eta_C = \tau_1\tau_2$. So the group is abelian. ∎

Corollary 30.22 If σ is a glide reflection with center c and τ is a translation with center c, then $\sigma\tau = \tau\sigma$.

Theorem 30.23 If τ is a translation with center c, line l is perpendicular to c at P, and σ is any isometry, then

(a) $\tau\eta_P$ is the halfturn about the midpoint of P and τP

(b) $\tau\rho_l$ is the reflection in the perpendicular bisector of P and τP

(c) $\sigma\eta_P\sigma^{-1}$ is the halfturn about σP.

Proof Let line m be perpendicular to c at M, the midpoint of P and τP. (a) Since $\tau = \eta_M\eta_P$, then $\tau\eta_P = \eta_M$. (b) Since $\tau = \rho_m\rho_l$, then $\tau\rho_l = \rho_m$. (c) $\sigma\eta_P\sigma^{-1}$ is an involutory, even isometry fixing σP. ∎

Around the frieze of a building there is often a pattern formed by the repetition of some figure over and over again. The essential property of an ornamental frieze pattern is that it is invariant under some "smallest translation." Other symmetries are often evident as well. Of course, there is an infinite variety in the subject matter for such patterns. However, by discounting the subject matter and considering only the symmetries under which such patterns are invariant, we shall see that there are essentially only seven possible types of ornamental frieze patterns.

DEFINITION 30.24 A group of isometries that fix line c and whose translations, together with the identity, form an infinite cyclic group is a *frieze group* with *center* c.

Let τ be a translation with center c. We shall determine all frieze groups F with center c and whose translations, together with the identity, form the cyclic group generated by τ. There will be seven of them. For each group we shall have a frieze pattern having that group as its group of symmetries. We shall also state in italics criteria which will distinguish those patterns with the given group of symmetries. The following notation will be used throughout. Suppose A is a point on c. For the moment A is arbitrary, but we shall be more specific in some cases. Let $A_i = \tau^i A$. So $A_o = A$. Since $\tau^n A_i = \tau^{i+n}A$, every translation in F must take A_i to some A_j. Let M be the midpoint of A and A_1

and $M_i = \tau^i M$. So M_i is the midpoint of A_i and A_{i+1} and the midpoint of A_0 and A_{2i+1}. One possibility for F is just the group generated by τ. Let $F_1 = \langle \tau \rangle$. *A frieze pattern having F_1 as its group of symmetries has no point of symmetry, has no line of symmetry, and is not fixed by a glide reflection.* See Figure 30.4.

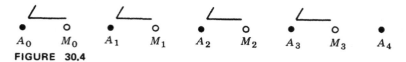

$A_0 \qquad M_0 \qquad A_1 \qquad M_1 \qquad A_2 \qquad M_2 \qquad A_3 \qquad M_3 \qquad A_4$

FIGURE 30.4

The only even isometries that fix c are the identity, the translations with center c, and the halfturns with center on c. Suppose F contains a halfturn. In this case we suppose A is picked to be the center of a halfturn in F. Since τ and η_A are in F, then $\tau^n \eta_A$ is in F for each integer n. Hence (Theorem 30.23a), F contains $\tau^{2m} \eta_A$, which is the halfturn about A_m, and F contains $\tau^{2m+1} \eta_A$, which is the halfturn about M_m. Now suppose P is the center of some halfturn in F. Then the translation $\eta_P \eta_A$ is in F. So $\eta_P \eta_A A = A_n$ for some n. Then $\eta_P A = A_n$, and P is the midpoint of A and A_n. Hence F contains exactly those halfturns that have center A_m and those that have center M_m. Let $F_2 = \langle \tau, \eta_A \rangle$. Since $\tau \eta_A$ is an involution, then $\tau \eta_A = \eta_A \tau^{-1}$. So every element in F_2 is of the form τ^i or $\eta_A \tau^i$. Every element in F_2 is of the form $\eta_A^j \tau^i$. Also, $F_2 = \langle \eta_A, \eta_M \rangle$ since $\eta_M \tau = \eta_A$. *A frieze pattern having F_2 as its group of symmetries has a point of symmetry but no line of symmetry.* See Figure 30.5.

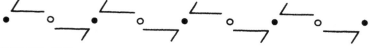

FIGURE 30.5

If F contains only even isometries, then F must be one of F_1 or F_2. In general, F must contain F_1 or F_2. The other possibilities for F are obtained by augmenting F_1 or F_2 with odd isometries. We first consider adding reflections. A reflection in line l fixes c iff $l = c$ or $l \perp c$. Let $F_1^1 = \langle \tau, \rho_c \rangle$. Since $\tau \rho_c = \rho_c \tau$ (Theorem 29.17), then F_1^1 is abelian and every element is of the form $\rho_c^j \tau^i$. If $n \neq 0$, then F_1^1 contains the glide reflection $\rho_c \tau^n$, which takes A to A_n. *A frieze pattern having F_1^1 as its group of symmetries has no point of symmetry and the center is a line of symmetry.* See Figure 30.6.

FIGURE 30.6

Let $F_2^1 = \langle \tau, \eta_A, \rho_c \rangle$. Since ρ_c commutes with both τ and η_A, then every element of F_2^1 is of the form $\rho_c^k \eta_A^j \tau^i$. If $n \neq 0$, then F_2^1 contains the glide reflection $\tau^n \rho_c$, which takes A to A_n. Also (Theorem 30.23b), F_2^1 contains $\tau^{2m} \eta_A \rho_c$, which is the reflection in the line perpendicular to c at A_m, and F_2^1 contains $\tau^{2m+1} \eta_A \rho_c$, which is the reflection in the line perpendicular to c at M_m. If a is the line perpendicular to c at A, then $F_2^1 = \langle \tau, \rho_a, \rho_c \rangle$. *A frieze pattern having F_2^1 as its group of symmetries has a point of symmetry and the center is a line of symmetry.* See Figure 30.7.

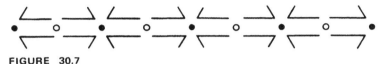

FIGURE 30.7

Suppose F does not contain a halfturn but does contain the reflection in a line a that is perpendicular to c. In this case, we suppose A is on a. Then (Theorem 30.23b), F contains $\tau^{2m} \rho_a$, which is the reflection in the line perpendicular to c at A_m, and F contains $\tau^{2m+1} \rho_a$, which is the reflection in the line perpendicular to c at M_m. Assume F contains another reflection ρ_l. Then $l \neq c$ since the halfturn $\rho_c \rho_a$ is not in F. So $l \perp c$. Then F contains the translation $\rho_l \rho_a$, which must take A to A_n for some n. So $\rho_l A = A_n$ for some n with $n \neq 0$, and l is perpendicular to c at some A_m or at some M_m. Therefore, F must contain exactly those reflections in lines perpendicular to c at A_m for each m and those reflections in lines perpendicular to c at M_m for each m. We have now considered all possible cases of adding reflections to F_1. Let $F_1^2 = \langle \tau, \rho_a \rangle$ where a is perpendicular to c at A. Since $\tau \rho_a = \rho_a \tau^{-1}$, every element of F_1^2 is of the form $\rho_a^j \tau^i$. F_1^2 does not contain ρ_c but does contain the reflections in the lines that are perpendicular to c at A_m or M_m. *A frieze pattern having F_1^2 as its group of symmetries has no point of symmetry, has a line of symmetry, but the center is not a line of symmetry.* See Figure 30.8.

FIGURE 30.8

Now suppose F does contain a halfturn and the reflection in a line p. If $p = c$, p is perpendicular to c at A_m, or p is perpendicular to c at M_m, then we are back to F_2^1. To obtain something new, we must suppose p is off each A_m and off each M_m. Let p be perpendicular to c at P. Since $\tau^n \rho_p$ and $\tau^{-n} \rho_p$ with $n > 0$ are the reflections in the lines perpendicular to c at the two points on c that are of distance nAM from P (Theorem 30.23b), we may suppose $A - P - M$ without loss of generality. F contains the halfturn $\rho_p \eta_A \rho_p$ about $\rho_p A$ (Theorem 30.23c). Since the only permissible centers of halfturns are the A_m and the M_m and since $A - P - M$, then we must have $\rho_\nu A = M$. Hence F contains ρ_p where p is the perpendicular bisector of A and M. If line a is perpendicular to c at A, then F cannot contain both ρ_p and ρ_a as the translation $\rho_p \rho_a$ would take A to M, which is impossible. Also, since $\rho_p \rho_a = \rho_p \rho_c \eta_A$, F cannot contain both ρ_p and ρ_c. We have now considered all possible cases of adding reflections to F_2. Let $F_2^2 = \langle \tau, \eta_A, \rho_p \rangle$ where p is the perpendicular bisector of A and M. F_2^2 contains the glide reflection $\rho_p \eta_A$ which takes A to M. Let $\gamma = \rho_p \eta_A$. Since $\tau = \gamma^2$ and $\rho_p = \gamma \eta_A$, then $F_2^2 = \langle \gamma, \eta_A \rangle$. F_2^2 does not contain ρ_c. *A frieze pattern having F_2^2 as its group of symmetries has a point of symmetry, has a line of symmetry, but the center is not a line of symmetry.* See Figure 30.9, but ignore the dots.

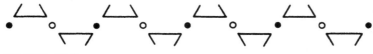

FIGURE 30.9

We have considered all possibilities for F that do not necessarily contain a glide reflection. Now suppose F contains the glide reflection σ. Then σ has center c and σ^2 is a translation with center c. We have two cases: $\sigma^2 = \tau^{2n}$ and $\sigma^2 = \tau^{2n+1}$ for some integer n. Suppose $\sigma^2 = \tau^{2n}$. Since σ and τ commute (Corollary 30.22), then $(\tau^{-n}\sigma)^2$ is the identity. So the odd, involutory isometry $\tau^{-n}\sigma$ must be ρ_c. Hence $\sigma = \tau^n \rho_c$. In this case F contains ρ_c and $\rho_c \tau^m$ for each integer m. If F does not contain a halfturn, then we are back to F_1^1; if F contains a halfturn, then we are back to F_2^1. Now suppose $\sigma^2 = \tau^{2n+1}$. Then $(\tau^{-n}\sigma)^2$ is τ. Let $\gamma = \tau^{-n}\sigma$. Then γ is an odd isometry whose square is τ. Hence γ must be the unique glide reflection with center c that takes A to M (Theorem 30.10). Since $\gamma^{2m} = \tau^m$ and $\gamma^{2m+1} = \tau^m \gamma$, the glide reflections in F are exactly those of the form $\tau^m \gamma$. Let $F_1^3 = \langle \gamma \rangle$ where γ is the glide reflection with center c such that $\gamma^2 = \tau$. *A frieze pattern having F_1^3 as its group of symmetries has no point of symmetry, has no line of symmetry, but is fixed by a glide reflection.* See Figure 30.10, but ignore the dots.

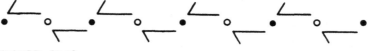

FIGURE 30.10

Suppose F contains isometries in addition to those generated by the glide reflection γ with center c where $\gamma^2 = \tau$. Since the square of the translation $\rho_c \gamma$ is τ, then $\rho_c \gamma$ is not in $\langle \tau \rangle$. So ρ_c cannot be in F. If F contains ρ_l with $l \perp c$, then F contains the halfturn $\rho_l \gamma$. If F contains a halfturn, then F must contain η_A. In this case, F contains η_A and the glide reflection γ such that $\gamma^2 = \tau$. Hence F is F_2^2. We have now run out of possibilities. F must be one of the seven groups given above.

```
R  R  R         E E E      A A A          D  D  D
  R  R  R        E E E      A A A          D  D  D


      N  N  N        O O O      H  H  H
        N  N  N      O O O      H  H  H
```
FIGURE 30.11

Theorem 30.25 Let F be a frieze group with center c whose translations, together with the identity, form the group generated by the translation τ. If F contains a halfturn, suppose F contains η_A; if F contains a reflection in a line perpendicular to c, suppose F contains ρ_a with $a \perp c$. Let γ be the (unique) glide reflection with center c such that $\gamma^2 = \tau$. Then F is exactly one of the seven groups defined as follows:

$$F_1 = \langle \tau \rangle, \qquad F_1^1 = \langle \tau, \rho_c \rangle, \qquad F_1^2 = \langle \tau, \rho_a \rangle, \qquad F_1^3 = \langle \gamma \rangle,$$

$$F_2 = \langle \tau, \eta_A \rangle, \qquad F_2^1 = \langle \tau, \eta_A, \rho_c \rangle, \qquad F_2^2 = \langle \gamma, \eta_A \rangle.$$

The seven types of ornamental frieze patterns are illustrated in Figure 30.11.

If P is a point and τ_1 and τ_2 are translations such that $P, \tau_1 P$, and $\tau_2 P$ are not collinear, then a group of isometries whose translations are exactly those in $\langle \tau_1, \tau_2 \rangle$ is called a *wallpaper group*. In the sense that there are exactly seven frieze groups, there are exactly seventeen wallpaper groups for the Euclidean plane. (See the references cited in Exercises 30.10, 30.20, and 30.21.) Patterns corresponding to many of these groups were known to ancient Egyptians and Chinese. However, all seventeen of the groups were known to the Moors as is shown by the ornamental patterns decorating the Alhambra in Granada. This has been considered one of the greatest mathematical achievements of ancient times. The study of wallpaper groups makes a nice topic for independent study or for a seminar.

30.3 EXERCISES

30.1 Theorem 30.7.

30.2 A nonempty set of points in the interior of a circle cannot have two points of symmetry.

● **30.3** Using the notation in the text for a regular polygon with group of symmetries D_n, show that the reflection in $\overleftrightarrow{CV}_i$ is $\sigma^{2i}\rho$. What is $\sigma^{2i+1}\rho$?

30.4 What is the group of symmetries for the Greek cross of Figure 30.1?

● **30.5** Find all possible groups of symmetries that fix a given line but do not contain a translation. What could be on the frieze of a building that corresponds to each of these groups?

● **30.6** Two mathematicians looking at the frieze of a building disagree on the frieze group. Give an example of such a frieze pattern.

30.7 Find the frieze group for each of the following patterns:

(a) DDDDDDDD
   DDDDDDDD ,

(b) E E E E
   E E E E ,

(c) XXXXXXXX
   XXXXXXXX ,

(d) ZZZZZZZZ
   ZZZZZZZZ ,

(e) XXXXXXXX
   ZZZZZZZZ ,

(f) XZXZXZXZ
   ZXZXZXZX ,

(g) X X X X
    X X X X ,

(h) Z Z Z Z
    Z Z Z Z ,

(i) X X X X
    Z Z Z Z ,

(j) YYYYYYYY
   YYYYYYYY .

● **30.8** True or False?

(a) No letter in the alphabet has exactly one point of symmetry and exactly one line of symmetry.

(b) There exists a nonempty set of points with two lines of symmetry but no point of symmetry.

(c) An n-gon has exactly 1, exactly n, or exactly $2n$ symmetries.

(d) A dihedral group of order greater than 2 is generated by two reflections.

(e) A finite cyclic group of isometries is generated by a rotation or by the identity.

(f) If α and β are isometries and $\alpha^2 = \beta^2$, then either $\alpha = \beta$ or $\alpha = \beta^{-1}$.

(g) If P is a point of symmetry for set S of points, then P is in S.

(h)   Given translation τ, in the Euclidean plane there are many glide reflections whose square is τ but in the Bolyai–Lobachevsky plane there is exactly one.

(i)   If τ is a translation with center l and line m is perpendicular to l at P, then $\tau\eta_P$ is the halfturn about τP and $\tau\rho_m$ is the reflection in τm.

(j)   An ornamental frieze pattern having no point of symmetry, having a line of symmetry, and which is fixed under a glide reflection has a frieze group F_1^1 as its group of symmetries.

30.9   Find the frieze group for each of the following patterns:

(a)   DDDD,   (b)   XXXX,   (c)   ZZZZ,   (d)   YYYY,

(e)   SSSS,   (f)   UUUU,   (g)   OOOO,   (h)   IIII,

(i)   QQQQ,   (j)   AAAA,   (k)   BBBB,   (l)   CCCC.

30.10   Read *Symmetry* by Hermann Weyl (Princeton, 1952).

30.11   Read *Paper Folding for the Mathematics Class* by Donovan A. Johnson (National Council of Teachers of Mathematics, 1957) or *Geometric Exercises in Paper Folding* by Sundara Row (Dover, 1966).

30.12   If S is the set of images of point A under the elements of some group G of isometries, then G may be a proper subgroup of the group of symmetries of S.

30.13   Give an example of a rotation σ such that $\langle\sigma\rangle$ is an infinite cyclic group.

30.14   What is the group of symmetries for the curve in the Cartesian plane with equation: (i) $y = \cos x$, (ii) $y = \cosh x$, (iii) $y = \tan x$, (iv) $y = \tanh x$?

30.15   What is the group of symmetries for the set of all points $(x, 0)$ in the Cartesian plane with x rational?

30.16   A frieze group F_2^2 is generated by two involutions; a frieze group F_2^1 is generated by three involutions.

30.17   Frieze groups F_1 and F_1^3 are isomorphic. Frieze groups F_2, F_1^2, and F_2^2 are isomorphic. Frieze groups F_1^1 and F_2^1 are not isomorphic.

30.18   If α and β are involutions such that $\langle\beta\alpha\rangle$ is finite, then $\langle\alpha, \beta\rangle$ is a dihedral group. Conversely, every dihedral group contains such involutions.

30.19 If group G of isometries contains an odd isometry, then there exists a one-to-one correspondence between the set of even isometries in G and the set of odd isometries in G.

30.20 Read Chapter II of *Geometry and the Imagination* by D. Hilbert and S. Cohn-Vossen (Chelsea, 1956).

*30.21 Read Part One of *Regular Figures* by L. Fejes Toth (Pergamon, 1964).

*30.22 Give an example of a group G generated by α and β, but every element of G is not of the form $\beta^j \alpha^i$.

*30.23 What are the polygons with all sides congruent and all angles congruent that are not regular polygons?

*30.24 Find all groups of isometries that fix a given line.

*30.25 Is every group of isometries the group of symmetries for some set of points?

*30.26 What are the finite groups of isometries in Euclidean three-space?

*30.27 Establish that there are exactly seventeen wallpaper groups for the Euclidean plane.

● ***30.28** What are the wallpaper groups for the Bolyai – Lobachevsky plane?

GRAFFITI

CHAPTER 31

Horocircles

31.1 LENGTH OF ARC

For our proof of the fundamental formula of Bolyai–Lobachevsky geometry we shall need a lemma concerning ratios of *lengths* of certain *arcs* of horocircles. After extending the domain of definition for the critical function Π to the set of reals, the rest of this section is devoted to proving this lemma, our Theorem 31.17. For this extension, $\Pi(0)$ will be defined to be $\pi/2$ since $\Pi(x)$ approaches $\pi/2$ as x approaches 0. Then $(0, \pi/2)$ will be made a point of symmetry for the graph of Π in the Cartesian plane. So the midpoint of $(x, \Pi(x))$ and $(-x, \Pi(-x))$ will be $(0, \pi/2)$. See Figure 31.1. Then for all real x, we will have $\Pi(x) + \Pi(-x) = \pi$.

DEFINITION 31.1 $\Pi(0) = \pi/2$, and $\Pi(-x) = \pi - \Pi(x)$ for $x > 0$.

Theorem 31.2 Π is a strictly decreasing, continuous function on the reals. $\Pi(0) = \pi/2$, $\lim_{x \to x} \Pi(x) = 0$, and $\lim_{x \to -x} \Pi(x) = \pi$.

Proof Corollary 24.17. ■

DEFINITION 31.3 If l and m are two parallel lines and S is either a point or a set of points, then S is **between** l and m if S is on the halfplane of l that contains m and on the halfplane of m that contains l.

If A and B are two points of horocircle $\mathscr{C}$ with center b, then $\widehat{AB}$ is the union of $\{A, B\}$ and the points of $\mathscr{C}$ that are between the line in b

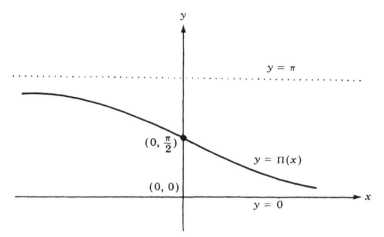

FIGURE 31.1

through A and the line in b through B, int $(\widehat{AB}) = \widehat{AB} \setminus \{A, B\}$, $\overline{AB}$ is the chord of $\mathscr{C}$ **subtended** by $\widehat{AB}$, and $\widehat{AB}$ is the **arc** of $\mathscr{C}$ **subtended** by $\overline{AB}$. The **interior** of $\widehat{AB}$ is int $(\widehat{AB})$. If b is the horopencil determined by $\overrightarrow{AC}$ and $\mathscr{C}$ is the horocircle through A with center b, then $\overrightarrow{AC}$ is a **radius** of $\mathscr{C}$. If line t intersects any set S of points such that $S \setminus t$ is on a halfplane of t, then t is a **tangent** of S.

The notation "$\widehat{AB}$" is somewhat inadequate since two points A and B lie on exactly two horocircles (determined by opposite rays on the perpendicular bisector of $\overline{AB}$). However, we shall assume that the reader always makes the *correct* choice of meaning. For example, in the next theorem it is tacitly assumed that $\widehat{AP}$, $\widehat{PB}$, and $\widehat{AB}$ are arcs of the same horocircle.

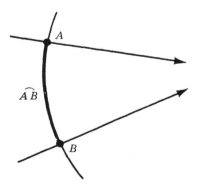

FIGURE 31.2

Theorem 31.4 Let $\overset{\frown}{AB}$ be on horocircle $\mathscr{C}$. If point P is in int $(\overset{\frown}{AB})$, then $\overset{\frown}{AP} \cap \overset{\frown}{PB} = \{P\}$ and $\overset{\frown}{AP} \cup \overset{\frown}{PB} = \overset{\frown}{AB}$.

Proof Let $\mathscr{C}$ have center b. Let $\overrightarrow{AC}, \overrightarrow{BD}$, and $\overrightarrow{PQ}$ be radii of $\mathscr{C}$. Then $\overleftrightarrow{PQ}$ lies between $\overleftrightarrow{AC}$ and $\overleftrightarrow{BD}$. Since $\overleftrightarrow{AC}, \overleftrightarrow{PQ}$, and $\overleftrightarrow{BD}$ are also three lines in b, then the set of points between $\overleftrightarrow{AC}$ and $\overleftrightarrow{BD}$ is the union of the three mutually disjoint sets: $\overleftrightarrow{PQ}$, the set of points between $\overleftrightarrow{AC}$ and $\overleftrightarrow{PQ}$, and the set of points between $\overleftrightarrow{PQ}$ and $\overleftrightarrow{BD}$ (Exercise 31.1). The theorem now follows from the definition of an arc of a horocircle. ∎

We are not bothering to state formally all the *obvious* consequences that follow from our definitions and theorems, particularly those consequences that are almost restatements of previous results. (For example, if point A is on horocircle $\mathscr{C}$, then the radius of $\mathscr{C}$ through A is unique.) Of course, we must avoid using any theorem that is *obviously true* but whose proof is not immediate. (For example, if $\overset{\frown}{AB}$ is on a horocircle with radii $\overrightarrow{AC}$ and $\overrightarrow{BD}$, then $\overline{DE}$ intersects $\overset{\frown}{AB}$ when $E-A-C$.) There are two principal reasons for this self-imposed avoidance. Mathematicians have learned that some of the so-called obvious theorems are the most difficult to prove while others turn out to be, in fact, *false*.

Let $\mathscr{C}$ be a horocircle with center b. We know that any line intersects $\mathscr{C}$ in at most two points, that every line of b contains exactly one point of $\mathscr{C}$, and that every point of $\mathscr{C}$ is on exactly one line of b. The next theorem is another elementary fact that we shall use again and again without reference.

Theorem 31.5 Let $\overrightarrow{AC}$ be a radius of horocircle $\mathscr{C}$. If point B is off $\overleftrightarrow{AC}$, then B is on $\mathscr{C}$ iff $m\angle CAB = \Pi(AB/2)$.

Proof Let $\mathscr{C}$ have center b. Let M be the midpoint of $\overline{AB}$. Then $AM = \frac{1}{2}AB$. Also, by definition of Π, $m\angle CAB = \Pi(AM)$ iff the perpendicular bisector of $\overline{AB}$ is in b. Further, by definition of $\mathscr{C}$, point B is on $\mathscr{C}$ iff the perpendicular bisector of $\overline{AB}$ is in b. ∎

In particular, if B is on $\mathscr{C}$ in the preceding theorem, then $\angle CAB$ is necessarily acute.

Theorem 31.6 Let A, B, P be three points on horocircle $\mathscr{C}$ with radius $\overrightarrow{AC}$. Then, P is in int $(\overset{\frown}{AB})$ iff P and C are on opposite sides of $\overleftrightarrow{AB}$. Also, $\overset{\frown}{AB}$ is the set of all points on $\mathscr{C}$ that are off the halfplane of $\overleftrightarrow{AB}$ containing C.

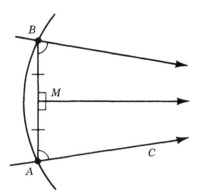

FIGURE 31.3

Proof Let $\overrightarrow{BD}$ and $\overrightarrow{PQ}$ be radii of $\mathscr{C}$. (See Figure 31.4.) Assume P is in int $(\widehat{AB})$ but P and C are on the same side of $\overleftrightarrow{AB}$. Then 2π is the sum of $m\angle QPA$, $m\angle QPB$, and $m\angle APB$. This is a contradiction since $\angle QPA$ and $\angle QPB$ are both acute. Conversely, assume P and C are on opposite sides of $\overleftrightarrow{AB}$ but P is not in int $(\widehat{AB})$. If A and P are on opposite sides of $\overleftrightarrow{BD}$, then the sum of $m\angle ABD$ and $m\angle DBP$ is greater than π; if B and P are on opposite sides of $\overleftrightarrow{AC}$, then the sum of $m\angle BAC$ and $m\angle CAP$ is greater than π. In either case, we have a contradiction since the sum of the measures of two acute angles is less than π. Therefore, P is in int $(\widehat{AB})$ iff P and C are on opposite sides of $\overleftrightarrow{AB}$. The last statement in the theorem now follows. ■

Theorem 31.7 Let A, B, P be three points on horocircle $\mathscr{C}$ with radius $\overrightarrow{AC}$ such that B and P are on the same side of $\overleftrightarrow{AC}$. Then, P is in int $(\widehat{AB})$ iff $m\angle CAP > m\angle CAB$. Also, P is in int $(\widehat{AB})$ iff $AP < AB$.

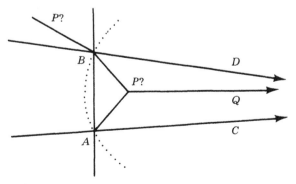

FIGURE 31.4

Proof Each of the following statements is equivalent to the next. (1) P is in int $(\widehat{AB})$. (2) P and C are on opposite sides of $\overleftrightarrow{AB}$. (3) $m\angle CAP$ is greater than $m\angle CAB$. (4) $\Pi(AP/2) > \Pi(AB/2)$. (5) $AP < AB$. ∎

Theorem 31.8 Let $\mathscr{C}$ be a horocircle with radius $\overrightarrow{AC}$. Then every circle with center A intersects $\mathscr{C}$ in exactly two points, one in each halfplane of $\overleftrightarrow{AC}$.

Proof Suppose $r > 0$. Point P is on $\mathscr{C}$ and on the circle with center A and radius r iff $m\angle CAP = \Pi(AP/2)$ and $AP = r$. By the Angle-Segment Construction Theorem, there are exactly two points satisfying the conditions, one on each side of $\overleftrightarrow{AC}$. ∎

The little theorem just proved can be used to parameterize a horocircle by the use of lengths of chords.

Theorem 31.9 Let $\overrightarrow{AC}$ be a radius of horocircle $\mathscr{C}$. Let H be a halfplane of $\overleftrightarrow{AC}$. Let $P_0 = A$. For positive real number r, let P_r be the point on $\mathscr{C}$ such that $P_0 P_r = r$ and P_r is in H, and let P_{-r} be the point on $\mathscr{C}$ such that $P_0 P_{-r} = r$ and P is off H. Then the mapping $f : \mathbf{R} \to \mathscr{C}$, where $f(x) = P_x$, is a bijection and $\widehat{P_a P_b} = \{P_x | a \le x \le b\}$ when $a < b$. Further, given real numbers r and ε with $\varepsilon > 0$, there exists a positive real δ such that $|r - x| < \delta$ implies $P_r P_x < \varepsilon$.

Proof Mapping f is a bijection (Theorem 31.8), and P_{-r} is the image of P_r under the reflection in $\overleftrightarrow{AC}$. Suppose $0 < c < d$. Then $\widehat{P_0 P_c}$ is $\{P_x | 0 \le x \le c\}$, and $\widehat{P_0 P_d}$ is $\{P_x | 0 \le x \le d\}$ (Theorem 31.7). Since P_c

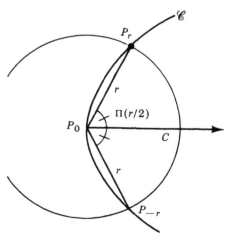

FIGURE 31.5

is in int $(\widehat{P_oP_d})$, then the intersection of $\widehat{P_oP_c}$ and $\widehat{P_cP_d}$ is $\{P_c\}$ and the union is $\widehat{P_oP_d}$ (Theorem 31.4). Hence $\widehat{P_cP_d}$ must be $\{P_x|c \leqq x \leqq d\}$. It follows that, for any two real numbers a and b, int $(\widehat{P_aP_b})$ is the set of all points P_x such that x is between a and b.

(If P_r is *close* to P_x, then r is *close* to x as $|r-x| \leqq P_rP_x$ by the Triangle Inequality. The last statement in the theorem is the converse of this. That is, we are to prove that r *close* to x implies P_r is *close* to P_x. Recall that $|r-x| < \delta$ iff $r-\delta < x < r+\delta$.)

The last statement in the theorem holds for $r=0$ (Theorem 31.7 with $\delta = \varepsilon$). The statement holds for all r by symmetry if it holds for positive r. We may suppose $\varepsilon < r$. Now let P_c and P_d be the two points on $\mathscr{C}$ that are on the circle with center P_r and radius ε such that P_c is on $\widehat{P_oP_r}$. (See Figure 31.6.) Let α be the minimum of $\Pi(c/2) - \Pi(r/2)$ and $\Pi(r/2) - \Pi(d/2)$. Then $\alpha > 0$ since α is the minimum of $m\angle P_cP_oP_r$ and $m\angle P_rP_oP_d$. Now since the function Π is continuous at r, there exists a positive δ such that $|r-x| < \delta$ implies $|\Pi(r/2) - \Pi(x/2)| < \alpha$. Then $\Pi(c/2) > \Pi(x/2) > \Pi(d/2)$ and $c < x < d$. So P_x is on $\widehat{P_cP_r}$ or $\widehat{P_rP_d}$. In either case, we have $|r-x| < \delta$ implies $P_rP_x < \varepsilon$. ∎

We next prove a couple of theorems about the tangents of a horocircle. The first is an analogue of Theorem 20.4.

Theorem 31.10 If a line is perpendicular to a radius of a horocircle at its vertex, then the line is a tangent of the horocircle. Conversely, every tangent of a horocircle is perpendicular to some radius at its vertex. A tangent of a horocircle contains exactly one point of the horocircle.

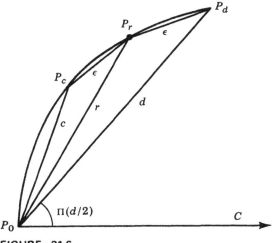

FIGURE 31.6

Proof Let $\overrightarrow{AC}$ be a radius of horocircle $\mathscr{C}$ with center b. Certainly $\overleftrightarrow{AC}$ is not a tangent of $\mathscr{C}$, since $\overleftrightarrow{AC}$ is a line of symmetry for $\mathscr{C}$. Let line n be perpendicular to $\overleftrightarrow{AC}$ at A. Let P be any point on $\mathscr{C}$ except A. Since $\angle CAP$ is acute, then P and C are on the same side of n. Thus n is a tangent of $\mathscr{C}$ and contains exactly one point of $\mathscr{C}$.

Let l be any line through A different from $\overleftrightarrow{AC}$ and from n. We may suppose $l = \overleftrightarrow{AE}$ where $\angle CAE$ is acute. There is a unique point P on $\overrightarrow{AE}$ such that $\Pi(AP/2) = m\angle CAE$. Thus P is on $\mathscr{C}$. So l intersects $\mathscr{C}$ in exactly the two points A and P. Let s and t be positive reals such that $s < AP < t$. Let S and T be the points on $\mathscr{C}$ and on the same side of $\overleftrightarrow{AC}$ as P such that $AS = s$ and $AT = t$. (See Figure 31.7.) Since $\Pi(AP/2)$ is between $\Pi(s/2)$ and $\Pi(t/2)$, we have $m\angle CAP$ is between $m\angle CAS$ and $m\angle CAT$. Hence S and T are points of $\mathscr{C}$ on opposite sides of l. Therefore, l is not a tangent of $\mathscr{C}$, and n is the unique tangent of $\mathscr{C}$ through A. ∎

Theorem 31.11 Let A and B be two points on horocircle $\mathscr{C}$. The tangents of $\mathscr{C}$ at A and B intersect iff $\Pi(AB/2) > \pi/4$ and are horoparallel iff $\Pi(AB/2) = \pi/4$. If the two tangents do intersect at a point Q, then int $(\overparen{AB})$ is on int $(\triangle AQB)$.

Proof Let $\overrightarrow{AC}$ and $\overrightarrow{BD}$ be radii of $\mathscr{C}$. Let M be the midpoint of $\overline{AB}$. Let the perpendicular bisector of $\overline{AB}$ intersect $\mathscr{C}$ at N. Let $\overleftrightarrow{AE}$ and $\overleftrightarrow{BF}$ be tangents of $\mathscr{C}$ with E and F on the same side of $\overleftrightarrow{AB}$ as N. (See Figure

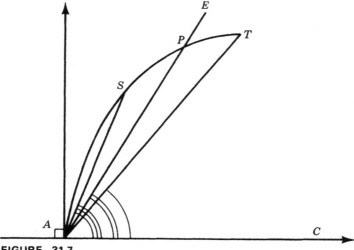

FIGURE 31.7

31.8.) Now, $m\angle MAE = \pi/2 - \Pi(AM)$. Hence $\overrightarrow{AE}|\overleftrightarrow{MN}$ iff $\Pi(AM) = \pi/4$, and $\overleftrightarrow{AE}$ intersects $\overleftrightarrow{MN}$ iff $\Pi(AM) > \pi/4$. Since $\overleftrightarrow{MN}$ is a line of symmetry for $\mathscr{C}$, it follows that $\overrightarrow{AE}|\overrightarrow{BF}$ iff $\Pi(AM) = \pi/4$ and that $\overleftrightarrow{AE}$ intersects $\overleftrightarrow{BF}$ iff $\Pi(AM) > \pi/4$.

Suppose $\overleftrightarrow{AE}$ does intersect $\overleftrightarrow{BF}$ at point Q. Then $\overrightarrow{AE}$ must intersect $\overrightarrow{BF}$ at Q. So Q and N are on the same side of $\overleftrightarrow{AB}$. Thus int $(\widehat{AB})$ is on the same side of $\overleftrightarrow{AB}$ as Q. Since $\overleftrightarrow{AQ}$ and $\overleftrightarrow{BQ}$ are tangents of $\mathscr{C}$, then int $(\widehat{AB})$ is on the same side of $\overleftrightarrow{AQ}$ as B and on the same side of $\overleftrightarrow{BQ}$ as A. Therefore int $(\widehat{AB})$ is on int $(\triangle AQB)$. ∎

We want to talk about the *length* of an arc of a horocircle. Of course we must define *length* first. Then there will be the problem of showing that the length exists, i.e., the length is a finite number. We must avoid the kind of nonsense of comparing infinite numbers that is illustrated in Section 23.4. A nice Euclidean curve that points out some of the difficulties is the Snowflake Curve of Exercise 31.18. As we have said before, the definition of a unicorn does not imply the existence of a unicorn.

DEFINITION 31.12 Let $\widehat{AB}$ be on horocircle $\mathscr{C}$ with center b. Then $|\widehat{AB}|$ is the least upper bound of all numbers $\Sigma_{i=0}^{i=n} T_i T_{i+1}$ where $\{T_0, T_{n+1}\} = \{A, B\}$, T_i is on $\mathscr{C}$, and T_i is between the line in b through T_{i-1} and the line in b through T_{i+1} for $i = 1, 2, \ldots, n$. If $|\widehat{AB}|$ exists, then $|\widehat{AB}|$ is called the *length* of $\widehat{AB}$. We say $\widehat{AB}$ is *longer* than $\widehat{CD}$ when $|\widehat{AB}| > |\widehat{CD}|$.

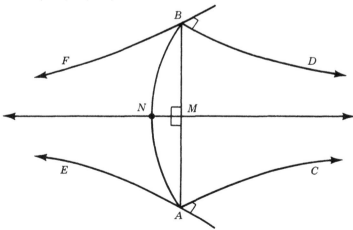

FIGURE 31.8

We first show that if chord $\overline{AB}$ of a horocircle is short enough, then $|\widehat{AB}|$ exists. In particular, if $\Pi(AB/2) > \pi/4$, then $|\widehat{AB}|$ exists.

Theorem 31.13   If the tangents at two points A and B on a horocircle intersect at a point Q, then $|\widehat{AB}| \leqq AQ + QB$.

Proof   Let $|\widehat{AB}|$ be on horocircle $\mathscr{C}$ with center b. We suppose the tangents at A and B intersect at Q. Let n be any positive integer. Let $T_o = A$, $T_{n+1} = B$, and $Q_o = Q$. For $i = 1, 2, \ldots, n$ let T_i be on $\mathscr{C}$ such that T_i is between the line in b through T_{i-1} and the line in b through T_{i+1}.

Since T_1 is on int $(\widehat{T_oB})$, then T_1 is in int $(\triangle T_oQ_oB)$, (Theorem 31.11). So, by Crossbar, $\overrightarrow{T_oT_1}$ intersects $\overline{Q_oB}$ at a point Q_1 such that Q_o-Q_1-B and $T_o-T_1-Q_1$. Likewise, since T_2 is in int $(\widehat{T_1B})$, then T_1 is in int $(\triangle T_1Q_1B)$. So, by Crossbar, $\overrightarrow{T_1T_2}$ intersects Q_1B at a point Q_2 such that Q_1-Q_2-B and $T_1-T_2-Q_2$. (Note that the last two sentences are obtained from the previous two by adding 1 to each subscript. See Figure 31.9.) We continue to define points Q_i in the same fashion for $i = 3, 4, \ldots, n$ such that $Q_{i-1}-Q_i-B$ and $T_{i-1}-T_i-Q_i$. Finally, let $Q_{n+1} = B$. Now the following inequalities follow from the Triangle Inequality:

$$T_oT_1 + (T_1Q_1) < T_oQ_o + Q_oQ_1,$$
$$T_1T_2 + (T_2Q_2) < (T_1Q_1) + Q_1Q_2,$$
$$T_2T_3 + (T_3Q_3) < (T_2Q_2) + Q_2Q_3,$$
$$\cdots$$
$$T_{n-1}T_n + (T_nQ_n) < (T_{n-1}Q_{n-1}) + Q_{n-1}Q_n,$$
$$T_nT_{n+1} + 0 \qquad < (T_nQ_n) \qquad + Q_nQ_{n+1}.$$

Adding the inequalities, we obtain

$$\Sigma \, T_iT_{i+1} < T_oQ_o + Q_oQ_{n+1} = AQ + QB.$$

Therefore, $|\widehat{AB}|$ exists and $|\widehat{AB}| \leqq AQ + QB$.   ∎

When $n = 1$, the proof above reduces to Euclid's Proposition I.21, our Theorem 18.16. It is true that $|\widehat{AB}|$ is strictly less than $AQ + QB$ in the theorem. However, we shall not need this result to show that $|\widehat{AB}|$ always exists.

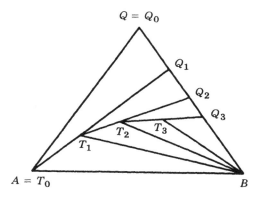

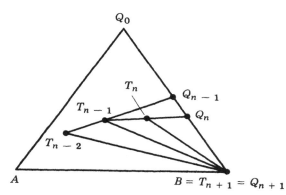

FIGURE 31.9

Theorem 31.14 If A and B are any two points on horocircle $\mathscr{C}$, then $|\widehat{AB}|$ exists and $|\widehat{AB}| > AB$. Congruent chords of one or more horocircles subtend congruent arcs. The longer of two noncongruent chords of one or more horocircles subtends a longer arc.

Proof Since any two horocircles are congruent (Theorem 29.2) and since the perpendicular bisector of a chord of a horocircle is a line of symmetry for that horocircle, it follows that there is an isometry (in fact, exactly two) mapping point P' on horocircle $\mathscr{C}'$ to point P on horocircle $\mathscr{C}$ with $\mathscr{C}'$ going to $\mathscr{C}$. Thus all the statements in the theorem follow immediately once it is established that every arc has a length.

Suppose $\widehat{AB}$ is on horocircle $\mathscr{C}$ with center b. Let $\Pi(r/2) = \pi/3$. If $AB \leqq r$, then $\Pi(AB/2) > \pi/4$ and $|\widehat{AB}|$ exists by the preceding theorem. So suppose $AB > r$. Let B_0 be on $\widehat{AB}$ and on the circle with center A and radius r. Define the sequence of points B_n for $n = 1, 2, 3, \cdots$ by letting B_n be the image of A under the reflection in the line of b through

B_{n-1}. Now B_n is on $\mathscr{C}$ and $AB_{n-1} < AB_n$ because B_{n-1} is in int $(\widehat{AB_n})$. Further, $|\widehat{AB_n}| = 2^n|\widehat{AB_0}|$. So $|\widehat{AB}|$ exists if there is some integer m such that B is on $\widehat{AB_m}$. In other words, we are done if $AB_m > AB$ for some m. (It is conceivable that the numbers AB_n remain bounded even though the numbers $|\widehat{AB_n}|$ do not.) Assume the set of all numbers AB_n has least upper bound s. There is a point S on $\mathscr{C}$ such that $AS = s$ with B and S on the same side of the radius through A. Also, there is a point T on $\widehat{AS}$ such that $ST = r$. Now, if some B_n is on $\widehat{TS}$, then S is on $\widehat{AB_{n+1}}$, contradicting the fact that s is an upper bound. Also, if no B_n is on $\widehat{TS}$, then s is an upper bound but not the least upper bound. In either case, we have a contradiction, and $|\widehat{AB}|$ always exists. ∎

Now that we know every arc on a horocircle has a length, we wish to show every positive real is the length of some arc.

Theorem 31.15 Given point A on horocircle $\mathscr{C}$ and $s > 0$, there exists a point P on $\mathscr{C}$ such that $|\widehat{AP}| = s$.

Proof We shall use the notation of Theorem 31.9. So $A = P_0$. Let $g(0) = 0$, and let $g(x) = |\widehat{P_0 P_x}|$ for nonzero real x. Since g is an unbounded function on the reals and $g(0) = 0$, the theorem follows from the intermediate value theorem provided g is continuous.

Suppose $\varepsilon > 0$. To show g is continuous at 0, we need to find a positive δ such that $0 < |x| < \delta$ implies $|\widehat{P_0 P_x}| < \varepsilon$. Let Q be a point such that $\overline{AQ} \perp \overline{AC}$ and $AQ = \varepsilon/4$. Then $\overleftrightarrow{AQ}$ is a tangent of $\mathscr{C}$. Let B be the image of A under the reflection in the line through Q lying in the center of $\mathscr{C}$. Then B is on $\mathscr{C}$, $\overleftrightarrow{QB}$ is a tangent of $\mathscr{C}$, and $QB = \varepsilon/4$. So $AQ + QB = \varepsilon/2$. Also, $|\widehat{AB}| \leq AQ + QB$. Further, for $0 < |x| < AB$, we have $|\widehat{P_0 P_x}| < |\widehat{AB}|$ since $P_0 P_x < AB$. Therefore, taking $\delta = AB$, we have $0 < |x| < \delta$ implies $|\widehat{P_0 P_x}| < \varepsilon$.

Suppose $\varepsilon > 0$ and $r \neq 0$. Then $|g(r) - g(x)| = |\widehat{P_r P_x}|$. To show g is continuous at r we need to find a positive δ such that $|r - x| < \delta$ implies $|\widehat{P_r P_x}| < \varepsilon$. By the preceding paragraph, there exists a positive η such that $0 < P_r P_x < \eta$ implies $|\widehat{P_r P_x}| < \varepsilon$. However, for $\eta > 0$ there exists a positive δ such that $|x - r| < \delta$ implies $P_r P_x < \eta$ (Theorem 31.9). Therefore, $|x - r| < \delta$ implies $|\widehat{P_r P_x}| < \varepsilon$. So g is a continuous function. ∎

DEFINITION 31.16 Let $\mathscr{C}$ be horocircle with center b. Point P is in int $(\mathscr{C})$, the **interior** of $\mathscr{C}$, if P is in the interior of some radius of $\mathscr{C}$.

The *exterior* of $\mathscr{C}$ is the set of points off both $\mathscr{C}$ and int $(\mathscr{C})$. Let $\mathscr{C}'$ be a horocircle concentric with $\mathscr{C}$. Let A and B be two points on $\mathscr{C}$. Let the lines in b through A and B, respectively, intersect $\mathscr{C}'$ at A' and B', respectively. Then, with respect to $\mathscr{C}$ and $\mathscr{C}'$, point A **corresponds** to point A', $\overline{AB}$ **corresponds** to $\overline{A'B'}$, $\widehat{AB}$ **corresponds** to $\widehat{A'B'}$ and AA' is the **distance between** $\mathscr{C}$ and $\mathscr{C}'$.

That the distance between concentric horocircles is well defined is included in Theorem 28.2.

Theorem 31.17 There is a positive constant k such that if $\mathscr{C}$ and $\mathscr{C}'$ are any concentric horocircles with $\mathscr{C}'$ on int $(\mathscr{C})$, $\widehat{AB}$ on $\mathscr{C}$ corresponds to $\widehat{A'B'}$ on $\mathscr{C}'$, and x is the distance between $\mathscr{C}$ and $\mathscr{C}'$, then $|\widehat{A'B'}| = |\widehat{AB}|e^{-x/k}$.

Proof Suppose point C on $\mathscr{C}$ corresponds to point C' on $\mathscr{C}'$ where B is in int $(\widehat{AC})$. We first show that $|\widehat{AC}|\,/\,|\widehat{AB}|$ is equal to $|\widehat{A'C'}|\,/\,|\widehat{A'B'}|$ by providing the absolute value of the difference of these numbers is less than every positive number. Suppose $\varepsilon > 0$. Let n be any positive integer such that $2^n\varepsilon > 1$. There exists a point P on $\widehat{AB}$ such that $|\widehat{AB}| = 2^n|\widehat{AP}|$. There exists a point Q in int $(\widehat{AC})$ such that $|\widehat{AC}| = m|\widehat{AP}| + |\widehat{QC}|$ with $|\widehat{QC}| \leq |\widehat{AP}|$ for some integer m (Theorem 31.15). Let P' and Q' be the points on $\mathscr{C}'$ that correspond to P and Q, respectively. (See Figure 31.10.) Then (Exercise 31.3), we have

$$|\widehat{A'B'}| = 2^n|\widehat{A'P'}|, \qquad |\widehat{A'C'}| = m|\widehat{A'P'}| + |\widehat{Q'C'}|,$$

and

$$|\widehat{Q'C'}| \leq |\widehat{A'P'}|.$$

Then,

$$\left| \frac{|\widehat{AC}|}{|\widehat{AB}|} - \frac{|\widehat{A'C'}|}{|\widehat{A'B'}|} \right| = \frac{1}{2^n} \left| \frac{|\widehat{QC}|}{|\widehat{AP}|} - \frac{|\widehat{Q'C'}|}{|\widehat{A'P'}|} \right| < \frac{1}{2^n} < \varepsilon.$$

So, $\dfrac{|\widehat{AC}|}{|\widehat{AB}|} = \dfrac{|\widehat{A'C'}|}{|\widehat{A'B'}|}$; therefore, $\dfrac{|\widehat{AB}|}{|\widehat{A'B'}|} = \dfrac{|\widehat{AC}|}{|\widehat{A'C'}|}$.

It follows from the last equation that, given two concentric horocircles $\mathscr{C}$ and $\mathscr{C}'$ with $\mathscr{C}'$ on int $(\mathscr{C})$, the ratio of the length of an arc on $\mathscr{C}$ to the length of the corresponding arc on $\mathscr{C}'$ can depend only on

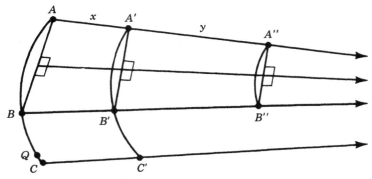

FIGURE 31.10

the distance between $\mathscr{C}$ and $\mathscr{C}'$. If x is the distance between $\mathscr{C}$ and $\mathscr{C}'$, we denote this ratio by $f(x)$.

For positive numbers x and y, we may suppose $AA' = x$ and $A'A'' = y$ where $A - A' - A''$. Let $\mathscr{C}''$ be the horocircle through A'' that is concentric with $\mathscr{C}$ and $\mathscr{C}'$. Then $\mathscr{C}''$ is on int $(\mathscr{C})$ and on int $(\mathscr{C}')$, and the distance between $\mathscr{C}$ and $\mathscr{C}''$ is $x + y$. (See Figure 31.10.) Let B'' on $\mathscr{C}''$ correspond to B on $\mathscr{C}$. Then

$$f(x) = \frac{|\widehat{AB}|}{|\widehat{A'B'}|}, \qquad f(y) = \frac{|\widehat{A'B'}|}{|\widehat{A''B''}|}, \qquad \text{and} \qquad f(x+y) = \frac{|\widehat{AB}|}{|\widehat{A''B''}|}.$$

Since $\frac{1}{2}A'B' > \frac{1}{2}A''B''$, then $f(x+y) > f(x)$. Thus f is a strictly increasing function such that $f(x+y) = f(x)f(y)$ for positive x and y.

Let $f(1) = a$. So $a > 1$. Computing $f(1/n + 1/n + \cdots + 1/n)$ once where there are n terms in the sum and again where there are m terms in the sum, we obtain $f(m/n) = a^{m/n}$ for positive integers m and n. Hence $f(r) = a^r$ for all positive rationals r. Since f is strictly increasing, we must have $f(x) = a^x$ for all positive reals x. To finish the proof and follow standard notation, we let $a = e^{1/k}$. ∎

DEFINITION 31.18 The *distance scale* is the positive constant k in Theorem 31.17.

By Theorem 31.17 each model of Σ has associated with it a positive constant k. If we multiply all distances of a given model by a positive constant t, we obtain a model of Σ, but the constant associated with this new model would be tk. (Replacing x by tx in the theorem requires replacing k by tk.) All we would be doing is changing the *scale* for distance. That is, if $(\mathscr{P}, \mathscr{L}, d, m)$ is a model of Σ with distance scale k, t is a positive constant, and $d'(P, Q) = t\, d(P, Q)$ for all points

P and Q, then $(\mathscr{P}, \mathscr{L}, d', m)$ is a model of Σ with distance scale tk. If we take $t = 1/k$, then $(\mathscr{P}, \mathscr{L}, d', m)$ is a model of Σ with distance scale 1. Using the Cayley–Klein Model and varying the value of t, we see that there exist models of Σ with any given positive distance scale. So, in the theory, the numerical value of the distance scale is not determined by the axioms.

31.2 HYPERBOLIC FUNCTIONS

Combinations of $\frac{1}{2}(e^x - e^{-x})$ and $\frac{1}{2}(e^x + e^{-x})$ appear so often in analysis that these expressions have been given the special names in our next definition.

DEFINITION 31.19 For a real number x

$$\sinh x = \frac{e^x - e^{-x}}{2}, \qquad \cosh x = \frac{e^x + e^{-x}}{2}, \qquad \tanh x = \frac{\sinh x}{\cosh x},$$

$$\sinh x \, \text{csch} \, x = \cosh x \, \text{sech} \, x = \tanh x \, \coth x = 1 \qquad \text{if} \quad x \neq 0.$$

Also, $\text{arcsinh} \, x = y$ if $\sinh y = x$; $\text{arccosh} \, x = y$ if $\cosh y = x$ and $y \geq 0$; $\text{arctanh} \, x = y$ if $\tanh y = x$.

The first six of the nine functions defined above are called the *hyperbolic functions*. Although the geometric connection between the hyperbolic functions and a hyperbola is slight (Exercise 31.16), this is the reason for the name. The name and notation for these functions is due to Lambert. The usual trig functions are often called the *circular functions* because of their close connection with circles. There is a good deal of algebraic similarity between the hyperbolic functions and the usual trig functions. For this reason, the hyperbolic functions are often called the *hyperbolic trig functions*. The full name of the function sinh is *hyperbolic sine*, and the other hyperbolic functions are similarly named.

The hyperbolic functions are often omitted from a calculus course because one can do without them. They are, after all, only *abbreviations*. From a practical point of view, we shall find these abbreviations indispensable for studying hyperbolic geometry. (Hyperbolic geometry was so named after the hyperbolic functions were so named.)

Graphs of sinh, cosh, tanh, and arctanh may be found in Figures 2.2 and 2.3.

Theorem 31.20 Let x and y be real numbers. Then

(a) $\cosh^2 x - \sinh^2 x = 1$, (b) $1 - \tanh^2 x = \operatorname{sech}^2 x$,

(c) $\cosh x + \sinh x = e^x$, (d) $\cosh x - \sinh x = e^{-x}$,

(e) $\sinh (x + y) = \sinh x \cosh y + \cosh x \sinh y$,

(f) $\sinh (x - y) = \sinh x \cosh y - \cosh x \sinh y$,

(g) $\cosh (x + y) = \cosh x \cosh y + \sinh x \sinh y$,

(h) $\cosh (x - y) = \cosh x \cosh y - \sinh x \sinh y$,

(i) $\tanh (x + y) = \dfrac{\tanh x + \tanh y}{1 + \tanh x \tanh y}$,

(j) $\sinh x + \sinh y = 2 \sinh \dfrac{x+y}{2} \cosh \dfrac{x-y}{2}$,

(k) $\sinh x - \sinh y = 2 \cosh \dfrac{x+y}{2} \sinh \dfrac{x-y}{2}$,

(l) $\cosh x + \cosh y = 2 \cosh \dfrac{x+y}{2} \cosh \dfrac{x-y}{2}$,

(m) $\cosh x - \cosh y = 2 \sinh \dfrac{x+y}{2} \sinh \dfrac{x-y}{2}$,

(n) $\sinh \dfrac{x}{2} = \pm \sqrt{\dfrac{\cosh x - 1}{2}}$, (o) $\cosh \dfrac{x}{2} = \sqrt{\dfrac{\cosh x + 1}{2}}$,

(p) $\tanh \dfrac{x}{2} = \pm \sqrt{\dfrac{\cosh x - 1}{\cosh x + 1}} = \dfrac{\cosh x - 1}{\sinh x} = \dfrac{\sinh x}{\cosh x + 1}$,

(q) $e^{-x} = \dfrac{\operatorname{sech} x}{1 + \tanh x} = \dfrac{1 - \tanh x}{\operatorname{sech} x}$,

(r) $\operatorname{arcsinh} x = \ln (x + \sqrt{x^2 + 1})$,

(s) $\operatorname{arccosh} x = \ln (x + \sqrt{x^2 - 1})$, $x \geq 1$,

(t) $\operatorname{arctanh} x = \tfrac{1}{2} \ln \dfrac{1+x}{1-x}$, $|x| < 1$.

Proof Each of the twenty formulas can be proved independently by simple substitution, using Definition 31.19. Of course, only (a), (c), (e), (g) and the two identities $\sinh -x = -\sinh x$ and $\cosh -x = +\cosh x$ should be proved this way. The others of the first seventeen follow from these. It is easy to derive the last three formulas. For example, suppose arccosh $x = y$. Then cosh $y = x \geq 1$. So $e^y + e^{-y} = 2x$. Multiplying both sides of this equation by e^y, we obtain a quadratic equation in e^y. Solving for e^y by the quadratic formula, we obtain the solution $e^y = x \pm \sqrt{x^2 - 1}$. Then $y = \pm \ln (x + \sqrt{x^2 - 1})$ and (s) follow by taking the logarithm of both sides of the solution. ∎

31.3 EXERCISES

31.1 If $\overleftrightarrow{AC}$, $\overleftrightarrow{PQ}$, and $\overleftrightarrow{BD}$ are in a horopencil and P is between $\overleftrightarrow{AC}$ and $\overleftrightarrow{BD}$, then the set of all points between $\overleftrightarrow{AC}$ and $\overleftrightarrow{BD}$ is the union of the three mutually disjoint sets: $\overleftrightarrow{PQ}$, the set of all points between $\overleftrightarrow{AC}$ and $\overleftrightarrow{PQ}$, and the set of all points between $\overleftrightarrow{PQ}$ and $\overleftrightarrow{BD}$.

● **31.2** Every point on a tangent of a cycle but off the cycle is on another tangent of the cycle. No point off a set of points can be on three tangents of that set.

31.3 Let $\overarc{AB}$ on horocircle $\mathscr{C}$ correspond to $\overarc{A'B'}$ on horocircle $\mathscr{C}'$. If point P is in int $(\overarc{AB})$, $|\overarc{AB}| = 2|\overarc{AP}|$, and P' on $\mathscr{C}'$ corresponds to P, then $|\overarc{A'B'}| = 2|\overarc{A'P'}|$. Now verify the step indicated in the proof of Theorem 31.17.

● **31.4** Given two nonperpendicular lines, there exists a third line perpendicular to one of the two lines and horoparallel to the other.

● **31.5** True or False?

(a) If point A is on horocircle $\mathscr{C}$, then there exists a unique point C such that $\overrightarrow{AC}$ is a radius of $\mathscr{C}$.

(b) Line l intersects horocircle $\mathscr{C}$ at exactly one point iff l is a tangent of $\mathscr{C}$.

(c) If point B is in int $(\overarc{AC})$, then $\angle ABC$ is acute.

(d) If point A is in int $(\overarc{BC})$, then $\angle ABC$ is acute.

(e) The tangents at two points A and B on a horocircle are hyperparallel iff $\Pi(AB/2) < \pi/4$.

(f) $\coth^2 x - \operatorname{csch}^2 x = 1$ for nonzero real x.

(g) $\sinh -x = -\sinh x$ but $\cosh -x = +\cosh x$ for all real x.

(h) Two points determine a horocircle.

(i) If horocircle $\mathscr{C}'$ is on the interior of horocircle $\mathscr{C}$, then a chord of $\mathscr{C}$ does not intersect $\mathscr{C}'$.

(j)   If $\overparen{AB}$ on horocircle $\mathscr{C}$ corresponds to $\overparen{A'B'}$ on horocircle $\mathscr{C}'$, then $|\overparen{AB}| \geqq |\overparen{A'B'}|$.

31.6   If $\overparen{AB}$ and $\overparen{EF}$ are equal arcs on horocircle $\mathscr{C}$, then $\{A, B\} = \{E, F\}$.

31.7   Analogue of Theorem 31.10 for hypercircles.

31.8   Suppose $\overline{AB}$ is a longest side of $\triangle ABC$. The A, B, C are on a circle, a horocircle, or a hypercircle iff $m\angle A$ is respectively greater than, equal to, or less than $\Pi(AC/2) - \Pi(AB/2)$.

31.9   Formulas (e) and (g) of Theorem 31.20.

31.10 $\sinh(a+b) + \sinh(a-b) = 2 \sinh a \cosh b$,

$\sinh(a+b) - \sinh(a-b) = 2 \cosh a \sinh b$,

$\cosh(a+b) + \cosh(a-b) = 2 \cosh a \cosh b$,

$\cosh(a+b) - \cosh(a-b) = 2 \sinh a \sinh b$.

● **31.11**   Formulas (j), (k), (l), and (m) of Theorem 31.20.

● **31.12**   Formulas (n), (o), and (p) of Theorem 31.20

31.13   Formulas (r) and (t) of Theorem 31.20.

31.14   A line can be hyperparallel but cannot be horoparallel to each of the three lines that contain a side of a given triangle.

31.15   Let l, m, n be three lines. If n is between l and m and l is between m and n, then m is between l and n.

31.16   Let (x', y') be a point in the first quadrant of the Cartesian plane. Let l be the x-axis and m the line through $(0, 0)$ and (x', y'). If (x', y') is on the circle with equation $x^2 + y^2 = 1$, then $x' = \cos\theta$ and $y' = \sin\theta$ where θ is twice the area of the region bounded by l, m, and the circle. If (x', y') is on the hyperbola with equation $x^2 - y^2 = 1$, then $x' = \cosh\theta$ and $y' = \sinh\theta$ where θ is twice the area of the region bounded by l, m, and the hyperbola.

31.17   The interior of a horocircle is a convex set.

31.18   A *snowflake curve*, due to Helge von Koch (1870–1924), is defined as the limit of the following sequence of curves. C_0 is an equilateral triangle. For positive integer n, C_{n+1} is obtained from C_n by constructing an equilateral triangle (not intersecting the interior of C_n) on the middle third of each side of C_n and then deleting those middle thirds. If C_0 has perimeter $3s$, the resulting snowflake curve bounds an area of $2\sqrt{3}\ s^2/5$. However, what can be said about arclength with respect to a snowflake curve?

*31.19 Generalize Euclid's Proposition I.21 for a finite number of points in the interior of a triangle.

GRAFFITI

$$\sin (x+y) = \sin x \cos y + \cos x \sin y,$$

$$\sin (x-y) = \sin x \cos y - \cos x \sin y,$$

$$\cos (x+y) = \cos x \cos y - \sin x \sin y,$$

$$\cos (x-y) = \cos x \cos y + \sin x \sin y.$$

$$\sin \frac{x}{2} = \pm \sqrt{\frac{1-\cos x}{2}}, \qquad \cos \frac{x}{2} = \pm \sqrt{\frac{1+\cos x}{2}},$$

$$\tan \frac{x}{2} = \frac{\sin x}{1+\cos x}.$$

$$\sin x + \sin y = 2 \sin \frac{x+y}{2} \cos \frac{x-y}{2},$$

$$\sin x - \sin y = 2 \sin \frac{x-y}{2} \cos \frac{x+y}{2},$$

$$\cos x + \cos y = 2 \cos \frac{x+y}{2} \cos \frac{x-y}{2},$$

$$\cos x - \cos y = -2 \sin \frac{x+y}{2} \sin \frac{x-y}{2}.$$

$$\int_1^e \frac{dx}{x} = 1, \qquad e = \sum_{n=0}^{\infty} \frac{1}{n!} = \lim_{n \to \infty} (1 + 1/n)^n.$$

$$\sinh x = \sum_{n=0}^{\infty} \frac{x^{2n+1}}{(2n+1)!}, \qquad \sin x = \sum_{n=0}^{\infty} (-1)^n \frac{x^{2n+1}}{(2n+1)!},$$

$$\cosh x = \sum_{n=0}^{\infty} \frac{x^{2n}}{(2n)!}, \qquad \cos x = \sum_{n=0}^{\infty} (-1)^n \frac{x^{2n}}{(2n)!},$$

$$\operatorname{arctanh} x = \sum_{n=0}^{\infty} \frac{x^{2n+1}}{2n+1}, \qquad \arctan x = \sum_{n=0}^{\infty} (-1)^n \frac{x^{2n+1}}{2n+1}.$$

$$i^2 + 1 = 0,$$

$$\sin z = \frac{e^{iz} - e^{-iz}}{2i}, \qquad \cos z = \frac{e^{iz} + e^{-iz}}{2},$$

$$e^{iz} = \cos z + i \sin z,$$

$$e^{i\pi} + 1 = 0.$$

CHAPTER 32

The Fundamental Formula

32.1 TRIGONOMETRY

A certain constant S is useful in developing the trigonometry.

DEFINITION 32.1   If $\Pi(p) = \pi/4$, then S is the constant such that chords of length $2p$ subtend arcs of length $2S$ on a horocircle.

In Figure 32.1, $\overset{\frown}{AB}$ has length S. Figure 32.2 is essentially the same as Figure 32.1 but from a different perspective. Since longer chords of a horocircle subtend longer arcs, the next theorem is just a restatement of Theorem 31.11.

Theorem 32.2   Let A and B be two points on horocircle $\mathscr{C}$. Then the

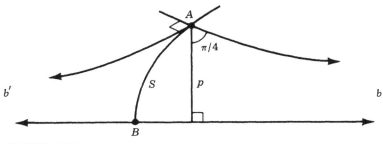

FIGURE 32.1

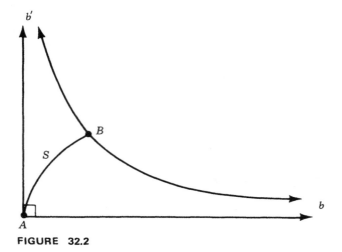

FIGURE 32.2

tangent to $\mathscr{C}$ at A and the line containing the radius of $\mathscr{C}$ through B intersect iff $|\widehat{AB}| < S$ and are horoparallel iff $|\widehat{AB}| = S$.

Suppose $\Pi(p) = \pi/4$. A change in the distance scale k would change the value of p. (See Figure 32.3.) So p depends on k. Since S is defined above in terms of p, then the value of S also depends on k. As it turns out, in developing the formulas for trigonometry the constant S always nicely cancels out at the end. So we don't really care what the actual value of S is. However, the distance scale k is not so obliging. In fact, k pops up all over the place, especially as a denominator. We are going to avoid this nuisance by assuming $k = 1$. In other words, we are going to develop the trigonometry for the *special case $k = 1$*. From the special case we will be able to deduce the desired results for arbitrary distance scale.

DEFINITION 32.3 Unless specifically stated otherwise, we assume

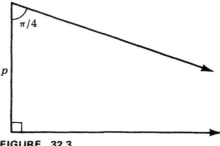

FIGURE 32.3

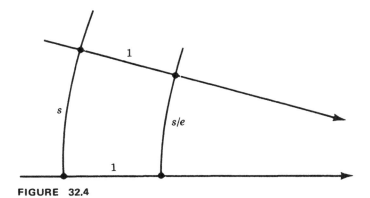

FIGURE 32.4

the distance scale is 1. Thus, "Suppose the distance scale is 1" is tacitly assumed to be a part of the statement of each of the remaining theorems that contain no other reference to the distance scale.

We have tentatively assumed the distance scale is such that a
ratio of the lengths of corresponding arcs on two horocircles is the
number e when the distance between the horocircles is 1. (See Figure 32.4). The number π is involved in our choice of scale for angle
measure, and the number e is now involved in our choice of scale for
distance. We are presently in somewhat the same situation as the student of high school trigonometry who fails to understand why radian
measurement may be preferred over degree measurement. Of course,
using radian measurement avoids having the constant $\pi/180$ appear-
ing all over the place in calculus. It is not obvious we have picked the
best scale for distance. Since distance and angle measure are related
by the critical function, it might seem more natural to pick the scale
such that p is 1 when $\Pi(p) = \pi/4$. Indeed, many earlier geometers did
exactly that. However, in that case, the constant arcsinh 1, which
equals $\ln(1 + \sqrt{2})$, is ubiquitous. Later, with the advantage of hind-
sight, it will be clear that we have made the most convenient choice
for the distance scale.

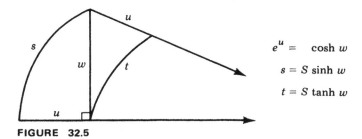

$$e^u = \cosh w$$
$$s = S \sinh w$$
$$t = S \tanh w$$

FIGURE 32.5

The content of the next theorem is contained in Figure 32.5.

Theorem 32.4 Let $\widehat{AB}$ on horocircle $\mathscr{C}$ have length s. Let C be the foot of the perpendicular from B to the radius of $\mathscr{C}$ through A. Let $\widehat{CD}$ correspond to $\widehat{AB}$ and $AC = u$. Then $e^u = \cosh BC$, $|\widehat{AB}| = S \sinh BC$, and $|\widehat{CD}| = S \tanh BC$.

Proof Let $|\widehat{AB}| = s$, $|\widehat{CD}| = t$, and $BC = w$. Let $\mathscr{C}$ have center b. $\overrightarrow{AC}$ and $\overrightarrow{BD}$ are radii of $\mathscr{C}$ and $BD = u$. Let $\widehat{CD}$ be on horocircle $\mathscr{C}'$. Since $\overleftrightarrow{BC}$ is tangent to $\mathscr{C}'$ at C and intersects $\overleftrightarrow{BD}$, we have $t < S$, (Theorem 32.2). Let E and F be the points on $\mathscr{C}'$ such that $|\widehat{CE}| = |\widehat{CF}| = S$ with D on $\widehat{CE}$. Then $|\widehat{DE}| = S - t$ and $|\widehat{DF}| = S + t$. (See Figure 32.6.)

From $m\angle BDC = \pi - \Pi(CD/2)$, we know $\angle BDC$ is an obtuse angle of $\triangle BDC$. So $w > u$. Let G and H be the two points on $\overleftrightarrow{BD}$ such that $BG = BH = w$ with D on $\overline{BG}$. Then $DG = w - u$ and $DH = w + u$.

Let I and J be the points such that $\widehat{GI}$ corresponds to $\widehat{DF}$ and $\widehat{HJ}$ corresponds to $\widehat{DE}$. Since $\overleftrightarrow{BC}$ is tangent to $\mathscr{C}'$ at C and $|\widehat{CF}| = |\widehat{CE}| = S$, then $\overrightarrow{BC}|\overrightarrow{IF}$ and $\overrightarrow{CB}|\overrightarrow{EJ}$ by definition of S. Let b_2 be the horopencil containing $\overrightarrow{BC}$ and $\overrightarrow{IF}$; let b_3 be the horopencil containing $\overrightarrow{BC}$ and $\overrightarrow{EJ}$. Since $m\angle CBD = \Pi(w)$, then the perpendicular to $\overline{BG}$ at

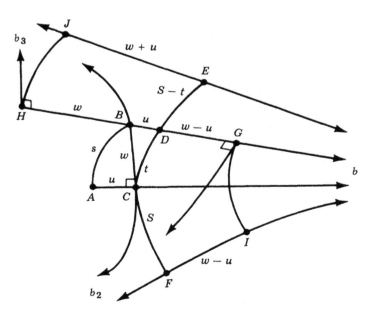

FIGURE 32.6

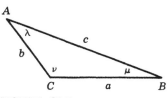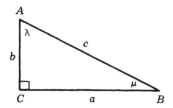

FIGURE 32.7

G is in b_2 and the perpendicular to $\overline{BH}$ at H is in b_3. So we have $|\widehat{GI}| = |\widehat{HJ}| = S$.

Now, applying the fundamental lemma on the ratio of lengths of corresponding arcs of concentric horocircles, we have

$$S + t = |\widehat{DF}| = |\widehat{GI}|\,e^{w-u} = Se^{-u}e^{w},$$

$$S - t = |\widehat{DE}| = |\widehat{HJ}|\,e^{-(w+u)} = Se^{-u}e^{-w},$$

$$t = |\widehat{CD}| = |\widehat{AB}|\,e^{-u} = se^{-u}.$$

Adding the first two equations, we obtain $e^u = \cosh w$. Then, subtracting the first two equations, we obtain $t = S \tanh w$. Finally, from the third equation, we have $s = S \sinh w$. ■

Corollary 32.5 A chord of length $2w$ of a horocircle subtends an arc of length $2S \sinh w$. If $\Pi(p) = \pi/4$, then $p = \operatorname{arcsinh} 1 = \operatorname{arccosh} \sqrt{2} = \ln (1 + \sqrt{2})$.

Proof The first statement follows directly from the theorem. If we take $w = p$, then we must have $2S = 2S \sinh p$. So $\sinh p = 1$ and $\cosh p = \sqrt{2}$. Adding these two equations, we have $e^p = 1 + \sqrt{2}$. ■

Given $\triangle ABC$, you are probably used to having a be the length of the side opposite $\angle A$. Also, in Euclidean geometry, α is often supposed to be $m\angle A$. In hyperbolic geometry a different notation is usually used. We still have $a = BC$, but α is $\Pi(a)$ while $m\angle A$ is λ. See Figure 32.7.

DEFINITION 32.6 Given $\triangle ABC$, we say we are using **standard notation** if we suppose

$$\alpha = \Pi(a) = \Pi(BC), \qquad \lambda = \Pi(l) = m\angle A,$$

$$\beta = \Pi(b) = \Pi(AC), \qquad \mu = \Pi(m) = m\angle B,$$

$$\gamma = \Pi(c) = \Pi(AB), \qquad \nu = \Pi(n) = m\angle C,$$

and $s = (a+b+c)/2$, regardless of the value of the distance scale.

Using standard notation for $\triangle ABC$, we have $\nu = \pi/2$ and $n = 0$ when $\angle C$ is right. Further, $n < 0$ iff $\angle C$ is obtuse (Definition 31.1). So there exists a segment of length n iff $\angle C$ is acute.

Theorem 32.7 If $\triangle ABC$ has a right angle at C, then, with standard notation,

(a) $\cosh c = \cosh a \cosh b$, (b) $\cosh c = \sinh l \sinh m$,

(c) $\sinh c = \sinh a \cosh l$, (c') $\sinh c = \sinh b \cosh m$,

(d) $\tanh b = \tanh c \tanh l$, (d') $\tanh a = \tanh c \tanh m$,

(e) $\sinh b = \tanh a \sinh l$, (e') $\sinh a = \tanh b \sinh m$,

(f) $\cosh a = \cosh m \tanh l$, (f') $\cosh b = \cosh l \tanh m$.

Proof Let b be the horopencil determined by $\overrightarrow{CA}$. Let $\mathscr{C}_1$ be the horocircle through B with center b. Let $\overleftrightarrow{CA}$ intersect $\mathscr{C}_1$ at R. Since $\angle C$ is right, we have $R - C - A$. Let $RC = r$. Let L be such that $B - A - L$ and $AL = l$. Let $\mathscr{C}_2$ be the horocircle through L with center b. Since $\Pi(l) = m \angle A$, then $\overleftrightarrow{BL}$ is tangent to $\mathscr{C}_2$ at L. Let points P and Q on $\mathscr{C}_2$ correspond to B and R, respectively, on $\mathscr{C}_1$. Let $AQ = q$. Then $BP = r + b + q$. (See Figure 32.8.) Let $s_1 = |\widehat{BR}|$, $s_2 = |\widehat{PQ}|$, and $s_3 = |\widehat{QL}|$. We have (Theorems 31.17 and 32.4)

(i) $s_1 = s_2 e^{r+b+q}$, (ii) $e^r = \cosh a$,

(iii) $s_1 = S \sinh a$, (iv) $e^q = \cosh l$,

(v) $s_3 = S \tanh l$, (vi) $e^{r+b+q} = \cosh (c + l)$,

(vii) $s_2 + s_3 = S \tanh (c + l)$.

From (i), using (iii) to substitute for s_1, using (v) and (vii) to

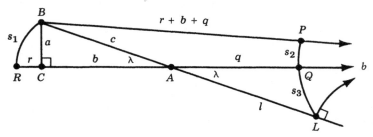

FIGURE 32.8

substitute for s_2, and using (vi) to substitute for e^{r+b+q}, we obtain (c). (The actual computations are left for Exercise 32.1.) Equation (c') follows from (c) by symmetry.

Now $e^b = [\cosh (c+l)]/[\cosh a \cosh l]$ by (ii), (iv), and (vi). Then, using (c) and substituting for e^b in $\cosh b = \frac{1}{2}(e^b + e^{-b})$ and $\tanh b = (e^{2b}-1)/(e^{2b}+1)$, after some time we obtain (a) and (d), respectively. By symmetry, we also have (d'). Using (c) and (d') to substitute in the expression $\cosh l \tanh m$, we obtain $\cosh b$ by (a). Thus (f) and (f') hold. Dividing (c) by (a), we have an expression for $\tanh c$ which when substituted in (d) gives (e). So (e') also holds. Finally, using (f) and (f') to substitute in (a), we obtain (b). ∎

In Euclidean geometry the famous equation $c^2 = a^2 + b^2$ gives the relation between the lengths of the hypotenuse and the legs of a right triangle. Here we have $\cosh c = \cosh a \cosh b$. Equations (c) through (f') in Theorem 32.7 are also analogues of familiar formulas from Euclidean geometry. Of course (b) has no analogue in Euclidean geometry. Why? We now derive our first hyperbolic analogues of the Euclidean *law of sines* and *law of cosines*.

Theorem 32.8   With standard notation for $\triangle ABC$,

$$\frac{\operatorname{sech} l}{\sinh a} = \frac{\operatorname{sech} m}{\sinh b} = \frac{\operatorname{sech} n}{\sinh c}$$

and

$$\cosh c = \cosh a \cosh b - \sinh a \sinh b \tanh n.$$

Proof   Let D be the foot of the perpendicular from A to $\overleftrightarrow{BC}$. Let $AD = h$. If $\angle C$ is not obtuse, then $\sinh b = \sinh h \cosh n$ (by (c) of Theorem 32.7). If $\angle C$ is obtuse, then $\sinh b = \sinh h \cosh t$ where $\Pi(t) = \pi - \Pi(n)$. (See Figure 32.9.) For the second case, $t = -n$. Since $\cosh n = \cosh -n$, we have $\sinh b = \sinh h \cosh n$ in any case. Likewise, whether $\angle B$ is obtuse or not, we have $\sinh c = \sinh h \cosh m$. So

$$\frac{\sinh b}{\sinh c} = \frac{\cosh n}{\cosh m} = \frac{\operatorname{sech} m}{\operatorname{sech} n}.$$

The first equation in the theorem now follows by symmetry.

Let $CD = d$. If neither $\angle B$ nor $\angle C$ is obtuse, then $\cosh DB = \cosh (a-d)$ and $\tanh d = \tanh b \tanh n$. If $\angle B$ is obtuse, then $\cosh DB = \cosh (a-d)$ and $\tanh d = \tanh b \tanh n$. If $\angle C$ is obtuse, then $\cosh BD = \cosh (a+d)$ and $\tanh d = \tanh b \tanh -n = -\tanh b \tanh n$. Since $\cosh c = \cosh h \cosh DB$ and $\cosh b = \cosh h \cosh d$, we have

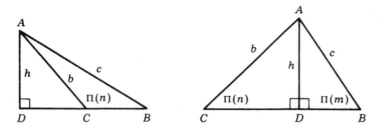

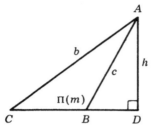

FIGURE 32.9

$$\cosh c = \frac{\cosh b \cosh DB}{\cosh d} = \frac{\cosh b \cosh (a \pm d)}{\cosh d}$$

$$= [\cosh b][\cosh a \cosh d \pm \sinh a \sinh d]/[\cosh d]$$

$$= \cosh a \cosh b \pm \sinh a \cosh b \tanh d$$

$$= \cosh a \cosh b - \sinh a \cosh b \tanh b \tanh n$$

$$= \cosh a \cosh b - \sinh a \sinh b \tanh n. \quad \blacksquare$$

Theorem 32.9 *Hyperbolic Pythagorean Theorem* Given $\triangle ABC$ with standard notation, $\angle C$ is right iff $\cosh c = \cosh a \cosh b$.

Proof Each of the following is equivalent to the next: $\cosh c = \cosh a \cosh b$, $\sinh a \sinh b \tanh n = 0$, $\tanh n = 0$, $n = 0$, $\nu = m\angle C = \pi/2$. $\quad \blacksquare$

Suppose $m\angle BAC = \lambda = \Pi(l) < \pi/2$. By (d) of Theorem 32.7, for Figure 32.10 we have $\tanh s = \tanh x \ \tanh l$. It follows that $s/x = [\text{arctanh } (\tanh x \tanh l)]/x$. As point P moves along int $(\overrightarrow{AB})$ we know s/x does not remain constant (Theorem 22.23). Since the defect of *small* triangles is *close* to 0, it seems that the hyperbolic trigonometry must be *close* to Euclidean trigonometry for such triangles. One might *hope* that the limit of s/x as x approaches 0 be $\cos \lambda$. If you remember

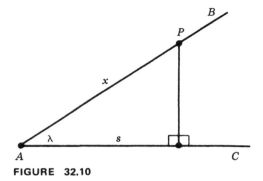

FIGURE 32.10

L'Hospital's rule from calculus, you can easily check that the limit is actually tanh l. So as a good *guess* we might have $\cos \lambda = \tanh l$. If this is correct, then $\lambda = \arccos (\tanh \Pi^{-1}(\lambda))$. This explains the first line in the proof of our next theorem, where we *prove* our guess is correct. (As usual, $r = \arccos s$ iff $\cos r = s$ and $0 \leqq r \leqq \pi$.)

Theorem 32.10 *The Fundamental Formula* For all real x,

$$\cos \Pi (x) = \tanh x.$$

Proof Let $f(\lambda) = \arccos \tanh \Pi^{-1}(\lambda)$ for $0 < \lambda < \pi$. The range of f coincides with its domain. Since f is a composite of three continuous functions, f is itself continuous. Further, $f(\pi/2) = \pi/2$. If $\lambda = \Pi(l)$ for real number l, then $\cos f(\lambda) = \tanh l$. It follows that then $\sin f(\lambda) = \operatorname{sech} l$. We shall prove $f(\lambda) = \lambda$.

First, suppose λ and μ are each positive numbers less than $\pi/2$. Let $\Pi(l) = \lambda$ and $\Pi(m) = \mu$. Let h be any positive number less than l and less than m. Let C and D be any two points such that $CD = h$. Let A and B be points on opposite sides of $\overleftrightarrow{CD}$ such that $m\angle DCB = \lambda$ and $m\angle DCA = \mu$. Since $\Pi(h) > \lambda$ and $\Pi(h) > \mu$, the perpendicular to $\overleftrightarrow{CD}$ at D intersects both $\overrightarrow{CA}$ and $\overrightarrow{CB}$. So we may suppose $\overleftrightarrow{AB}$ is per-

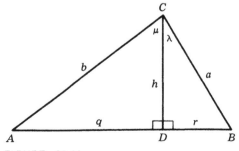

FIGURE 32.11

pendicular to $\overleftrightarrow{CD}$ at D. (See Figure 32.11. Note we are not using the standard notation for $\triangle ABC$.) Let $AD=q$ and $DB=r$. So $AB=r+q$. Since $0<\lambda+\mu<\pi$, there is a real t such that $\Pi(t)=\lambda+\mu$. Then, since $\cosh(r+q)=\cosh a\cosh b-\sinh a\sinh b\tanh t$, we have

$$\cos f(\lambda+\mu)=\tanh t=\frac{\cosh a\cosh b-\cosh(r+q)}{\sinh a\sinh b}$$

$$=\coth a\coth b-\frac{\cosh r\cosh q}{\sinh a\sinh b}-\frac{\sinh r\sinh q}{\sinh a\sinh b}.$$

We shall evaluate the last three terms separately. From $\tanh h=\tanh a\quad\tanh l=\tanh a\quad\cos f(\lambda)$ and $\tanh h=\tanh b\quad\tanh m=\tanh b\cos f(\mu)$, we have $\coth a\coth b=\coth^2 h\cos f(\lambda)\cos f(\mu)$.

From $\quad\dfrac{\cosh r}{\sinh a}=\dfrac{\cosh r\cosh h}{\sinh a\cosh h}=\dfrac{\cosh a}{\sinh a\cosh h}$

$$=\frac{\tanh h}{\tanh a\sinh h}=\frac{\tanh l}{\sinh h}=\frac{\cos f(\lambda)}{\sinh h}$$

and the similar equation $\dfrac{\cosh q}{\sinh b}=\dfrac{\cos f(\mu)}{\sinh h}$,

we have $\quad\dfrac{\cosh r\cosh q}{\sinh a\sinh b}=\operatorname{csch}^2 h\cos f(\lambda)\cos f(\mu)$.

From $\quad\dfrac{\sinh r}{\sinh a}=\dfrac{\sinh r}{\sinh r\cosh l}=\operatorname{sech} l=\sin f(\lambda)$

and the similar equation $\dfrac{\sinh q}{\sinh b}=\sin f(\mu)$,

we have $\quad\dfrac{\sinh r\sinh q}{\sinh a\sinh b}=\sin f(\lambda)\sin f(\mu)$.

So $\quad\cos f(\lambda+\mu)=\cos f(\lambda)\cos f(\mu)-\sin f(\lambda)\sin f(\mu)$

$$=\cos(f(\lambda)+f(\mu)).$$

Hence $\quad f(\lambda+\mu)=f(\lambda)+f(\mu)\quad$ when $\quad 0<\lambda<\pi/2$

and $\quad 0<\mu<\pi/2$.

Let $f(1)=c$. Let s and t be positive integers such that $s/t<\pi$. So $s/2t<\pi/2$. Evaluating $f(1/2t+1/2t+\cdots+1/2t)$ when there are $2t$ terms in the sum and again when there are s terms in the sum, we obtain $f(s/2t) = sc/2t$. Then $f(s/2t + s/2t) = sc/t$. So $f(\lambda) = \lambda c$ for all positive rationals λ less than π. Since f is continuous, then $f(\lambda) = \lambda c$ for all positive real λ less than π. Finally, since $f(\pi/2) = \pi/2$, we have $c = 1$ and $f(\lambda) = \lambda$ for all λ in the domain of f. ∎

Corollary 32.11 For any nonzero real number x,

$$\sin \Pi(x) = \operatorname{sech} x, \qquad \cos \Pi(x) = \tanh x, \qquad \tan \Pi(x) = \operatorname{csch} x,$$

$$\csc \Pi(x) = \cosh x, \qquad \sec \Pi(x) = \coth x, \qquad \cot \Pi(x) = \sinh x.$$

Proof $\sin \Pi(x) = +(1 - \cos^2 \Pi(x))^{1/2} = \operatorname{sech} x$ for all x, since $0 < \Pi(x) < \pi$. The remaining equations follow easily from the first two. ∎

Corollary 32.12 For any real number x,

$$\tan \frac{\Pi(x)}{2} = e^{-x} \qquad \text{and} \qquad \Pi(x) = 2 \arctan e^{-x}$$

Proof By the preceding corollary, we have

$$\tan \frac{\Pi(x)}{2} = \frac{\sin \Pi(x)}{1 + \cos \Pi(x)} = \frac{\operatorname{sech} x}{1 + \tanh x} = e^{-x}. \quad \blacksquare$$

Corollary 32.13 If $\triangle ABC$ has a right angle at C, then, with standard notation,

(a) $\cosh c = \cosh a \cosh b$, (b) $\cosh c = \cot \lambda \cot \mu$,

(c) $\sin \lambda = \dfrac{\sinh a}{\sinh c}$, (c') $\sin \mu = \dfrac{\sinh b}{\sinh c}$,

(d) $\cos \lambda = \dfrac{\tanh b}{\tanh c}$, (d') $\cos \mu = \dfrac{\tanh a}{\tanh c}$,

(e) $\tan \lambda = \dfrac{\tanh a}{\sinh b}$, (e') $\tan \mu = \dfrac{\tanh b}{\sinh a}$,

(f) $\cosh a = \dfrac{\cos \lambda}{\sin \mu}$, (f') $\cosh b = \dfrac{\cos \mu}{\sin \lambda}$,

(g) $\tan \dfrac{1}{2} \delta \triangle ABC = \tanh \dfrac{a}{2} \tanh \dfrac{b}{2}$,

(h) $\sin \delta \triangle ABC = \dfrac{\sinh a \sinh b}{1 + \cosh a \cosh b}$.

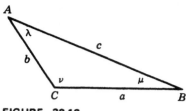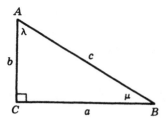

FIGURE 32.12

Proof Except for the last two, the equations follow directly from Theorem 32.7. Then, using only (f), (f′), and identities for tanh $(x/2)$ and tan $(x/2)$, we have

$$\tanh^2 \frac{a}{2} \tanh^2 \frac{b}{2} = \frac{(\cosh a) - 1}{(\cosh a) + 1} \frac{(\cosh b) - 1}{(\cosh b) + 1}$$

$$= \frac{1 - \sin (\lambda + \mu)}{1 + \sin (\lambda + \mu)} \frac{\cos (\lambda - \mu)}{\cos (\lambda - \mu)}$$

$$= \frac{1 - \cos \delta \triangle ABC}{1 + \cos \delta \triangle ABC}$$

$$= \tan^2 \tfrac{1}{2} \delta \triangle ABC.$$

Finally, from formulas (c), (c′), (d), and (d′), we have

$$\sin \delta \triangle ABC = \cos \lambda \cos \mu - \sin \lambda \sin \mu$$

$$= (\sinh a \sinh b)(\cosh c - 1)/(\sinh^2 c)$$

$$= (\sinh a \sinh b)/(1 + \cosh c). \quad \blacksquare$$

Corollary 32.14 *Hyperbolic Law of Sines* With standard notation for $\triangle ABC$,

$$\frac{\sin \lambda}{\sinh a} = \frac{\sin \mu}{\sinh b} = \frac{\sin \nu}{\sinh c}.$$

Proof Theorem 32.8 and sech $x = \sin \Pi(x)$. $\quad \blacksquare$

Corollary 32.15 *Hyperbolic Law of Cosines* With standard notation for $\triangle ABC$,

$$\cosh a = \cosh b \cosh c - \sinh b \sinh c \cos \lambda,$$

$$\cosh b = \cosh a \cosh c - \sinh a \sinh c \cos \mu,$$

$\cosh c = \cosh a \cosh b - \sinh a \sinh b \cos \nu$.

Proof Theorem 32.8 and $\tanh x = \cos \Pi(x)$. ■

Because AAA is a theorem in hyperbolic geometry, we can expect to find an equation that gives the length of a side of a given triangle in terms of the measures of the angles of the triangle. This is done in the next theorem, where only one of three similar formulas is stated.

Corollary 32.16 With standard notation for $\triangle ABC$,

$$\cosh c = \frac{\cos \lambda \cos \mu + \cos \nu}{\sin \lambda \sin \mu}.$$

Proof By the Hyperbolic Law of Sines and the Hyperbolic Law of Cosines, we can obtain the following identity:

$\sin \lambda \sin \mu \ (\sinh a \sinh b \sinh^2 c)$

$$= (\sin \lambda \sinh c)(\sin \mu \sinh c)(\sinh a \sinh b)$$

$$= \sinh^2 a \sinh^2 b \sin^2 \nu$$

$$= \sinh^2 a \sinh^2 b - (\sinh a \sinh b \cos \nu)^2$$

$$= (-1 + \cosh^2 a)(-1 + \cosh^2 b) - (\cosh a \cosh b - \cosh c)^2$$

$$= 1 - \cosh^2 a - \cosh^2 b - \cosh^2 c + 2 \cosh a \cosh b \cosh c.$$

Now, solving for each of $\cos \lambda$, $\cos \mu$, and $\cos \nu$ in the three equations of the Hyperbolic Law of Cosines and then substituting these solutions in the expression $\cos \lambda \cos \mu + \cos \nu$, we obtain the desired equation from the identity above. ■

Suppose $\triangle ABC$ has a right angle at C. With standard notation and distance scale 1, we have $\cosh c = \cosh a \cosh b$ and $\sin \lambda = (\sinh a)/(\sinh c)$. Now suppose the distance scale is k. Dividing all distances by k has the effect of changing the distance scale to 1. (See the remarks following Definition 31.18.) So, with distance scale k, it follows that

$\cosh (c/k) = [\cosh (a/k)][\cosh (b/k)]$

and

$\sin \lambda = [\sinh (a/k)]/[\sinh (c/k)]$

The other trigonometric formulas are obtained in the same way.

Theorem 32.17 Let the distance scale be k. Then with standard notation for $\triangle ABC$

$$\cos \Pi(x) = \tanh(x/k) \qquad \text{for all real } x,$$

$$\cos \nu = \frac{[\cosh(a/k)][\cosh(b/k)] - [\cosh(c/k)]}{[\sinh(a/k)][\sinh(b/k)]},$$

$$\cosh(c/k) = \frac{\cos \lambda \cos \mu + \cos \nu}{\sin \lambda \sin \mu}.$$

Proof Suppose $\llcorner\!\!\lrcorner ABCD$ is closed with $\angle C$ right and $a = BC$. Regardless of the distance scale, we have $\Pi(a) = m \angle ABC$ by the definition of Π, (Definition 24.11). If the distance scale is 1, then $m \angle ABC = 2 \arctan e^{-a}$ (Corollary 32.12). By the definition of Π, it follows that $m \angle ABC = 2 \arctan e^{-a/k}$ when the distance scale is k. So $\Pi(a) = 2 \arctan e^{-a/k}$ and $\cos \Pi(a) = \tanh(a/k)$ for $a > 0$. The first formula in the statement of the theorem now follows by the symmetry in the definition of Π for negative values (Definition 31.1). The remaining two formulas in the theorem follow immediately from the Hyperbolic Law of Cosines (Corollary 32.15) and its "dual" (Corollary 32.16). ∎

The formulas in Theorem 32.17 have some interesting implications. Let the terms involving sinh and cosh be replaced by their infinite power series (Section 31.4). Then taking the limit as k approaches infinity, we obtain the following three equations:

$$\Pi(x) = \pi/2, \qquad \cos \nu = \frac{a^2 + b^2 - c^2}{2ab}, \qquad \lambda + \mu + \nu = \pi.$$

These are equations for Euclidean geometry. So Euclidean geometry is a limiting case of hyperbolic geometry. To be flippant, "The Euclidean plane is a Bolyai–Lobachevsky plane with infinite distance scale." From a different perspective, we can conclude that Euclidean trigonometry is a very good approximation for small triangles. For example, if $\nu = \pi/2$ and the ratio c/k is small, then $c^2 = a^2 + b^2$ is a very good approximation to the relation of the lengths of the sides of the right triangle. Further, as we might guess from the remarks preceding the proof of the Fundamental Formula (Theorem 32.10), $\sin \lambda = a/c$ and $\cos \lambda = b/c$ are very good approximations for small right triangles.

32.2 COMPLEMENTARY SEGMENTS

The definitions of *complementary segments* and the involutory mapping *star* are each motivated by Figure 32.13 with $\overrightarrow{AD} \mid \overrightarrow{CE}$. In gen-

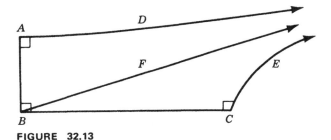

FIGURE 32.13

eral, the star mapping is dependent on the distance scale, as is Π.

DEFINITION 32.18 If $\overrightarrow{AB} \perp \overline{BC}$ and the perpendicular to $\overleftrightarrow{AB}$ at A is horoparallel to the perpendicular to $\overleftrightarrow{BC}$ at C, then $\overline{AB}$ and $\overline{BC}$ are *complementary segments.* For any positive real x the positive real x^* is defined by $\Pi(x^*) + \Pi(x) = \pi/2$, regardless of the value of the distance scale.

Theorem 32.19 If $\overline{AB} \perp \overline{BC}$, then $\overline{AB}$ and $\overline{BC}$ are complementary segments iff $BC = (AB)^*$.

Proof Suppose $\overline{AB} \perp \overline{BC}$, $\overline{AD} \perp \overline{AB}$, $\overline{EC} \perp \overline{BC}$, and $\overrightarrow{AD} | \overrightarrow{BF}$ with D, E, F in int $(\angle ABC)$. Then each of the following conditions is easily seen to be equivalent to the next:

(i) $\overline{AB}$ and $\overline{CD}$ are complementary segments,

(ii) $\overleftrightarrow{CE} | \overleftrightarrow{AD},$ (iii) $\overrightarrow{CE} | \overrightarrow{AD},$ (iv) $\overrightarrow{CE} | \overrightarrow{BF},$

(v) $\Pi(BC) = \pi/2 - \Pi(AB),$ (vi) $BC = (AB)^*.$ ∎

Theorem 32.20 For positive real x:

$\sinh x^* = \operatorname{csch} x,$ $\cosh x^* = \coth x,$ $\tanh x^* = \operatorname{sech} x,$

$\tanh (x^*/2) = e^{-x},$ and $x^* = 2 \operatorname{arctanh} e^{-x}.$

Proof $\sinh x^* = \cot \Pi(x^*) = \tan \Pi(x) = \operatorname{csch} x$ and $\cosh x^* = \csc \Pi(x^*)$ $= \sec \Pi(x) = \coth x$. The remaining equations follow immediately from these and the identity $\tanh (y/2) = (\sinh y)/(1 + \cosh y)$. ∎

The *trigonometry* for Lambert quadrilaterals is considered next. See Figure 32.14.

Theorem 32.21 Given $\square ABCD$, let $u = AD$, $v = AB$, $w = CD$, $z = BC$, and $\phi = m\angle C$. Then,

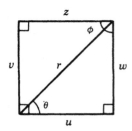 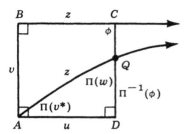

FIGURE 32.14

(a) $\sinh w = \sinh v \cosh z,$ (a') $\sinh z = \sinh u \cosh w,$

(b) $\tanh w = \cosh u \tanh v,$ (b') $\tanh z = \cosh v \tanh u,$

(c) $\sin \phi = \dfrac{\cosh v}{\cosh w} = \dfrac{\cosh u}{\cosh z},$

(d) $\cos \phi = \sinh u \sinh v = \tanh w \tanh z,$

(e) $\cot \phi = \tanh u \sinh w = \tanh v \sinh z.$

Proof   Let $\theta = m\angle CAD$ and $r = AC$. Then, $\sinh w = \sin\theta \ \sinh r = \cos(\pi/2 - \theta) \sinh r = \tanh v \cosh r$. Replacing $\cosh r$ in this equation by $\cosh v \cosh z$ or by $\cosh u \cosh w$, we obtain (a) and (b), respectively. Equation (a') and (b') follow by symmetry. Then,

$$\frac{\sin\phi}{\sinh BD} = \frac{\sin m\angle BDC}{\sinh z} = \frac{\cos m\angle ADB}{\sinh z} = \frac{\tanh u}{\tanh BD \sinh z}.$$

So, by (a') we have

$$\sin\phi = \frac{\tanh u \cosh BD}{\sinh z} = \frac{\tanh u \cosh u \cosh v}{\sinh u \cosh w} = \frac{\cosh v}{\cosh w}.$$

Equations (c) now follow by symmetry.

Since $z > u$, let Q be the intersection of $\overrightarrow{DC}$ and the circle with center A and radius z. So $\triangle AQD$ has a right angle at D. By (a'), (b'), and (c), respectively, we have

$$\sin m\angle AQD = \frac{\sinh u}{\sinh z} = \operatorname{sech} w,$$

$$\cos m\angle QAD = \frac{\tanh u}{\tanh z} = \operatorname{sech} v = \tanh v^{*},$$

$$\cosh QD \quad = \frac{\cosh z}{\cosh u} = \csc\phi.$$

Hence $m\angle AQD = \Pi(w)$, $m\angle QAD = \Pi(v^*)$, and $\Pi(QD) = \phi$. Equations (d) and (e) now follow from the formulas for a right triangle. (In particular, (d) follows from (e') and (d) of Theorem 32.7, and (e) follows from (e) and (c') of Theorem 32.7.) ■

Corollary 32.22   Given $\boxed{\text{L}}\,NMAB$ and $t > MN$, there is a unique point P on $\overrightarrow{MA}$ whose distance to $\overleftrightarrow{NB}$ is t.

Proof   By (a) of the theorem, it follows that P is uniquely determined on $\overrightarrow{MA}$ by $\cosh MP = (\sinh t)/(\sinh MN)$.   ■

Corollary 32.23   Suppose point A is off line l. Let B be the foot of the perpendicular from A to l and $\overleftrightarrow{BC} = l$. Let m be the line perpendicular to $\overleftrightarrow{AB}$ at A. Let D be the foot of the perpendicular from C to m. Then the circle with center A and radius BC intersects $\overline{CD}$ at a point Q such that $\overrightarrow{AQ}|\overrightarrow{BC}$.

Proof   In the notation used in the proof of the theorem, we have $\Pi(CD) < \phi = \Pi(QD)$ since $\overleftrightarrow{BC}$ and $\overleftrightarrow{AD}$ are hyperparallel. So Q is on $\overline{CD}$. Also, since $m\angle QAD = \Pi(v^*)$, then $m\angle QAB = \Pi(v) = \Pi(AB)$. Since $\angle ABC$ is right, we have $\overrightarrow{AQ}|\overrightarrow{BC}$.   ■

The circle in the preceding corollary intersects $\overrightarrow{CD}$ in another point R such that $\overleftrightarrow{AR}$ and $\overleftrightarrow{AQ}$ are the two lines through A that are horoparallel to $\overleftrightarrow{BC}$. See Figure 32.15. Therefore, we have a *construction* for horoparallels.

Looking back at Figure 32.14, we can see that $\triangle AQD$ exists iff $\boxed{\text{L}}\,ABCD$ exists. Now let's change the notation as follows: Let $u = a$, $v = m^*$, $w = l$, $z = c$, and $\phi = \Pi(b)$. Then $\triangle AQD$ of Figure 32.14 is labeled as in Figure 32.16, and $\boxed{\text{L}}\,ABCD$ of Figure 32.14 is labeled as in Figure 32.17. Therefore, there exists a right triangle having the

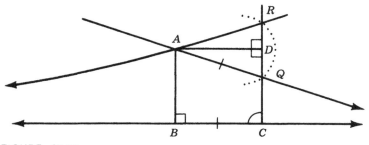

FIGURE   32.15

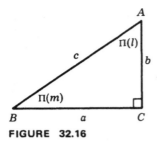

FIGURE 32.16

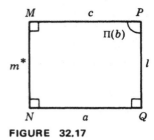

FIGURE 32.17

parameters indicated in Figure 32.16 iff there exists a Lambert quadrilateral having the parameters indicated in Figure 32.17.

We have just shown that the existence of right triangle $\triangle ABC$ having the parameters in the first row of Table 32.1 is equivalent to the existence of $\boxed{L}\, NMPQ$ having the parameters in the first row of Table 32.2. Suppose we have written down a list of rules by which one obtains the first row of one table from the first row of the other. Note that the columns of each table are arranged so that reading a row backwards determines a congruent figure. Now applying our rules to the first row of the second table read backwards, we obtain the second row of the first table. So the existence of the second triangle is equivalent to the existence of the first Lambert quadrilateral. Next apply the rules to the second row of the first table read backwards. Continuing in this fashion until we return to the original triangle (read backwards), we shall obtain both tables. (Although this may sound confusing, the actual computation is not difficult.) The following corollary is then established.

Corollary 32.24 The existence of a triangle having the parameters of any row in Table 32.1 implies the existence of a triangle having the parameters of any other row in Table 32.1 and the existence of a Lambert quadrilateral having the parameters of any row in Table 32.2. Further, the existence of a Lambert quadrilateral having the parameters of a row in Table 32.2 implies the existence of a triangle having the parameters of any row in Table 32.1.

It follows that the equations in Theorem 32.7 and Corollary 32.13 apply to the Lambert quadrilateral illustrated in Figure 32.17.

TABLE 32.1

$\triangle ABC$ $m\angle C = \pi/2$	BC	$m\angle B$	AB	$m\angle A$	AC
(1)	a	$\Pi(m)$	c	$\Pi(l)$	b
(2)	m^*	$\Pi(a^*)$	l	$\Pi(c)$	b
(3)	c^*	$\Pi(b^*)$	a^*	$\Pi(l)$	m^*
(4)	l^*	$\Pi(m)$	b^*	$\Pi(a^*)$	c^*
(5)	a	$\Pi(c)$	m	$\Pi(b^*)$	l^*

TABLE 32.2

$\boxed{L}\, NMPQ$	NM	MP	$m\angle P$	PQ	QN
(1)	m^*	c	$\Pi(b)$	l	a
(2)	c^*	l	$\Pi(m^*)$	a^*	b
(3)	l^*	a^*	$\Pi(c^*)$	b^*	m^*
(4)	a	b^*	$\Pi(l^*)$	m	c^*
(5)	b	m	$\Pi(a)$	c	l^*

32.3 EXERCISES

The distance scale is assumed to be 1.

32.1 Carry out the computation needed in the proof of Theorem 32.7.

32.2 For all real x and y,

$\tan (\Pi(x+y)/2) = \tan (\Pi(x)/2) \tan (\Pi(y)/2)$.

● **32.3** For all real x and y,

$$\sin \Pi(x \pm y) = \frac{\sin \Pi(x) \sin \Pi(y)}{1 \pm \cos \Pi(x) \cos \Pi(y)},$$

$$\cos \Pi(x \pm y) = \frac{\cos \Pi(x) \pm \cos \Pi(y)}{1 \pm \cos \Pi(x) \cos \Pi(y)}.$$

● **32.4** Lobachevsky's equations for $\triangle ABC$ in standard notation:

$$\sin \Pi(c) = \frac{\sin \Pi(l) \sin \Pi(m)}{\cos \Pi(l) \cos \Pi(m) + \cos \Pi(n)}$$

$$= \frac{\sin \Pi(a) \sin \Pi(b)}{1 - \cos \Pi(a) \cos \Pi(b) \cos \Pi(n)}.$$

● **32.5** With standard notation for $\triangle ABC$ and M the midpoint of $\overline{AB}$, $\cosh CM = (\cosh a + \cosh b)/(2 \cosh c/2)$.

● **32.6** Restate Theorem 32.10 and Theorem 32.20 without the assumption that the distance scale is 1.

● **32.7** True or False?

(a) If line m is between the two horoparallel lines l and n, then m is in the horopencil containing l and n.

(b) For all positive x, $\sinh x = \operatorname{csch} x^*$ and $\tanh x = \operatorname{sech} x^*$.

(c) For all positive x, $\tanh^2 x + \tanh^2 x^* = 1$.

(d) If $x \neq 0$, then $\tan \Pi(2x) = \frac{1}{2} \sin \Pi(x) \tan \Pi(x)$.

(e) For all x, $\tan \Pi(x/2) = e^{-x}$.

(f) For all x, $\cos \Pi(x/2) = (\cos \Pi(x))/(1 + \sin \Pi(x))$.

(g) For all positive x, $\sinh x^* = \tan \Pi(x)$ and $\tan x^* = \sin \Pi(x)$.

(h) $\Pi(1) < \Pi(\ln (1 + \sqrt{2})) = \pi/4$.

(i)   For $\triangle ABC$ in standard notation with $m\angle C = \pi/2$:

$\sin \Pi(c) = \sin \Pi(a) \sin \Pi(b),$

$\cos \Pi(a) = \cos \Pi(m) \cos \Pi(c).$

(j)   With standard notation for $\triangle ABC$:

$\cosh c = \cosh a \cosh b - \sinh a \sinh b \cos \nu,$

$\cos \nu = -\cos \lambda \cos \mu + \sin \lambda \sin \mu \cosh c.$

32.8 $\displaystyle\int_0^x \operatorname{sech} x \, dx = \pi/2 - \Pi(x).$

32.9 In the Cartesian plane, sketch the graph of f if $f(x) = x^*$.

32.10 A regular n-gon inscribed in a circle of radius R has sides of length a where $\sinh (a/2) = \sinh R \sin (\pi/n)$.

32.11 The radius of the circle inscribed in a regular n-gon with sides of length a is r where $\tanh (a/2) = \sinh r \tan (\pi/n)$.

● **32.12** The distance from the center of symmetry to a side of any regular 4-gon is always less than 1.

32.13 The circumference of a circle with radius r is $2\pi \sinh r$.

32.14 Let $\overline{AB}$ be a longest side of $\triangle ABC$. Then A, B, C are on a circle, a horocircle, or a hypercircle iff $\sinh \frac{1}{2}AB$ is respectively less than, equal to, or greater than $\sinh \frac{1}{2}BC + \sinh \frac{1}{2}CA$.

32.15 Let l and m be the tangents at points A and B of a cycle. If the cycle is a circle with radius r, then l and m intersect, are horoparallel, or are hyperparallel iff $\sinh (AB/2)$ is respectively less than, equal to, or greater than $\tanh r$. If the cycle is a horocircle, then l and m intersect, are horoparallel, or are hyperparallel iff $\sinh (AB/2)$ is respectively less than, equal to, or greater than 1. If the cycle is a hypercircle with radius r, then l and m intersect, are horoparallel, or are hyperparallel iff $\sinh (AB/2)$ is respectively less than, equal to, or greater than $\coth r$.

● **32.16** A circle whose diameter is at least $\ln 3$ cannot be the inscribed circle of any triangle.

32.17 Using L'Hospital's rule, prove

$$\lim_{x \to 0} \frac{\operatorname{arctanh} (\tanh l \tanh x)}{x} = \tanh l.$$

32.18 If $\boxed{S}\,ABCD$, $AB = r$, and $AD = q$, then the length of the arc with endpoints B and C on the hypercircle with center $\overleftrightarrow{AD}$ is $q\cosh r$. What is the length of an arc on the hypercircle subtended by a chord of length c?

32.19 If $dy/dx = \cos y$ with $|y| < \pi/2$ and $y = 0$ when $x = 0$, then $\tan y = \sinh x$. The function gd defined by $gd\,x = \arctan \sinh x$ for real x is called the *Gudermannian*. Sketch the graphs of gd and f where $f(x) = \pi/2 - gd\,x$.

32.20 A chord of length $\ln\,\tfrac{1}{2}(3 + \sqrt{5})$ subtends an arc of length S on a horocircle.

32.21 Given $\triangle ABC$ in standard notation with $n = 0$, $a = 1$, and $\Pi(l) = \pi/4$, find b, c, and m.

32.22 $\Pi(x/2) \neq \Pi(x)/2$ for all x; $(x/2)^* \neq x^*/2$ for $x > 0$.

32.23 With reference to Figure 32.14,

$$\cos \Pi(u) = \sin \Pi(v)\,\cos \Pi(z),$$

$$\sin \Pi(r) = \sin \Pi(u)\,\sin \Pi(w) = \sin \Pi(v)\,\sin \Pi(z),$$

$$\tanh^2 u + \tanh^2 v = \tanh^2 r.$$

32.24 Let $\overline{AB}$ be a chord of horocircle $\mathscr{C}$. Let $\overleftrightarrow{AC}$ be a tangent of $\mathscr{C}$ with B and C on the same side of the radius through a. If $m\angle BAC = \theta$, then $|\widehat{AB}| = 2S\tan\theta$. If $\overrightarrow{CB}$ contains a radius of $\mathscr{C}$ and $m\angle ACB = \phi$, then $|\widehat{AB}| = 2S\cos\phi$.

● **32.25** Use an isosceles right triangle with legs of length 2 to show that (one-half) the product of the length of a side of a triangle and the length of the altitude to that side depends, in general, on which side of the triangle is chosen.

● **32.26** One half the product of the hyperbolic sine of the length of a side of a triangle and the hyperbolic sine of the length of the altitude to that side is independent of which side of the triangle is chosen. (Why should this constant be called H, the *Heron* of the triangle?)

32.27 Using the formulas in the next exercise, find the constant of the preceding exercise in terms of the sides of $\triangle ABC$ in standard notation.

32.28 With standard notation for $\triangle ABC$,

$$\sin\frac{\lambda}{2} = \sqrt{\frac{\sinh\,(s - b)\,\sinh\,(s - c)}{\sinh b\,\sinh c}}$$

and

$$\cos\frac{\lambda}{2} = \sqrt{\frac{\sinh s \sinh (s-a)}{\sinh b \sinh c}}.$$

32.29 If r is the radius of the inscribed circle of $\triangle ABC$ in standard notation, then $\tanh r = \tan \frac{1}{2}\lambda \sinh (s-a)$.

32.30 For $\triangle ABC$ in standard notation, $\Pi(s-b)$ is greater than, equal to, or less than $\frac{1}{2}(\pi-\nu)$ iff $\Pi(s-c)$ is respectively greater than, equal to, or less than $\frac{1}{2}(\pi-\mu)$.

32.31 A cycle other than the inscribed circle of a given triangle that is tangent to the three lines containing the sides of the triangle is called an *escribed cycle* of the triangle. A triangle has three escribed cycles. Let $\mathscr{C}$ be the escribed cycle in int $(\angle A)$ of $\triangle ABC$ in standard notation. Let $q = \tan \frac{1}{2}\lambda \sinh s$. Then $\mathscr{C}$ is a circle, a horocircle, or a hypercircle iff q is respectively less than, equal to, or greater than 1. If $\mathscr{C}$ is a circle with radius r_a, then $\tanh r_a = q$; if $\mathscr{C}$ is a hypercircle with radius d_a, then $\coth d_a = q$.

32.32 A triangle with two escribed horocircles is isosceles. If $\triangle ABC$ has three escribed horocircles, then the triangle is equilateral with $\cosh AB = 3/2$, $\cos m\angle A = 3/5$, $\tanh r = 1/4$, and $\tanh R = 1/2$ where r and R are the radii of the inscribed and circumscribed circles, respectively.

32.33 Gauss' equations for $\triangle ABC$ with standard notation:

$$\sin\frac{\lambda+\mu}{2} \cosh\frac{c}{2} = \cos\frac{\nu}{2}\cosh\frac{a-b}{2},$$

$$\cos\frac{\lambda+\mu}{2} \cosh\frac{c}{2} = \sin\frac{\nu}{2}\cosh\frac{a+b}{2},$$

$$\sin\frac{\lambda-\mu}{2} \sinh\frac{c}{2} = \cos\frac{\nu}{2}\sinh\frac{a-b}{2},$$

$$\cos\frac{\lambda-\mu}{2} \sinh\frac{c}{2} = \sin\frac{\nu}{2}\sinh\frac{a-b}{2}.$$

32.34 Analogues of Heron's formula for $\triangle ABC$ with standard notation:

$$\sin\frac{\delta\triangle ABC}{2} = \frac{\sqrt{\sinh s \sinh (s-a) \sinh (s-b) \sinh (s-c)}}{2\cosh\dfrac{a}{2}\cosh\dfrac{b}{2}\cosh\dfrac{c}{2}},$$

$$\tan\frac{\delta\triangle ABC}{4}=\sqrt{\tanh\frac{s}{2}\tanh\frac{s-a}{2}\tanh\frac{s-b}{2}\tanh\frac{s-c}{2}}.$$

32.35 If $\triangle ABC$ in standard notation is inscribed in a circle of radius R, then

$$\tanh R \sin \tfrac{1}{2}\delta\triangle ABC = \tanh \tfrac{1}{2}a \tanh \tfrac{1}{2}b \tanh \tfrac{1}{2}c.$$

*32.36 *Mukhopadhyaya's pentagon:* Given $\triangle ABC$ in standard notation, there exists a pentagon with sides, in order, having length l, b^*, c, a^*, m.

*32.37 $S = 1$.

● ***32.38** Are there any hyperbolic Pythagorean triples?

GRAFFITI

Some Euclidean formulas for $\triangle ABC$ *where* K *is area,* R *is circumradius,* r *is inradius,* r_a *is radius of escribed circle with center in int* $(\angle A)$, *and* h_a *is the distance from* A *to* $\overleftrightarrow{BC}$:

$$K = \sqrt{s(s-a)(s-b)(s-c)} = rs = r_a(s-a)$$

$$= \sqrt{r\,r_a\,r_b\,r_c} = \frac{abc}{4R} = \tfrac{1}{2}a\,h_a,$$

$$\frac{1}{r} = \frac{1}{r_a} + \frac{1}{r_b} + \frac{1}{r_c} = \frac{1}{h_a} + \frac{1}{h_b} + \frac{1}{h_c},$$

$$r_a + r_b + r_c = r + 4R.$$

Formulas for $\triangle ABC$ *from* Euclidean spherical trigonometry:

$$\frac{\sin\lambda}{\sin a} = \frac{\sin\mu}{\sin b} = \frac{\sin\nu}{\sin c},$$

$$\cos c = +\cos a \cos b + \sin a \sin b \cos \nu,$$

$$\cos \nu = -\cos \lambda \cos \mu + \sin \lambda \sin \mu \cos c.$$

If $\nu = \pi/2$, then $\cos c = \cos a \cos b = \cot \lambda \cot \mu,$

$$\sin\lambda = \frac{\sin a}{\sin c}, \qquad \cos\lambda = \frac{\tan b}{\tan c}, \qquad \tan\lambda = \frac{\tan a}{\sin b}.$$

If z *is a complex number, then sin* $iz = i \sinh z$ *and cos* $iz = \cosh z$.

CHAPTER 33

Categoricalness and Area

33.1 ANALYTIC GEOMETRY

No one of the distance function d, the angle measure function m, or the distance scale k determines another of the three. For example, knowing m, we can determine whether $\overline{AB}$ is congruent to $\overline{CD}$ or not from AAA, but we cannot find the length of $\overline{AB}$ without knowing k. Likewise, knowing d, we can determine whether $\angle ABC$ is congruent to $\angle DEF$ or not from SSS, but we cannot find the measure of $\angle ABC$ without knowing k. Any two of d, m, k completely determines the third by the formulas in Theorem 32.17. In particular, since k is a constant, the distance scale is determined by any single equation $\cos m\angle PVQ = \tanh(VP/k)$ where $\angle PVQ$ is a critical angle for $\overline{VP}$. See Figure 33.1. We shall use this fact to show the Cayley–Klein Model has distance scale 1.

The Cayley–Klein Model (Section 23.2) has distance function h and angle measure function n. Let $0 < a < 1$. Let $P = (0, 0)$, $V = (0, a)$, $Q = (1/2, a/2)$, and $R = (1/2, 0)$. See Figure 33.2. Then $\overrightarrow{VQ} | \overrightarrow{PR}$. From the formula for angle measure we have $\cos n\angle VPR = 0$ and $\cos n\angle PVQ = a$. So $\angle VPR$ is right. Then $\angle PVQ$ is a critical angle for $\overline{VP}$. Thus, the model has distance scale 1 iff $a = \tanh h(V, P)$. From the formula for distance, we have $h(V, P) = \frac{1}{2}\ln[(1+a)/(1-a)]$. So $\tanh h(V, P) = a$, as desired. (See equation (t) of Theorem 31.20. Note the seemingly superfluous $\frac{1}{2}$ in the definition of the distance

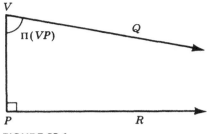

FIGURE 33.1

function h in Section 9.2 is just what is needed here.) The Cayley–Klein Model has distance scale 1.

Coordinatization of a Bolyai–Lobachevsky plane is considered next. Because a Lambert quadrilateral is not a rectangle, the introduction of coordinates in a Bolyai–Lobachevsky plane is a little more interesting than the introduction of coordinates in the Euclidean plane. In Figure 33.3 the distance scale is assumed to be 1.

DEFINITION 33.1   Let k be the distance scale. Suppose $\overrightarrow{OX} \perp \overrightarrow{OY}$, $\overleftrightarrow{OX}$ has coordinate system f such that $f(O)=0$ and $f(X)>0$, and $\overleftrightarrow{OY}$ has coordinate system g such that $g(O)=0$ and $g(Y)>0$. For arbitrary point P: let U and V be the feet of the perpendiculars from P to $\overleftrightarrow{OX}$ and $\overleftrightarrow{OY}$, respectively; let $u=f(U)$ and $v=g(V)$; let $r=OP$; let θ be a number such that $-\pi < \theta \leqq \pi$, $\tanh (u/k) = \cos \theta \tanh (r/k)$, and $\tanh (v/k) = \sin \theta \tanh (r/k)$; and let w be the real number such that $\tanh (w/k) = \cosh (u/k) \tanh (v/k)$. Then the ordered pair $\{\overrightarrow{OX}, \overrightarrow{OY}\}$ is a *frame* with *axes* $\overrightarrow{OX}$ and $\overrightarrow{OY}$. With respect to this frame, point P has *axial coordinates* (u,v), *polar coordinates* (r,θ), and *Lobachevsky coordinates* (u,w).

In the definition, θ is uniquely determined if $r \neq 0$. If $v>0$, then $\theta = m\angle XOP$, but $\theta = -m\angle XOP$ if $v<0$. If O, P_1, P_2 are three points,

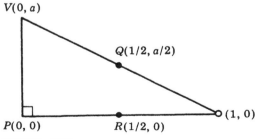

FIGURE   33.2

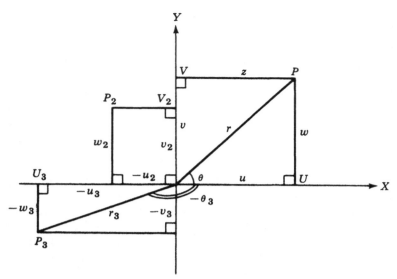

FIGURE 33.3

then $\cos(\theta_2 - \theta_1) = \pm 1$ iff the points are collinear and, otherwise, $\cos(\theta_2 - \theta_1) = \cos m\angle P_1 O P_2$. Although the coordinates are defined for arbitrary distance scale k, to save space in what follows below, we consider only the case $k = 1$.

Every point has a unique ordered pair of Lobachevsky coordinates, and, conversely, every ordered pair of real numbers is the pair of Lobachevsky coordinates for some unique point. If a is a nonzero real, then $u = a$ is an equation of a line but $w = a$ is an equation of a hypercircle. Further (Exercise 33.9), $e^{-u} = \tanh w$ is an equation of the line in the *first quadrant* that is horoparallel to both axes. Thus, in general, a line does not have a linear equation in Lobachevsky coordinates.

Every point has a unique ordered pair of axial coordinates. However, not every ordered pair of real numbers is a pair of axial coordinates. Let U and V be points on $\overleftrightarrow{OX}$ and $\overleftrightarrow{OY}$, respectively, with $V \neq 0$. By the definition of complementary segments, the perpendiculars to the axes at U and V will intersect iff $|u| < |v|^*$, which holds iff $\tanh^2 u < \tanh^2 |v|^*$. It follows that (u, v) *are the axial coordinates of a point iff* $\tanh^2 u + \tanh^2 v < 1$.

We shall now derive a formula for distance between points. Suppose O, P, Q are distinct points. Let P have axial coordinates (u, v) and polar coordinates (OP, θ); let Q have axial coordinates (u_2, v_2) and polar coordinates (OQ, θ_2). From the definition of θ, we have $\tanh^2 u + \tanh^2 v = \tanh^2 OP$, (Theorem 32.21). Thus

$$\cosh OP = \frac{1}{(1 - \tanh^2 u - \tanh^2 v)^{1/2}}.$$

Also,

$$\cos(\theta_2 - \theta) = \cos\theta\cos\theta_2 + \sin\theta\sin\theta_2$$

$$= \frac{\tanh u \tanh u_2 + \tanh v \tanh v_2}{\tanh OP \tanh OQ}.$$

Suppose O, P, Q are collinear. Then $\cos(\theta_2 - \theta) = -1$ and $PQ = OP + OQ$ iff $P\text{-}O\text{-}Q$, but $\cos(\theta_2 - \theta) = 1$ and $PQ = |OP - OQ|$ otherwise. From $\cosh PQ = \cosh(OP \pm OQ)$, we obtain

$$\cosh PQ = \cosh OP \cosh OQ - \sinh OP \sinh OQ \cos(\theta_2 - \theta)$$

$$= [\cosh OP \cosh OQ][1 - \tanh OP \tanh OQ \cos(\theta_2 - \theta)].$$

This last result also holds when O, P, Q are not collinear by the Hyperbolic Law of Cosines. Substitution now gives the desired formula. *If point P_1 has axial coordinates (u_1, v_1) and point P_2 has axial coordinates (u_2, v_2), then*

$$\cosh P_1 P_2 = \frac{1 - \tanh u_1 \tanh u_2 - \tanh v_1 \tanh v_2}{(1 - \tanh^2 u_1 - \tanh^2 v_1)^{1/2}(1 - \tanh^2 u_2 - \tanh^2 v_2)^{1/2}}.$$

DEFINITION 33.2 If point P has axial coordinates (u, v) with respect to a given frame, then P has **Beltrami coordinates** (x, y) with respect to that frame where k is the distance scale, $x = \tanh(u/k)$, and $y = \tanh(v/k)$.

We can now restate the two results obtained above in terms of Beltrami coordinates. Since there is no statement to the contrary, the hypothesis of the next theorem includes the assumption that the distance scale is 1 by Definition 32.3.

Theorem 33.3 With respect to a given frame:

(a) Every point has a unique ordered pair of Beltrami coordinates, and (x, y) is an ordered pair of Beltrami coordinates iff $x^2 + y^2 < 1$.

(b) If point P_1 has Beltrami coordinates (x_1, y_1) and point P_2 has Beltrami coordinates (x_2, y_2), then the distance $P_1 P_2$ is given by the following equivalent formulas:

$$\cosh P_1 P_2 = \frac{1 - x_1 x_2 - y_1 y_2}{(1 - x_1^2 - y_1^2)^{1/2}(1 - x_2^2 - y_2^2)^{1/2}},$$

$$\tanh P_1 P_2 = \frac{[(x_2 - x_1)^2 + (y_2 - y_1)^2 - (x_1 y_2 - x_2 y_1)^2]^{1/2}}{1 - x_1 x_2 - y_1 y_2},$$

$$P_1 P_2 = \frac{1}{2} \ln \frac{1 - x_1 x_2 - y_1 y_2 + [(x_2 - x_1)^2 + (y_2 - y_1)^2 - (x_1 y_2 - x_2 y_1)^2]^{1/2}}{1 - x_1 x_2 - y_1 y_2 - [(x_2 - x_1)^2 + (y_2 - y_1)^2 - (x_1 y_2 - x_2 y_1)^2]^{1/2}}.$$

(c) $Ax + By + C = 0$ is an equation of a line in Beltrami coordinates iff $A^2 + B^2 > C^2$, and every line has such an equation.

(d) Given $\angle PVQ$, if the Beltrami coordinates of P, V, and Q are respectively (x_1, y_1), (x_0, y_0), and (x_2, y_2), then

$$\cos m\angle PVQ =$$

$$\frac{(x_1 - x_0)(x_2 - x_0) + (y_1 - y_0)(y_2 - y_0) - (x_1 y_0 - y_1 x_0)(x_2 y_0 - y_2 x_0)}{[(x_1 - x_0)^2 + (y_1 - y_0)^2 - (x_1 y_0 - y_1 x_0)^2]^{1/2}[(x_2 - x_0)^2 + (y_2 - y_0)^2 - (x_2 y_0 - y_2 x_0)^2]}$$

(e) If $A_1 x + B_1 y + C_1 = 0$ and $A_2 x + B_2 y + C_2 = 0$ are equations of two intersecting lines in Beltrami coordinates and ψ is the measure of an angle in the union of these two lines, then

$$\cos \psi = \pm \frac{A_1 A_2 + B_1 B_2 - C_1 C_2}{(A_1^2 + B_1^2 - C_1^2)^{1/2}(A_2^2 + B_2^2 - C_2^2)^{1/2}}.$$

In particular, the two lines are perpendicular iff $A_1 A_2 + B_1 B_2 = C_1 C_2$.

(f) If (x_1, y_1) and (x_2, y_2) are the Beltrami coordinates of two distinct points, let $t_1 = (1 - x_1^2 - y_1^2)^{1/2}$ and $t_2 = (1 - x_2^2 - y_2^2)^{1/2}$. Then the midpoint of the two points has Beltrami coordinates

$$\left(\frac{x_1 t_2 + x_2 t_1}{t_1 + t_2}, \frac{y_1 t_2 + y_2 t_1}{t_1 + t_2} \right)$$

and the perpendicular bisector of the two points has an equation

$$(x_1 t_2 - x_2 t_1)x + (y_1 t_2 - y_2 t_1)y + (t_1 - t_2) = 0.$$

(g) If c is the distance from the point with Beltrami coordinates (a, b) to the line with equation $Ax + By + C = 0$, then

$$\sinh c = \frac{|Aa + Bb + C|}{(A^2 + B^2 - C^2)^{1/2}(1 - a^2 - b^2)^{1/2}}.$$

Proof The first two results are restatements of previous results that we obtained above in terms of axial coordinates. By result (a), the mapping α that sends the point with Beltrami coordinates (x, y) to the point (x, y) in the Cayley–Klein Model is a bijection. (In our plane, the Beltrami coordinates (x, y) are the *name* of a point; in the Cayley–Klein Model (x, y) *is* a point.) Since the distance scale is assumed to be 1, then α preserves distance by the third equation of result (b) and the formula for the distance function h in the Cayley–Klein Model. That is, $PQ = h(\alpha P, \alpha Q)$. It follows that α is a collineation from $(\mathscr{P}, \mathscr{L})$ onto M13. Result (c) then follows from our knowledge about equations of lines in the Cayley–Klein Model. Since the distance scale is assumed to be 1 and since α is a collineation that preserves distance, then α must preserve angles and angle measure (Corollary 32.15). Hence, α is an isomorphism onto the Cayley–Klein Model. Result (d) then follows from the formula for the angle measure function n in the Cayley–Klein Model. Since $(y_i - y_0)x - (x_i - x_0)y + (x_i y_0 - y_i x_0) = 0$ is an equation for the line through the two points with Beltrami coordinates (x_0, y_0) and (x_i, y_i), then result (e) follows directly from result (d). With the observation that $(B + bC)x - (A + aC)y + (bA - aB) = 0$ is an equation of the line through the point with Beltrami coordinates (a, b) and perpendicular to the line with equation $Ax + By + C = 0$, the proof of results (f) and (g) is left for Exercise 33.1. ∎

From the proof above, we conclude that every model of Σ with distance scale 1 is isomorphic to the Cayley–Klein Model. So any two models of Σ with distance scale 1 are isomorphic. It follows that any two models of Σ with the same distance scale are isomorphic. In particular, every model of Σ with distance scale k is isomorphic to (M13, kh, n) where (M13, h, n) is the Cayley–Klein Model. Suppose β is an isomorphism from (M13, kh, n) onto (M13, h, n). Let $\triangle ABC$ be an equilateral triangle in (M13, h, n). Then this triangle is also an equilateral triangle in (M13, kh, n) with the same vertices. Since β preserves distance, then $k\,h(A, B) = h(A', B')$ and $\triangle A'B'C'$ is an equilateral triangle in (M13, h, n) where $A' = \beta A$, $B' = \beta B$, and $C' = \beta C$. Since β preserves angle measure and the angle measure is the same for both models, then $\triangle ABC$ is congruent to $\triangle A'B'C'$ in (M13, h, n) by AAA. So $h(A', B') = h(A, B)$. Thus $k = 1$. It follows that if two models of Σ are isomorphic, then the two models must have the same distance scale. TWO MODELS OF Σ ARE ISOMORPHIC IFF THE TWO MODELS HAVE THE SAME DISTANCE SCALE.

Given any two models of the Bolyai–Lobachevsky plane there is a collineation from the first onto the second that preserves betweenness, congruence of segments, and congruence of angles. In fact, if $(\mathscr{P}_1, \mathscr{L}_1, d_1, m_1)$ and $(\mathscr{P}_2, \mathscr{L}_2, d_2, m_2)$ are models of Σ, then there exists a positive constant t such that $(\mathcal{P}_1, \mathcal{L}_1, td_1, m_1)$ and $(\mathcal{P}_2, \mathcal{L}_2, d_2, m_2)$ are isomorphic. So any two models of Σ are *almost* isomorphic. If a categorical axiom system is desired, we have only to add "The Normalization Axiom," which fixes the value of the distance scale in the theory. This is the only independent axiom that can be added and still have a consistent system. (There is such an axiom for each positive real number.) In applying our theory to physical space, the distance scale k would have to be an exceedingly large number in order to have a segment of length 1 correspond to a meter (or even a light year). On the other hand, it is convenient to suppose $k = 1$ in order to simplify formulas and calculations. By Definition 32.3, we are tacitly assuming The Normalization Axiom that states the distance scale is 1. However, we allow ourselves the opportunity of discarding this supposition and considering the general case whenever we like.

Three more coordinate systems will be mentioned. The distance scale is assumed to be 1. If point P has Beltrami coordinates (x, y), $t = 1 + (1 - x^2 - y^2)^{1/2}$, $p = x/t$, and $q = y/t$, then (p, q) are the *Poincaré coordinates* of P. Of course the Poincaré coordinates correspond to the Poincaré Model of the hyperbolic plane. Calculations using Poincaré coordinates can be simplified by using complex numbers. If point P has Poincaré coordinates (p, q) and z is the complex number $p + qi$, then (z) is the *Gauss coordinate* of P. As is true for Euclidean geometry, different coordinate systems are suited for different purposes. If point P has Beltrami coordinates (x, y) and $x^2 + y^2 = \tanh^2 r$, then $(x \cosh r, y \cosh r, \cosh r)$ are the *Weierstrass coordinates* of P. With the notation of Figure 33.3, point P has Weierstrass coordinates $(\sinh z, \sinh w, \cosh r)$. So (a, b, c) are the Weierstrass coordinates of a point iff $a^2 + b^2 - c^2 = -1$ and $c \geqq 1$. Lines have linear equations in Weierstrass coordinates. If points P_1 and P_2 have Weierstrass coordinates (a_1, b_1, c_1) and (a_2, b_2, c_2), respectively, then $\cosh P_1 P_2 = c_1 c_2 - a_1 a_2 - b_1 b_2$. It is the Weierstrass coordinates that are most often employed in advanced work.

33.2 AREA

In three-dimensional Euclidean geometry there exist solid tetrahedrons P and Q with equal altitudes and bases of equal area such that it is impossible to cut up P into solid tetrahedrons that can be reassembled to form a solid congruent to Q. In the Euclidean plane the analogous situation is quite different. The following result holds there: If P and Q are triangular regions of equal area, then P can be cut up into triangular regions that can be reassembled to form a triangular region congruent to Q. We shall say such regions are *equivalent by triangulation.* The result above is a special case of *Bolyai's Theorem for Absolute Geometry* (Theorem 33.16).

In this section we first make precise the "equality" of polygonal regions that was tacitly introduced by Euclid in his Proposition I.35. (See Section 11.1. His "equality" is our "equivalent by triangulation.") Secondly, it is shown that in any reasonable application of the word "area" to the Bolyai–Lobachevsky plane the *area* of a polygonal region must be proportional to the defect. Finally, Bolyai's Theorem for Absolute Geometry is proved.

The definitions and theorems of this section hold for the Euclidean plane provided: (1) The word "defect" is replaced throughout by the word "area," (2) The first sentence in Definition 33.4 is omitted, and (3) All the material after the statement of Theorem 33.14 through the statement of Definition 33.15 is replaced by the trivial proof of Theorem 33.14 that is applicable to the Euclidean plane.

DEFINITION 33.4 The *defect* of a convex polygon P with n sides is the positive difference between $(n-2)\pi$ and the sum of the measures of the n angles of P. The union of a convex polygon and its interior is a *convex polygonal region* having the same *vertices, sides, interior,* and *defect* as the convex polygon. If Q is a convex polygon or a convex polygonal region, then int (Q) and δQ are the interior of Q and the defect of Q, respectively. The union of a triangle and its interior is a *triangular region.* A *polygonal region* R is the union of a positive number of triangular regions $T_1, T_2, \ldots, T_n$ such that int (T_i) and int (T_j) are disjoint for $i \neq j;$ further, if $T = \{T_1, T_2, \ldots, T_n\}$ and δT is the sum of the defects of the triangular regions in T, then the set T is a *triangulation* of R and the number δT is the *defect* of the triangulation T. If S and T are triangulations of the same polygonal region such that each triangular region in S lies in some triangular region of T, then S is a *subtriangulation* of T.

Since a triangle is a convex polygon, a triangular region is a convex polygonal region. The sides of a convex polygonal region and any point in the interior determine triangular regions that can be used to show a convex polygonal region is indeed a polygonal region. See Figure 33.4, where we note the defect of such a *star triangulation* is equal to the defect of the convex polygonal region itself. In general, a polygonal region can be quite complicated. See Figure 33.5. Since polygonal regions do not have unique triangulations, we cannot define the defect of an arbitrary polygonal region to be the defect of some particular triangulation until it is shown that all the triangulations of a polygonal region have the same defect. Figure 33.6 illustrates three different triangulations of one polygonal region. Studying the figure,

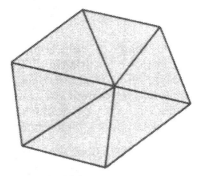

FIGURE 33.4

we see that the triangular regions of the third triangulation can be combined in one way to form the triangular regions of the first and in another way to form the triangular regions of the second. That is, the third triangulation is a subtriangulation of the others. After stating a lemma that has an obvious proof, we shall show that every two triangulations of a given polygonal region have such a common subtriangulation.

Theorem 33.5 If line l with halfplanes H and K intersects the interior of convex polygonal region R, then $R \cap (H \cup l)$ and $R \cap (K \cup l)$ are convex polygonal regions whose union is R and whose defects add up to δR.

Theorem 33.6 Any two triangulations of a polygonal region have a common subtriangulation. Any two triangulations of a polygonal region have the same defect.

Proof Let $A = \{T_1, T_2, \ldots, T_n\}$, $B = \{T_{n+1}, T_{n+2}, \ldots, T_{n+m}\}$, and $L = \{l_1, l_2, \ldots, l_r\}$ where A and B are two triangulations of polygonal region R and L is the set of all lines containing the sides of the tri-

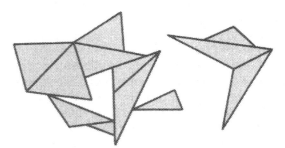

FIGURE 33.5

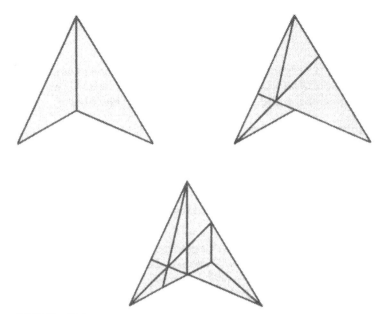

FIGURE 33.6

angular regions in A and B. $(r < 3n + 3m.)$ For $i = 1, 2, \ldots, n + m$, if l_1 has halfplanes H and K and intersects T_i, replace the convex polygonal region T_i in A or B by the two sets $T_i \cap (H \cup l)$ and $T_i \cap (K \cup l)$. We now have sets A_1 and B_1, each of which has the property that it is a set of convex polygonal regions whose union is R and whose interiors are disjoint. Further, by the preceding theorem, the sum of the defects of the convex polygonal regions in A_1 is equal to δA, and the sum of the defects of the convex polygonal regions in B_1 is equal to δB. (In general, A_1 and B_1 are not triangulations.) In the same way, starting with A_1 and B_1 and using line l_2, we obtain sets A_2 and B_2 having these same properties. Continuing in the same fashion for each of the lines in L in turn, we have sets A_r and B_r with these properties. By the definition of L, it follows that $A_r = B_r$. Therefore, $\delta A = \delta B$. Let $A_r = \{C_1, C_2, \ldots, C_s\}$ and P_i be a point in int (C_i). Suppose convex polygonal region C_i has t sides. Then C_i has a (star) triangulation $\{C_{i1}, C_{i2}, \ldots, C_{it}\}$ where each of the triangular regions C_{ij} has a vertex P_i and a side in common with C_i. Let T be the union of all such sets C_{ij} for $1 \leq i \leq s$. Then T is a subtriangulation of A and of B. ∎

Because of this theorem, we are now in a position to define the defect of *any* polygonal region. Also, the definition of the defect of a convex polygonal region is consistent with the following definition.

DEFINITION 33.7   If R is a polygonal region with triangulation T, then the **defect** of R is δT; δR is the defect of R.

The next definition gives a special case of *piecewise congruence*. This is the essence of the popular decomposition puzzles: Given two figures, can one be cut up and reassembled to form the other?

DEFINITION 33.8   If $\{T_1, T_2, \ldots, T_n\}$ and $\{S_1, S_2, \ldots, S_n\}$ are triangulations of polygonal regions R_1 and R_2, respectively, such that T_i is congruent to S_i for $i = 1, 2, \ldots, n$, then we say R_1 and R_2 are **equivalent by triangulation** and write $R_1 \equiv R_2$.

Theorem 33.9   Equivalence by triangulation is an equivalence relation on the set of polygonal regions.

Proof   Let R_1, R_2, R_3 be polygonal regions. That $R_1 \equiv R_1$ and that $R_1 \equiv R_2$ implies $R_2 \equiv R_1$ are trivial observations. Suppose $R_1 \equiv R_2$ and $R_2 \equiv R_3$. We need to show $R_1 \equiv R_3$. Since $R_1 \equiv R_2$, then R_1 and R_2 have triangulations $\{P_1, P_2, \ldots, P_n\}$ and $\{S_1, S_2, \ldots, S_n\}$, respectively, such that P_i is congruent to S_i; since $R_2 \equiv R_3$, then R_2 and R_3 have triangulations $\{S_{n+1}, S_{n+2}, \ldots, S_{n+m}\}$ and $\{Q_1, Q_2, \ldots, Q_m\}$, respectively, such that S_{n+j} is congruent to Q_j. Let T be a common subtriangulation of these triangulations of R_2. Suppose T is $\{T_1, T_2, \ldots, T_t\}$. It follows that R_1 has a triangulation $\{P_1', P_2', \ldots, P_t'\}$ such that P_i' is congruent to T_i. Likewise, it follows that R_3 has a triangulation $\{Q_1', Q_2', \ldots, Q_t'\}$ such that Q_i' is congruent to T_i. Since congruence of triangles is a transitive relation, we must have P_i' is congruent to Q_i' for $i = 1, 2, \ldots, t$. Hence $R_1 \equiv R_3$. ∎

Corollary 33.10   If two polygonal regions are equivalent by triangulation, then the polygonal regions have the same defect. In particular, two triangular regions that are equivalent by triangulation have the same defect.

So if a triangular region with defect d_0 is cut up and reassembled to form another triangular region, the new triangular region also has defect d_0. Conversely, given two triangular regions with the same defect, can each be cut up and reassembled to form the other? As we shall see in the next three theorems, the answer is "yes."

Theorem 33.11   Given $\triangle ABC$, let D and E be the midpoints of $\overline{AB}$ and $\overline{AC}$, respectively. Let G and H be the feet of the perpendiculars to $\overleftrightarrow{DE}$ from B and C, respectively. Then $\boxed{S}\,GBCH$. Further, if T is the union of $\triangle ABC$ and its interior and S is the union of $\boxed{S}\,GBCH$ and its interior, then $T \equiv S$. So $\delta\boxed{S}\,GBCH = \delta\triangle ABC$.

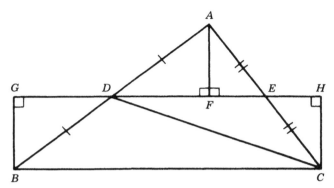

FIGURE 33.7

Proof Let F be the foot of the perpendicular from A to $\overleftrightarrow{DE}$. Either D, G, F are distinct or coincident. If D, G, F are distinct, then $\triangle BDG \simeq \triangle ADF$ by SAA. In either case, $BG = AF$ and $DG = DF$. Either E, H, F are distinct or coincident. If E, F, H are distinct, then $\triangle AEF \simeq \triangle CEH$ by SAA. In either case, $AF = CH$ and $EF = EH$. Then $BG = AF = CH$ and $\boxed{S} GBCH$. (See Figure 33.7.) Also, $\eta_D \eta_E H = \eta_D F = G$. So $GH = 2DE$. Since one of $\angle B$ or $\angle C$ must be acute, we may suppose $\angle B$ is acute without loss of generality. Let D_0 be such that E is the midpoint of D and D_0. Then define D_{i+1} to be such that $G - D_{i+1} - D_i$ and $D_{i+1}D_i = GH$. So $D_1 = D$. (See Figure 33.8.) By Archimedes' axiom, there is an integer n such that $D_{n+1} = H$ or $G - D_{n+1} - H$. Let Q_i be the polygonal region determined by $\square BD_{i+1}D_iC$. Since $\triangle ADE \cong \triangle CD_0E$, we have $T \equiv Q_0$, $BD_1 = CD_0$, and $m\angle GD_1B = m\angle GD_0C$. So $\triangle BD_1D_2 \cong \triangle CD_0D_1$, and we have $Q_0 \equiv Q_1$, $BD_2 = CD_1$, and $m\angle GD_2B = m\angle GD_1C$. Likewise, $Q_i \equiv Q_{i+1}$ for $i = 1, 2, \ldots, n-1$. (A triangular region is chopped off one side of Q_i and replaced on the other to form Q_{i+1}.) It follows that $T \equiv Q_n$. However, $Q_n \equiv S$ since $\triangle BGD_{n+1} \cong \triangle CHD_n$. Therefore, $T \equiv S$ and so T and S have the same defect. ∎

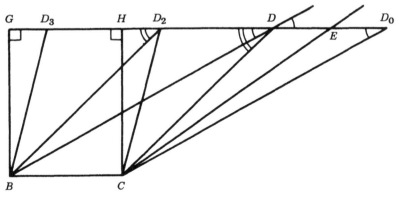

FIGURE 33.8

Theorem 33.12 Given $\triangle ABC$ and $x > AB$, there exists a point P such that $PB = x$ and the union of $\triangle PBC$ and its interior is equivalent by triangulation to the union of $\triangle ABC$ and its interior.

Proof Let T be the triangular region determined by $\triangle ABC$. Let D, E, G, H be as in the preceding proof. Let S be the polygonal region determined by $\boxed{S}GBCH$. Then $T \equiv S$. Since $\frac{1}{2}x > BD \cong BG$, there exists a point M on $\overleftrightarrow{DE}$ such that $BM = \frac{1}{2}x$. Let P be the point such that M is the midpoint of P and B. So $PB = x$. Let N be the midpoint of P and C. Let R be the triangular region determined by $\triangle PBC$. By the preceding theorem, since $\overleftrightarrow{MN}$ is the unique line through M that is perpendicular to the perpendicular bisector of $\overline{BC}$, then $\overleftrightarrow{MN} = \overleftrightarrow{DE}$ and $R \equiv S$. Therefore $T \equiv R$. ∎

Theorem 33.13 Triangular regions with the same defect are equivalent by triangulation.

Proof Since triangular regions determined by congruent equilateral triangles are obviously equivalent by triangulation, suppose R and S are the triangular regions determined by $\triangle ABC$ and $\triangle DEF$, respectively, where $DE > AB$. Suppose R and S both have defect d_0. By the preceding theorem, there exists a triangular region T determined by $\triangle PBC$ such that $PB = DE$ and $R \equiv T$. So $\delta T = d_0$. Now each of T and S is equivalent by triangulation to a polygonal region determined by a Saccheri quadrilateral with defect d_0 and upper base of length DE (Theorem 33.11). Since any two such Saccheri quadrilaterals are congruent (Theorem 26.8), then $T \equiv S$. Therefore $R \equiv S$. ∎

Suppose there is an *area function* α defined on the set of polygonal regions. If R is a polygonal region, then αR is the *area* of R. There are two essential properties α must have if *area* is to have any sort of meaning that agrees with our usual connotation. First, αR must be a positive real numbers. So we suppose α is a *positive* function. Secondly, if polygonal region R has triangulation $\{T_1, T_2, \ldots, T_n\}$, then αR must be the sum of the αT_i. So we suppose α is *additive*. Then α is determined once we know αT for every triangular region T. For convenience of notation, if T is the triangular region determined by $\triangle ABC$, we define $\alpha\triangle ABC$ to be αT. (Defining $\alpha\triangle ABC$ to be one-half the base times the height is out of the question, since, in general, there are three different such numerical products, one for each "base" of $\triangle ABC$.) Since we know triangular regions with the same defect are equivalent by triangulation, then triangles with the same defect must have the same area. We shall use this fact to show that area must be directly proportional to defect. We shall also use the following lemma.

Theorem 33.14 Given $\triangle RST$ and $0 < x < \delta\triangle RST$, there exists a unique point P on $\overline{ST}$ such that $\delta\triangle RSP = x$.

Proof If $\angle ACB$ is right, then $\tanh \frac{1}{2}AC \tanh \frac{1}{2}BC = \tan \frac{1}{2}\delta\triangle ABC$. The theorem follows from at most two applications of this formula. ■

Let $\triangle ABC$ and $\triangle PQR$ be any triangles. Suppose, first, $\delta\triangle PQR$ is less than $\delta\triangle ABC$. Then there is a unique point D on $\overline{BC}$ such that $\triangle ABD$ has defect $\delta\triangle PQR$. Let ε be any positive number. Let n be any positive integer such that $2^{n-1}\varepsilon > 1$. Then there exist points S and T on $\overline{BC}$ and positive integer t such that $\delta\triangle ABD = 2^n\delta\triangle ABS$ and $\delta\triangle ABC = t\delta\triangle ABS + \delta\triangle ATC$ with $\delta\triangle ATC \leqq \delta\triangle ABS$. So we must also have $\alpha\triangle ABD = 2^n\alpha\triangle ABS$ and $\alpha\triangle ABC = t\alpha\triangle ABS + \alpha\triangle ATC$ with $\alpha\triangle ATC \leqq \alpha\triangle ABS$. With these equations, we then have

$$\left|\frac{\alpha\triangle ABC}{\alpha\triangle PQR} - \frac{\delta\triangle ABC}{\delta\triangle PQR}\right| = \left|\frac{\alpha\triangle ABC}{\alpha\triangle ABD} - \frac{\delta\triangle ABC}{\delta\triangle ABD}\right| \leqq \frac{1}{2^n}(1+1) < \varepsilon.$$

Since the only nonnegative real number less than every positive real is 0, then

$$\frac{\alpha\triangle ABC}{\alpha\triangle PQR} = \frac{\delta\triangle ABC}{\delta\triangle PQR} \quad \text{and} \quad \frac{\alpha\triangle ABC}{\delta\triangle ABC} = \frac{\alpha\triangle PQR}{\delta\triangle PQR}.$$

If $\delta\triangle ABC < \delta\triangle PQR$, then we obtain the same result by interchanging letters. Hence, in any case,

$$\alpha\triangle ABC = \left(\frac{\alpha\triangle PQR}{\delta\triangle PQR}\right)\delta\triangle ABC.$$

Let $g = (\alpha\triangle PQR)/(\delta\triangle PQR)$. Then for triangular region T defined by any triangle, we have $\alpha T = g\delta T$. Hence area must be directly proportional to defect. Since in any practical sense area ought to be related to the distance scale k and, to make certain calculations easier, it is convenient to choose g to be k^2. Therefore, not only have we shown that the following definition is reasonable, we have also shown that it is the only possible definition up to a constant of proportionality!

DEFINITION 33.15 The *area* αR of polygonal region R is $k^2\delta T$ where k is the distance scale.

In Euclidean geometry every polygonal region is equivalent by triangulation to a triangular region. This cannot be the case in the

Bolyai – Lobachevsky plane since the defect of a polygonal region may be arbitrarily large but the defect of a triangular region is less than π.

Suppose $A = \{A_1, A_2, \ldots, A_m\}$ and $B = \{B_1, B_2, \ldots, B_n\}$ where A and B are triangulations of polygonal regions P and Q, respectively, and $\delta P = \delta Q$. So the sum of the δA_i is equal to the sum of the δB_i. Bouncing back and forth and using Theorem 33.14, we can chop up the triangular regions in each of A and B to obtain subtriangulations C and D of A and B, respectively, such that $C = \{C_1, C_2, \ldots, C_r\}$, $D = \{D_1, D_2, \ldots, D_r\}$, and $\delta C_i = \delta D_i$ for $i = 1, 2, \ldots, r$. Figure 33.9 illustrates this for the case $m = 5$ and $n = 4$ with $r = 8$. (In the figure, D_1 and D_2 are cut off B_1 but there is not enough left to cut off a triangular region with defect δA_3. So D_3 is cut off A_3, etc.) With this observation, we can now prove *Bolyai's Theorem for Absolute Geometry*, which is attributed to Wolfgang Bolyai.

Theorem 33.16 *Bolyai's Theorem for Absolute Geometry* Two polygonal regions are equivalent by triangulation iff the polygonal regions have the same area.

Proof Let $S = \{S_1, S_2, \ldots, S_m\}$ and $T = \{T_1, T_2, \ldots, T_n\}$ where

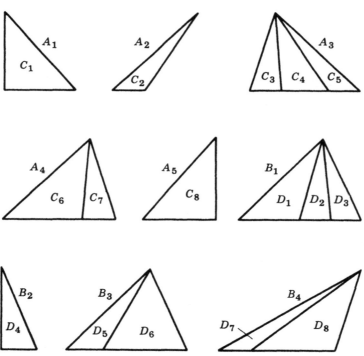

FIGURE 33.9

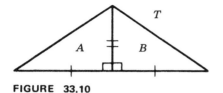 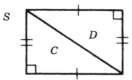

FIGURE 33.10

S and T are triangulations of polygonal regions P and Q, respectively. Suppose $\alpha P = \alpha Q$. Then $\delta P = \delta Q$. By the observation above, we may suppose $m = n$ and $\delta S_i = \delta T_i$ without loss of generality. Then, although S_i and T_i may not be congruent, we do know S_i is equivalent by triangulation to T_i, (Theorem 33.13). From $S_i \equiv T_i$ for all i, it follows that S and T have subtriangulations S' and T', respectively, such that $S' = \{S_{ij}\}$, $T' = \{T_{ij}\}$, and S_{ij} is congruent to T_{ij} for all i and j. Hence $P \equiv Q$. So $\alpha P = \alpha Q$ implies $P \equiv Q$. Conversely, if $P \equiv Q$, then $\delta P = \delta Q$ and $\alpha P = \alpha Q$, (Corollary 33.10). Therefore, $P \equiv Q$ iff $\alpha P = \alpha Q$. ∎

In Figure 33.10, the triangular regions A, B, C, D are supposed to be congruent. Then, since $T = A \cup B$ and $S = C \cup D$, we have T is equivalent by triangulation to S. However, in cutting up T to form S the "hypotenuses" of A and B are superimposed and the vertex of the right angle of A has to be "split" to form two vertices of S. This comes about because A and B are not disjoint and C and D are not disjoint. With this in mind, you might be led to the following definition. Set T of points is *equivalent by set decomposition* to set S of points if T is the union of disjoint sets T_i and S is the union of disjoint sets S_i such that T_i is congruent (Definition 19.13) to S_i for $i = 1, 2, \ldots, n$. This is a whole new ball game! See Exercises 33.25, 33.31, and 33.32. Can it be true in Euclidean geometry that a sphere the size of the sun is equivalent by set decomposition to a sphere the size of a pea?

33.3 EXERCISES

The distance scale is assumed to be 1.

33.1 Results (f) and (g) of Theorem 33.3.

● **33.2** If $Ax + By + C = 0$ with $C \neq 0$ is an equation of line l in Beltrami coordinates, then no point has Beltrami coordinates $(-A/C, -B/C)$ but $(x, y) = (-A/C, -B/C)$ is an algebraic solution to any equation of any line perpendicular to l.

● **33.3** If $A_1 x + B_1 y + C_1 = 0$ and $A_2 x + B_2 y + C_2 = 0$ are equations of two lines in Beltrami coordinates and $A_1B_2 = A_2B_1$, then the two lines are hyperparallel.

33.4 If $A_1x + B_1y + C_1 = 0$ and $A_2x + B_2y + C_2 = 0$ are equations of two lines in Beltrami coordinates, then the two lines intersect, are horoparallel, or are hyperparallel iff

$$\begin{vmatrix} A_1 & C_1 \\ A_2 & C_2 \end{vmatrix}^2 + \begin{vmatrix} B_1 & C_1 \\ B_2 & C_2 \end{vmatrix}^2 - \begin{vmatrix} A_1 & B_1 \\ A_2 & B_2 \end{vmatrix}^2$$

is respectively less than, equal to, or greater than 0.

• **33.5** If $A_ix + B_iy + C_i = 0$, $i = 1, 2, 3$, are equations of three lines in Beltrami coordinates, then the lines are in a brush iff

$$\begin{vmatrix} A_1 & B_1 & C_1 \\ A_2 & B_2 & C_2 \\ A_3 & B_3 & C_3 \end{vmatrix} = 0.$$

• **33.6** True or False?

(a) Points with Beltrami coordinates (x_0, y_0), (x_1, y_1), and (x_2, y_2) are collinear iff

$$\begin{vmatrix} x_0 & y_0 & 1 \\ x_1 & y_1 & 1 \\ x_2 & y_2 & 1 \end{vmatrix} = 0.$$

(b) $Ax + By + C > 0$ describes one halfplane of the line having equation $Ax + By + C = 0$ in Beltrami coordinates.

(c) $A_1x + B_1y + C_1 = 0$ and $A_2x + B_2y = 0$ are equations in Beltrami coordinates of two perpendicular lines if the equations are those of two perpendicular lines in the Cartesian plane.

(d) Given point P, then $(x^2 + y^2)^{1/2} = \tanh OP$ and $(1 - x^2 - y^2)^{1/2} = \operatorname{sech} OP$ iff P has Beltrami coordinates (x, y).

(e) $x^2 + y^2 = \tanh 1$ is an equation of the unit circle in Beltrami coordinates.

(f) $(x^2/1^2) + (y^2/\tanh^2 d) = 1$ is an equation of the locus of all points of positive distance d from $\overleftrightarrow{OX}$.

(g) $\cosh x - \sinh x \cos \theta \geq e^{-x}$ for real x and θ when $x > 0$.

(h) $\cosh \operatorname{arctanh} x = (1 - x^2)^{-1/2}$ for real x.

(i)   A point with Beltrami coordinates

$(\tanh u, (\tanh w)/(\cosh u))$

has Lobachevsky coordinates (u, w).

(j)   A point with Lobachevsky coordinates

$(\text{arctanh}\, x, \text{arctanh}\, [(1-x^2)^{-1/2}y])$

has Beltrami coordinates (x, y).

33.7   Exhibit the triangulations in the proof of Theorem 33.11 for the case $H-D-E-F$ when $HD < DE$ and when $DE < HD < GH$.

33.8   Triangulate the Greek cross, Figure 30.1, with all angles of measure θ, to form the polygonal region determined by a Gersonides quadrilateral.

33.9 $e^{-u} = \tanh w$ and $e^u = \cosh w$ are, respectively, equations in Lobachevsky coordinates of the line in the first quadrant horoparallel to both axes and of the horocircle with radius $\overrightarrow{OX}$.

33.10   Give an algebraic description of the pencil with center O having Beltrami coordinates $(x_0\, y_0)$ and of the horopencil determined by $\overrightarrow{OP}$ where P has Beltrami coordinates (x_1, y_1).

33.11   Using Beltrami coordinates, find an equation of the circle with radius r and center with coordinates (x_0, y_0), and find an equation of each hypercircle with center having equation $Ax + By + C = 0$ and of distance d from its center.

33.12   Find an equation in Beltrami coordinates of the horocircle with radius $\overrightarrow{PQ}$ where P and Q have coordinates (x_0, y_0) and (x_1, y_1), respectively.

33.13 $(1-x_0^2-y_0^2)^{1/2} > e^{-t}(1-x_1^2-y_1^2)^{1/2}$ if (x_0, y_0) and (x_1, y_1) are, respectively, the Beltrami coordinates of two points P and Q with $t = PQ$.

33.14   Every cycle has an equation in Beltrami coordinates that is of the form $(1-x^2-y^2)^{1/2} = ax + by + c$. The cycle is a circle iff $-1 < a^2 + b^2 - c^2 < 0$ and $c > 0$. The cycle is a horocircle iff $a^2 + b^2 - c^2 = 0$ and $c > 0$. The cycle is a hypercircle iff $0 < a^2 + b^2 - c^2$. What sets of points have an equation obtained by squaring both sides of the equation above?

33.15 $C(p^2 + q^2) + 2Ap + 2Bq + C = 0$ is an equation of a line in Poincaré coordinates iff $A^2 + B^2 > C^2$, and every line has such an equation.

33.16 Every cycle has an equation in Poincaré coordinates that is of the form $(c+1)(p^2+q^2)+2ap+2bq+(c-1)=0$ where a, b, c are as in Exercise 33.14.

33.17 Verify the result indicated in Figure 27.9 for the Poincaré Model.

33.18 If $(1-x^2-y^2)^{1/2}=c[1-(-a/c)x-(-b/c)y]$ is an equation of a cycle in Beltrami coordinates, what algebraic significance do $(-a/c, -b/c)$ and $ax+by+c=0$ have?

33.19 Reflect on our various uses of the words "center" and "concentric."

33.20 A distance formula in Gauss coordinates is

$$\tanh\frac{P_0P_1}{2}=\left|\frac{z_1-z_0}{1-z_1\bar{z}_0}\right|.$$

33.21 Find an equation in Beltrami coordinates of the locus of all points equidistant from a nonincident point and line and an equation of the locus of all points such that the sum (difference) of the distances from two fixed points is a constant. Now consider a *focus-directrix-curve* analogous to a conic in the Cartesian plane.

33.22 Find an equation in Beltrami coordinates of the locus of all points P such that $\angle APB$ is right where A and B are two fixed points.

33.23 The set of the sets of all points satisfying an equation in Beltrami coordinates of the form $Ax^2+Bxy+Cy^2+Dx+Ey+F=0$ is fixed under the isometries.

33.24 If S and T are the polygonal regions determined by $\boxed{S}\,ABCD$ and $\boxed{S}\,EFGH$, respectively, where $\angle B \simeq \angle F$ and $BC=2FG$, then give triangulations of S and T that show S and T are equivalent by triangulation.

33.25 Read "'A Paradox, A Paradox, A Most Ingenious Paradox'" by L. M. Blumenthal in *The American Mathematical Monthly*, Vol. 47 (1940), pp. 346–353.

33.26 Read *Recreational Problems in Geometry, Dissections and How to Solve Them* by H. Lindgren (Dover, 1972).

*33.27 Find a trisection point of $\overline{OT}$ where T has Beltrami coordinates $(t, 0)$.

*33.28 Are $(M13, h)$ and $(M13, kh)$ isomorphic for $k>0$?

*33.29 What are the equations of the sets of points that are equidistant from two cycles?

*33.30   Read Part One (83 pages) of *Theory of Functions of a Complex Variable* by C. Caratheodory (Chelsea, 1964).

*33.31   Read "On the Congruence of Sets and their Equivalence by Finite Decomposition" by W. Sierpinski in *Congruence of Sets and Other Monographs* (Chelsea, n.d.).

*33.32   Read *Unsolved and Unsolvable Problems in Geometry* by H. Meschkowski (Oliver and Boyd, 1966).

*33.33   Read *The Banach-Tarski Paradox* by Stan Wagon (Oxford University Press, 1985).

GRAFFITI

Suppose the distance from an equidistant curve in the absolute plane *to a center of the equidistant curve is* r. *Since the ratio of the arclength of an arc on this curve to the length of the projection of the arc onto this center depends only on* r, *denote this ratio by* E_r. *Let* O_r *denote the circumference of a circle with radius* r. *In 1878 Joseph deTilly gave the following formulas for* $\triangle ABC$ *in standard notation when* $\angle C$ *is right:*

$$O_a = O_c \sin \lambda, \qquad E_a = \frac{\cos \lambda}{\sin \mu}, \qquad E_c = E_a E_b.$$

All of absolute trigonometry can be obtained from de Tilly's formulas. In particular, the Absolute Pythagorean Theorem can be expressed by the equation

$$(O_a)^2 E_b(E_a + E_b E_c) + (O_b)^2 E_a(E_b + E_a E_c) = (O_c)^2(E_c + E_a E_b)$$

for $\triangle ABC$ *in standard notation when* $\angle C$ *is right. The Absolute Law of Sines was given earlier by Bolyai:*

$$\frac{\sin \lambda}{O_a} = \frac{\sin \mu}{O_b} = \frac{\sin \nu}{O_c}$$

for $\triangle ABC$ *in standard notation.*

CHAPTER 34

Quadrature of the Circle

34.1 CLASSICAL THEOREMS

About 100 A.D. Menelaus of Alexandria extended a then well-known lemma to spherical triangles in his *Sphaerica,* which is extant in an Arabic translation. In plane geometry this lemma is known as *Menelaus' Theorem.* The well-known lemma that is now called *Pappus' Theorem* may be found in Pappus' *Collection.* It is quite likely that both Menelaus' Theorem and Pappus' Theorem were known to Euclid, since both appear in the *Collection* as lemmas that are useful for an understanding of the now lost book *Porisms* by Euclid.

In 1639 the book *Brouillon Project* by Girad Desargues (1591 – 1661) unsuccessfully introduced projective geometry. Written in bizarre language and employing methods that thoroughly broke with tradition, the great book was not accepted by the mathematical community — even labeled "dangerous and unsound." *Desargues' Theorem,* which is the cornerstone of the foundations of projective geometry, does not actually appear in the *Brouillon Project.* Desargues published his famous theorem on perspective triangles in an appendix to a book on perspective by his friend, the engraver Abraham Bosse, in 1648. The significance of Desargues' Theorem and Pappus' Theorem to projective geometry was recognized only in the nineteenth century.

In 1678 Giovanni Ceva published the forgotten Menelaus' Theorem along with the theorem that now bears his name. *Ceva's Theorem* and Menelaus' Theorem are twins. Why two thousand years separates the formulation of these two theorems is a mystery.

Statements and proofs of the theorems mentioned above are indicated later in this section, where we prove the hyperbolic analogues of these classical theorems.

DEFINITION 34.1 Given $\triangle ABC$, if (i) D, B, C are three collinear points, (ii) E, A, C are three collinear points, and (iii) F, A, B are three collinear points, then we say "D, E, F are **Menelaus points** for $\triangle ABC$," noting that the order of the letters in the phrase is significant. A line through a vertex of a triangle is a **Cevian** for that triangle.

Theorem 34.2 Suppose D, E, F are Menelaus points for $\triangle ABC$. If D, E, F are collinear, then

$$\frac{\sinh AF}{\sinh FB} \cdot \frac{\sinh BD}{\sinh DC} \cdot \frac{\sinh CE}{\sinh EA} = 1.$$

Proof Let $[P, Q, R]$ denote $(\sinh PQ)/(\sinh QR)$ for any three points P, Q, R. Suppose D, E, F are on line l. Let A', B', C' be the feet of the perpendiculars to l from A, B, C, respectively. Then (Corollary 32.13c or trivially when l is perpendicular to a side of $\triangle ABC$), each of the following equation holds: $[A', A, F] = [B', B, F]$, $[B', B, D] = [C', C, D]$, and $[C', C, E] = [A', A, E]$. The desired equation now follows by elementary algebra from multiplying these three equations together. ∎

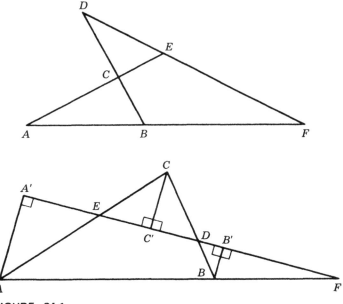

FIGURE 34.1

We should note that if "sinh" is consistently deleted from the statement of Theorem 24.2 and its proof, then we have a theorem and its proof for the Euclidean plane. The next theorem is a special case that shows the converse of the theorem above does not hold.

Theorem 34.3 Suppose D, E, F are Menelaus points for $\triangle ABC$. If Cevians $\overleftrightarrow{AD}$, $\overleftrightarrow{BE}$, $\overleftrightarrow{CF}$ are concurrent, then

$$\frac{\sinh AF}{\sinh FB} \cdot \frac{\sinh BD}{\sinh DC} \cdot \frac{\sinh CE}{\sinh EA} = 1.$$

Proof Suppose the Cevians are concurrent at point P. (See Figure 34.2.) By hypothesis, we have that C, P, F are Menelaus points for $\triangle ABD$ and that E, P, B are Menelaus points for $\triangle DCA$. The desired equation now follows by elementary algebra from two applications of the previous theorem. ∎

For the statement of the theorems of Menelaus and Ceva we shall need the idea of *directed distance*.

DEFINITION 34.4 Let F be a fixed set of coordinate systems that contains exactly one coordinate system for each line. If P is a point on $\overleftrightarrow{AB}$ and f is the coordinate system in F for $\overleftrightarrow{AB}$, then AP is the **directed distance** from A to P where $AP = f(P) - f(A)$. (Which particular choice

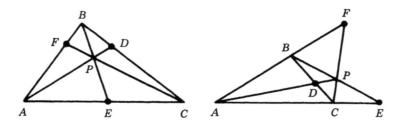

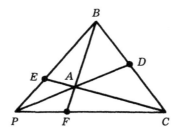

FIGURE 34.2

is made for F is immaterial as all of our results will be independent of the choice.)

So $AP = \pm AP$ and $AP = -PA$ for any points A and P. Our use of directed distance will be restricted to expressions of the form that occur in the following lemma.

Theorem 34.5 If $\langle P, Q, R \rangle$ denotes $(\sinh PQ)/(\sinh QR)$ for any three collinear points P, Q, R, then $\langle P, Q, R \rangle$ is positive iff $P-Q-R$ and, further, $\langle P, Q, R \rangle = \langle P, Q', R \rangle$ implies $Q = Q'$.

Proof By hypothesis, $\langle P, Q, R \rangle = (\sinh (q-p))/(\sinh (r-q))$ where P, Q, R have coordinates p, q, r, respectively, with respect to some coordinate system. So each of the following is equivalent to the next: $\langle P, Q, R \rangle$ is positive; $q-p$ and $r-q$ have the same sign; q is between p and r; and $P-Q-R$. Further, since $P \neq R$, it then follows (Theorem 31.20m) that each of the following implies the next: $\langle P, Q, R \rangle = \langle P, Q', R \rangle = \cosh (PQ + Q'R) - \cosh (PQ - Q'R) = \cosh (PQ' + QR) - \cosh (PQ' - QR)$; $PQ + Q'R = \pm(PQ' + QR)$; $q = q'$; and $Q = Q'$. ∎

It is easy to check that the lemma above holds for the Euclidean plane if "sinh" is consistently deleted from its statement. (The second part follows from the fact that $(p-r)(q-q') = 0$ and $p \neq r$ implies $q = q'$.) Then, by consistently deleting "sinh" from the statement of our next theorem and its proof, we obtain the statement of *Menelaus' Theorem* for the Euclidean plane with its proof!

Theorem 34.6 *Hyperbolic Menelaus' Theorem*   If D, E, F are Menelaus points for $\triangle ABC$, then D, E, F are collinear iff

$$\frac{\sinh AF}{\sinh FB} \cdot \frac{\sinh BD}{\sinh DC} \cdot \frac{\sinh CE}{\sinh EA} = -1.$$

Proof Suppose D, E, F are on line l. Then (Theorem 34.2) the left-hand side of the equation is equal to $+1$ or -1. Since l intersects $\triangle ABC$ exactly twice or not at all by PASCH, then exactly one or else each of the three terms in the left-hand side of the equation is negative. In either case, the product of three terms must be -1.

Conversely, suppose the equation holds. Since exactly one or else each of the three terms on the left-hand side is negative, then either exactly two or none of the points D, E, F are on $\triangle ABC$. In either case, it follows that at least one of $\overleftrightarrow{DE}$, $\overleftrightarrow{EF}$, or $\overleftrightarrow{DF}$ intersects each of the three lines that contains a side of $\triangle ABC$. By symmetry, we may suppose $\overleftrightarrow{DE}$ intersects $\overleftrightarrow{AB}$ at a point F' without loss of generality. So, in the notation of the lemma, the product of $\langle A, F', B \rangle$, $\langle B, D, C \rangle$, and $\langle C, E, A \rangle$ must be -1 by the first part of the theorem. Thus $\langle A, F', B \rangle = \langle A, F, B \rangle$, and $F = F'$ by the lemma. Therefore, F is on $\overleftrightarrow{DE}$. ∎

Theorem 34.7 *Hyperbolic Ceva's Theorem* If D, E, F are Menelaus points for $\triangle ABC$, then the Cevians $\overleftrightarrow{AD}$, $\overleftrightarrow{BE}$, $\overleftrightarrow{CF}$ are concurrent iff each pair intersect and

$$\frac{\sinh AF}{\sinh FB} \cdot \frac{\sinh BD}{\sinh DC} \cdot \frac{\sinh CE}{\sinh EA} = +1.$$

Proof Suppose the Cevians are concurrent at point P. Since C, P, F are Menelaus points for $\triangle ABD$ and since E, P, B are Menelaus points for $\triangle DCA$, then the first part of the theorem follows by two applications of the previous theorem.

Conversely, suppose the equation holds. Since either all of the three terms on the left-hand side of the equation are positive or else exactly one of the terms is positive, then either D, E, F are all on $\triangle ABC$ or else exactly one of D, E, F is on $\triangle ABC$. By symmetry, we may suppose D is on $\triangle ABC$ without loss of generality. Suppose $\overleftrightarrow{BE}$ intersects $\overleftrightarrow{CF}$ at a point P. Since either E and F are both on or else both off $\triangle ABC$, it follows that P is either in int $(\angle BAC)$ or else in the interior of the vertical angle of $\angle BAC$. In either case, $\overleftrightarrow{AP}$ intersects $\overline{BC}$ at a point D' by Crossbar. Thus D', E, F are Menelaus points for $\triangle ABC$ such that the Cevians $\overleftrightarrow{AD'}$, $\overleftrightarrow{BE}$, $\overleftrightarrow{CF}$ are concurrent. By the first part of the theorem (and the lemma Theorem 34.5), it then follows that $D' = D$. Therefore, $\overleftrightarrow{AD}$ is on P. ∎

Ceva's Theorem for the Euclidean plane can be stated as follows: If D, E, F are Menelaus points for $\triangle ABC$, then the Cevians $\overleftrightarrow{AD}$, $\overleftrightarrow{BE}$, $\overleftrightarrow{CF}$ are parallel or concurrent iff $AF \cdot BD \cdot CE = FB \cdot DC \cdot EA$. The proof of the "only if" part is left for Exercise 34.2. To prove the "if" part, we take the second paragraph of the proof of Theorem 34.7 just as it stands and add the following. "Now suppose $\overleftrightarrow{BE}$ and $\overleftrightarrow{CF}$ are parallel. Let l be the (unique) line through A that is parallel to $\overleftrightarrow{BE}$. Then l intersects $\overleftrightarrow{BC}$ at a point D''. So D'', E, F are Menelaus points for $\triangle ABC$ such that the Cevians $\overleftrightarrow{AD''}$, $\overleftrightarrow{BE}$, $\overleftrightarrow{CF}$ are parallel. By the first part of the theorem (and the lemma Theorem 34.5), it follows that $D'' = D$. Therefore $\overleftrightarrow{AD} = l$, finishing the proof." After seeing Ceva's Theorem, one might conjecture that the proposition "If D, E, F are Menelaus points for $\triangle ABC$, then the Cevians $\overleftrightarrow{AD}$, $\overleftrightarrow{BE}$, $\overleftrightarrow{CF}$ are in a brush iff the equation in Theorem 34.7 holds," is a theorem for the Bolyai–Lobachevsky plane. However, the conjecture is false (Exercise 34.15).

The Hyperbolic Menelaus' Theorem gives a necessary and sufficient condition for the collinearity of points. The Hyperbolic Ceva's Theorem gives a necessary and sufficient condition for the concurrency of lines. Both of these theorems relate incidence and distance. The remaining theorems in this section are pure incidence theorems.

Theorem 34.8 Suppose $\triangle ABC$ and $\triangle A'B'C'$ have no vertex in common, point A'' is the intersection of $\overleftrightarrow{BC}$ and $\overleftrightarrow{B'C'}$, point B'' is the intersection of $\overleftrightarrow{AC}$ and $\overleftrightarrow{A'C'}$, and point C'' is the intersection of $\overleftrightarrow{AB}$ and $\overleftrightarrow{A'B'}$. If $\overleftrightarrow{AA'}$, $\overleftrightarrow{BB'}$, $\overleftrightarrow{CC'}$ are concurrent, then A'', B'', C'' are collinear.

Proof   Let $\overleftrightarrow{AA'}$, $\overleftrightarrow{BB'}$, $\overleftrightarrow{CC'}$ be concurrent at point V. If V is a vertex of $\triangle ABC$ or of $\triangle A'B'C'$, then A'', B'', C'' are not distinct and the result is trivial. So suppose V is off the triangles. (See Figure 34.3.) The collinear points B', A', C'' are Menelaus points for $\triangle ABV$; the collinear points C', B', A'' are Menelaus points for $\triangle BCV$; and the collinear points A', C', B'' are Menelaus points for $\triangle CAV$. From the Hyperbolic Menelaus' Theorem, we have three equations which when multiplied together such that one side is $(-1)^3$ give us a fourth equation. Now A'', B'', C'' are Menelaus points for $\triangle ABC$ and, again by the Hyperbolic Menelaus' Theorem, are collinear iff this fourth equation holds.   ∎

Desargues' Theorem is the statement of Theorem 34.8 as applied

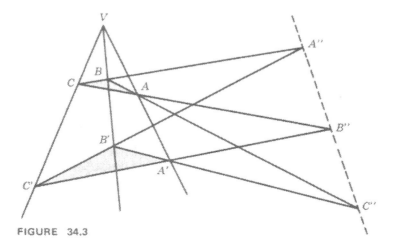

FIGURE   34.3

to the real projective plane. The real projective plane is obtained by augmenting the Euclidean plane so that every two lines intersect (see Section 4.2). Thus two lines in the real projective are in a unique *pencil*. Desargues' Theorem implies several theorems for the Euclidean plane. One of these is the statement of our Theorem 34.8, which has as a proof the proof above with the word "Hyperbolic" consistently deleted. Another is the so-called *Little Desargues' Theorem* in Exercise 34.4.

Desargues' Theorem implies a theorem for the Bolyai–Lobachevsky plane that is more general than our Theorem 34.8. Our proof of this generalization is something of a *tour de force* in that, besides the Bolyai–Lobachevsky plane, the proof employs the Euclidean plane, the real projective plane, and real projective three-space. Real projective three-space is obtained by augmenting Euclidean three-space so that every two lines in a plane intersect. We may suppose, since the value of the distance scale is immaterial for our purposes, that a Bolyai–Lobachevsky plane is the Cayley–Klein Model (Section 33.1). This geometry is embedded in the Euclidean plane (Section 23.2). The Euclidean plane is embedded in the real projective plane, which is embedded in the real projective three-space. Utilizing the *nesting* of these geometries, we have a powerful tool for proving incidence theorems for the Bolyai–Lobachevsky plane. This "nesting method" is illustrated in the proof of our next theorem.

Theorem 34.9 *Hyperbolic Desargues' Theorem* Suppose $\triangle ABC$ and $\triangle A'B'C'$ have no vertex in common, point A'' is the intersection of $\overleftrightarrow{BC}$ and $\overleftrightarrow{B'C'}$, and point B'' is the intersection of $\overleftrightarrow{AC}$ and $\overleftrightarrow{A'C'}$. Then $\overleftrightarrow{AA'}$, $\overleftrightarrow{BB'}$, $\overleftrightarrow{CC'}$ are in a brush iff $\overleftrightarrow{AB}$, $\overleftrightarrow{A'B'}$, $\overleftrightarrow{A''B''}$ are in a brush.

Proof Let $\triangle ABC$, $\triangle A'B'C'$, A'', B'', and C'' be in the real projective plane P and satisfy all the hypotheses of the statement of Theorem 34.8. (We are going to prove Desargues' Theorem for the real projective plane by using real projective three-space. The idea is that every Desargues' figure (Figure 34.3 in a plane) is a projection of a three-dimensional Desargues' figure (Figure 34.3 with the plane of $\triangle ABC$ different from the plane of $\triangle A'B'C'$).) Let V be the point of concurrency of $\overleftrightarrow{AA'}$, $\overleftrightarrow{BB'}$, $\overleftrightarrow{CC'}$. Let S be a point in real projective three-space that is off the plane P. (See Figure 34.4.) Let D be any point such that D, A, S are three collinear points. The two coplanar lines $\overleftrightarrow{SA'}$ and $\overleftrightarrow{VD}$ intersect at a point D'. The two coplanar lines $\overleftrightarrow{DB}$ and $\overleftrightarrow{D'B'}$ intersect at a point F. The two coplanar lines $\overleftrightarrow{DC}$ and $\overleftrightarrow{D'C'}$ intersect at a point G. Let Q be a plane containing the three points S, F, G. Then $Q \neq P$ since point S is off plane P. Now $\triangle BCD$ and $\triangle B'C'D'$ are in different

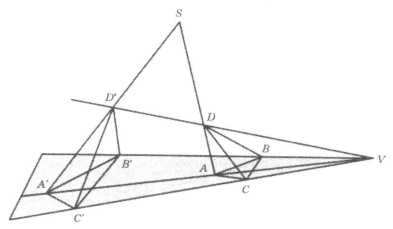

FIGURE 34.4

planes, and points A'', F, G are collinear since each of A'', F, G is in both of these planes. (Two intersecting planes intersect in a unique line.) Likewise, since $\triangle ACD$ and $\triangle A'C'D'$ are in different planes, the points S, F, B'' are collinear; and, since $\triangle ABD$ and $\triangle A'B'D'$ are in different planes, the points S, G, C'' are collinear. Therefore, since plane Q contains the three points F, G, S, then points A'', B'', C'' are in plane Q. Since A'', B'', C'' are also in plane P and $P \neq Q$, then A'', B'', C'' must be collinear as each is in the line of intersection of the planes P and Q. We have now proved Desargues' Theorem; that is, we have proved the statement of our Theorem 34.8 for the real projective plane.

We always suppose below that $\triangle ABC$ and $\triangle A'B'C'$ have no vertex in common, that point A'' is the intersection of $\overleftrightarrow{BC}$ and $\overleftrightarrow{B'C'}$, and that point B'' is the intersection of $\overleftrightarrow{AC}$ and $\overleftrightarrow{A'C'}$. We can restate the conclusion of Desargues' Theorem (for the real projective plane) as follows: If $\overleftrightarrow{AA'}$, $\overleftrightarrow{BB'}$, $\overleftrightarrow{CC'}$ are in a pencil, then $\overleftrightarrow{AB}$, $\overleftrightarrow{A'B'}$, $\overleftrightarrow{A''B''}$ are in a pencil (see Figure 34.5, ignoring the shading). In this form it is easy to see that Desargue's Theorem implies its converse: If $\overleftrightarrow{AB}$, $\overleftrightarrow{A'B'}$, $\overleftrightarrow{A''B''}$ are in a pencil, then $\overleftrightarrow{AA'}$, $\overleftrightarrow{BB'}$, $\overleftrightarrow{CC'}$ are in a pencil. To prove this we have only to apply Desargues' Theorem to $\triangle A''BB'$ and $\triangle B''AA'$ (see Figure 34.5 with its shading). Thus, for the real projective plane, we have: Lines $\overleftrightarrow{AA'}$, $\overleftrightarrow{BB'}$, $\overleftrightarrow{CC'}$ are in a pencil iff lines $\overleftrightarrow{AB}$, $\overleftrightarrow{A'B'}$, $\overleftrightarrow{A''B''}$ are in a pencil. Since the pencils in the real projective plane correspond to the pencils and parallel pencils in the Euclidean plane, then we have also demonstrated the following proposition for the Euclidean plane: Lines $\overleftrightarrow{AA'}$, $\overleftrightarrow{BB'}$, $\overleftrightarrow{CC'}$ are either in a pencil or a parallel pencil iff lines $\overleftrightarrow{AB}$, $\overleftrightarrow{A'B'}$, $\overleftrightarrow{A''B''}$ are either in a pencil or a parallel pencil. (This

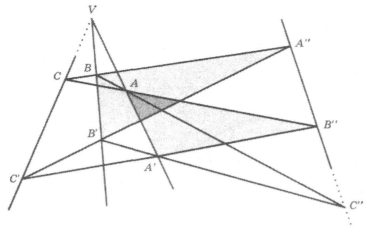

FIGURE 34.5

proposition can be proved by imitating the first part of the proof but using only Euclidean three-space. However, such a proof would be very long because of all the possible cases.) Further, the pencils and parallel pencils in the Euclidean plane correspond to the brushes in the Cayley–Klein Model. (It is not difficult to show that the brushes in the Cayley–Klein Model are exactly the intersections of the pencils and parallel pencils in the Euclidean plane with the set of subsets of the points in the Cayley–Klein Model. See Exercise 33.5.) Hence, for the Cayley–Klein Model we have the conclusion: Lines $\overleftrightarrow{AA'}$, $\overleftrightarrow{BB'}$, $\overleftrightarrow{CC'}$ are in a brush iff lines $\overleftrightarrow{AB}$, $\overleftrightarrow{A'B'}$, $\overleftrightarrow{A''B''}$ are in a brush. This final result must apply to the Bolyai–Lobachevsky plane in general. ∎

Theorem 34.10 Suppose A, B, C, D, E, F are six points such that points A, C, E are on one line, points B, D, F are on another line, lines $\overleftrightarrow{AB}$ and $\overleftrightarrow{DE}$ intersect at point L, lines $\overleftrightarrow{BC}$ and $\overleftrightarrow{EF}$ intersect at point M, and lines $\overleftrightarrow{CD}$ and $\overleftrightarrow{FA}$ intersect a point N. Further, suppose lines $\overleftrightarrow{AB}$ and $\overleftrightarrow{CD}$ intersect at point P, lines $\overleftrightarrow{CD}$ and $\overleftrightarrow{EF}$ intersect at point Q, and lines $\overleftrightarrow{EF}$ and $\overleftrightarrow{AB}$ intersect at point R. Then points L, M, N are collinear.

Proof (See Figure 34.6.) The points P, Q, R are not collinear. The collinear points F, A, N are Menelaus points for $\triangle PQR$. The collinear points B, C, M are Menelaus points for $\triangle QRP$. The collinear points D, E, L are Menelaus points for $\triangle RPQ$. The collinear points D, B, F are Menelaus points for $\triangle RQP$. The collinear points E, C, A are Menelaus points for $\triangle PRQ$. From the Hyperbolic Menelaus' Theorem, we

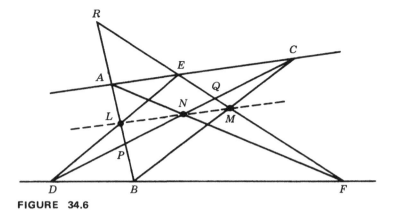

FIGURE 34.6

have five equations which when multiplied together such that one side is $(-1)^5$ give us a sixth equation. Now M, L, N are Menelaus points for $\triangle PQR$ and, again by the Hyperbolic Menelaus' Theorem, are collinear iff this sixth equation holds. ∎

Rather than the proposition in Exercise 34.12, we choose the following theorem to be called the Hyperbolic Pappus' Theorem. *Pappus' Theorem* is the statement of Theorem 34.11 as applied to the real projective plane. See Figure 34.7.

Theorem 34.11 *Hyperbolic Pappus' Theorem* If A, B, C, D, E, F are six points such that points A, C, E are on one line, points B, D, F are on another line, lines $\overleftrightarrow{AB}$ and $\overleftrightarrow{DE}$ intersect at point L, lines $\overleftrightarrow{BC}$ and $\overleftrightarrow{EF}$ intersect at point M, and lines $\overleftrightarrow{CD}$ and $\overleftrightarrow{FA}$ intersect at point N, then L, M, N are collinear.

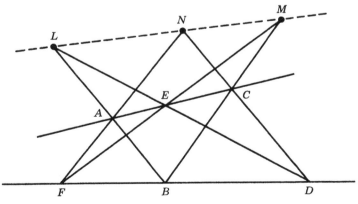

FIGURE 34.7

Proof Exercise 34.8. ■

Finally, we mention a theorem that can be proved easily by the nesting method. The theorem is not a classical theorem in that there is no analogue for the Euclidean plane!

Theorem 34.12 There exists a collineation that maps all the points onto the interior of any given circle.

Proof Exercise 34.1. ■

34.2 CALCULUS

Elementary calculus is assumed in this section. Derivatives, definite integrals, and series are from the theory of functions of a real variable and are independent of geometry. These same tools that you applied to problems in the Euclidean plane will now be applied to analogous problems in the Bolyai–Lobachevsky plane. As in your calculus class, it is assumed here that the concept of area for polygonal regions can be generalized to apply to other regions. If you have not read Section 33.2 on area, then replace "area" by "defect" in the sequel. We shall be just as informal as the average calculus book in deriving our results and in not questioning whether these results are independent of choice of coordinatization. From Section 33.1 on analytic geometry, only Definition 31.1 is essential since it is the Lobachevsky coordinates and the polar coordinates that are most useful for our applications. Otherwise, the necessary formulas are to be found in Chapter 32.

Our use of series is limited to the following observation. If t is a *very small* number, then t is a very good approximation for $\sin t$, for $\tan t$, for $\sinh t$, and for $\tanh t$. Applying this to the formula $\tanh \frac{1}{2}a \cdot \tanh \frac{1}{2}b = \tan \frac{1}{2}\delta \triangle ABC$ for $\triangle ABC$ in standard notation with $\angle C$ right, we see that when a and b are both very small the left-hand side is approximately $\frac{1}{4}ab$ and the right-hand side is approximately $\frac{1}{2}\delta \triangle ABC$. So $\frac{1}{2}ab$ is a good approximation for $\delta \triangle ABC$. Thus, the Euclidean formulas for area give very good approximations for area when applied to very small regions in the Bolyai–Lobachevsky plane. (This explains why k^2 is chosen for the constant of proportionality in Definition 33.15.) Further, as was noted at the end of Section 32.1, when a and b are both very small then $a^2 + b^2$ is a very good approximation for c^2.

Let s denote arc length along a curve with equation $w = f(u)$ in Lobachevsky coordinates. The differential of arc length ds is a function of f, du, and dw. In Figure 34.8, when $\triangle u$ and $\triangle w$ are very small,

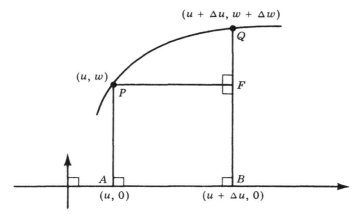

FIGURE 34.8

PQ^2 is a good approximation for ds^2. So $PF^2 + QF^2$ is a good approximation for ds^2. From the equation $\sinh PF = \sinh AB \cosh AP$, we deduce that PF is approximately $AB \cosh AP$, that is, $\triangle u \cosh w$. As $\triangle u$ approaches 0, then QF approaches $\triangle w$. Putting all this together, we conclude that

$$ds^2 = \cosh^2 w \, du^2 + dw^2$$

is the formula for differential of arc length in Lobachevsky coordinates.

We can find the length s of the arc from P to Q on the equidistant curve with equation $w = d$ where P and Q have coordinates $(0, d)$ and (q, d), respectively. See Figure 34.9. Since w is a constant, we have $ds^2 = \cosh^2 d \, du^2 + 0$. Hence,

$$s = \int_0^q \cosh d \, du = q \cosh d.$$

For another example, let's find the length s of the arc from O to Q on the horocircle with radius $\overrightarrow{OX}$ where O, X, Q have coordinates $(0, 0), (b, 0), (b, a)$, respectively, with $b > 0$. The horocircle has equation $e^u = \cosh w$ in Lobachevsky coordinates (Theorem 32.4). So $du = \tanh w \, dw$ and $ds^2 = \cosh^2 w \, dw^2$. Hence,

$$s = \int_0^a \cosh w \, dw = \sinh a.$$

From Theorem 32.4 we know that $s = S \sinh a$ where S is the constant introduced in Definition 32.1. Therefore, $S = 1$. (In general, $S = k$.)

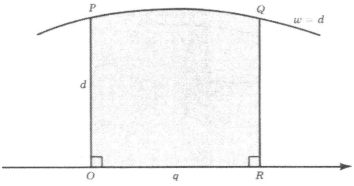

FIGURE 34.9

Equations $\tanh u = \cos\theta\,\tanh r$ and $\sinh w = \sin\theta\,\sinh r$ relate Lobachevsky coordinates (u, w) with polar coordinates (r, θ). We can take the differentials of both sides of each equation and find (Exercise 34.17) that

$$ds^2 = dr^2 + \sinh^2 r\, d\theta^2$$

is the formula for differential of arc length in polar coordinates.

In particular, when r is a constant we have a circle. Thus, it is easily seen that the circumference of a circle with radius r is $2\pi \sinh r$.

Still using polar coordinates, we then see that the length of the circular arc from P to Q. in Figure 34.10 is exactly $\triangle\theta \sinh r$. So when $\triangle\theta$ is very small, then PQ is approximately $\triangle\theta \sinh r$. (This result also follows directly from the formula $\sinh \frac{1}{2}PQ = \sin \frac{1}{2}\triangle\theta \sinh r$.)

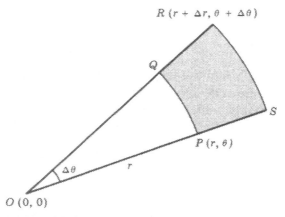

FIGURE 34.10

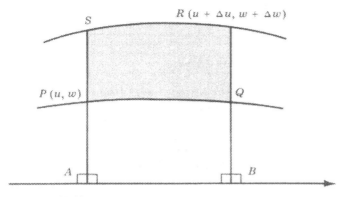

FIGURE 34.11

Since $PS = \triangle r$, when $\triangle r$ and $\triangle \theta$ are both very small, the shaded region in the figure approaches the interior of a rectangle and has approximate area $\sinh r \, \triangle r \, \triangle \theta$. Thus, we conclude that

$$dA = \sinh r \, dr \, d\theta$$

is the formula for differential of area in polar coordinates.
  In particular, if A is the area of a circle with radius r, then

$$A = 4 \int_0^{\pi/2} \int_0^r \sinh r \, dr \, d\theta = 4 \int_0^{\pi/2} (\cosh r - 1) \, d\theta$$
$$= 2\pi (\cosh r - 1) = 4\pi \sinh^2 \tfrac{1}{2}r.$$

  Returning to Lobachevsky coordinates, we know from our previous results that the length of the hypercircular arc from P to Q in Figure 34.11 is exactly $\triangle u \cosh w$. Letting $\triangle u$ and $\triangle w$ approach 0, we conclude by the same reasoning as above that

$$dA = \cosh w \, dw \, du$$

is the formula for differential of area in Lobachevsky coordinates. Thus, if A is the area of the region bounded by the curves with equation $w = f(u)$, $w = 0$, $u = a$, and $u = b$ where $f(u) \geqq 0$ for $a \leqq u \leqq b$, then

$$A = \int_a^b \int_0^w \cosh w \, dw \, du = \int_a^b \sinh f(u) \, du.$$

  For example, returning to Figure 34.9, we see that the area A

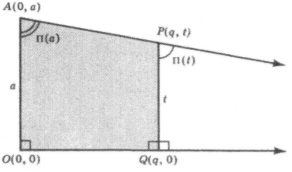

FIGURE 34.12

of the region between an arc of length $q \cosh d$ on a hypercircle of distance d from its center and the center is given by

$$A = \int_0^q \sinh d \, du = q \sinh d.$$

For another example, let's use calculus to compute the area A of $\square OAPQ$ in Figure 34.12. The hardest part of the problem is finding an equation for $\overleftrightarrow{AP}$. Perhaps the easiest way is to check that $\overleftrightarrow{AP}$ has equation $y = (1 - x) \tanh a$ in Beltrami coordinates and so has equation $\tanh w = e^{-u} \tanh a$ in Lobachevsky coordinates. Then, $du = -\text{sech}^2 w \coth w \, dw$ and

$$A = \int_0^q \sinh w \, du = \int_a^t -\text{sech} \, w \, dw$$

$$= [2 \arctan e^{-w}]_a^t = \Pi(t) - \Pi(a).$$

Further, letting t approach infinity in the result above, we have that the area of the region bounded by $\llcorner PAOQ$ is $\pi/2 - \Pi(a)$, as expected.

Finally, let A be the area of the region bounded by the horocircle with equation $e^u = \cosh w$ and the lines with equations $w = 0$ and $u = a$. See Figure 34.13. (We have previously shown that the length of the arc from O to Q is $\sinh a$.) Then

$$A = \int_0^b \sinh w \, du = \int_0^a \sinh w \tanh w \, dw$$

$$= \int_0^a (\cosh w - \text{sech} \, w) \, dw = [\sinh w + \Pi(w)]_0^a$$

$$= \sinh a + \Pi(a) - \pi/2.$$

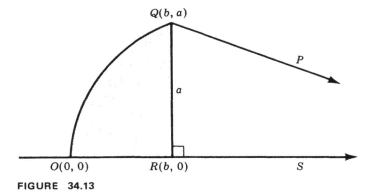

FIGURE 34.13

Putting our results together, we see that the area of the region bounded by the horocircle and the two radii $\overrightarrow{QP}$ and $\overrightarrow{OS}$ is equal to the length of the arc from O to Q.

34.3 CONSTRUCTIONS

The old games are the best games. One of the oldest is the *ruler and compass* game in Euclidean geometry. Masters of a game often give themselves handicaps, which they then try to overcome. Abul Wafa al-Buzjani (940–998) was such a master. He played the game with the restriction that his compass had a fixed opening. Such a tool is known as a *rusty compass* and is no limitation in the hands of a master player. In fact, a master player needs to use the rusty compass only once (the Poncelet-Steiner Theorem). On the other hand, a master player with a regular compass need not use the ruler at all (the Mohr-Mascheroni Theorem).

The ruler and compass game can be played on the Bolyai–Lobachevsky plane as well as on the Euclidean plane. In either case, we talk about *Euclidean constructions!* In Table 34.1, we give lists of equivalent *tools,* one list for each playing board. (We shall not prove the validity of the table here. See Exercise 34.31.) Some of the terminology is defined as follows. Given any two points, a *ruler* determines the line through the given points. Given any two points A and B, a *compass* determines the circle with center A and radius $\overline{AB}$. Given point P off line l, a *hypercompass* determines the hypercircle through P with center l. Given two points A and B, a *horocompass* determines the horocircle with radius $\overrightarrow{AB}$. A *rusty compass* is a compass that determines only circles having one fixed radius (distance), and a *rusty hypercompass* is a hypercompass that determines only hypercircles of one fixed radius (distance). (Since all horocircles are congruent, a *rusty horocompass* would be the same as a horocompass.) Given point *P* off line *l*, a *ruler with horoparallel edges* determines the lines through *P* that are horoparallel to *l* (see Theorem 26.19). A line is *constructed* if two points on the line are determined; a cycle is *constructed* if three points on the cycle are determined; and, with two exceptions, a point is *constructed* if it is the intersection of constructed lines and/or constructed circles. The exceptions are two points to begin with. You can't win a game you're not allowed to play. So we suppose that two points *P* and *Q* are given and defined to be *constructed*, where $PQ = 1$ in the Euclidean plane and $\Pi(PQ) = \pi/4$ in the Bolyai–Lobachevsky plane.

A *construction* in geometry is a list of statements that show how a desired locus can be determined in a finite number of steps. It is totally unnecessary to illustrate or to approximate this mathematical construction with some physical construction in an imperfect model such as a sheet of paper with dots representing points. Unnecessary, yes, but drawing this approximation is part of the fun for both amateurs and professionals. (Anyone who has carried out a long physical

TABLE 34.1

Equivalent Tools in the Euclidean Plane	Equivalent Tools in the Bolyai–Lobachevsky Plane
	Ruler and compass
Ruler and compass	Ruler and hypercompass
	Ruler and horocompass
Ruler and rusty compass	Ruler and rusty compass
	Ruler and rusty hypercompass
	Ruler, one circle with its center, and two lines from a horopencil
Ruler and one circle with its center	Ruler, one hypercircle with its center, and two lines from a horopencil not containing the center of the hypercircle
	Ruler, one horocircle with a radius, and two lines from a horopencil not containing the line on the radius of the horocircle
A ruler with parallel edges	A ruler with horoparallel edges
Compass	Compass and horocompass
	Hypercompass

construction on paper need not be told the meaning of the word "approximation.") It is always necessary to refer to the mathematical theory to actually prove the validity of the construction. Although our ruler and compass are not physical tools, children and amateurs (should) play the game as if they were, using such language as "draw the line . . ." rather than such language as "let l be the line . . . ." If "draw the line . . ." is ever to make sense in mathematical geometry, it is here. In fact, many conservative master players enjoy using this language to illustrate the venerable heritage the game has accrued throughout history, especially in the last century. These are matters of manners. If anyone wishes to "draw a horocircle," let her or him do so.

More important matters are illustrated by the following construction, due to Archimedes. Suppose we are given any acute angle with vertex V in the Euclidean plane. Let there be two marks on our ruler. With our compass, draw the circle with center V and radius equal to the distance between the two marks. Let the circle intersect the angle at two points A and B, and let the circle intersect $\overleftrightarrow{VA}$ at the two points A and C. See Figure 34.14. Now slide the ruler so that it passes through B, so that one of the two marks determines a point P on $\overrightarrow{AV}$, and so that the other mark determines a point Q on the circle. By considering the exterior angles of the isosceles triangles $\triangle BVQ$ and $\triangle PQV$, it is easy to see that $m\angle APB = \frac{1}{3} m\angle AVB$. This yields a construction for the trisection of any angle! Foul play? Yes and no. The construction is valid *if* you allow the *marked ruler*. However, then you would no longer be playing the ruler and compass game. (There are other games!) With a ruler we are only able to determine the points on the line through *two given points*. (It is only when you think of a ruler as a physical tool, which it is not, that it would make sense to say marking the ruler is against the rules of the game.) The ruler in Birkhoff's *ruler and protractor* (Section 14.1), upon which the axiom system for our geometry has been based, is very different from the ruler in the *ruler and compass* constructions. To alleviate some of the

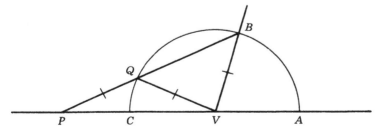

FIGURE 34.14

confusion caused by the traditional terminology, some people prefer using the terms *straightedge* or *unmarked ruler* to the term *ruler* when talking about Euclidean constructions.

The compass is analogous to the classical tool called a *pair of compasses,* which describes a circle but collapses when lifted off the paper, rather than analogous to the modern drawing tool called a compass, at least in the United States, and which can be used to draw a circle and to carry a distance. In Archimedes' construction above, the compass seems to be used to "carry a distance." Given points A, B, C with $B \neq C$, say that a *modern compass* determines the circle with center A and radius BC. It is not obvious that the ruler and compass is equivalent to the ruler and modern compass in absolute geometry. Nevertheless, this is exactly what Euclid proves in his first two propositions in the *Elements,* as may be seen by rereading his proofs of our Theorems 21.1 and 21.2.

As in any good game, you can't expect to win all the time. Let E be the set of real numbers that can be obtained in a finite number of steps by starting with the number 1 and using the operations of addition, subtraction, multiplication, division, and square root. Then E is a field. It can be shown that, given only a segment of length 1 in the Euclidean plane, we can construct with ruler and compass a segment of length x iff x is in the field E. (This is not too surprising when we recall that lines and circles have linear and quadratic equations, respectively, in the Cartesian plane.) Analogously, given only a segment of length arcsinh 1 (i.e., a segment of length p where $\Pi(p) = \pi/4$) in the Bolyai–Lobachevsky plane we can construct with ruler and compass a segment of length r iff tanh r is in the field E. Note that sinh r is in E iff tanh r is in E.

These results can be used to show that for the Euclidean plane the three great construction problems that were left unsolved by the Greeks cannot be solved. The problems are called *the trisection of an angle, the duplication of the cube,* and *the quadrature of the circle.* For the duplication of the cube, it is required to construct with ruler and compass the edge of a cube having twice the volume of a given cube. For the quadrature of the circle, it is required to construct with ruler and compass a regular quadrilateral having the same area as a given circle; since a regular 4-gon in Euclidean geometry is a square, this problem is also called *squaring the circle.*

The problem of the trisection of an angle by ruler and compass refers to arbitrary angles, since some angles can obviously be trisected. The usual way of showing this problem is unsolvable is to show that an angle of measure $\pi/3$ cannot be trisected by proving that it is impossible to construct an angle of measure $\pi/9$. Considering the side adjacent to an angle of measure $\pi/9$ in a right triangle with hypotenuse of length 1, we see that an angle of measure $\pi/9$ can be constructed in the Euclidean plane iff a segment of length $\cos \pi/9$ can be constructed. Since $2 \cos (\pi/9)$ is a root of the equation $x^2 - 3x - 1 = 0$ (let $\theta = \pi/9$ in the identity $\cos 3\theta = 4 \cos^3 \theta - 3 \cos \theta$) and since it can be shown that no root of this equation is in the field E, then it follows that an angle of measure $\pi/9$ cannot be constructed by ruler and compass. That the other two classical construction problems are also unsolvable by ruler and compass follows from the fact that neither the equation $x^3 = 2$ nor the equation $x^2 = \pi$ has a root in the field E. References to the proofs of the statements we have made may be found in Exercise 34.30. Actually proving the arguments outlined above is not trivial. It is no wonder that the Greeks were unable to solve these three problems.

Now suppose we are given a segment of length p in the Bolyai – Lobachevsky plane where $\Pi(p) = \pi/4$. So $p = \operatorname{arcsinh} 1$, since we are assuming the distance scale is 1. Considering right triangles with a leg of length p, we see that an angle of measure λ can be constructed iff a segment of length a can be constructed where $\tan \lambda = \tanh a$. Since $\tan \lambda$ is in the field E iff an angle of measure λ can be constructed in the Euclidean plane, then an angle of measure λ can be constructed in the Bolyai – Lobachevsky plane iff an angle of measure λ can be constructed in the Euclidean plane. Hence, the construction of a regular n-gon in one plane is possible iff it is possible in the other. In particular, the trisection of an angle problem is unsolvable in the Bolyai – Lobachevsky plane. If that doesn't surprise you, perhaps the next result will. Since $2 \sinh (p/3)$ is a root of the equation $x^3 + 3x - 2 = 0$ (let $3t = p$ in the identity $\sinh 3t = 4 \sinh^3 t + 3 \sinh t$) and since it can be shown this equation has no root in the field E, then it follows that a segment of length $p/3$ cannot be constructed by ruler and compass when given a segment of length p. Therefore, the trisection of an arbitrary *segment* by ruler and compass is impossible in the Bolyai – Lobachevsky plane!

Henceforth, we consider only *Euclidean constructions for the Bolyai – Lobachevsky plane*. We shall not assume the results of Table 34.1, but restrict ourselves to ruler and compass constructions. That we know there are unsolvable problems adds to the excitement of the game. Although a construction is often an outline of its own proof, sometimes a construction gives only the most subtle hint of its proof. Some geometers consider a construction to be incomplete if it does not incorporate a proof of its assertions. We have taken the view that the proof of a construction and the construction itself stand in the same relationship as the proof of a theorem and the theorem itself. As a theorem answers a question, so a construction answers a problem; and there is little difference between a question and a problem. The *elementary constructions* are the Euclidean constructions within the first twenty-eight propositions of Euclid's *Elements*. We assume the ele-

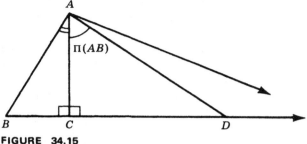

FIGURE 34.15

mentary constructions are known and begin numbering our problems with the following.

Problem 1: Given point P off line l, construct the lines through P that are horoparallel to l; construct a critical angle for a given segment.

The two problems in Problem 1 are equivalent by the elementary constructions. *Construction* 1, which answers these problems, is the statement of Corollary 32.23, and its proof is the proof of Corollary 32.23. So we have already solved Problem 1. Next, we turn to the converse problem.

Problem 2: Given an acute angle, construct a line perpendicular to one side and horoparallel to the other; construct a segment such that a given acute angle is a critical angle for that segment.

Construction 2: Given acute angle $\angle BAC$, suppose C is the foot of the perpendicular from B to $\overleftrightarrow{AC}$. Let point D be constructed on $\overleftrightarrow{BC}$ such that $m\angle CAD = \Pi(AB)$. (See Figure 34.15.) Then $\angle BAC$ is a critical angle for $\overline{AD}$.

Proof: Since $AB > AC$, then $\Pi(AB) < \Pi(AC)$. So point D exists (i.e., can be constructed) by Construction 1. Then

$$\cos m\angle BAC = \frac{\tanh AC}{\tanh AB} = \frac{\tanh AC}{\cos \Pi(AB)} = \tanh AD.$$

So $m\angle BAC = \Pi(AD)$, and $\angle BAC$ is a critical angle for $\overline{AD}$.   Q.E.F.

Problem 3: Construct the common perpendicular to two hyperparallel lines.

By Constructions 1 and 2, our proof of Theorem 26.16 provides both a construction for the common perpendicular to two given hyperparallel lines and a proof of that construction. Let's call this con-

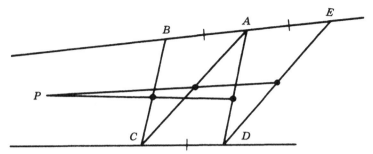

FIGURE 34.16

struction *Construction* 3a. So we may consider Problem 3 solved. However the following construction is of interest because it is shorter.

Construction 3b: Given two hyperparallel lines $\overleftrightarrow{PR}$ and $\overleftrightarrow{ST}$, let point Q be constructed on $\overleftrightarrow{ST}$ such that $\overleftrightarrow{PQ} \perp \overleftrightarrow{ST}$. Assuming $\overleftrightarrow{PQ}$ is not the common perpendicular to $\overleftrightarrow{PR}$ and $\overleftrightarrow{ST}$, we suppose $\angle QPR$ is acute. Let $\triangle ABC$ be constructed such that $\angle C$ is right, $\Pi(AC) = m\angle QPR$, and $m\angle CAB = \Pi(PQ)$. Let M be constructed on $\overrightarrow{PR}$ such that $PM = AB$. Then the perpendicular to $\overleftrightarrow{PR}$ at M is also perpendicular to $\overleftrightarrow{ST}$.

The proof of Construction 3b follows easily from Figures 32.17 and 32.16 with $l = PQ$. We shall give one more solution to Problem 3. The following construction is of special interest because it involves only elementary constructions. See Figure 34.16. However, the proof is not so elementary (Exercise 34.39).

Construction 3c: Given $\sqcup ABCD$ with $\overleftrightarrow{AB}$ and $\overleftrightarrow{CD}$ hyperparallel, we may suppose $CD = AB$. Let point E be constructed in int $(\overrightarrow{BA})$ such that $AE = AB$. Let l be the line through the midpoints of $\overline{AD}$ and of $\overline{BC}$; let m be the line through the midpoints of $\overline{AC}$ and of $\overline{ED}$. Then the line through the intersection of l and m that is perpendicular to $\overleftrightarrow{AB}$ is also perpendicular to $\overleftrightarrow{CD}$.

Construction 3c suggests the following problem.

Problem 4: Construct an isosceles closed biangle with vertex B that is equivalent to given closed biangle $\sqcup ABCD$.

A construction, hereafter referred to as *Construction* 4, that solves Problem 4 and that uses only elementary constructions can easily be gleaned from our proof of Theorem 24.3.

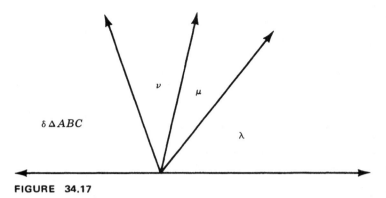

FIGURE 34.17

Problem 5: Construct a triangle having angles that are respectively congruent to three given angles.

By AAA a solution is essentially unique if it exists, and a solution exists iff the sum of the measures of the given angles is less than π. See Figure 34.17. If $\triangle ABC$ is the desired triangle, then the problem becomes trivial once BC is constructed. Therefore, the following construction solves the problem. See Figure 34.18.

Construction 5: Given $\angle L$, $\angle M$, $\angle N$ that are respectively congruent to the angles of some triangle, we may suppose that neither $\angle M$ nor $\angle N$ is larger than $\angle L$. Let $\angle SRQ$ be constructed congruent to $\angle L$ and such that $\angle N$ is a critical angle for $\overline{QR}$. Let P be the point constructed such that $S-R-P$ and $\angle M$ is a critical angle for $\overline{PR}$. Let lines p and q be constructed such that p is perpendicular to $\overleftrightarrow{PR}$ at P and q is perpendicular to $\overleftrightarrow{QR}$ at Q. Let the common perpendicular to p and q intersect p and q at points B and C, respectively. Then $\overline{BC}$ is

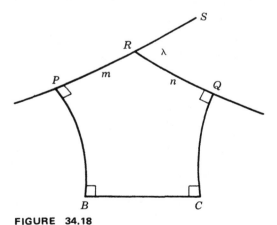

FIGURE 34.18

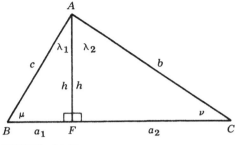

FIGURE 34.19

the side of a triangle $\triangle ABC$ such that $\angle A \simeq \angle L$, $\angle B \simeq \angle M$, and $\angle C \simeq \angle N$.

The proof of Construction 5 is not immediately evident from the statement of the construction. We need to verify that $\overline{BC}$ in Figure 34.18 exists and has the correct length. The missing links are provided by Figures 34.19 and 34.20 where $m\angle PRF = \Pi(l_1^*)$, $m\angle QRF = \Pi(l_2^*)$, $\lambda_1 = \Pi(l_1)$, $\lambda_2 = \Pi(l_2)$, $\lambda = \lambda_1 + \lambda_2$, $\mu = \Pi(m)$, and $\nu = \Pi(n)$. By Corollary 32.24 (row 1 of Table 32.1 and row 4 of Table 32.2), each of these two figures determines the validity of the other. Thus, $\overline{BC}$ in Figure 34.20 is determined by $\angle SRQ$, m, and n. Fortunately, $m\angle SRQ = \pi - \Pi(l_1^*) - \Pi(l_2^*) = \Pi(l_1) + \Pi(l_2) = \lambda$, and the mystery of Figure 34.18 disappears.

Figure 34.21 and Corollary 32.24 can be used to give a shorter solution to Problem 5 when one of the three given angles is a right angle. In general, we have to be given the acute angles since not all angles can be constructed by ruler and compass. Whether we can construct such triangles or not, the existence of such triangles is especially interesting in the case $\lambda = \pi/n$ and $\mu = \pi/m$ with n and m positive integers. Considering that the defect of a triangle is positive, we see that a necessary and sufficient condition for the existence of such triangles is

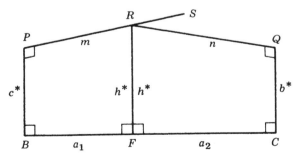

FIGURE 34.20

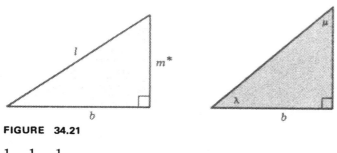

FIGURE 34.21

$$\frac{1}{n} + \frac{1}{m} < \frac{1}{2}.$$

With n a positive integer and $\lambda = \pi/n$, the triangles are the building blocks for the regular n-gons with angles of measure 2μ. (Compare Figures 34.21 and 34.22 when $n = 4$ and $\lambda = \pi/4$.) With m also a positive integer and $\mu = \pi/m$, these regular n-gons have angles of measure $2\pi/m$ and can be used to *tile* the plane, where each vertex of an n-gon is the common vertex of m such n-gons. Unlike the case for the Euclidean plane, there are infinitely many essentially different ways to tile the Bolyai–Lobachevsky plane with congruent regular n-gons. For example, taking $n = 4$ and $m = 8$, we see there exist regular 4-gons that can be used to tile the plane where each vertex of the 4-gons is the common vertex of 8 such 4-gons. (A figure is difficult to draw since the side of such a 4-gon would be very long relative to the size of the page.) There is something rather special about the example cited. Since, given any two points, we can construct angles of measure $\pi/4$ and angles of measure $\pi/8$ by elementary constructions alone, then we can construct the 4-gons in the example by ruler and compass. Although we have shown a lot more, in particular, we have shown that the following problem has a rather easy solution.

Problem 6: Construct a regular 4-gon with angles of measure $\pi/4$ (which then can be used to tile the plane).

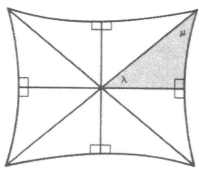

FIGURE 34.22

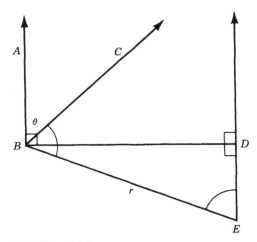

FIGURE 34.23

The equation $r = \Pi(\phi)$ determines a one-to-one correspondence between the constructable circles (with radius r) and the constructable acute angles (with measure ϕ). Another one-to-one correspondence is suggested by the following problem. Bolyai used the solution of this problem for a construction that gives this chapter its name.

Problem 7: Given either one of an acute angle of measure θ or else a segment of length r, construct the other such that $\tan \theta = 2 \sinh \frac{1}{2}r$.

The following construction only determines the segment given the angle. However, from this the converse construction is easily obtained. See Figure 34.23.

Construction 7: Given an acute angle $\angle ABC$ of measure θ, let point D be construction such that $\angle ABD$ is right, points D and C are on the same side of $\overleftrightarrow{AB}$, and $\angle CBD$ is a critical angle for $\overline{BD}$. Let point E be constructed such that $\llcorner\!CBED$ is closed and isosceles. If $\overline{BE}$ has length r, then $\tan \theta = 2 \sinh \frac{1}{2}r$.

Proof: Point D exists by Construction 2, and point E exists by Construction 4. Then $m\angle CBD = \pi/2 - \theta$, $m\angle BED = \Pi(\frac{1}{2}r)$, and

$$\tan \theta = \cot \Pi(BD)$$
$$= \sinh BD$$
$$= (\sin \Pi(\tfrac{1}{2}r))(\sinh r)$$
$$= (\operatorname{sech} \tfrac{1}{2}r)(2 \sinh \tfrac{1}{2}r \cosh \tfrac{1}{2}r)$$
$$= 2 \sinh \tfrac{1}{2}r. \qquad\qquad \text{Q.E.F.}$$

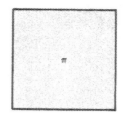

FIGURE 34.24

If you have not read Section 33.2 on area, replace "area" by "defect" in the sequel.

Either by the calculus in the preceding section or by the high school method of considering a circle as the limit of a regular n-gon, we know the area of a circle with radius r is $4\pi \sinh^2 \frac{1}{2}r$. Therefore, by Construction 7, if we can construct an acute angle with measure θ, then we can construct a circle with area $\pi \tan^2 \theta$. In particular, since we can construct an angle of measure $\pi/4$, then we can construct a circle with area π.

Problem 8: Construct (using ruler and compass alone) a circle and a regular quadrilateral that have the same area.

By our remarks above, we can construct a circle with area π. Further, we know we can construct a regular 4-gon with area π, since we can easily solve Problem 6. Hence, we can solve Problem 8, although the problem is unsolvable in the Euclidean plane. We have *squared the circle in the Bolyai–Lobachevsky plane!*

34.4 EXERCISES

All exercises refer to the Bolyai–Lobachevsky plane unless otherwise indicated. The distance scale is assumed to be 1.

34.1 Theorem 34.12.

34.2 Finish the proof of Ceva's Theorem for the Euclidean plane.

• 34.3 Prove the following proposition is false for each of the three types of brushes: If A, B, C, A', B', C' are six points such that $\overleftrightarrow{AA'}$, $\overleftrightarrow{BB'}$, $\overleftrightarrow{CC'}$ are in a brush, $\overleftrightarrow{AC}\|\overleftrightarrow{A'C'}$, and $\overleftrightarrow{BC}\|\overleftrightarrow{B'C'}$, then $\overleftrightarrow{AB}\|\overleftrightarrow{A'B'}$.

34.4 For the Euclidean plane, prove the proposition in Exercise 34.3. (When the brush is a parallel pencil the proposition is called the *Little Desargues' Theorem.*)

34.5 For the Euclidean plane, prove the statement of Theorem 34.11 assuming $\overleftrightarrow{AB}$, $\overleftrightarrow{CD}$, $\overleftrightarrow{EF}$ are parallel.

34.6 For the Euclidean plane, prove the statement of Theorem 34.11 assuming $\overleftrightarrow{AB}$ is parallel to $\overleftrightarrow{CD}$ but is not parallel to $\overleftrightarrow{EF}$.

• **34.7** For the Euclidean plane, prove the statement of Theorem 34.11

• **34.8** Theorem 34.11.

34.9 The following proposition is a theorem for the Euclidean plane but is a false statement for the Bolyai–Lobachevsky plane: If A, B, C, D, E, F are six points such that points A, C, E are on one line, points B, D, F are on another line, lines $\overleftrightarrow{AB}$ and $\overleftrightarrow{DE}$ are parallel, and lines $\overleftrightarrow{BC}$ and $\overleftrightarrow{EF}$ are parallel, then lines $\overleftrightarrow{CD}$ and $\overleftrightarrow{FA}$ are parallel.

34.10 How might Desargues' Theorem be used to know where to dig tunnels on opposite sides of a mountain so that the tunnels will meet to form a straight tunnel?

34.11 For the real projective plane, prove Pappus' Theorem.

34.12 Given $\triangle ABC$ and $\triangle A'B'C'$ with no vertex in common and such that point C is on $\overleftrightarrow{A'B'}$, point C' is on $\overleftrightarrow{AB}$, point A'' is the intersection of $\overleftrightarrow{BC}$ and $\overleftrightarrow{B'C'}$, and point B'' is the intersection of $\overleftrightarrow{AC}$ and $\overleftrightarrow{A'C'}$, then lines $\overleftrightarrow{AB'}$, $\overleftrightarrow{A'B}$, $\overleftrightarrow{A''B''}$ are in a brush.

34.13 Suppose two lines intersect off your paper in a drawing. How can you draw a line through a given point P that also passes through the inaccessible point of intersection?

• **34.14** Let P be the set of all brushes. Define a set L whose elements are subsets of P such that (P, L) is a projective plane, as defined by Axiom System 2 in Section 4.2, with set of *points* P and set of *lines* L.

34.15 Disprove: If D, E, F are Menlaus points for $\triangle ABC$ and the Cevians $\overleftrightarrow{AD}$, $\overleftrightarrow{BE}$, $\overleftrightarrow{CF}$ are in a brush, then the equation in the statement of Theorem 34.7 holds.

34.16 In Definition 34.4, the existence of the set F does not depend on the *axiom of choice*.

34.17 Verify the formula for ds^2 in polar coordinates.

• **34.18** In Lobachevsky coordinates, if two points P and O have coordinates $(0, a)$ and $(0, 0)$, respectively, what is the equation of the perpendicular to $\overleftrightarrow{PO}$ at P?

34.19 Use calculus to find the area of a Lambert quadrilateral.

34.20 Suppose a horocircle passes through the center of a circle with radius r. What are the lengths of the intercepted arcs? What are the areas of the regions bounded by the two cycles?

34.21 The area of the region bounded by each of three mutually horoparallel lines not in one brush is π. What is the area of the region bounded by the four lines in Figure 26.2, where lines are assumed horoparallel when possible? Generalize.

34.22 If $m\angle POQ = \pi/3$ and $r = OP = OQ$, find the areas of the regions bounded by $\overline{PQ}$ and the circle through P with center O. Show that $PQ > r$.

34.23 In *A New Theory of Parallels* the author of *Alice in Wonderland* uses the following axiom in place of Euclid's Parallel Postulate: "In every Circle, the inscribed equilateral Hexagon is greater [in area] than any one of the Segments which lie outside it." Dodgson's "Segments" are the smaller regions in Exercise 34.22. Show that the axiom is false in the hyperbolic plane.

*34.24 Find formulas for ds^2 and dA in Beltrami coordinates.

34.25 Beginning with only two points that are defined to be constructed points, give a Euclidean construction for an angle of measure $\pi/4$.

34.26 Draw figures that illustrate the elementary constructions.

● **34.27** Give a Euclidean construction for an angle of measure $\pi/6$.

● **34.28** Outline a Euclidean construction for a circle and a regular quadrilateral that both have area A with $A \neq \pi$.

34.29 Give a Euclidean construction for the point P in Corollary 32.22, assuming a segment of length t is given.

34.30 For the unsolvability of the classical construction problems, read the following two classic paperbacks written for the high school teacher. The new classic is N. D. Kazarinoff's *Ruler and the Round* (Prindle, Weber, and Schmidt, 1970), which contains a very interesting bit on Gauss on page 30. The 1895 classic is F. Klein's *Famous Problems of Elementary Geometry* (Dover, 1956).

34.31 For the equivalent tools in the Euclidean plane, read Chapter 4 of H. Eves' *A Survey of Geometry* (Allyn and Bacon, 1972). For references to the equivalent tools in the hyperbolic plane, see Section 15.8 of H. S. M. Coxeter's *Non-Euclidean Geometry* (University of Toronto Press, 1957).

34.32 Read the 1961 Blaisdell paperbacks *Geometrical Constructions using Compass Only* by A. N. Kostovskii and *The Ruler in Geometrical Constructions* by A. S. Smogorzhevskii.

34.33 Read *Geometric Exercises in Paper Folding* by T. S. Row (Dover, 1966).

34.34 The mirror is the only *tool* needed for the Euclidean constructions.

● **34.35** The ruler alone is not sufficient for the Euclidean constructions.

34.36 Give a Euclidean construction for the tangents to a given circle through a given point in the exterior of the circle.

34.37 The smallest regular 4-gon that can be used to tile the plane with regular 4-gons has area $2\pi/5$.

34.38 There exist exactly four noncongruent polygons that can be used to tile the plane with congruent polygons of area π.

34.39 Prove Construction 3c.

*__34.40__ Give a Euclidean construction for an angle of measure $\pi/5$.

Hints and Answers

CHAPTER 1

1.1   i: p any false statement; ii: q any true statement.
1.2   Use contrapositives, e.g.: (ii′) If $a \leqq b$, then $c \leqq d$.
1.4 $[a]$ contains all integers $5k + a$ where k is an integer. There are 5 equivalence classes: $[0], [1], [2], [3], [4]$.
1.5   FTFTT − FFTFT.
1.6   All points; the empty set.

CHAPTER 2

2.1   All 1-1; only f_1 and f_4 onto.
2.2 $2x − 1, 2(x − 1), 2x^3 − 1, 8(x − 1)^3, 2(x^3 − 1)$.
2.3   from, into, of, at, to, under, onto, one-to-one, on.
2.4 $A = B = \{1, 2\}, D = \{(1, 1), (2, 2), (1, 2)\}, f(1)$?.
2.6   TTFFF − TFFTF.
2.7 $n^n, n!, n!, n!$.
2.9 $D = \mathbf{Z}, C = \mathbf{R}, f(x) = x$.
2.10 $D = \mathbf{R}, f(x) = −x$.
2.12   e.g.: $f(x) = x + 2$ if $x < 0$ and $f(x) = x − 3$ if $x \geqq 0$.

CHAPTER 3

3.4   1 from 03; 2 from 01, 3 is 02, 4 through 8 from definition of $>$; 9 from 7 and 8; 10 from 2, 5, and 8.
3.5   FTTTT − FFFTF.
3.6 $f(n) = n + 1$.

3.10 Not a or b.

3.13 Historical precedence.

CHAPTER 4

4.2 $\mathscr{P}$ points of Euclidean plane, $\mathscr{L}$ Euclidean lines through some fixed point in the Euclidean plane, usual incidence; Euclidean plane; Euclidean line.

4.3 Axiom 1: old P and old Q on unique old line and both off l_x; P_n and P_m on l_x, the only line with two old pencils on it; P_n and old Q on old line through Q and parallel to n. Axiom 2': if old l parallel to old m then both through only P_i; if old P on old l and old m, then no other new point on both old l and old m; P_m is only new point on old m and l_x.

4.4 TFTFT−TFTFF.

4.6 Affine plane and Axiom 4: Every line has at most two points on it.

4.7 $\mathscr{P}$ the set of points on a Euclidean sphere, $\mathscr{L}$ the set of circles on the sphere, usual incidence.

4.8 Let $(\mathscr{P}_3, \mathscr{L}_3, \mathscr{F}_3)$ be derived from E in Figure 4.4. Take $l, O, E,$ P', P as in Figure 4.4. Then P', P_m, m, l_x in $(\mathscr{P}_3, \mathscr{L}_3, \mathscr{F}_3)$ correspond, respectively, to P, Euclidean line in l through O and parallel to m, plane through m and O, l.

CHAPTER 5

5.1 See Figure 4.2.

5.2 (x, y) to (x, y) if $x \leqq 1$ and (x, y) to $(x-1, y)$ if $x > 2$ determines a collineation onto M1.

5.4 $x = 1$; $y = -1$; $y = -2x + 5$; $y = x$ if $x \leqq 0$ and $y = \frac{1}{2}x$ if $x > 0$; $y = 4x - 3$ if $x \leqq 0$ and $y = 2x - 3$ if $x > 0$; $3y = 8x + 5$ if $x \leqq 0$ and $3y = 4x + 5$ if $x > 0$.

5.5 TTFFT−FTTFT.

5.7 (x, y) to (x, y^3) gives a collineation from M1 to M9.

5.8 Cardinality argument.

5.10 Compare with $(\mathscr{P}_2, \mathscr{L}_2, \mathscr{F}_2)$ of Section 4.2.

5.18 Use Gauss plane and Exercise 3.22.

CHAPTER 6

6.2 For $y = mx + b$, take $f((x, mx + b)) = x(1 + m^2)^{1/2}$.

6.3 First sentence of Section 6.3.

6.6 Use Exercise 6.5 on M1.

6.7 TFTTF – TTFFT.

6.14 Taking $P = 2$ and $Q = 6$ in the Gauss plane, then $PQ = 4$ and $\overleftrightarrow{PQ} = \mathbf{R}$.

6.15 All except M2.

CHAPTER 7

7.4 Draw the streets and avenues through the two points; consider equation in each of the nine regions in the resulting tick-tack-toe figure.

7.5 Theorem 7.7.

7.7 TFTTF – FTTFF.

CHAPTER 8

8.4 Let H_1 and H_2 be such sets. Let $A - V - B$ with $l = \overleftrightarrow{AB}$ and $A \in H_1$. Show $P \in$ int $(\overrightarrow{VB})$ implies $P \notin H_1$ but $P \in H_2$. So $B \in H_2$. Then, $Q \in$ int $(\overrightarrow{VA})$ implies $Q \notin H_2$ but $Q \in H_1$.

8.5 Cartesian line through A and B in Figure 8.6 contains A and B but not W.

8.6 TFFTT – TFTFF.

8.8 Same as Cartesian plane.

8.9 A off $\overline{PQ}$ in Figure 7.5.

8.11 Take M as A in Figure 7.5 with $x_1 = y_1$ and $x_2 = y_2$.

CHAPTER 9

9.2 Figures 9.15 and 9.16.

9.4 Figure 9.16.

9.5 All points off both $\overleftrightarrow{VA}$ and $\overleftrightarrow{VB}$. (3, 3) is not in inside.

9.6 FTFTF – TTFTT.

CHAPTER 10

10.2 Tetrahedron, octahedron, hexahedron, icosahedron, and dodecahedron.

10.3 FTTTT – TFFFT.

CHAPTER 12

12.3 Use Theorem 12.14.
12.5 TTFTF − TTFTF.
12.7 See Section 8.2.
12.9 Yes.
12.15 This is Sylvester's original conjecture.
12.17 False.

CHAPTER 13

13.1 A, E, F on a side of $\overleftrightarrow{CD}$; B and D on opposite sides of $\overleftrightarrow{AC}$ as are B and F.
13.3 Although distance is not Euclidean distance, betweenness for points in the model is the same as Euclidean betweenness.
13.4 Use Sylvester's Theorem.
13.5 TTTTF − FFFFF.

CHAPTER 14

14.4 $x_1 < x_2 < x_3$ iff $\sqrt{3}x_1 < \sqrt{3}x_2 < \sqrt{3}x_3$.
14.5 Lines with nonzero slopes n_1 and n_2 are perpendicular iff $n_1 n_2 = -3$.
14.6 FTFFF − FTFTF.
14.7 If $P = (-1, 1)$, then $m' \angle AVP$ is $\pi/3, 2\pi/3$.

CHAPTER 15

15.1 Veblen's axiom system is categorical, every model being isomorphic to the set of Cartesian points with the conventional betweenness relation. However, since Taxicab geometry and Cartesian geometry share the same points and betweenness relation, Veblen's system does not fully describe what is usually called Euclidean geometry.

CHAPTER 16

16.2 With $m \angle A = m \angle A'$ and $m \angle B = m \angle B'$, take D on $\overrightarrow{BC}$ to build $\triangle ABD$ as a copy of $\triangle A'B'C'$.

16.5 E.g., let $V = (0, 0)$, $A = (0, 5)$, and $B = (4, 3)$. A mirror map for l would take A to B, B to A, and V to V. But $VA \neq VB$.
16.6 TFTTF − TTFFT.
16.8 Fold paper so desired points coincide.
16.10 (x, y) to $(2x, 2y)$ for Cartesian plane; $(M10, t)$.

CHAPTER 17

17.3 See Exercise 17.6(g).
17.4 ASA.
17.6 FTTFF − TTTFT.
17.7 No: $(M1, d, m')$.
17.9 Right angles at H in Figure 17.6.
17.13 In Cartesian three-space project distances from plane $z = 2y$ to the $x - y$-plane.

CHAPTER 18

18.4 See Theorem 18.12.
18.5 FTTTT − TFFFF.
18.10 $A - D - E$ such that $2AD = AE - AC$; compare with Theorem 18.16.

CHAPTER 19

19.1 Corollary of Theorem 19.9.
19.3 (x, y) to $(2x, 2y)$; show slope is preserved.
19.4 1; 2 or 6; 6.
19.6 TFTTF − TFTFT.
19.7 From slope argument $A(y - y') = B(x - x')$; midpoint of (x, y) and (x', y') is on l; solve two equations in two unknowns.
19.12 (x, y) to $(a_1 x + b_1 y + c_1, a_2 x + b_2 y + c_2)$.
19.13 $a_1 = \pm b_2 = \cos \theta$ and $a_2 = \mp b_1 = \sin \theta$ in Answer 19.12.

CHAPTER 20

20.3 Use Triangle Inequality.
20.5 TTFTF − TFFTF.

20.9 $c > 0$ and one (i) greater than, (ii) equal to, (iii) less than sum of other two; (iv): $a - b = c = 0$.
20.12 Use contrapositives of statements in hypothesis.

CHAPTER 21

21.3 Theorem 21.8.
21.4 Not true for the Cartesian plane when B and C on opposite sides of $\overleftrightarrow{AD}$.
21.5 TTFFF − FFFTF.
21.6 Assumes congruent segments on $\overrightarrow{BA}$ have congruent projections on $\overrightarrow{BC}$.

CHAPTER 22

22.4 Look at Theorem 23.1 only as a last resort.
22.5 Convex; angles acute or right; opposite sides parallel and congruent; obtained by reflecting a Saccheri quadrilateral in lower base.
22.6 FFTFF − FTFFT.
22.7 Theorem 18.17.
22.8 Exercise 21.6.
22.12 Theorem 22.17 and Theorem 18.16.
22.14 Four cases: $G = C = K$; $G \neq C = K$; $G = K, B - C - G$; and $G = K, B - G - C$.
22.16 See Theorem 33.11.

CHAPTER 23

23.1 For Y implies V use idea of Theorem 23.1.
23.3 Theorem 23.6.
23.4 FTTFT − TFTFT.
23.8 Not under the Hypothesis of the Acute Angle.
23.12 There is a positive number r such that the distance from C_n to $\overleftrightarrow{AB}$ is greater than r for all n.

CHAPTER 24

24.2 Proposition Y of Theorem 23.7.
24.5 FFTFT − FFFFF.

24.9 In proof of Theorem 24.20, let $AD = s$ and $A - F - B$ such that $m\angle ACF = \pi/2$ to obtain $\triangle ACF$.

CHAPTER 25

25.1 The rub is in defining *area* in the first place.
25.4 f must preserve the interior of any angle.
25.5 Possibly no planes.

CHAPTER 26

26.1 See Figure 26.2. For $n > 1$, $\Pi(d_0) = \pi/(n+1)$.
26.2 Either part of Figure 26.2.
26.3 Consider $\llcorner BADC$ and $\llcorner ABCD$.
26.5 TFFFF $-$ TFFTF.
26.12 $\angle BAE$ critical angle for $\overline{AE}$, $A - E - C$, and $\overline{CD} \perp \overline{AC}$.
26.18 Exercise 22.16.

CHAPTER 27

27.1 PQ big enough in Figure 26.1 so that l and m are hyperparallel.
27.2 Figure 23.7.
27.4 Theorem 27.2; p in proof of Theorem 26.3.
27.5 Only g false.
27.10 $44\infty\infty\infty - 21101$.
27.11 Nine cases, considering position of c.

CHAPTER 28

28.1 Only Theorem 28.6 is substantially different.
28.2 AAA.
28.3 $\rho_m \rho_l = (\rho_m \rho_n)(\rho_n \rho_l)$.
28.4 Corollary 28.13.
28.5 TTTTT $-$ FFFFF.

CHAPTER 29

29.1 The perpendicular bisectors of two chords $\overline{PQ}$ and $\overline{QR}$ of a hypercircle cannot both be perpendicular to each of two lines.

29.2   Corollary 28.13, making a wise choice.

29.4   With $\eta_c = \rho_e \rho_d$ where $d \perp \overleftrightarrow{AB}$, you also have a proof that the product of three halfturns is a halfturn in the Euclidean plane.

29.5   Theorem 29.13; $\rho_l \eta_M$ with M the midpoint.

29.6   TFTTF – TTFTF.

29.8 $\rho_d \rho_b \rho_c \rho_a = \rho_f \rho_d \rho_a \rho_e = \rho_f \rho_a \rho_d \rho_e = \eta_Q \eta_P$.

29.9   Involutory $\sigma \eta_P \sigma^{-1}$ fixes point Q iff $Q = \sigma P$.

29.20   If σ is odd with center c, then $n = c$; if σ is even, then $\sigma P = \sigma \rho_l P$ for all points P on l.

CHAPTER 30

30.3 $\sigma^i \rho \sigma^{-i}$ and Theorem 29.8.

30.5   1881, 1883, 1961, 1984, MM.

30.6   One is colorblind.

30.8   TTTTF – FFTFT.

30.28   See Exercise 30.21.

CHAPTER 31

31.2   Proof of Theorem 31.15.

31.4   Theorem 26.19.

31.5   FFFTT – TTFFF.

31.11   In Exercise 31.10: $x = a + b$, $y = a - b$.

31.12 $\cosh x = \cosh (x/2 + x/2)$.

CHAPTER 32

32.3 $\sin \Pi(x + y) \cosh (x + y) = 1$.

32.4   Use Corollary 32.11 in Corollaries 32.16 and 32.15.

32.5 $\cos (\pi - \theta) = -\cos \theta$.

32.6 $\sinh ((x^*)/k) = \operatorname{csch} (x/k)$.

32.7   Only (a) and (e) false.

32.12   By Exercise 32.5 or 32.11, less than $\ln (1 + \sqrt{2})$.

32.16   If $\Pi(r) = \pi/3$, then $2r = \ln 3$.

32.25   Exercise 32.12 and $\cosh^2 2 < \cosh 2^2$.

32.26 H is a multiple of the constant in Corollary 32.14.

32.38 a, b, c integers: no; but if $a^2 + b^2 < c^2 < (a + b)^2$ then ta, tb, tc for some real t.

CHAPTER 33

33.2   See Figure 23.7.
33.3 $B_1 x = A_1 y$.
33.5   Expand by third row.
33.6   TTFFF – TTTTT.

CHAPTER 34

34.3 A'' and B'' outside Cayley–Klein Model in Figure 34.5.
34.7   Theorem 34.10, Exercise 34.5, Exercise 34.6.
34.8   Corollary of Exercise 34.7.
34.14   E.g., a horopencil and all hyperpencils with center in the horo-pencil.
34.18   Theorem 32.21b with $v = a$.
34.27   The hypotenuse of an isosceles right triangle with legs of length p is very interesting.
34.28   Exercise 34.27.
34.35   Midpoints cannot be constructed.

Notation Index

Index

*Page numbers in italics indicate an element of the theory, usually a definition or a theorem.

Undergraduate Texts in Mathematics

(continued from page ii)

Undergraduate Texts in Mathematics

Troutman: Variational Calculus and Optimal Control. Second edition.

Valenza: Linear Algebra: An Introduction to Abstract Mathematics.

Whyburn/Duda: Dynamic Topology.

Wilson: Much Ado About Calculus.

Lightning Source UK Ltd.
Milton Keynes UK
UKOW05n0029090317
296199UK00014B/86/P